Springer Series in
# CLUSTER PHYSICS

Springer-Verlag Berlin Heidelberg GmbH

Springer Series in
# CLUSTER PHYSICS

Julius Jellinek (Ed.)

# Theory of Atomic and Molecular Clusters

## With a Glimpse at Experiments

With 169 Figures and 34 Tables

 Springer

Dr. Julius Jellinek
Chemistry Division
Argonne National Laboratory
9700 S. Cass Avenue
Argonne, IL 60439
USA

*Series Editors:*

Professor A. W. Castleman, Jr.
(*Editor-in-Chief*)
Department of Chemistry
The Pennsylvania State University
152 Davey Laboratory
University Park, PA 16802, USA

Professor R. Stephen Berry
Department of Chemistry
The University of Chicago
5735 South Ellis Avenue
Chicago, IL 60637, USA

Professor Dr. Hellmut Haberland
Albert-Ludwigs-Universität Freiburg
Fakultät für Physik
Hermann-Herder-Strasse 3
D-79104 Freiburg, Germany

Professor Dr. Joshua Jortner
School of Chemistry, Tel Aviv University
Raymond and Beverly Sackler
Faculty of Sciences
Ramat Aviv, Tel Aviv 69978, Israel

Dr. Tamotsu Kondow
Toyota Technological Institute
Cluster Research Laboratory
East Tokyo Laboratory
Genesis Research Institute Inc.
Futamata 717-86
Ichikawa, Chiba 272-0001, Japan

ISSN 1437-0395
ISBN 978-3-540-62000-6     ISBN 978-3-642-58389-6 (eBook)
DOI 10.1007/978-3-642-58389-6
Library of Congress Cataloging-in-Publication Data applied for.

This work is subject to copyright. All rights are reserved, whether the whole or part of the material is concerned, specifically the rights of translation, reprinting, reuse of illustrations, recitation, broadcasting, reproduction on microfilm or in any other way, and storage in data banks. Duplication of this publication or parts thereof is permitted only under the provisions of the German Copyright Law of September 9, 1965, in its current version, and permission for use must always be obtained from Springer-Verlag. Violations are liable for prosecution under the German Copyright Law.

© Springer-Verlag Berlin Heidelberg 1999
Originally published by Springer-Verlag Berlin Heidelberg New York in 1999

The use of general descriptive names, registered names, trademarks, etc. in this publication does not imply, even in the absence of a specific statement, that such names are exempt from the relevant protective laws and regulations and therefore free for general use.

Typesetting: Camera-ready copy by the authors
Cover concept: eStudio Calamar Steinen
Cover production: *design & production* GmbH, Heidelberg
SPIN: 10553576     57/3144/Tr - 5 4 3 2 1 0 – Printed on acid-free paper

# Preface

The emergence and spectacularly rapid evolution of the field of atomic and molecular clusters are among the most exciting developments in the recent history of natural sciences. The field of clusters expands into the traditional disciplines of physics, chemistry, materials science, and biology, yet in many respects it forms a cognition area of its own. The identifying attributes of this area reflect the specificity of the objects, subjects, problems, and issues it addresses, and the concepts, methodologies, techniques, and tools it utilizes. All these ultimately relate to and are defined by a single common characteristic – the finite size of the targeted systems.

The term "finite" represents a broad range of sizes – from small (a few to a few tens of atoms and molecules) to large (thousands of atoms and molecules). This variety of sizes and the size-dependence of the properties would alone be sufficient to make the cluster field a broad and challenging research area. The variety of elements one can use (and nature uses) to form clusters and the different types of interatomic interactions in clusters of different elements and materials only enhance the diversity and the richness of the field.

To unravel, understand, and describe cluster properties one often has to invoke complementary concepts and techniques, and, what is most challenging and stimulating, develop new ones. This book presents a "cross section" of theoretical approaches and their applications in studies of different cluster systems. The last three articles provide a "glimpse" at experimental cluster research.

A book of this size can give only a partial account of the activity in an expanding field, but the collection offered is representative. It covers atomic and molecular clusters, homogeneous (one-component) and heterogeneous (two-component). The systems discussed range from weakly (van der Waals) bonded, through hydrogen- and covalently bonded, to semiconductor and metallic clusters. The properties of isolated clusters as well as those interacting with molecules, each other, surfaces, and electromagnetic field are described and analyzed. The discussion includes electronic features, geometric structures and structural transitions, phases and phase changes, peculiarities of intracluster dynamics (electronic, nuclear, and the coupling between the two), fragmentation processes, chemical reactivity, etc. The theoretical approaches presented involve high-level electronic structure computations, more approximate electronic structure treatments, use of semiempirical potentials, dynamical and statistical analyses, and illustrate the utility of both classical and quantum mechanical concepts. The contributions are written by experts in the respective areas, and represent, for the most part, an exposition of the author's own research.

Argonne, December 1998                                          Julius Jellinek

# Contents

# Phases and Phase Changes of Small Systems

R. Stephen Berry

Department of Chemistry and the James Franck Institute, The University of Chicago,
5735 South Ellis Avenue, Chicago, Illinois 60637, USA

**Abstract.** Atomic and molecular clusters exhibit a variety of phase-like forms and phase changes that differ from those of bulk matter. It is possible to relate some–but not all–of these to corresponding phases and phase transitions of bulk matter. Clusters are attractive vehicles for studying phases and phase changes because they are susceptible to the analytic and computational methods applicable to small systems, and they allow the study of size dependence, even to quite large systems. Among the properties that distinguish phases and phase changes of small systems are bands of temperature and pressure within which two *or more* phase-like forms may coexist–not just curves of coexistence. This coexistence is dynamic, like that of chemical isomers coexisting. Moreover clusters may exhibit phase-like forms that do not exist for bulk matter. These properties of existence and coexistence are the consequence of the small differences between the free energies of clusters in different phase-like forms. Theory predicts that the bands of coexistence should have sharp boundaries, due to the disappearance of local stability of each phase-like form, implying that, in the bulk limit, the two branches of the spinodals should have sharp limits of temperature and pressure. A necessary condition that a species of cluster exhibit a particular phase is that such clusters must reside in the corresponding region of configuration and phase space long enough to establish equilibrium-like properties characteristic of that phase.

## 1 Introduction

Finite systems, especially small finite systems on the scale of atomic and molecular clusters, exhibit some of the solid-like and liquid-like behavior of bulk solids and liquids. Because we can treat small systems in considerable detail, and study how they approach the many-particleΠ limit as they grow, we can use them to give us new insights into the phases and phase transitions of bulk matter. However small systems have many interesting phase-like properties that do not extend to bulk matter, properties that have been recognized but, for the most part, not yet exploited. This discussion will explore the phase-like properties of finite systems in terms of which properties extend simply to bulk matter and which are specific to small systems. It will also point out areas where further research is waiting to be done.

The phase-like properties of clusters and nanoscale particles lend themselves to study by simulation and by analytic theory. Until now, a few experimental studies have demonstrated specific phase-like forms of these species (Bartell, 1992, Bartell, et al., 1991b, Bartell, et al., 1988, Buck, et al., 1993, Goldstein, et al., 1990, Martin, et al., 1994, Valente, et al., 1984a, Valente, et al., 1984b), but very little has come from the laboratory to elucidate the nature of the equilibrium or the transitions between these forms(Bartell, et al., 1992, Bartell, et al., 1991a, Dibble, et al., 1992, Ellert, et al., 1995, Schmidt, et al., 1997). Optical excitation has recently proved to be a powerful tool for probing phase changes. Observations of the differences in optical response of sodium clusters (Ellert, et al., 1995) opened this approach. Then the same group selected charged socium clusters of a single size, thermalized these clusters and then photodissociated the clusters, using the fragmentation pattern as a measure of the internal energy of the equilibrated clusters. A sharp increase in the fraction of small-cluster fragments was taken as a signal of melting (Schmidt, et al., 1997). They then carried out similar experiments with clusters of various sizes, and determined tht there is no apparent regularity in the size dependence of the appearance temperature of a liquid phase (Schmidt, et al., 1998). Another approach that has made it possible to study phase changes experimentally is microcalorimetry; by allowing very small particles on surfaces to act as their own bolometers, Allen and coworkers have been able to determine heats of fusion, e.g. of large clusters of tin atoms (Lai, et al., 1996). The relevant experimental methods have included electron diffraction, which probes structure directly, and bolometric, spectroscopic and mass-spectroscopic studies, such as the intensity distributions of mass peaks generated by photoionization and photodissociation.

The first fundamental concept one must grasp to understand the thermodynamics of small systems and especially the description of their phase-like behavior is the idea that thermodynamic equilibrium of clusters and nanoscale particles must be the dynamic equilibrium of an ensemble. One may consider the state of a single, isolated particle as well, as being in a kind of equilibrium state that traces a path through its phase space. Such an equilibrium state may truly be a stable mechanical state of an isolated dynamical system, but the path of the system may be very intricate, most likely aperiodic and, we believe, because clusters and nanoscale particles are complicated dynamical systems, mostly chaotic and ergodic. Such systems have no internal, separable modes of motion, and hence lend themselves to descriptions involving some kind of averaging. Because they are presumably ergodic, such a description could emerge from the long time history of a single system, or from the instantaneous state of an ensemble of many, many such systems.

As in all of statistical thermodynamics, the ensemble of choice depends on the conditions of interest: microcanonical, for a constant-energy system; canonical for a constant-temperature system; grand-canonical for a system at constant chemical potential; isobaric, for a system at constant pressure, and so on. Histories of single systems based on computer simulations often yield desired information, dynamical as well as equilibrium, regarding changes of phase or about the phases of clusters themselves. Such histories come from molecular dynamics (MD) simulations, numerical solutions of the equations of Newton's Second Law if the simulations are classical, of the time-dependent Schrödinger (or quantum-mechanical Liouville) equation if the simulations are quantum-mechanical. The dynamical information from MD is limited, because computer-induced, random errors destroy the mechanical reversibility of the integration. In practice, such simulations retain full reversibility for 500 to 10,000 time steps with usual degrees of precision, and lose at least one or two significant figures with each doubling of the length of the trajectory. This means that very slow processes cannot be reproduced reliably with MD simulations as we now carry them out. Nevertheless MD simulations do reveal short-time, high-frequency dynamics. Long MD simulations should be thought of as stochastic models retaining short-time correlations.

The alternative approach to simulation is through Monte Carlo (MC) methods, which explore the system's phase space but use no dynamics to link one step to the next. This is therefore a way to sample an ensemble at an instant, the alternative to a time history and equally valid, for any ergodic system. But because they use no dynamics, MC methods also yield no dynamics. Both methods have been useful, molecular dynamics perhaps a bit more because it gives both dynamical and thermodynamical information.

"Phase transitions in small systems are gradual, not sharp." This is a commonplace that still sometimes appears in discussions of clusters and nanoparticles. In one sense this is correct (Hill, 1963, Hill, 1964). However these changes have a kind of precision and sharpness of their own, which we can find if we explore their nature in a bit of detail. To make our language precise, we shall refer to the changes of phase exhibited by bulk matter as "phase transitions," and their analogues for finite systems, as closely as they come, as "phase changes." The reason is that the changes of phase for small systems are not the same as the phase transitions of bulk systems. We shall see how we can follow the emergence of bulk transitions from the phase changes of small systems. The phase changes of small systems cannot be classified according to "order" in the Ehrenfest pattern of cataloguing according to which derivative of the energy or entropy is the lowest to vanish at the point of the transition. Moreover, as we shall see, the Gibbs Phase Rule loses its meaning because the distinction between "phase" and "component" becomes unclear for small systems.

Another important distinction between the phases of bulk matter and the phase-like forms of clusters and nanoparticles is that there are many varieties of the latter which do not persist in the limit of very large systems. Table 1 lists a menagerie of such forms, with examples or possible examples.

We begin by examining the simplest, best-studied forms of clusters, the solid-like and liquid-like forms, and the passages between these forms. This will inform us regarding the way a first-order transition occurs, and about limits on metastability and the spinodals of bulk matter. Then we go on to the more exotic phase-like forms of clusters and to the question of coexistence of these forms. We conclude by pointing out some of the most challenging open questions.

**Table 1.** A Menagerie of Phase-Like Forms that Clusters May Exhibit

| Phase-Like Form | Example |
| --- | --- |
| solid | any cluster but $He_n$, at low enough T |
| soft solid (or "fluxional cluster") | 6-particle metal clusters (Sawada, et al., 1989); $Au_{55}$(Sawada, et al., 1992). |
| liquid | $Ar_7$ (Berry, et al., 1988, Briant, et al., 1975) |
| surface-melted | $Ar_{55}$ (Cheng, et al., 1991, Cheng, et al., 1992a, Kunz, et al., 1993, Kunz, et al., 1994, Nauchitel, et al., 1980) |
| core-melted | possibly $Ga_n$ or $(H_2O)_n$ (Kunz, et al., 1994) |
| glassy or amorphous | $Ar_n$ or mixed rare-gas clusters; $(KCl)_n$ (Amini, et al., 1979) |
| "restricted liquid" | $Li_8^+$ (Jellinek, et al., 1994) |

# 2 Solid and Liquid Clusters and Their Equilibria

## 2.1 Solid-like and Liquid-like Forms of Clusters

At sufficiently low temperatures or energies, all clusters with the exception of those of helium and possibly of hydrogen molecules behave like solids. Their component atoms or molecules undergo nearly-harmonic, small-amplitude vibrations around the equilibrium sites to which they are bound. In simulations, the Lindemann criterion (Lindemann, 1910, Lindemann, 1912), that the relative root-mean-square deviation of nearest-neighbor distances be less than about 0.1, is satisfied (Berry, et al., 1988). Very little diffusion occurs; the mean square displacement of the particles with time is nearly zero (Berry, et al., 1988). The velocity autocorrelation functions have no very-low-frequency components, meaning that there are no very soft, diffusive modes of motion in these clusters (Berry, et al., 1988). Many clusters exhibit well-ordered geometries, but many of these are not geometries consistent with periodic lattices. For example many kinds of atomic clusters have solid-like structures based on icosahedra. Clusters of no more than a few thousand rare-gas atoms generally take on variations of icosahedral geometries in the lowest-energy or global-minimum structures on their potential surfaces, although there are a few sizes for which the global minima are close-packed; much larger clusters have close-packed, lattice-based structures. The precise way this change occurs as the number of component atoms increases is not yet understood, but how it happens and at what size depend on the range of the interatomic forces (Doye, et al., 1995c, Wales, et al., 1995). There is sound experimental evidence for such structures (Bartell, 1986, Bartell, et al., 1988, Farges, et al., 1983, Farges, et al., 1986, Lee, et al., 1987, Raoult, et al., 1989, Torchet, et al., 1990a, Torchet, et al., 1990b, Valente, et al., 1984a, Valente, et al., 1984b). Simulated cold clusters show pair distribution functions with the sharp peaks of successive shells of neighbors, like an ordered solid, and angular distributions likewise are characteristic of solid-like structures. For example icosahedral clusters show angular distributions with negligible amplitude at 90°, an angle that simply does not occur for triples of neighbors in that structure (Quirke, et al., 1984).

At higher temperatures, many kinds of clusters show liquid-like behavior in simulations (Berry, et al., 1988). The particles exhibit mean square displacements that increase linearly as functions of time, corresponding to well-defined diffusion coefficients, until the displacements reach the linear dimension of the cluster. The Lindemann criterion parameter is typically well above 0.10 for warm enough clusters. The velocity autocorrelation function has a significant contribution from very-low-frequency modes, which are the soft modes of a liquid; these show clearly in the Fourier transforms. The pair distribution function and the angular distribution function have the broad form characteristic of liquids. Experimentally, a few instances of liquid clusters have been identified (Bartell, et al., 1989). Simulations first revealed such forms (Amini, et al., 1979, Briant, et al., 1975, Cotterill, et al., 1973, Cotterill, 1975, Damgaard Kristensen, et al., 1974, Etters, et al., 1975, Etters, et al., 1977b, Kaelberer, et al., 1977, Lee, et al., 1973, McGinty, 1973) and implied one of the conditions for such liquid-like forms to be recognizable forms of clusters: the liquid had to persist long enough for the system to establish at least vibrational equilibrium in that form. Many of these same simulation studies indicated that clusters of certain sizes, e.g. $Ar_7$ (the smallest of the Lennard-Jones clusters to show this), $Ar_{13}$ and $Ar_{19}$, all modeled by pairwise Lennard-Jones potentials, and others such as $Au_6$ (now the smallest to show solid-liquid equilibrium), $Au_7$, $Au_{13}$ (Garzn, 1991), and other gold clusters, all modeled with Gupta potentials (Jellinek, 1991; Garzn, 1993) can exhibit well-defined dynamic equilibrium between their solid and liquid forms. The same was confirmed and shown for $Ar_{55}$ soon after (Nauchitel, et al., 1980).

Some aspects of the dynamic equilibrium ion those simulations led to the speculation by Briant and Burton that the solid-liquid change might even be a first-order transition. This was puzzling, since dogma had it that first-order transitions are properties only of bulk matter and cannot occur in small systems.

Certain clusters exhibit, in some range of energy or temperature, a floppiness or fluxional character that allows them to explore only a limited set of potential minima, and not become liquid in the sense of exploring their entire potential energy surface. One is the 6-particle cluster modeled by the Gupta potential, a system studied by Sawada and Sugano (Sawada, 1987); they have also examined other metal clusters in this manner(Sawada, et al., 1992). The tetramers and pentamers of alkali halides exhibit such behavior; they have a kind of nonrigid, planar, phase-like form which can pass readily between open rings and "ladders" or rectangles (Luo, et al., 1987, Rose, et al., 1992). Still another exotic phase-like form of cluster is the very flexible $Li_8^+$ ion, which has a region of its potential surface in which it is extremely flexible and liquid-like, with the one qualification that one atom, at the center of the cluster, cannot participate in the permutational motions that mix all the other seven and make them liquid-like

(Jellinek, et al., 1994). We call this kind of system, in which most but not all the component particles are mobile, a "restricted liquid".

## 2.2 Equilibrium of Solid and Liquid Clusters

We may think of the solid-like cluster as an ordinary, near-rigid molecule with small-amplitude vibrations, and the liquid-like cluster as a sort of nonrigid, fluxional molecule. From this perspective, it is natural to construct a quantum-statistical model to infer and rationalize the temperatures at which these two forms may coexist (Berry, et al., 1984a, Berry, et al., 1984b, Jellinek, et al., 1986). This requires making a model based on the postulate that there is at least some temperature at which the two forms are both locally stable, in this sense: the free energy, expressed as a function of temperature, pressure (or density), and a nonrigidity parameter that serves like an order parameter, has two minima as a function of the nonrigidity parameter at some temperature and pressure. This condition, plus the dynamic condition of long-enough persistence of each phase, are sufficient to describe coexistence of the solid and liquid forms of the cluster. The next step in the logic implies that if such coexistence occurs at all for clusters, it occurs not at a single point or along a single curve in the space of two such variables as pressure and temperature; coexistence of solid and liquid clusters, if it occurs, occurs within a *band* of the space of thermodynamic variables of the system (Beck, et al., 1988a, Beck, et al., 1987, Beck, et al., 1988b, Berry, et al., 1988, Berry, et al., 1984a, Berry, et al., 1984b, Davis, et al., 1988, Davis, et al., 1987).

The reasons for this are straightforward. To begin, the densities of states of solid and liquid forms can be represented as functions of $\gamma$, the nonrigidity or order parameter, at any given temperature. The density of solid-like states is invariably higher at low temperatures, but the density of liquid-like states rises considerably faster with T than that of the solid clusters. This means that the free energy of the solid is lower, at low temperatures, than that of the liquid and is a monotonically increasing function of the nonrigidity parameter there. The only minimum in $F(T, \gamma)$ occurs for some small value of $\gamma$, the nonrigidity parameter, in the solid-like range. However as the temperature increases, the greater density of states for high degrees of nonrigidity, i.e. for large $\gamma$, reduces the free energy for nonrigid forms of the cluster, relative to more rigid forms; the free energy $F(T,\gamma)$ becomes less strongly monotonic and, eventually at a sufficiently high temperature $T_f$, develops a point of zero slope, i.e.

$$\left[ \frac{\partial F(T,\gamma)}{\partial \gamma} \right]_{T_f} = 0$$

at some value of $\gamma$. At temperatures above $T_f$, $F(T,\gamma)$ has two minima, one in the solid-like range and one in the liquid-like range of $\gamma$. But as the temperature continues to increase and the density of states of the liquid-like form becomes larger and larger with respect to that of the solid-like form, the curve of $F(T,\gamma)$

as a function of $\gamma$ continues to tip, more and more, toward the liquid-like side until the system reaches a temperature $T_m$ at which the minimum in $F(T,\gamma)$ near the solid-like end of the scale turns into just a flat spot. At temperatures above $T_m$, $F(T,\gamma)$ has only a single minimum, and that is in the liquid-like region. This is illustrated in Fig.1.

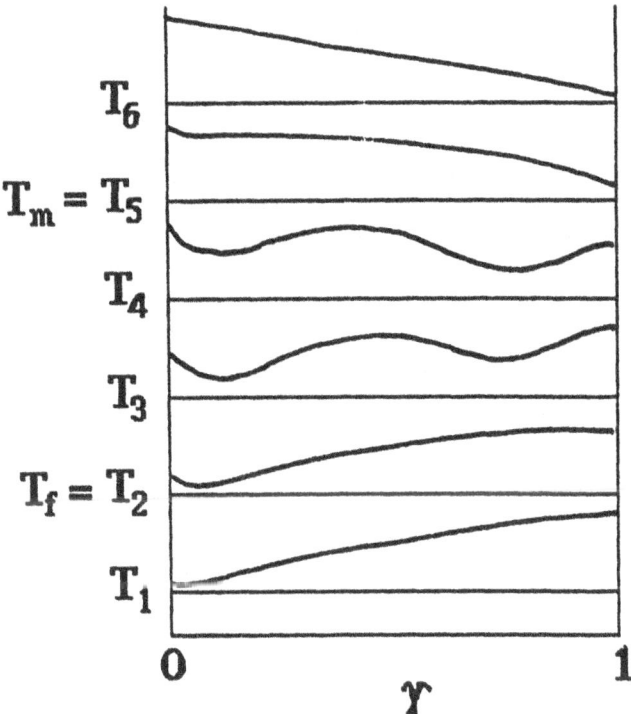

**Fig. 1.** The free energy $F(T,\gamma)$ as a function of the nonrigidity parameter $\gamma$, for six temperatures, increasing from $T_1$ through $T_6$. Below $T_2$, $F(T,\gamma)$ has only one minimum. near $\gamma = 0$, in the solid-like end of the scale. At $T_2 = T_f$, $F(T,\gamma)$ develops a point of zero slope near the nonrigid limit, i.e. near $\gamma = 1$. Between $T_2 = T_f$ and $T_5 = T_m$, $F(T,\gamma)$ has two minima, that for lower $\gamma$ corresponding to a locally stable solid-like form and that for the higher $\gamma$, to a locally stable liquid. At $T_5 = T_m$, the free energy has only one minimum and one other point of zero slope and zero second derivative, as a function of $\gamma$. Above $T_5 = T_m$, the free energy has only one minimum, corresponding to the one stable form.

This argument implies that between the lower limit $T_f$ and the upper limit $T_m$, the solid and liquid forms of the cluster may coexist, all at the single pressure for which the curves were constructed. Hence at each pressure there must be a range of temperature within which the free energy has two minima, therefore two locally stable forms, therefore two coexisting forms. The relative amounts of these two forms is fixed by a chemical equilibrium constant,

$$K_{eq} = exp[-\Delta F(T, \gamma)/kT].$$

The argument also implies that $K_{eq}$ should have two discontinuities, one at $T_f$ and another at $T_m$, for each pressure. However the relative amounts of solid and liquid, fixed as the ratio $K_{eq}$, vary smoothly between these limits, with the ratio of liquid to solid increasing with T (Berry, et al., 1988, Berry, et al., 1984a, Berry, et al., 1984b). The analytic argument has been well supported by simulations, both constant-energy and constant-temperature, and by simulations based on both molecular dynamics and Monte Carlo methods (Amar, et al., 1986, Beck, et al., 1988a, Beck, et al., 1987, Beck, et al., 1988b, Berry, et al., 1988, Blaisten-Barojas, et al., 1986, Blaisten-Barojas, et al., 1987, Davis, et al., 1988, Davis, et al., 1987, Garzon, et al., 1989, Honeycutt, et al., 1987, Sawada, 1987, Sawada, et al., 1989, Sawada, et al., 1992). One characteristic signature is a bimodal form for the distributions of various properties. One used frequently as such an index is the distribution of (short-term) mean kinetic energies, i.e. of mean vibrational temperatures, in isoergic dynamic simulations of the cluster's evolution. Likewise, another such signature is a bimodal distribution of total energies and potential energies in isothermal molecular dynamics simulations, if the conditions of temperature and pressure correspond to the range of coexistence. Outside the coexistence range, the distributions are unimodal. The ratio of the fraction of the time the cluster spends as a solid to that spent as a liquid is just the equilibrium constant $K_{eq}$.

The relation between the phase change between solid and liquid clusters and the solid-liquid first-order phase transition of bulk matter now becomes apparent from this argument. To see this relation, it is easier to think in terms of a transformation of $K_{eq}$, specifically the equilibrium distribution function

$$D_{eq} = (K_{eq}-1)/(K_{eq} + 1),$$

rather than of $K_{eq}$ itself. The reason is that $K_{eq}$ varies from zero to infinity, with the value 1 when the amounts of solid and liquid are equal, while the function $D_{eq}$ varies between -1 if the system is all solid and +1 if it is all liquid. At low temperatures, below $T_f$, $D_{eq}$ is a constant -1; at $T_f$, it shows a discontinuity, and rises to some finite value greater than -1. Between $T_f$ and $T_m$, $D_{eq}$ rises monotonically and smoothly, presumably going through zero and up from negative to positive values. Then, at $T_m$, it has another discontinuity, and rises to its high-temperature limit of +1. Thus, between these two limiting temperatures, solid and liquid clusters coexist in a canonical ensemble; alternatively, a single cluster, at constant temperature, passes back and forth between solid-like and liquid-like forms, so long as that temperature lies between $T_f$ and $T_m$. As we shall see in the context of finite analogues of second-order transitions, it is possible that a phase-like form of a cluster might be stable and present in detectable quantities, yet, under no conditions, have a chemical potential equal to or lower than that of any other phase-like form.

If the cluster is small, then the discontinuities in $D_{eq}$ are detectably large and the transition from negative to positive values is gradual. As the cluster gets larger, the magnitudes of the discontinuities at $T_f$ and $T_m$ decrease, the sigmoidal curve of $D_{eq}$ becomes steeper and more abrupt, and the values of $D_{eq}$ remain close to -1 and +1 until $D_{eq}$ comes close to $T_{eq}$, the value of T at which $D_{eq} = 0$, where almost all its change in value occurs. In other words, the discontinuities get smaller and smaller and the continuous change of equilibrium constant becomes sharper and sharper. There is still an open question of whether fluctuations destabilize the undercooled liquid or superheated solid so much that the discontinuities in $D_{eq}$ are observable (Fisher, et al., 1985, Fisher, et al., 1986, Privman, et al., 1983). If the cluster approaches macroscopic size, the discontinuities are immeasurably small and the continuous change of $D_{eq}$ from very near -1 to very near +1 takes place so abruptly that it is in effect discontinuous at $T_{eq}$! That is how a first-order melting and freezing transition emerges from its counterpart in small systems. Fig.2 illustrates the behavior of $D_{eq}$ schematically for three sizes of clusters; the figure is drawn to show all three clusters with the same $T_{eq}$, but this is only done here to emphasize the evolution of the shape of $D_{eq}$ with N. In reality, $T_{eq}$ can be expected to vary with N.

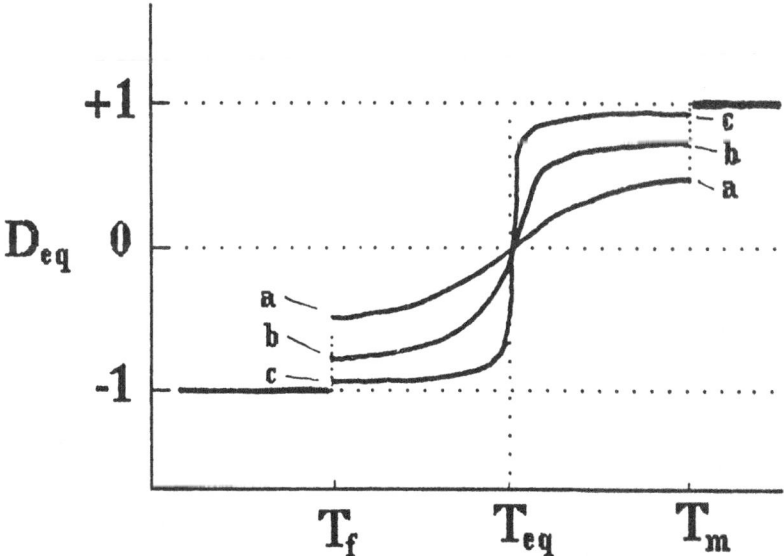

**Fig. 2.** Schematic curves of $D_{eq}$ for a) a small cluster, b) a cluster of intermediate size, and c) a large cluster.

A variety of *sufficient* conditions were found in which clusters can coexist in solid and liquid phase-like forms, both in simulations (Beck, et al., 1988a, Beck, et al., 1987, Beck, et al., 1988b, Blaisten-Barojas, et al., 1986, Briant, et al., 1975, Etters, et al., 1977a, Etters, et al., 1975, Etters, et al., 1977b, Honeycutt, et al., 1987, Jellinek, et al., 1986, Kaelberer, et al., 1977, McGinty, 1973) and analytically (Berry, et al., 1984a, Wales, et al., 1990). The analytic results follow from a model in which the nonrigidity is expressed by a free energy which is a polynomial function of the density of defects. The model is phenomenological insofar as the exact nature of the defects need not be specified. From this, a statistical theory, quantum or classical, leads to the sufficient conditions that the free energy have two minima within some finite range of temperature. Furthermore, provided the defects either attract each other or lower the vibrational frequencies of the cluster, the interval between the two minima persists as $N$ becomes infinite. In other words, clusters of substances whose free energy has such a dependence on defects have solid and liquid forms whose phase changes merge smoothly into conventional first-order phase transitions as the clusters grow to become bulk matter.

More recently, *necessary* conditions were also found for the coexistence of solid and liquid (or any other two dense, phase-like) forms of clusters (Wales, et al., 1994; Lynden-Bell et al., 1994; Doye et al., 1995b). These conditions are expressed as the requirements to assure that the distribution of short-time mean temperatures of a microcanonical ensemble, or the distribution of short-time mean free energies of a canonical ensemble have a bimodal distribution. Two sets of necessary conditions are that there be three points of inflection in the canonical distribution (for bistability in a microcanonical ensemble) or in the grand canonical distribution (for bistability in a canonical ensemble), respectively. The next section addresses the question of what is meant by a "short-time mean" temperature or free energy.

One intriguing aspect of phase coexistence of clusters appears in the caloric curves, the representations of the mean temperature as a function of energy, for an isoergic system, or of the mean energy as a function of temperature for an isothermal system. The slope of a curve of $T$ vs. $E$, subject to constraints, is of course the heat capacity of the system under the conditions of those constraints. Such curves have been constructed for many years (cf. Briant, et al., 1975). These typically are nearly linear in the regions of single phases but in the coexistence regions, may be flat or even, in the case of curves based on microcanonical distributions, have negative slope (Labastie, et al., 1990). The s-shapes of such curves as found from molecular dynamics simulations, at least for $Ar_{13}$ and $Ar_{55}$, match caloric curves based on model partition functions, provided those partition functions include the anharmonic contributions to the potential surface (Doye, J. P. K., et al., 1995a). It seems that a region of negative slope, corresponding to a region of negative heat capacity, is sufficient for coexistence of two phases of clusters. However it is not *necessary* for the caloric curve to have a region of negative slope in order to have dynamic bistability; an outstanding example is the $Ar_{13}$ cluster (cf. Doye, et al., 1995b).

## 2.3   Time Scales and Phases of Small Systems

The possibility of observing solid-like and liquid-like clusters depends, as we have seen, on the individual clusters spending long enough intervals in one phase-like form to establish equilibrium-like properties characteristic of that phase. This requires that the mean time spent in one such phase be long relative not only to the vibrational period of the components, but also to the time required for the system to establish well-defined mean-square displacements $\langle d^2(t) \rangle$ of its particles, which means it must establish well-defined diffusion coefficients $D$ since these are directly related:

$$D = (1/6)[d \langle d^2(t) \rangle / dt]$$

in three dimensions.

The system must also establish a stable velocity autocorrelation function for a time long enough for that autocorrelation function to have a stable Fourier transform, which means that the distribution of vibrational frequencies must be stable. Clusters even as small as $Ar_7$ do exhibit this behavior. However many other small clusters, such as $Ar_{12}$ and $Ar_{14}$, do not; they pass between solid-like and liquid-like forms too frequently to establish such properties for purposes, for example, of infrared spectroscopy. In their infrared or Raman spectra, a canonical ensemble of $Ar_7$, $Ar_{13}$ or $Ar_{19}$ would look like a collection of solid particles at low temperatures, like a collection of liquid particles at high temperatures (but below temperatures at which the particles would vaporize within the time scale of the observation), and at intermediate temperatures, like a mixture of solid and liquid particles. By contrast, a similar ensemble of clusters of $Ar_{12}$, $Ar_{14}$ or $Ar_{17}$ would look like a collection of slush balls at those intermediate temperatures. If the probe were very much slower than infrared spectroscopy, for example if it were a radio-frequency probe or some other means requiring microseconds or milliseconds, then the observer would see the long-time average behavior—on that time scale—of all the clusters in the ensemble, and would therefore interpret the behavior of the clusters as slush-like, whatever the size of the clusters. This is because the time scale for residence in solid-like and liquid-like forms for $Ar_7$, $Ar_{13}$ and $Ar_{19}$ is of order 50-500 ps, long relative to vibrational line widths but very short compared with times for absorption or emission of a radio-frequency photon. What we mean by "equilibrium" of these clusters depends on the time scale we intend for their description. On a short time scale, they exhibit traditional, distinguishable phases, but on a long time scale, they transform smoothly from solid through slush to liquid.

It is precisely because the time scales of the dynamics of clusters fall into ranges we can span with our experimental methods that we become sensitized to these fundamental ideas. The same concepts lurk even in some of the most elementary-seeming concepts of thermodynamics, but there, they are easy to overlook because the readily-ignorable time scales do not force us to confront them. For example the notion of a "reversible adiabatic process" would be an oxymoron, were it not possible to separate the time scale for its internal thermal equilibration from the much longer time scale for loss of heat to the surroundings (Woods, 1975).

Time scales are not only important for distinguishing what we mean by a "state" and "equilibrium". They are also important because what we can extract from a simulation may depend on the time scales of the computation as well as of the dynamics it seeks to simulate. At the short-time end of the scale, it is important to ask whether or not the results one seeks require accurate representation of the vibrations. If so, then the time steps must be significantly shorter than the vibrational period. This is an obvious conclusion. Speeding up a molecular dynamics computation by using long time steps can behazardous, if one intends to extract dynamical information. Yet it is at the long end of the time scale that inferences are more likely to be misled. For example one may in principle infer from isothermal molecular dynamics simulations the free energy differences between two forms of a cluster in dynamic equilibrium. If the system is ergodic, then the equilibrium constant is the ratio of the times spent in the two forms. However the duration of a simulation long enough to provide a stable value for that ratio may be very long indeed, compared with most current simulations. Moreover it is also important for molecular dynamics to recognize another time scale mentioned previously, which is associated not with the system but with the computation: this is the time scale over which the computation remains mechanically reversible. Typically, this is about 5,000 to 10,000 time steps. Reversing dynamics calculations longer than these typically reveals a loss of significant figures if one tries to recover the initial conditions. This means that long molecular dynamics calculations are not really following reversible, Hamiltonian dynamics, even though they may remain isoergic to within very narrow limits. Rather, they become a kind of slowly-randomized on-shell or isoergic walk with strong correlations for short times but none for very long times.

## 2.4   Phase Rule, Maxwell Construction and Phase Diagrams

The behavior of solid-liquid equilibrium of clusters seems at first sight to contradict the Phase Rule for the number of stable phases $p$, for a system with $c$ components and $n$ degrees of freedom, $n = c - p + 2$, and the Maxwell construction argument against any second phase existing except along the binodal and the "equal-area" tie-line. In a sense this is correct, but for a reason that makes the behavior of clusters fall outside the realm where these two traditional ideas apply. The Maxwell construction argues that between the binodal and the spinodal, only the phase of lower chemical potential $\mu$ can be stable because the free energy $N\mu$ of the less stable phase would be enormously higher, effectively infinitely higher, than the free energy of the more stable phase. Clusters, by contrast with bulk matter, have values of N small enough that the free energy of the less stable form may be not so very much higher than that of the more stable form, so that the ratio

$$K_{eq} = exp[-\Delta F(T, \gamma)/kT]$$

is still large enough even near the spinodal that both forms may be detected in an ensemble or in the time history of a single system, if it is in the coexistence region.

We use the Phase Rule when we implicitly distinguish phases from components in chemical equilibrium by tacitly supposing that the free energy difference between phases is effectively $\pm\infty$ at all points where it is not zero, while the free energy difference between components may take on any value. Hence if our systems are clusters and not bulk matter, the former supposition does not apply, and the distinction between phases and components is lost–so the Phase Rule is irrelevant to ensembles or time histories of clusters.

Thus far, the discussion has neglected effects of pressure on the behavior of clusters; we have discussed only simulations at constant pressure, usually either not specified or zero. Of course pressures may vary, and with them, the temperature dependence of free energy relationships. The full phase behavior of clusters may be simulated (Cheng, et al., 1992b) by using the Nosé method (Nosé, 1984a, Nosé, 1984b, Nosé, 1991) or now, the stochastic method (Kast, 1995, Kast, 1996) to maintain constant temperature and the Anderson method (Andersen, 1980) to keep the simulated system isobaric. The results show how the distribution $D_{eq}$ between solid and liquid forms changes with pressure as well as with temperature. Furthermore these results give us enough insight to construct phase diagrams for clusters and nanoparticles that are somewhat different from–and extensions of–conventional phase diagrams (Cheng, et al., 1992b) . In particular, the phase diagram for a cluster requires, in addition to the traditional thermodynamic variables such as pressure p and temperature T, an additional variable which we most naturally choose to be $D_{eq}$, the distribution function introduced previously. The phase diagram for solid-liquid equilibrium of a *bulk* system in this representation contains no more information than that in a conventional plot of vapor pressure vs. temperature; off the curve along which $\mu_{vap} = \mu_{liq}$, the system is either all solid with $D_{eq} = -1$ (at low temperatures) or all liquid with $D_{eq} = +1$ (at high temperatures) and the transition between these two values of $D_{eq}$ is so abrupt that it appears to be a discontinuity. The phase diagram thus consists of a half-plane with $D_{eq} = -1$ and a half-plane with $D_{eq} = +1$, and the two are separated along the curve of the equilibrium vapor pressure.

A small cluster presents a much richer diagram. In the solid-liquid phase diagram of the small cluster, there are indeed discontinuities in $D_{eq}$, but not at $p(T_{eq})$. These discontinuities occur where $D_{eq}$ is truly discontinuous, along the curves of $p(T_f)$ and $p(T_m)$. These discontinuities separate, respectively, a half-plane where $D_{eq} = -1$ from a curved surface that joins that plane at the lowest fractional values of $D_{eq}$, and then, that curved surface at the highest fractional values of $D_{eq}$ with the half-plane where $D_{eq} = +1$. The new information in such a diagram is in the curved surface between the two half-planes. Fig. 3 shows schematic representations of the two cases just described. As yet, no such diagram has been presented for any real system, although there have been steps toward such a construction (Cheng, et al., 1992b; Doye, et al., 1995b; Vekhter, 1997).

14

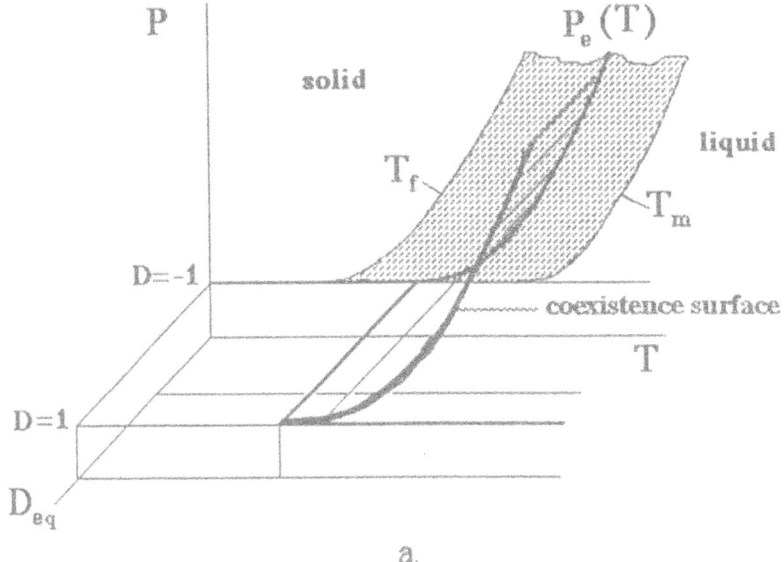

a.

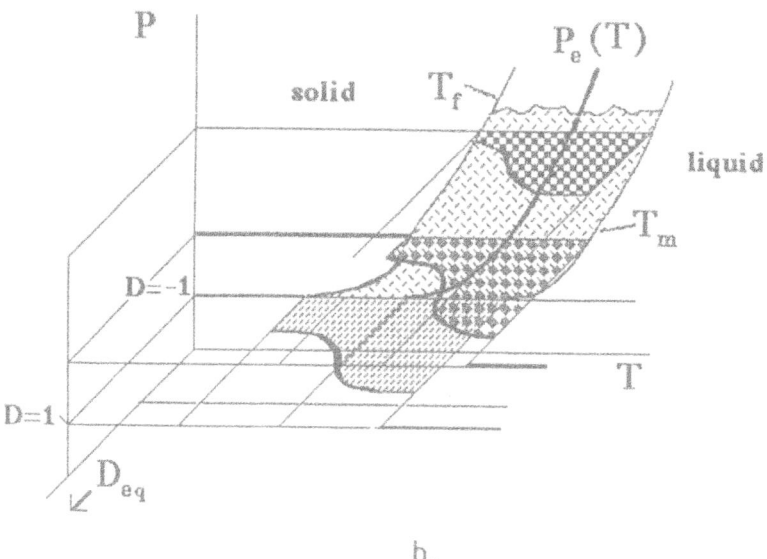

b.

**Fig. 3.** Schematic phase diagrams for solid-liquid equilibrium of finite systems: a) a large system, and b) a system of moderate size, e.g. $N$ of about 20 or 100. The three axes correspond to pressure p(vertical), temperature T (horizontal, in the plane of this surface), and $D_{eq}$, the distribution (projecting horizontally out of the plane of this surface).

## 2.5  Limits on Metastability of Bulk Phases

This line of reasoning has one important implication for bulk matter, specifically for the spinodal and for the metastable superheated solid and undercooled or supercooled liquid. The logic based on the densities of states of solid and liquid forms of finite systems implied the existence of the limiting temperatures $T_f$, below which the liquid has no local stability, and $T_m$, above which the solid has no local stability. We also reviewed conditions for the coexistence of two free energy minima, each responsible for the stability of one phase, and for the persistence of two such minima as $N$ grows arbitrarily large. If a system with very large $N$ is not at its $T_{eq}$, then what is the role of the higher-energy minimum of the free energy $F(T, \gamma)$? To see this, suppose we cool an ensemble of very large liquid systems and, instead of allowing the ensemble to come to thermodynamic equilibrium, we keep all of it in the vicinity of its local, liquid-like minimum, as an ensemble of undercooled liquid systems. This local stability can be maintained at temperatures below $T_{eq}$ just so long as the local minimum in the free energy $F(T, \gamma)$ is there. If that minimum disappears, then so does local stability. Hence the liquid branch of the spinodal exists down only to the temperature $T_f$, and the solid branch of the spinodal, up only to the temperature $T_m$. There are limits to the temperatures, then, beyond which metastable solid and liquid phases cannot exist. In other words, at any pressure, there are temperature limits to the two branches of a family of spinodal curves.

## 3  "Surface-Melted" Clusters and Coexistence of Multiple Phases

Surface melting of rare-gas clusters seemed plausible to Briant and Burton (Briant, et al., 1975), when they carried out molecular dynamics simulations of small systems, and appeared unambiguously in the simulations of the 55-atom Lennard-Jones cluster by Nauchitel and Pertsin (Nauchitel, et al., 1980). It is now well established in simulations of metal clusters as well, with various representations of the potential (Jellinek, 1995, 1996). A softening, sometimes called "premelting," occurs in somewhat smaller clusters, e.g. of Ni (Garzn, 1992), at temperatures a bit below that at which homogeneous liquid appears. This however is probably not simply related to surface-melting, for reasons that will become clear in the next paragraph. The inference that the surface is liquid and the core, solid comes from standard diagnostics–diffusion coefficient, velocity autocorrelation function and pair distribution function, for example–with the particles comprising the cluster kept in separate categories, corresponding to the layers of the cluster. Furthermore snapshots of clusters with liquid surfaces show unstructured, amorphous outer layers and ordered, polyhedral or crystalline cores. Typically, in these snapshots, a few atoms float in the region just outside the outer layer of the cluster. Lennard-Jones clusters of about 45 or more particles display this behavior within bands of temperature beginning just below and somewhat narrower than the bands of liquid-solid coexistence.

The surface-melted cluster was assumed to have a polyhedral core and an amorphous, swarm-like liquid coat–until animations revealed something quite different (Cheng, et al., 1991, Cheng, et al., 1992a). Animations constructed from dynamics simulations of such clusters as $Ar_{147}$ (or, more precisely, the Lennard-Jones cluster "$LJ_{147}$") show that in the "surface-melted" state, a) the cores are indeed solid-like polyhedra in this range of sizes, b) most of the particles of the outer layer execute large-amplitude, highly anharmonic, complex oscillations, and c) a few particles, about 1 in 30 of those in the outer layer, have come out of the surface and float relatively freely around the cluster's surface. These "floaters" carry the large-amplitude, low-frequency motion and contribute the values of the numerical diagnostics that imply that the surface is liquid-like. The other particles that remain in the surface, undergoing large-amplitude vibrations, move in a highly collective manner, oscillating around a well-defined polyhedral equilibrium structure. In other words, the surface-melted state, while so distinct and self-organized that it behaves as much like a distinct phase as do the normal solid and liquid forms, is not at all like a conventional liquid coating a conventional solid. This is especially significant in the context of nanoparticles because it implies that the "melted surface" of the cluster in the surface-melted, phase-like form *does not provide nuclei to initiate normal, homogeneous melting.* This in turn implies that the standard argument, that materials whose liquids wet their solid forms cannot be superheated, is not applicable to clusters, and therefore that clusters may, at least in some cases, be superheated. This may be useful in the fabrication and annealing of nanoscale particles.

Lennard-Jones clusters of about 45 or more atoms simulated at constant energy indeed exhibit bimodal and multimodal distributions of short-time-average mean kinetic energies (times long with respect to the vibrational period, short with respect to dwell times in a given phase-like form)–that is, of mean vibrational temperatures; in isothermal simulations, they show bimodal or multimodal distributions of short-time-average or even of instantaneous potential or total energy (Kunz, et al., 1993, Kunz, et al., 1994). The plateaus in these bimodal and multimodal distributions are not all of the same kind. Just above the lowest temperatures where the distributions are unimodal and the clusters are solid, the bimodal distributions correspond to clusters that are sometimes solids and sometimes in their surface-melted phase. At still higher temperatures the distributions become trimodal, then bimodal again, and finally, at the highest temperatures before evaporation dominates the simulations, unimodal. In the trimodal region, the three plateaus correspond to solid, surface-melted and homogeneously-melted phases. The bimodal region above that is the region of dynamic coexistence of surface-melted and liquid phases, and the high-temperature, unimodal region is that of the homogeneous liquid.

We can express the partition function of such clusters in terms of contributions from their cores, from their surfaces and from their floaters, and include the floater-surface and surface-core interactions. These contributions can be expressed in terms of parameters reflecting the energy required to produce defects and floaters (Kunz, et al., 1993, Kunz, et al., 1994). With such partition functions, we can search for the limits of the conditions of stability of each phase-like form, just as with the solid-liquid equilibrium. It is convenient to study the results in the form of a kind of phase diagram, such as that in Fig. 4. This Figure was constructed with parameters that made it correspond to the diagram found from MD simulations of $Ar_{55}$ and $Ar_{147}$. It reveals a region of two-phase equilibrium of the solid and surface-melted forms, a region of equilibrium of three phases, solid, surface-melted and liquid, a region of two-phase equilibrium of surface-melted cluster and solid-like cluster, and a region in which only the liquid is stable.

Not all plausible diagrams of this type have precisely these regions of stability or mutual stability. It is possible to have non-overlapping regions of bistability, for example (Kunz, et al., 1993, Kunz, et al., 1994). There need not be a tristable region, although the Lennard-Jones system does show one. In fact, with suitable and plausible parameters, this theoretical framework predicts that there could be substances that could have, instead of a surface-melted phase, a core-melted or frozen-surface phase. Naturally one condition for such behavior is that the liquid be denser than the solid. The likely candidate clusters for a frozen-surface phase are those of gallium, indium and even water. Clusters in this state would be much like ice cubes that have not yet frozen through, but have liquid centers.

## 4   Some Unsolved and Open Questions

Most of the discussion thus far has dealt with phases and phase changes that are now moderately well understood. However the study of phase-like forms of clusters and nanoparticles can hardly be said to be complete. We have already pointed out that the role of fluctuations in metastable, superheated or under-cooled systems is still uncertain, at least with regard to whether they would mask the discontinuies of the distribution $D_{eq}$. There are some others that deserve mention.

One is the question of what the finite-system analogue of a second-order phase transition would be (Proykova, et al., 1997). The true second-order transition of bulk matter has only a single stable minimum with respect to the order parameter, at any temperature. The value of the order parameter at which the free energy is a minimum changes with the system's temperature, and this value moves from a region in which one phase is stable to a region in which the other phase is stable. An analogue of this second-order transition for finite systems might be one which has only a single minimum with respect to the order parameter, for all sizes of clusters. This would be the most obvious formal analogue.

But there is another possibility: the system might have two minima for very small or even moderate-size systems, but these minima might converge to a single, common, stable value as $N$ grows very large. This can indeed happen if the defects repel each other, or if they raise the frequencies of the normal modes of the cluster.

It may even be that both possibilities occur, and that there are two kinds of second-order transition, one emerging from each of the two conditions. One type would emerge from a single minimum for all $N$, and the other, from the coalescence of two minima in the free energy for small $N$, as the cluster grows large. This question is under study at the time of this writing (Proykova, et al., 1997). The second possibility raises a subsidiary question, of whether the convergence occurs at some finite $N$ or only as $N$ goes to infinity. Still another, related possibility is that the two minima approach but do not quite merge for very large $N$. This case is clearly a prototype for a weak first-order transition, associated with a very small but nonzero latent heat.

In small systems, for which multiple phase coexistence is possible, an intermediate phase may appear as an intermediate between the high- and low-temperature forms and yet, under no conditions, be the most stable phase. If this occurs, the intermediate phase would be found in clusters but not in bulk matter, where its free energy would everywhere be above that of one or another phase. Such a case may occur in clusters of $TeF_6$, which exhibit an intermediate, monoclinic phase between two others found both in clusters and bulk $TeF_6$ (Proykova, et al., 1997).

One other open issue deserves mention. It appears from still-unpublished results that clusters may exhibit solid-liquid critical points. The conditions for this to occur are entirely consistent with the finite, nonperiodic nature of the cluster. The stability condition for two phases is, in effect, the condition that the change of entropic contribution to the free energy difference of the phases be matched by the change in energy or enthalpy. If this condition is met at two points then two phases are in equilibrium. This can be put in terms of two curves that cross at a point. If the two crossing points were to become a single tangent point, that tangent point would be a limit, beyond which the two phases could no longer be distinguished as equilibrium forms of the substance. In short, the point of tangency would be the critical point. It appears at present that there is no logical barrier to the existence of such a point for clusters.

It is even possible to argue that there might be solid-liquid critical points for some bulk systems, on the following grounds. The traditional argument that they cannot exist for bulk systems is the symmetry ground, that there can be no continuous transition from the discrete symmetry of the crystalline solid to the continuous translational and rotational symmetry of the amorphous liquid. However as the temperature of a solid increases, so does the density of its defects, so that establishing the translational symmetry of a hot solid requires averaging over successively longer lengths to establish that average discrete translational symmetry. Likewise, as the pressure increases on a liquid, it becomes more and more ordered. This requires averaging over longer and longer lengths as the pressure increases, in order to establish the continuous translational symmetry of the liquid. If these lengths increase only slowly with N, then the symmetry arguments will be valid. However if they increase rapidly enough, or if they were even to diverge as temperature or pressure increase, then the symmetry argument would be inapplicable and there could perfectly well be a solid-liquid critical point for a macroscopic system.

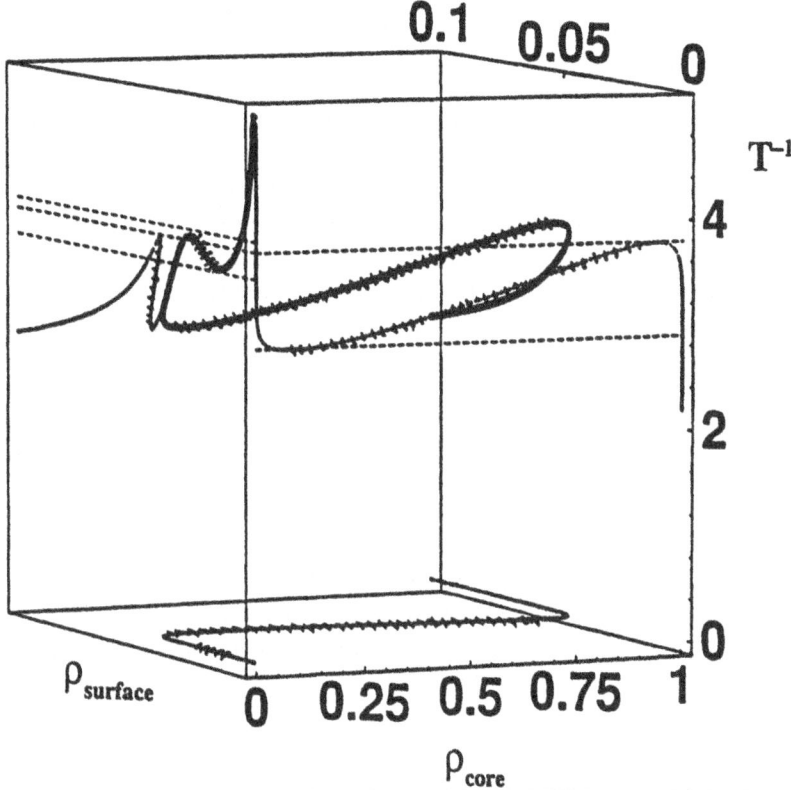

**Fig. 4.** A type of phase diagram for a cluster indicating regions of stability and co-existence for three phase-like forms: solid, surface-melted and liquid the vertical axis measures $T^{-1}$, and the two horizontal axes, the densities $\rho_{surf}$ and $\rho_{core}$, of defects in the surface and core, respectively. The heavy curve is the curve of stability, like that of a van der Waals gas in a pressure-volume plot. The lighter curves are its projection onto the three planes of the graph. Wherever the curve's slope is downward, the system is stable; where it slopes up, the system fails to satisfy a local stability condition. The regions of unstable equilibrium are cross-hatched. Each downward branch corresponds to a stable phase, whose limits are the points where the curve's direction reverses. Thus, in this figure, the solid is stable from the lowest temperatures (highest part of the heavy curve) in the region where there are very few defects in the surface or the core; the surface-melted form is stable only over a finite band of temperature because the curve reverses its direction twice, in the vicinity of very low densities of bulk defects but moderately high densities of surface defects; and the homogeneous liquid is stable in the region of high densities of surface and bulk defects.

# 5  Summary

The phase-like behavior of small clusters leads us not only to a rich variety of new phenomena such as "phases" that do not exist as such in bulk matter, but also to deepened insights into thermodynamics of materials. The solid-liquid equilibria of ensembles of clusters reveal, through the the finite bands of their coexistence regions, the basis of the sharp coexistence curves of bulk solids and liquids, and of the limits of metastability. The coexistence of multiple phase-like forms of clusters and nanoparticles clarifies some of the limits on concepts of "phase" and "component", on the Phase Rule, and on the Maxwell construction normally used to explain the "tie-line" of the solid-liquid phase diagram, with its discontinuities of slope.

Furthermore, in attempting to elucidate the phase-like behavior of clusters, we are forced to reexamine the role of time scales in the thermodynamic interpretation of phenomena when it is possible to make observations on different time scales. Concepts assumed without scrutiny in the context of bulk matter become elusive, requiring care and precision, when we confront them in the context of systems so small that different time scales do not necessarily separate neatly. Careful consideration of time scales is important not only conceptually but technically as well, especially when we deal with molecular dynamics simulations.

Clusters may exhibit coexistence of more than two phases over finite bands of temperature and pressure. This is because, as with solid-liquid equilibria, the various phase-like forms differ from one another by relatively small amounts of free energy, so that detectable amounts of several forms may be present under conditions in which they have nonzero differences in their chemical potentials. Among these forms are not only solid and liquid, but also surface-melted and possibly core-melted (frozen-shell) clusters. This is an example of a situation in which "phase" and "component" cannot be separated.

Open questions remain, such as the nature of the finite analogue or analogues of second-order phase transitions. Another is the question of how the transition from polyhedral to lattice-based structures occur in those clusters, such as the Lennard-Jones, which are polyhedral for small sizes but close-packed in the bulk crystal. Still another is the question of the structures and phases of molecular clusters, of the extent to which these take on the structures of their bulk counterparts even at small sizes, and of what kinds of phase transitions they may show—clearly an issue closely related to that of the second-order transitions. Finally, the phase behavior of clusters is a subject in which theory and simulation have far outpaced experiment; we can hope that recognition of these ideas will serve as stimuli for new laboratory studies of the phase behavior of clusters and nanoparticles.

# 6 Acknowledgments

The author would like to express his debt to his many coworkers whose efforts have made it possible to bring this subject to its present level. The research described here that was carried out at The University of Chicago has been supported by Grants from the National Science Foundation.

# 7 References

Amar, F. and Berry, R. S. (1986), "The onset of nonrigid dynamics and the melting transition in $Ar_7$," J. Chem. Phys. **85**, 5943-54.

Amini, M. and Hockney, R. W. (1979), "Computer Simulation of Melting and Glass Formation in a Potassium Chloride Microcrystal," J. Non-Cryst. Sol. **31**, 447-452.

Andersen, H. C. (1980), "Molecular dynamics simulations at constant pressure and/or temperature," J. Chem. Phys. **72**, 2384-2393.

Bartell, L. S. (1986), "Diffraction Studies of Clusters Generated in Supersonic Flow," Chem. Rev. **86**, 492-505.

Bartell, L. S. (1992), "Inference of Cluster Phase from Considerations of Homogeneous Nucleation in an Evaporative Ensemble," J. Phys. Chem. **96**, 108-111.

Bartell, L. S. and Chen, J. (1992), "Structure and Dynamics of Molecular Clusters. 2. Melting and Freezing of $CCl_4$ Clusters," J. Phys. Chem. **96**, 8801-8808.

Bartell, L. S. and Dibble, T. S. (1991a), "Electron Diffraction Studies of the Kinetics of Phase Changes in Molecular Clusters. Freezing of $CCl_4$ in Supersonic Flow," J. Phys. Chem. **95**, 1159-1167.

Bartell, L. S. and Dibble, T. S. (1991b), "Kinetics of phase changes in large molecular clusters," Z. Phys. D **20**, 255-257.

Bartell, L. S., Harsanyi, L. and Valente, E. J. (1989), "Phases and Phase Changes of Molecular Clusters Generated in Supersonic Flow," J. Phys. Chem. **93**, 6201-6205.

Bartell, L. S., Sharkey, L. R. and Shi, X. (1988), "Electron Diffraction and Monte Carlo Studies of Liquids. 3. Supercooled Benzene," J. Am. Chem. Soc. **110**, 7006-7013.

Beck, T. L. and Berry, R. S. (1988a), "The interplay of structure and dynamics in the melting of small clusters," J. Chem. Phys. **88**, 3910-3922.

Beck, T. L., Jellinek, J. and Berry, R. S. (1987), "Rare gas clusters: Solids, liquids, slush and magic numbers," J. Chem. Phys. **87**, 545-554.

Beck, T. L., Leitner, D. M. and Berry, R. S. (1988b), "Melting and phase space transitions in small clusters: Spectral characteristics, dimensions and K-entropy," J. Chem. Phys. **89**, 1681-1694.

Berry, R. S., Beck, T. L., Davis, H. L. and Jellinek, J. (1988), "Solid-Liquid Phase Behavior in Microclusters," in *Evolution of Size Effects in Chemical Dynamics, Part 2*, Prigogine, I. and Rice, S. A., ed. (John Wiley and Sons, New York), 75-138.

Berry, R. S., Jellinek, J. and Natanson, G. (1984a), "Melting of clusters and melting," Phys. Rev. A **30**, 919-931.

Berry, R. S., Jellinek, J. and Natanson, G. (1984b), "Unequal freezing and melting temperatures for clusters," Chem. Phys. Lett. **107**, 227.

Blaisten-Barojas, E. and Levesque, D. (1986), "Molecular-dynamics simulation of silicon clusters," Phys. Rev. B **34**, 3910-3916.

Blaisten-Barojas, E. and Levesque, D. (1987), "A Molecular Dynamics Study of Silicon Clusters," in *Physics and Chemistry of Small Clusters*, Jena, P., Rao, B. K. and Khanna, S. N., ed. (Plenum Press, New York), 157-168.

Briant, C. L. and Burton, J. J. (1975), "Molecular dynamics study of the structure and thermodynamic properties of argon microclusters," J. Chem. Phys. **63**, 2045-2058.

Buck, U., Schmidt, B. and Siebers, J. G. (1993), "Structural transitions and thermally averaged infrared spectra of small methanol clusters," J. Chem. Phys. **99**, 9428-9437.

Cheng, H.-P. and Berry, R. S. (1991), "Surface Melting and Surface Diffusion on Clusters," Proc. Materials Res. Soc. **206**, 241-252.

Cheng, H.-P. and Berry, R. S. (1992a), "Surface melting of clusters and implications for bulk matter," Phys. Rev. A **45**, 7969-7980.

Cheng, H.-P., Li, X., Whetten, R. L. and Berry, R. S. (1992b), "Complete statistical thermodynamics of the cluster solid-liquid transition," Phys. Rev. A **46**, 791-800.

Cotterill, R. M., Damgaard Kristensen, W., Martin, J. W., Peterson, L. B. and Jensen, E. J. (1973), Comput. Phys. Comm. **5**, 28.

Cotterill, R. M. J. (1975), Phil. Mag. **32**, 1283-1288.

Damgaard Kristensen, W., Jensen, E. J. and Cotterill, R. M. J. (1974), "Thermodynamics of small clusters of atoms: A molecular dynamics simulation," J. Chem Phys. **60**, 4161-4169.

Davis, H. L., Beck, T. L., Braier, P. A. and Berry, R. S. (1988), "Time Scale Considerations in the Characterization of Melting and Freezing in Microclusters," in *The Time Domain in Surface and Structural Dynamics*, Long, G. J. and Grandjean, F., ed. (Kluwer Academic Publishers, Dordrecht) 535-549.

Davis, H. L., Jellinek, J. and Berry, R. S. (1987), "Melting and freezing in isothermal $Ar_13$ clusters," J. Chem. Phys. **86**, 6456-6469.

Dibble, T. S. and Bartell, L. S. (1992), "Electron Diffraction Studies of the Kinetics of Phase Changes in Molecular Clusters. 2. Freezing of $CH_3Cl_3$," J. Phys. Chem. **96**, 2317.

Doye, J. P. K. and Wales, D. J. (1995a), "Calculation of thermodynamic properties of small Lennard-Jones clusters incorporating anharmonicity," J. Chem. Phys. **102**, 9659-9672.

Doye, J. P. K. and Wales, D. J. (1995b), "An order parameter approach to coexistence in atomic clusters," J. Chem. Phys. **102**, 9673-9688.

Doye, J. P. K., Wales, D. J. and Berry, R. S. (1995c), "The effect of the range of the potential on the Structures of Clusters," J. Chem. Phys. **103**, 3061-3070.

Ellert, C., Schmidt, M., Schmitt, C., Reiners, T. and Haberland, H. (1995), "Temperature Dependence of the Optical Response of Small, Open Shell Sodium clusters," Phys. Rev. Lett. **75**, 1731-1734.

Etters, R. D., Danilowicz, R. and Kaelberer, J. (1977a), "Metastable states of small rare gas crystallites," J. Chem. Phys. **67**, 4145-4148.

Etters, R. D. and Kaelberer, J. B. (1975), "Thermodynamic properties of small aggregates of rare-gas atoms," Phys. Rev. A **11**, 1068-1079.

Etters, R. D. and Kaelberer, J. B. (1977b), "On the character of the melting transition in small atomic aggregates," J. Chem. Phys. **66**, 5112-5116.

Farges, J., deFeraudy, M. F., Raoult, B. and Torchet, G. (1983), "Noncrystalline structure of argon clusters. I. Polyicosahedral structure of $Ar_N$ clusters, $20 < N < 50$," J. Chem. Phys. **78**, 5067-5080.

Farges, J., deFeraudy, M. F., Raoult, B. and Torchet, G. (1986), J. Chem. Phys. **84**, 3491.

Fisher, M. E. and Privman, V. (1985), "First-order transitions breaking O(n) symmetry: finite-size scaling," Phys. Rev. B **32**, 447-464.

Fisher, M. E. and Privman, V. (1986), "First-Order Transitions in Spherical Models: Finite-Size Scaling," Commun. Math. Phys. **103**, 527-548.

Garzn, I. L., Borja, M. A. and Blaisten-Barojas, E. (1989), "Phenomenological model of melting in Lennard-Jones clusters," Phys. Rev. B **40**, 4749-4759.

Garzn, I. L. and Jellinek, J. (1991), "Melting of gold microclusters," Z. Phys. D **20** 235-238.

Garzn, I. L. and Jellinek, J. (1992), "Melting of Nickel Clusters," in *Physics and Chemistry of Finite Systems: From Clusters to Crystals, Vol.1*. Jena, P. et al., eds. (Kluwer, Dordrecht), pp. 405-410.

Garzn, I. L. and Jellinek, J. (1993), "Peculiarities of structures and meltinglike transition in gold clusters," Z. Phys. D **26**, 316-318.

Goldstein, A. N., Colvin, V. L. and Alivisatos, A. P. (1990), "Observation of Melting in 30Å Diameter CdS Nanocrystals," Proc. Materials Res. Soc. **206**, 271-274.

Hill, T. L. (1963) *The Thermodynamics of Small Systems, Part 1* (W. A. Benjamin, New York).

Hill, T. L. (1964) *The Thermodynamics of Small Systems, Part 2* (W. A. Benjamin, New York).

Honeycutt, J. D. and Andersen, H. C. (1987), "Molecular Dynamics Study of Melting and Freezing of Small Lennard-Jones Clusters," J. Phys. Chem. **91**, 4950-4963.

Jellinek, J., Beck, T. L. and Berry, R. S. (1986), "Solid-liquid phase changes in simulated isoenergetic $Ar_{13}$," J. Chem. Phys. **84**, 2783-2794.

Jellinek, J. and Garzn, I. L. (1991), "Structural and dynamical properties of transition metal clusters," Z. Phys. D **20**, 239-242.

Jellinek, J., Bonacic-Koutecky, V., Fantucci, P. and Wiechert, M. (1994), "Ab initio HF SCF study of structure and dynamics of $Li_8$," J. Chem. Phys. **101**, 10092-10100.

Jellinek, J., (1995) "Structure, Melting, and Reactivity of Nickel Clusters from Numerical Simulations," in *The Synergy Between Dynamics and Reactivity at Clusters and Surfaces*, Farrugia, L. J., ed. (Kluwer, Dordrecht), pp. 217-240.

Jellinek, J., (1996) "Theoretical Dynamical Studies of Metal Clusters and Cluster-Ligand Systems," in *Metal-Ligand Interactions*, Russo, N. and Salahub, D., eds. (Kluwer, Dordrecht), pp. 325-360.

Kaelberer, J. B. and Etters, R. D. (1977), "Phase transitions in small clusters of atoms," J. Chem. Phys. **66**, 3233-3239.

Kast, S. M., Nicklas, L., Bär, H.-J. and Brickmann, J., "Constant Temperature Molecular Dynamics Simulations by Means of a Stochastic Collision Model. I. Noninteracting Particles," (1994) J. Chem. Phys. **100**, 566.

Kast, S. M. and Brickmann, J., (1996), "Constant temperature molecular dynamics simulations by means of a stochastic collision model. II. The harmonic oscillator," J. Chem. Phys. **104**, 3732.

Kunz, R. E. and Berry, R. S. (1993), "Coexistence of multiple phases in finite systems," Phys. Rev. Lett. **71**, 3987-3990.

Kunz, R. E. and Berry, R. S. (1994), "Multiple phase coexistence in finite systems," Phys. Rev. E **49**, 1895-1908.

Labastie, P. and Whetten, R.L. (1990), "Statistical Thermodynamics of the Cluster Solid-Liquid Transition," Phys. Rev. Lett. **65**, 1567.

Lai, S.L., Guo, J.Y., Petrova, V., Ramanath, G., and Allen, L.H. (1996), "Size-Dependent Melting Properties of Small Tin Particles: Nanocalorimetric Measurements," Phys. Rev. Lett. **77**, 99-102.

Lai, S.L.,Ramanath, G., Allen, L.H., Infante, P., and Ma, Z., (1997), "Heat capacity of Sn nanostructures via a thin-film scanning calorimeter," Appl. Phys. Lett. **70**, 43.

Lee, J. K., Barker, J. A. and Abraham, F. F. (1973), "Theory and Monte Carlo simulation of physical clusters in the imperfect vapor," J. Chem. Phys. **58**, 3166-3180.

Lee, J. W. and Stein, G.D. (1987), "Structure Change with Size of Argon Clusters Formed in Laval Nozzle Beams," J. Phys. Chem. **91**, 2450-2457.

Lindemann, F. A. (1910), Phys. Z. **11**, 609.

Lindemann, F. A. (1912), Engineering **94**, 515.

Luo, J., Landman, U. and Jortner, J. (1987), "Isomerization and Melting of Small Alkali- Halide Clusters," in *Physics and Chemistry of Small Clusters*, Jena, P., Rao, B. K. and Khanna, S. N., ed. (Plenum Press, New York, N. Y.), 201.

Lynden-Bell, R. M. and Wales, D. J. (1994), "Free Energy Barriers to Melting in Atomic Clusters," J. Chem. Phys. **101**, 1460

Martin, T. P., Näher, U., Schaber, H. and Zimmerman, U. (1994), "Evidence for a size dependent melting of sodium clusters," J. Chem. Phys. **100**, 2322.

McGinty, D. J. (1973), "Molecular dynamics studies of the properties of small clusters of argon atoms," J. Chem. Phys. **58**, 4733.

Nauchitel, V. V. and Pertsin, A. J. (1980), "A Monte Carlo study of the structure and thermodynamic behaviour of small Lennard-Jones clusters," Mol. Phys. **40**, 1341.

Nosé, S. (1984a), Mol. Phys. **52**, 255.

Nosé, S. (1984b), "A unified formulation of the constant temperature molecular dynamics methods," J. Chem. Phys. **81**, 511-519.

Nosé, S. (1991), "Constant Temperature Molecular Dynamics Methods," Prog. Theor. Phys. Supplement **103**, 1-46.

Privman, V. and Fisher, M. E. (1983), "Finite-Size Effects at First-Order Transitions," J. Stat. Phys. **33**, 385-417.

Proykova, A. and Berry, R. S. (1997), "Analogues in Clusters of Second-Order Transitions," Z. Phys. D (in press).

Quirke, N. and Sheng, P. (1984), "The Melting Behavior of Small Clusters of Atoms," Chem. Phys. Lett. **110**, 63-66.

Raoult, B., Farges, J., DeFeraudy, M. F. and Torchet, G. (1989), "Comparison between icosahedral, decahedral and crystalline Lennard-Jones models containing 500 to 6000 atoms," Phil. Mag. B **60**, 881-906.

Rose, J. P. and Berry, R. S. (1992), "Towards elucidating the interplay of structure and dynamics in clusters: Small KCl clusters as models," J. Chem. Phys. **96**, 517-538.

Sawada, S. (1987), "Dynamics of Transition Metal Clusters," in *Microclusters*, Sugano, S., Nishina, Y. and Ohnishi, S., ed. (Springer-Verlag, Berlin), 211-217.

Sawada, S. and Sugano, S. (1989), Z. Phys. D **12**, 189.

Sawada, S. and Sugano, S. (1992), "Structural fluctuations of $Au_{55}$ and $Au_{147}$: Substrate effect," Z. Phys. D **24**, 377-384.

Schmidt, M., Kusche, R., Kronmller, W., von Issendorff, B. and Haberland, H. (1997), "Experimental Determination of the Melting Point and Heat Capacity for a Free Cluster of 139 Sodium Atoms," Phys. Rev. Lett. **79**, 99-102.

Schmidt, M., Kusche, R., von Issendorff, B. and Haberland, H. (1998), "Irregular variations in the melting point of size-selected atomic clusters," Nature **393**, 238-240.

Torchet, G., de Feraudy, M. F., Raoult, B., Farges, J., Fuchs, A. H. and Pawley, G. S. (1990a), "Cluster model for the monoclinic to cubic transition in $SF_6$ clusters," J. Chem. Phys. **92**, 6768-6774.

Torchet, G., Farges, J., deFeraudy, M. F. and Raoult, B. (1990b), "Electron Diffraction Studies of Clusters Produced in a Free Jet Expansion," in *The Chemical Physics of Atomic and Molecular Clusters*, Scoles, G., ed. (North-Holland, Amsterdam), 513-542.

Valente, E. J. and Bartell, L. S. (1984a), "Electron diffraction studies of supersonic jets. VI. Microdrops of benzene," J. Chem. Phys. **80**, 1451-1457.

Valente, E. J. and Bartell, L. S. (1984b), "Electron diffraction studies of supersonic jets. VII. Liquid and plastic crystalline carbon tetrachloride," J. Chem. Phys. **80**, 1458-1461.

Vekhter, B. and Berry, R. S. (1997), "Phase coexistence in clusters: An experimental isobar and an elementary model," J. Chem. Phys. **106**, 6456-6459.

Wales, D. J. and Berry, R. S. (1990), "Freezing, melting, spinodals and clusters," J. Chem. Phys. **92**, 4473-4482.

Wales, D. J. and Berry, R. S. (1994), "Coexistence in Finite Systems," Phys. Rev. Lett. **73**, 2875-2878.

Wales, D. J. and Doye, J. P. K. (1995), "Coexistence and phase separation in clusters: From the small to the not-so-small regime," J. Chem. Phys. **103**, 3061-3070.

Woods, L. C. (1975), *Thermodynamics of Fluid Systems* (Clarendon, Oxford).

# Physics of Clusters and Cluster Assemblies

P. Jena, S. N. Khanna, and B. K. Rao
Physics Department, Virginia Commonwealth University
Richmond, VA 23284-2000, USA
E-mail: brao@gems.vcu.edu

**Abstract**. Atomic clusters with a specific size and composition can be viewed as "super-atoms" whose chemistry mimics that of the atoms in the periodic table. A novel class of materials with tailored properties can then be synthesized by assembling these clusters in a variety of ways. This review deals with the physics and chemistry of unusually stable clusters and electronic structure and energetics of crystals where these clusters form the building blocks. Also discussed are the electronic structure and magnetic properties of clusters supported on metallic substrates.

## 1. Introduction

The discovery of magic numbers in alkali metal clusters started a new era in the physics of clusters. Knight *et al.* [1] observed that clusters of Na atoms containing 2, 8, 20, 40...atoms were abundant in the mass spectra. They concluded that the unusual stability of these clusters might have the same origin as the magic numbers in nuclei. In analogy with the nuclear shell structure, they attributed the magic numbers in alkali metal clusters to be due to electronic shell structures. Imagine that a cluster can be thought of as a spherical object where the ionic charges are distributed uniformly throughout the cluster. If the density of this positive charge distribution is $n_0$, the radius, R of the cluster is given by the relation

$$4\frac{\pi}{3}R^3 n_0 = N \qquad (1)$$

where N is the total number of positive charges. These charges are compensated by N valence electrons, thus leaving the cluster electrically neutral. The distribution of the positive charges can be written as

$$n_+(\vec{r}) = n_0\theta(\vec{R}-\vec{r}) , \qquad (2)$$

where $\theta(\vec{R}-\vec{r}) = 1$ if $r \le R$ and $\theta(\vec{R}-\vec{r}) = 0$ if $r > R$. The distribution of N-electrons can then be calculated by taking Eq. (2) as the external potential and following the standard quantum mechanical procedure. A qualitative understanding of the electron energy levels responding to the charge distribution in Eq. (2) can be achieved by considering a simple square well potential,

$$V(\vec{r}) = V_0\theta(\vec{R}-\vec{r}) \tag{3}$$

The energy levels of the electrons confined in this square well are quantized. A schematic plot of these levels is given in Fig. 1. Note that clusters containing 2, 8, 20, 40,...electrons have closed electronic shells. Since the Na atom is monovalent, the number of valence electrons in the cluster equals the number of atoms in the cluster. Thus, the magic numbers in the mass spectra of Na clusters can be attributed to electronic shell closure effect. The total energy calculations of these clusters also indicate that the clusters of 2, 8, 20, 40,...Na atoms are energetically more preferred over their neighboring clusters. A study of the ionization potential [2] of these clusters as a function of size illustrates that the magic clusters are characterized by large ionization potentials that drop abruptly as one increases the cluster size by an additional atom.

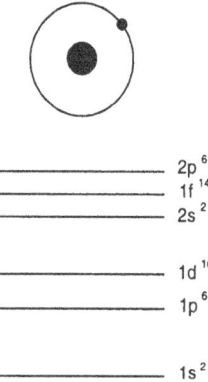

Fig. 1. Energy level structure in a "super-atom" with the cluster core.

This characteristic of the ionization potential is also seen in the atoms [3]. The ionization potentials of atoms are quite large for the rare-gas atoms and drop abruptly as one goes to the next atom in the periodic table. Note that the rare-gas atoms derive their stability and chemical inertness from the electronic shell closures. This analogy points to the expectation that atomic clusters could be made to mimic the chemistry of the atoms. Since the magic Na clusters are expected to be chemically inert, the interaction between two such clusters is expected to be weak. The number of valence electrons of a cluster can be easily altered either by changing their size or composition in such a way that it may have the same electronic configuration as another atom in the periodic table. The ability to alter the composition and size at will provides an unprecedented opportunity to design clusters as "super-atoms". If such clusters can be assembled to form new materials, it will usher in a new era in materials science as we approach the twenty first Century.

   The task of achieving this goal is, of course, enormous. The two major obstacles are: clusters are metastable and have a tendency to coalesce when assembled. They must be prevented from doing so. Secondly, one must find a cost effective way to produce well characterized mono-dispersed clusters in large quantities. This review only deals with the first problem. The two common ways to prevent clusters from coalescing is either to produce clusters that are so stable that they resist the temptation to coalesce and/or to keep them isolated by coating the clusters with a surfactant and/or depositing them on substrates at low enough densities so they remain in as-deposited form. In section II we discuss the design of clusters that mimic the chemistry of atoms. The electronic band structures and cohesive energies of crystals composed of these designer clusters are discussed in Section III. Section IV deals with the energetics and electronic structure of supported clusters. A summary of our conclusions are given in section V.

## 2. Clusters as "Super-Atoms"

While the jellium model of a cluster provides a glimpse of how a cluster's electronic configuration can be tailored to mimic the electronic structure of an atom, it should be emphasized that real clusters are not like a jellium. The geometry of a cluster need not be spherical, and more often than not, they are Jahn-Teller distorted [4]. In addition, the electronic structure of even simple metal clusters may not be free-electron-like, and the orbital degeneracies of the cluster electrons are not governed by the $2(2\ell+1)$ rule as is the case with atoms. Thus, what is needed at first is a determination of the equilibrium geometry of a cluster and how the electronic structure is influenced by the special symmetry imposed by the cluster topology.

Using currently available first principles theoretical procedure and computer architecture, it is possible to determine the equilibrium geometry, binding energy, and electronic structure and properties of compound clusters of a few dozen atoms. We first describe the theoretical procedure briefly and discuss the energetics of a typical set of clusters to demonstrate that real clusters, not just those modelled by a jellium, can mimic the chemistry of atoms and thus can be viewed as "super-atoms".

The starting point of our calculational procedure is to describe the one electron orbitals for a cluster in terms of a linear combination of atomic orbitals [5], namely,

$$\psi_i(\vec{r}) = \sum_\mu C_{i\mu} \phi_\mu(\vec{r}-\vec{R}_\mu) \tag{4}$$

Here $R_\mu$ is the position vector of the $\mu$th atom. The coefficients, $C_{i\mu}$ of linear combination are determined by solving the variational equation,

$$[-\nabla^2 + V(\vec{r})] \; \psi_i(\vec{r}) = E_i \psi_i(\vec{r}) \; . \tag{5}$$

where $V = V_{ion} + V_H + V_{xc}$. Eq. (5) has been expressed here in Rydberg atomic units. The quantities $V_H$ and $V_{ion}$ are the electrostatic potentials arising from the Coulomb interaction between electrons and nuclei respectively. $V_{xc}$ is the exchange-correlation contribution to the potential and is approximated by the local density approximation [6] to the density functional theory. Improvements to this potential can be made by either including gradient corrections [7] to the exchange-correlation energy or by calculating the exchange-correlation potential through more rigorous quantum chemical schemes [5]. However, such procedures are computationally demanding and the local density approximation in many situations provides a reliable description. The total energy of a cluster is calculated by using the approximation,

$$E = \sum_{i\sigma} \langle \psi_i^{\sigma} | -\nabla^2 + V_{ion} | \psi_i^{\sigma} \rangle + \frac{1}{2} \iint \frac{\rho(\vec{r})\rho(\vec{r}')}{|\vec{r}-\vec{r}'|} d^3r \, d^3r'$$

$$+ \int \epsilon_{xc}(\vec{r})\rho(\vec{r}) d^3r + \frac{1}{2} \sum_{i,j} \frac{Z_i Z_j}{|\vec{R}_i - \vec{R}_j|} \quad . \tag{6}$$

Here $\rho$ is the electron density and $Z_i$ is the nuclear charge on the ith nucleus.

To optimize the geometries of a cluster, we start with an arbitrary configuration and calculate the total energy self-consistently using Eqs. (4) - (6). The forces at each atomic site are then computed by taking the gradient of the energy. Each of the atoms is then displaced along the path of steepest descent and the total energy recalculated at the new configuration. These steps are repeated until the forces at each atomic site vanish. To ensure that we have arrived at the global minimum energy configuration and are not stuck at a local minimum, the calculations are repeated with different initial configurations. It is also necessary to compare the energies for different spin multiplicities. All these variations make the computation of the ground state very complex, particularly for large clusters. It is, therefore, often necessary that the geometries of large clusters are optimized subjected to certain symmetry constraints. There are other methods, such as the simulated annealing and the genetic algorithms that are also currently being applied for optimizing geometries.

Here we discuss the electronic structure and energetics of some Al-based clusters with icosahedric geometries [8-10]. These include: $Al_{13}$, $BAl_{12}$, $CAl_{12}$, $NAl_{12}$, $KAl_{13}$, and $BAl_{12}K$. We begin with the $Al_{13}$ cluster. Since Al is trivalent, the number of valence electrons in $Al_{13}$ is 39. If these electrons could be confined to a square well potential, the outer shell of the cluster will have the $2p^5$ configuration (see Fig. 1). The cluster would need one extra electron to close its outermost shell. In analogy with the periodic table of elements, the chemistry of the $Al_{13}$ should then be that of a chlorine atom. This indeed seems to be the case [9]. We have optimized the geometry of the $Al_{13}$ cluster [8]. It is a perfect icosahedron (see Fig. 2). In this structure there are two characteristic bond lengths - the bond between the central atom and a surface atom, and the bond between surface atoms. The former is about 5 % shorter than the latter. The bond distances, binding energy/atom, ionization potential and electron affinity of this cluster are given in Table 1. Note that the electron affinity of $Al_{13}$ is nearly identical with the electron affinity of Cl atom which is 3.6 eV. Thus, $Al_{13}$ should be a reactive cluster. Castleman and co-workers [11] have observed this to be the case. While $Al_{13}$ readily reacted with oxygen, $Al_{13}^-$ with closed electronic shells was found to be chemically inert. Note that rare-gas atoms with closed electronic shells are also chemically inert. Thus $Al_{13}$ and $Al_{13}^-$ clusters can be thought of as being chemically equivalent to Cl and Ar.

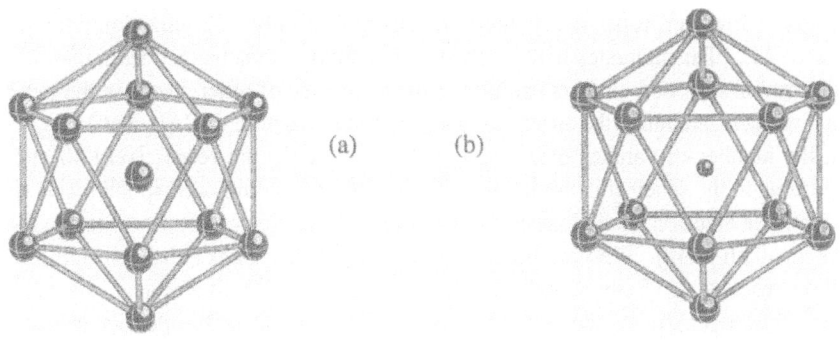

(a)          (b)

**Fig. 2.** Icosahedral geometries of (a) $Al_{13}$ and (b) $XAl_{12}$ (X = B, C, N) clusters.

**Table 1.** Binding energies, inter-atomic distances, ionization potentials, and electron affinities of Al-based clusters. The numbers marked by a * indicate the radial distances of the K atom.

| Cluster | Binding Energy/ atom (eV) | Inter-atomic distances (a,b) (Å) | Ionization Potential (eV) | Electron Affinity (eV) |
|---|---|---|---|---|
| $Al_{13}$ | 2.82 | (2.68, 2.82) | 7.2 | 3.8 |
| $KAl_{13}$ | 3.04 | (2.67, 2.80) 4.87* | 6.5 | |
| $CAl_{12}$ | 3.16 | (2.53, 2.66) | 7.0 | |
| $NAl_{12}$ | 2.97 | (2.56, 2.69) | 5.8 | |
| $BAl_{12}$ | 3.09 | (2.52, 2.65) | 7.0 | 3.6 |
| $(BAl_{12})K$ | 3.66 | (2.54, 2.67) 4.78* | 6.4 | |

The stability of $Al_{13}$ could be further enhanced without changing its chemistry. Consider for example $BAl_{12}$ cluster [10]. Since B is trivalent like Al, the $BAl_{12}$ cluster is isoelectronic with $Al_{13}$. However, the size of the B-atom is smaller than that of the Al-atom. Consequently, the B-Al distance is shorter than the Al-Al distance. Thus in a

BAl$_{12}$ cluster (see Fig. 2), the surface bond between Al atoms will be shorter than that in the Al$_{13}$ cluster. Consequently, this will lead to an enhancement in the binding energy of the BAl$_{12}$ cluster over the Al$_{13}$ cluster. To examine if this is true, we have computed the equilibrium geometry, binding energy, ionization potential and electron affinity of BAl$_{12}$. The results are given in Table 1. Note that BAl$_{12}$ is more strongly bound than Al$_{13}$ by 3.43 eV. However, it has nearly the same electron affinity and ionization potential as those of Al$_{13}$. This indicates that the chemistry of BAl$_{12}$ is unchanged from that of Al$_{13}$.

To see if Al based clusters can mimic the chemistry of other atoms in the periodic table, we have examined the energetics and electronic structure of CAl$_{12}$ and NAl$_{12}$. Note that the number of valence electrons in CAl$_{12}$ and NAl$_{12}$ are respectively 40 and 41. According to the schematic plot in Fig. 1, these electrons would correspond respectively to complete electron shell closure (as in Ar) or a single electron on the outermost shell (like an alkali atom). We first discuss CAl$_{12}$. This cluster gains enhanced stability due to three reasons: (1) Being icosahedric, it is compact with closed atomic shells. (2) Like BAl$_{12}$ cluster, CAl$_{12}$ reduces the structural strain inherent in the perfect icosahedric structure. (3) The complete electronic shell closure renders CAl$_{12}$ chemical inertness and stability. Results of the binding energy and ionization potential listed in Table 1 bear these out. CAl$_{12}$ cluster is 1.0 eV more bound than BAl$_{12}$ and 4.4 eV more bound than Al$_{13}$. Study of the interaction of CAl$_{12}$ with hydrogen also shows that CAl$_{12}$ is less reactive than Al$_{13}$. NAl$_{12}$, on the other hand, is less stable than CAl$_{12}$ (see Table 1). Its ionization potential is 5.8 eV which is similar to that of the Na atom (ionization potential of Na is 5.14 eV). This suggests that the chemistry of NAl$_{12}$ can be similar to that of an alkali atom just like the chemistry of Al$_{13}$ was found to be similar to that of a halogen atom.

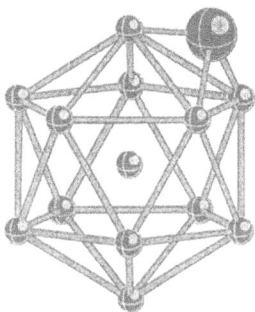

**Fig. 3.**    An Al$_{13}$K cluster. The K atom is shown by the larger sphere occupying the three fold coordinated hollow site.

This similarity can be further elucidated by considering the energetics of KAl$_{13}$ cluster [9]. Note that K is electropositive. Since Al$_{13}$ like Cl is electronegative, the interaction between K and Al$_{13}$ should be ionic in nature. This, if true, would be exciting on its own merit since both K and Al are metallic elements and one would not have expected that two metallic elements could be bound by an ionic interaction. We have optimized the structure of $KAl_{13}$. The equilibrium structure is found to be the one where K is attached to a three-fold surface site of an icosahedric $Al_{13}$ cluster (see Fig. 3). The binding energy of $KAl_{13}$ measured against the energy of K and $Al_{13}$ cluster is 3.0 eV. This is about five times larger than the binding energy of the KAl dimer. This large difference is due to the fact that the bonding between K and $Al_{13}$ is ionic while that between K and Al is "metallic." This is further proved by carrying out the Mulliken charge analysis of the $KAl_{13}$ cluster. We find that the electronic structure of the cluster consists of $K^+$ and $Al_{13}^-$ ion. The distance between the centers of these two charge distributions obtained from our self consistent calculation is 9.20 a.u. It is interesting to note that the electrostatic attraction between two equal but opposite unit charges separated by this distance is 3.0 eV which is almost identical to the binding energy of $KAl_{13}$ noted above. This analysis reconfirms our earlier expectation that clusters can be suitably designed to mimic the chemistry of the atoms.

Finally, we should point out that both $CAl_{12}$ and $KAl_{13}$ clusters contain the same number of valence electrons, namely 40. Yet their stabilities are governed by separate mechanisms: For $CAl_{12}$ it is the electronic and atomic shell closure that is at the root while for $KAl_{13}$ it is the ionic bond. These illustrations point to the enormous possibilities clusters offer for the design of novel materials. In the next two sections we discuss some of these possibilities.

## 3. Crystals of Clusters

In this section we discuss a new class of materials where clusters serve as building blocks much in the same way as atoms do in ordinary materials. These cluster assembled materials (CAM) are expected to exhibit novel properties for several reasons. (1) In contrast to conventional materials, the CAM's are characterized by two length scales - the intra-cluster and inter-cluster distances. (2) Similarly, the bonding within the cluster and between the clusters could be different. (3) The energy bands formed by the overlap of cluster wave functions would be very different from those in conventional solids where the atomic overlap guides the energy band structure. (4) The phonon spectrum will be richer not only because of the richness of electronic structure but also because a cluster has more degrees of freedom than an individual atom, thus affecting the number of modes. That this is indeed the case is already evident from the fullerides [12] where the $C_{60}$ clusters composed of covalently bonded C atoms are held together via Van der Waals forces. It is now known that the electronic, optical and other properties of the fullerides are completely different from the graphite, diamond, or amorphous Carbon, the other solids made of C atoms. To give an example, fullerides intercalated with graphite have a high superconducting transition temperature [13] and the fullerides show non linear optical behavior. We would like to emphasize that cluster assembled materials are different from molecular solids such as ice. Clusters and cluster assembled materials are metastable while molecules such as $H_2O$ that form the building blocks of ice are stable.

In order to carry over the fulleride experience to metallic clusters, one needs to design clusters which are stable and therefore will survive as they are assembled to form solids. The electronic and geometric golden rules outlined in section I, can provide a way. In this section, we discuss three classes of crystals, namely (1) crystals of model clusters, (2) crystals made up of the $CAl_{12}$ clusters, and (3) crystals of $KAl_{13}$ or $(BAl_{12})Cs$ clusters. While $CAl_{12}$ has a closed electronic shell and therefore, may be expected to form a weakly interacting solid, the $KAl_{13}$ or $(BAl_{12})Cs$ are ionically bonded and are therefore expected to form ionic solids.

## 3.1. Crystals of Model Clusters

As discussed before, the electronic structure and stability of simple metal clusters can be fairly well understood within a Jellium picture. The clusters with filled electronic levels resemble inert gases and are therefore expected to interact weakly when assembled. It is with this in mind that Manninen et al. [14] calculated the band structure of a solid composed of Jellium spheres with filled electronic shells. In such a model the self consistent potential is nearly constant within the sphere and can be represented by a square well of the form given in eq. (3) where $V_0 = - \phi - \epsilon_F$. Here $\phi$ is the work function and $\epsilon_F$ is the Fermi energy. R is the radius of the cluster related to the number of electrons N via $R = r_s N^{1/3}$ where $r_s$ is the electron density parameter. Manninen et al. assumed $r_s = 2.07$ and $V_0 = -0.586$ a. u. which correspond to metallic Al. Further, since the van der Waals materials have a tendency to form closed packed structures they considered a face centered cubic lattice of Jellium spheres each containing 40 valence electrons as is the case in $CAl_{12}$. The electronic band structure was obtained by using a plane wave basis of about 2000 plane waves. In this model, it is not possible to optimize the inter-cluster separations since Jellium spheres are only models and one has to include repulsive interactions to calculate the total energy. Therefore, Manninen et al. carried out calculations for various lattice constants.

Fig. 4 shows the progression of one electron levels as the Jellium spheres are brought together from a large separation. As in case of atomic solids, as the spheres are brought together, the individual electron levels broaden to form bands. To put things in perspective, note that the inter-cluster Al-Al separation in $CAl_{12}$ is around 2.70 Å. It is likely that the inter-cluster Al-Al distance in the cluster will be larger than intra-cluster distance. With this in mind, a 3.18 Å inter-cluster Al-Al distance would correspond to a lattice constant of 12.18 Å. A 2.75 Å separation would correspond to a lattice constant of 11.64 Å. Fig. 4 shows that in both these cases, the corresponding solid would be have a gap at the Fermi energy.

Fig. 5 shows the band structure of a cluster solid obtained by Manninen et al. corresponding to a lattice constant of 11.64 Å. Since the cluster has 40 electrons, the Fermi energy lies between 2p and 1g shells (see Fig. 1). It is interesting to note that all the angular momentum states form broad bands. This is different from atomic solids where the bands become progressively narrower as one goes from s and p to d, f and g levels. The energy bands, however, do go through maxima and minima at the high symmetry points in the Brillouin zone. The smallest gap occurs at $\Gamma$-point but small gaps also occur at L and X points. Manninen et al. also calculated the effective mass of electrons at the bottom the conduction and at the top of the valence band to be 0.42 eV and 0.47 eV respectively. This combined with a gap of around 0.5 eV indicates that the resulting solid would be a semiconductor. It is clear that this result is based on the premise that the clusters maintain their identity in going to the solid state. Is $CAl_{12}$ the right choice ? Only *ab initio* calculations with real clusters can provide the answer.

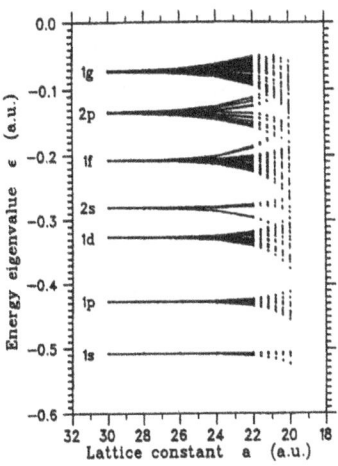

**Fig. 4.**  Energy eigen values of cluster material as a function of the lattice constant. The different energy bands are denoted by the single electron states in a single potential well. The Fermi energy lies between 2p and 1g bands.

## 3.2. Crystal of $Al_{12}C$ Clusters

To examine whether a solid made up of $CAl_{12}$ would be stable, Seitsonen *et al.* [15] have recently studied the stability of an isolated $CAl_{12}$ cluster and a solid composed of such clusters. Their studies used pseudopotentials for the Al and C cores and employed the Car-Parrinello method [16] based on the density functional theory. For an isolated cluster, they obtained bond lengths and binding energy close to those previously obtained by Khanna and Jena. Note that the C atom in $CAl_{12}$ has a coordination of 12 which is unusual for a C atom. To test if raising the temperature may break the outer shell of Al atoms into a group of (e.g. 4 and 8) giving C an $sp^3$ coordination, they carried out molecular dynamics calculations at elevated temperatures. The results showed that the C centered cluster is stable up to around 1000 K, well beyond the melting point of bulk Al which is around 930 K confirming that it is indeed a very stable structure and that the electronic and geometric shell closure rules do have some merit.

Seitsonen *et al.* then examined the possibility of forming a fcc solid composed of icosahedral $CAl_{12}$ units. They studied the total energy of the solid as a function of the lattice constant. The energy decreased monotonically as the lattice constant was reduced excluding the possibility of a metastable minimum. To test the dynamic stability, they carried out a Molecular Dynamics simulation starting from ideal $CAl_{12}$ clusters arranged on a fcc lattice with a lattice parameter of about 10.58 Å. The clusters, unfortunately, broke within 0.2 ps and the temperature of the system stabilized to around 1000 K. The pair correlation function of the final system showed the presence of $CAl_4$ units solvated in liquid Al.

There may be several reasons as to why a crystal of $CAl_{12}$ clusters is not metastable. Firstly, the charge density around the cluster is not as well localized as in e.g. $C_{60}$. The other is that C may prefer a four-fold coordination. This indicates that another choice of atoms with stronger inter-cluster interaction and more localized density need to be explored. It is interesting to point out that earlier simulations [17] on two isolated $CAl_{12}$ interacting units had indicated that the clusters undergo only minimal distortion when put together.

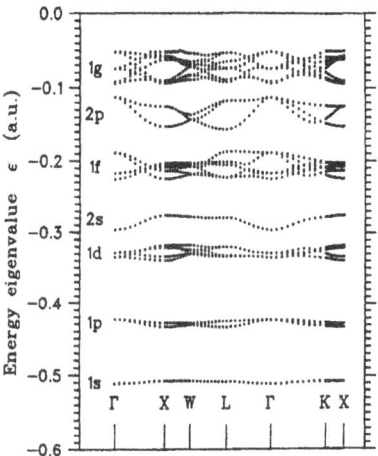

**Fig. 5.** Energy bands of a fcc cluster solid. The lattice constant is 11.64 Å.

## 3.3. Ionic Solids Using Metallic Clusters

A different class of CAM can be envisioned by taking clusters which are analog of $KAl_{13}$ [9]. Although this cluster has 40 valence electrons, it is held together by ionic coupling as mentioned before. The possibility of an ionic solid using $KAl_{13}$ has been examined recently by Liu *et al.* [18] These authors have carried out electronic structure calculations within the density functional approximation using ultra soft pseudopotentials proposed by Vanderbilt [19]. The exchange correlation effects were incorporated via Ceperly-Alder [20] form for the exchange correlation potential. The electronic equation was solved via a preconditioned gradient minimization scheme and the atomic positions were optimized via a Newtonian damping procedure. A 20 Ry cut off was used for the plane waves.

The conventional ionic solids take a NaCl (fcc) or the CsCl (bcc) structures depending on the size of the constituent atoms. Liu *et al.* [18] considered body centered cubic and the face centered cubic structure and calculated the total energies as a function of lattice parameter. Further, the clusters were allowed to change their structure as the lattice constant was changed. This was done by optimizing the energy as a function of $l_1$ and $l_2$ as shown in Fig. 6 for the case of CsCl structure. The cluster can change from the cuboctahedric to icosahedric shape by a change from $l_1 = l_2$ to $l_1 = 1.618\, l_2$.

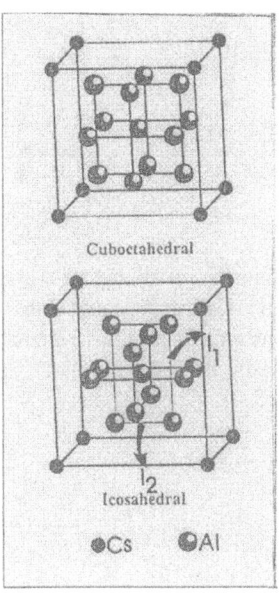

**Fig. 6.** Schematic picture of a prototype CsCl solid formed out of $Al_{13}$ (cuboctahedral or icosahedral) clusters and K atoms.

Fig. 7 shows the binding energy of the $KAl_{13}$ solid as a function of volume for the NaCl and CsCl structures. It is clear that the CsCl structure is far more stable than the NaCl structure. This was expected in view of the dissimilarity in size between the cluster and alkali atom since ionic solids of dissimilar size atoms prefer CsCl structure. Fig. 8 shows the calculated energy of a fcc $KAl_{13}$ lattice containing icosahedral (hollow circles) and cuboctahedral (filled circles). At large volumes, the $Al_{13}$ clusters remain icosahedral as in the gas phase. However, as the volume is reduced, the $Al_{13}$ clusters change from the icosahedral to cuboctahedral form. The change is driven by the enhanced inter-cluster interaction as well as the crystal field of K atoms which has an $O_h$ symmetry since the K atoms are arranged on a cubic lattice. At the equilibrium volume, the shortest Al-Al distance was between the Al atoms not in the same $Al_{13}$ cluster but belonging to neighboring cells. The binding energy per unit cell was 11.6 eV and a calculation of the elastic constants showed that the lattice will be stable against distortions.

As pointed out before, the stability of $Al_{13}$ can be enhanced by replacing the central Al by a B atom. Also, the interaction between the clusters can be reduced by choosing a larger alkali ion. With these in mind, Ashman *et al.* [10] have recently studied the

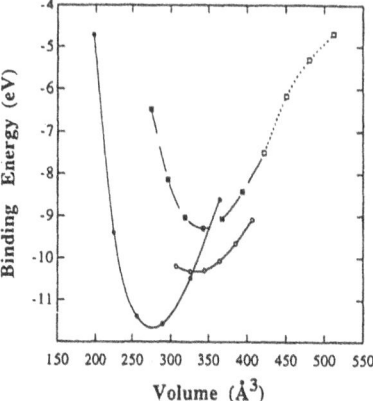

**Fig. 7.** Cohesive energy as a function of volume for a crystal of $KAl_{13}$. The solid and the hollow circles correspond to CsCl arrangement with cuboctahedral and icosahedral $Al_{13}$ clusters respectively. The squares are a NaCl arrangement with icosahedral $Al_{13}$.

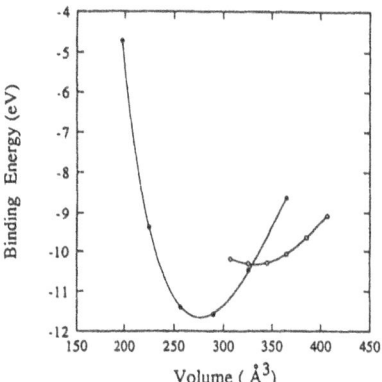

**Fig. 8.** Cohesive energy as a function of volume for $(BAl_{12})Cs$ solid. The solid and the hollow circles correspond to CsCl arrangement with cuboctahedral and icosahedral $BAl_{12}$ clusters respectively.

stability of $(BAl_{12})Cs$ using the same approach as above. Fig. 9 shows their calculated binding energy as a function of volume for the icosahedral and cuboctahedral units. Starting at the large volume, the energy curve goes through the minimum as the volume is decreased. The $BAl_{12}$ clusters maintain an icosahedral form (the calculated points are marked by the hollow circles). Upon further compression, a second minimum where the $BAl_{12}$ clusters transform to the cuboctahedral shape is obtained. The second minimum is marginally more stable than the initial minima but is protected by a barrier. This shows that unlike the case of KAl$_{13}$, a (BAl$_{12}$)Cs solid with icosahedral clusters would be metastable. To examine the properties of the new solid, Ashman *et al.* calculated the electronic density of states of the resulting solid. Their results are shown in Fig. 10. Note that there is an appreciable density of states at the Fermi energy (chosen as the zero of energy) suggesting that the new solid will be conducting.

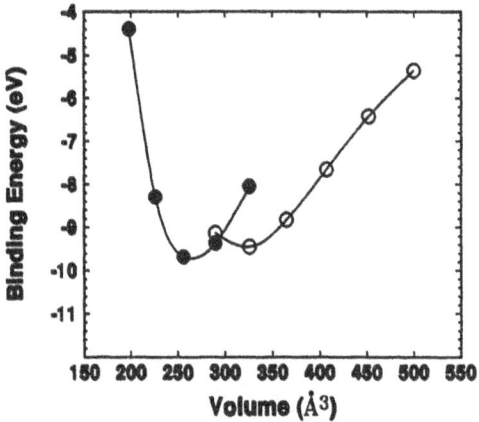

**Fig. 9.** The cohesive energy of (Al$_{12}$B) Cs crystal as a function of lattice volume. The crystal structure was taken to be the CsCl type and Al$_{12}$B cluster was allowed to assume either icosahedric or cuboctahedric shape. The solid and hollow circles correspond to cuboctahedral and icosahedral BAl$_{12}$ clusters respectively.

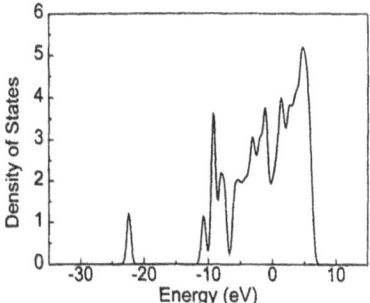

**Fig. 10.** The electron density of states of (Al$_{12}$B)Cs crystal.

The above results indicate two things. (1) That it is indeed possible to make an ionically bonded CAM which is metastable and (2) that if the building blocks are highly stable as is the case for $BAl_{12}$, the clusters may maintain their identity. It is clear that what we have discussed above is just an illustration. We hope, however, that it will stimulate further theoretical and experimental work on the formation of these new materials.

## 4. Supported Clusters

Recent experiments show that novel materials (e. g. thin films [21], nanowires [22]) can be grown through controlled deposition of atomic clusters on surfaces. The methods of cluster deposition include thin film deposition [23], multiple expansion cluster source [24], pulsed arc cluster ion source [25], and liquid metal ion source [26]. The substrates vary from thin film polymers [27], silica [28, 29], silicon wafer, graphite [30], rare gas matrices [31], alkali halide crystals [23, 32], and amorphous carbon surfaces [33] to various metal surfaces [34]. Self-assembly of gold clusters deposited on a substrate have led to the development of nanocircuit elements [35].

An understanding of the nature of the cluster-surface interaction is important for the development of suitable materials through cluster deposition. The important questions that need to be investigated are as follows: Do the size and geometry specific properties of the clusters change upon deposition on a substrate ? Do the clusters migrate on the surface ? If so, do they form islands by coalescing later ? Is there any specific cluster size that resists coalescence ? Do the atoms find preferential sites on the surface to adhere to ? Some experiments already indicate that the substrate often influences the properties of the deposited clusters. For example, growth of platinum film on Pt (111) surface [36] shows that below a certain temperature, an abundance of $Pt_7$ cluster is observed in stead of the steady growth of a film. Similar experiments [37] with Rh clusters on Rh (100) surface also show preferential growth of specific sized clusters. In another experiment, when gold atoms are field evaporated on gold surface [38], it is seen that local atomic relaxation and restructuring of the surface take place to relieve the stress produced by the deposition and the deposited gold clusters lose their identity in the resultant surface.

Theoretical studies of the stability and electronic structure of clusters on surfaces are difficult for a number of reasons. (1) The surface may reconstruct with or without the cluster. (2) The geometry of the deposited cluster can be modified from its gas phase structure if the interaction between the cluster and the substrate is strong. (3) The orientation of the surface can also influence the geometry of the cluster. While *ab initio* calculations that take into account the difference in the electronic and atomic structure of the cluster and the substrate are difficult, valuable insight can be obtained by considering model systems. In the following we discuss three different cases to illustrate the properties of supported clusters: (a) relative stability of metal clusters confined to two dimensions, (b) molecular dynamics calculations aimed at obtaining globally optimized geometries on relaxed substrates, and (c) magnetic properties of transition metal clusters on surfaces. We discuss these systems separately.

## 4.1. Stability of Clusters in Two Dimensions

For small clusters it is expected that their equilibrium geometries will conform to the underlying atomic structure of the substrate [39]. The important question that arises is whether the relative stabilities of clusters confined to two dimensions (no electronic interaction with the substrate being considered) would be different from that of clusters formed in the gas phase. To study this we have considered $Li_n$ and $Al_n$ clusters containing up to 11 atoms deposited on (100) and (111) surfaces of the respective crystals [40]. We have assumed that the interatomic distances are the same as that of the underlying lattice. We than calculate the minimum energy configurations of deposited clusters using the self-consistent field linear combination of atomic orbitals - molecular orbital (SCF-LCAO-MO) method [5]. We have used the Hartree-Fock method for computing the electrostatic and exchange contributions to the potential and Möller-Plesset fourth order perturbation theory (MP4) [5] for the correlation contribution. The Gaussian 94 software [41] has been used for this purpose.

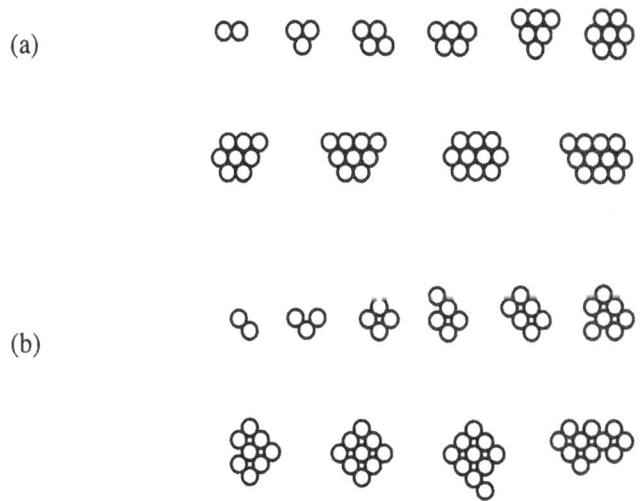

**Fig. 11.** Structures of lowest energy clusters on the (a) fcc (111) plane and (b) fcc (100) plane.

The equilibrium geometries of clusters corresponding to the (111) and (100) planes of an fcc lattice (e. g. Al) are given in Figs. 11(a) and Fig. 11(b) respectively. The energetics for these are presented in Fig. 12 where one observes distinct peaks at n = 4, 7, and 10 for clusters on the (111) plane. The corresponding geometries in Fig. 11(a) are close packed structures with the maximum number of bonds possible for that size. These sizes correspond to close atomic packing as observed in experiment [36] on Pt clusters deposited on Pt (111) surface. Note that $Al_7$ has 21 electrons and is not expected to show unusual stability in the gas phase according to the electronic shell filling criterion [1].

The marked stability of this cluster confined to a plane demonstrates that the properties of clusters deposited on surfaces follow a different rule than those in the gas phase. Fig. 12 also shows that when the cluster is deposited on the (100) surface of an fcc lattice, the nature of the stability changes completely and $Al_7$ becomes less stable than the adjacent sizes.

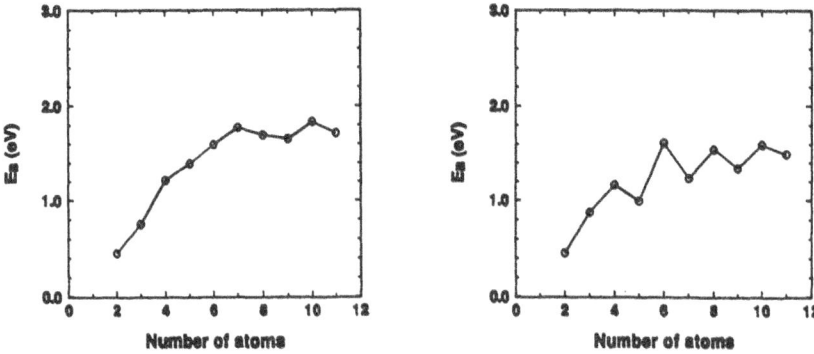

**Fig. 12.** Plot of $E_B$ as a function of cluster size for $Al_n$ clusters on the (111) plane (left) and the (100) plane (right) of the fcc lattice. $E_B = (nE_1 - E_n)/n$ where $E_n$ is the total energy of the clusters with n atoms.and $E_1$ is the energy of a free atom.

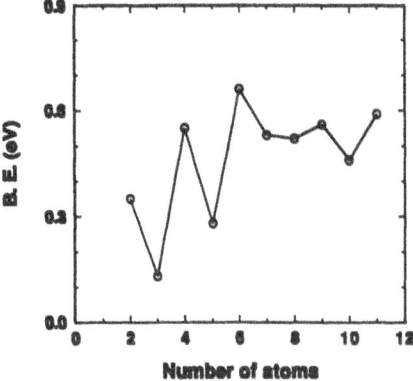

**Fig. 13.** Plot of $E_B$ as a function of cluster size for $Li_n$ clusters on the (111) plane of the fcc lattice.

To examine if the electronic structure of a cluster is important in determining its stability in two dimensions, we have studied $Li_n$ clusters arranged in the same shape as dictated by the (111) surface of an fcc lattice. The results are given in Fig. 13. Note that $Li_8$ is not magic as it is in the gas phase. Moreover, the energetics of $Li_n$ clusters on the fcc (111) surface is not like that of the $Al_n$ clusters on the same surface. Thus, the electronic structure of the clusters and that of the surface are both responsible for the final behavior of the clusters when deposited on a substrate.

To verify if the above behavior persists when a substrate is present, the same calculation was repeated with Li atoms deposited sequentially on a model (111) surface formed out of Li atoms. In this case, the energies were calculated self-consistently including the atoms of the surface. The close-packed geometry of $Li_4$ (Fig. 11(a)) showed extra stability as before.

We have also studied the interactions of pre-formed clusters of lithium atoms deposited on the (100) surface of lithium [42]. In this case, we have optimized the geometry of the deposited cluster while keeping the surface unchanged. The deposited cluster showed signs of melting and becoming a part of the surface with the same symmetry. This symmetry is different from that of the cluster before the interaction began. This result is comparable to what was observed in the case of gold clusters deposited on gold surface [38]. The similarity of the interactions among the intracluster atoms and the intrasurface atoms is definitely the underlying reason for this behavior.

## 4.2. Molecular Dynamics Simulation of Supported Clusters

Most of the above calculations only took into account the effect of cluster stability if the cluster atoms are restricted to a plane. However, as mentioned above, the situation would be more complicated if the interaction of the substrate with the cluster is strong and if the surface undergoes relaxation as the cluster is deposited. These effects can be better studied provided the interatomic potentials are well known. Here we describe three examples.

### $Be_{55}$ cluster on Be(0001) surface

Using a many-body potential [43] that was developed by a global fitting procedure to the potential energy hyper surface of small clusters as well as to the bulk cohesive energies, lattice constant, and elastic constants, we have studied the dynamics of $Be_{55}$ interacting with Be (0001) surface [44]. In this case, the surface was represented by a slab of 420 atoms in 12 layers. In order to simulate a semi-infinite crystal, the two bottom layers were kept rigid. Periodic boundary conditions were imposed on the directions parallel to the surface. After testing out the characteristics of melting of this slab, we modeled the substrate for the cluster-surface interaction by a slab containing 504 Be atoms arranged in three layers forming the (0001) surface. During the dynamics, the atoms of the top two layers were allowed to move. The results are summarized in Fig. 14. We note that as the cluster is deposited even at 0 K, there is significant perturbation of the surface in addition to a noticeable change in the cluster geometry. The cluster spreads out, but remains three dimensional in nature. As the temperature of the substrate is increased, the cluster spreads even more, eventually tending to become planar. This is consistent with the experimental results on metal clusters deposited on the surface of the same metal [38].

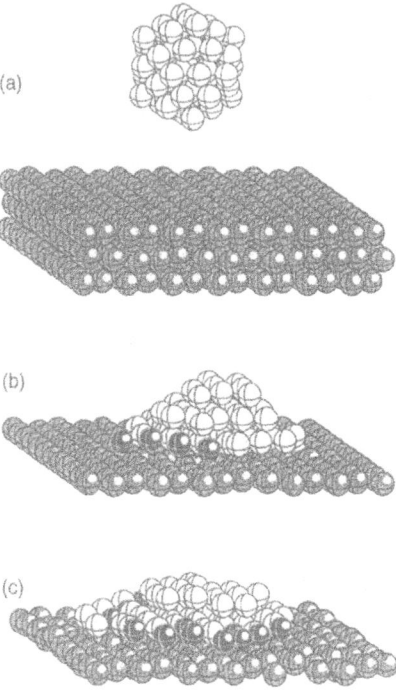

**Fig. 14.** Interaction of $Be_{55}$ cluster with Be surface showing the situations (a) before interaction, (b) at 0 K, and (c) at 700 K. The white spheres and the gray spheres represent the atoms and the substrate respectively. The black spheres show the positions of the surface atoms which have left the surface to wet the cluster.

## $Na_8$ cluster on Na and NaCl substrates

Recently Häkkinen and Manninen [45, 46] have calculated the dynamics of $Na_8$ on NaCl and Na substrates using quantum molecular dynamics approach. Note that $Na_8$ is a magic cluster in the gas phase. It is very stable and is expected to be chemically inert. The question is if $Na_8$ remains magic on a substrate (i. e. if its structure changes). The authors found that $Na_8$ clusters, in interacting with the (100) surface of metallic sodium, lose their identities immediately and the atoms take their epitaxial positions on the adsorption layer. However, when these clusters interact with the NaCl surface, the geometry of $Na_8$ clusters do not change.

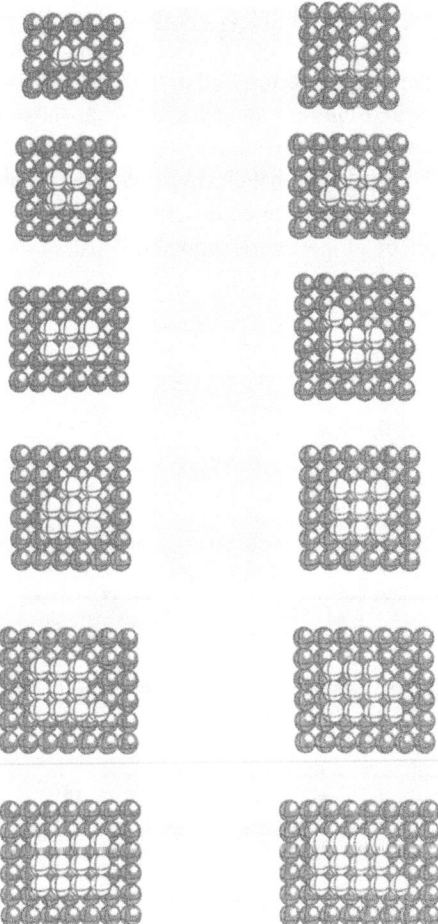

**Fig. 15.** Optimized geometries of $Ag_n$ clusters deposited on Ag (001) surface.

## $Ag_n$ clusters on Ag (001) surface

The above calculations examined what happens to the geometry of a given cluster when deposited on a substrate. Recently, we have studied systematically [47] the relative stabilities of $Ag_n$ clusters on Ag (001) substrate for $2 \leq n \leq 14$ using molecular dynamics simulation with a tight-binding form for the interatomic potential. We recall that relative stabilities of $Ag_n$ clusters in the gas phase show the same pattern as that of the alkali metal clusters, i. e. they have magic numbers at 2, 8, 20, ... . In Fig. 15 we give the optimized geometries of the deposited clusters. Note that all these clusters assume planar configuration even though both the cluster and the surface atoms are allowed all degrees of freedom. To study the relative stability of the supported clusters, we plot the second derivative of the binding energies in Fig. 16. The maxima correspond to energetically preferred clusters while the minima correspond to the less preferred ones. Note that the magic numbers of supported clusters are different from those in the gas phase and resemble the magic numbers for a two dimensional array of clusters as given in Fig. 11(b). The conclusion, therefore, is that clusters deposited on an interacting substrate have different geometries and binding energies. In the following we discuss the role of the substrate on the magnetic properties of supported clusters.

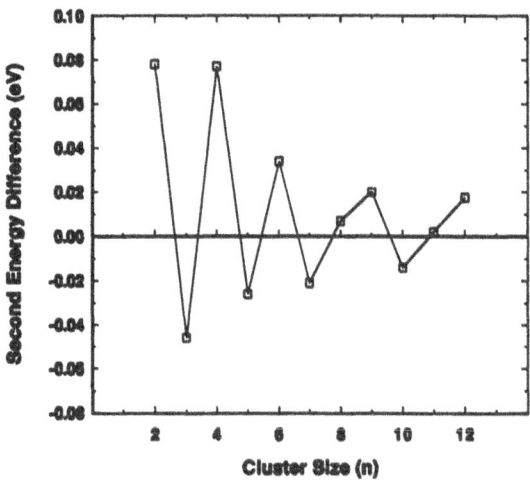

**Fig. 16.** Second difference of binding energy as a function of cluster size with substrate relaxation.

## 4.3. Magnetism of Supported Transition Metal Clusters

Magnetic moments of clusters in the gas phase have been studied extensively [48, 49]. Unlike in the bulk, the magnetism of clusters can be strongly modified by changing their size, composition, and geometry. Thus, clusters of nonmagnetic elements can be magnetic [48, 50] while magnetic moments of ferromagnetic elements can be enhanced by almost a factor of three [51] for certain sizes of clusters. This ability to control the magnetic character of clusters has given rise to the hope that clusters can be a unique source for molecular magnets and can be ideal candidates for high density magnetic storage. For such applications, clusters have to deposited on substrates. Therefore, it is necessary to understand the role of substrates on the magnetism of supported clusters.

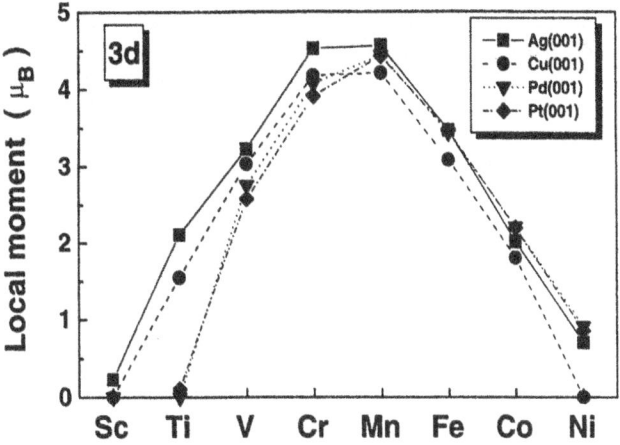

**Fig. 17.** Magnetic moments of 3d adatoms on different substrates.

Recently the magnetic properties of 3d, 4d, and 5d clusters on metallic substrates [52, 53] have been studied using the KKR-Green's function method [54-56]. Here we briefly discuss some of the important results. In Fig. 17 we plot the magnetic moments of 3d adatoms on different metallic substrates. Note that most of these atoms carry a finite magnetic moment [53, 55, 577] with those in the middle of the series carrying the largest moment. The moments also depend, to some extent, on the substrate.

In Fig. 18 we show the magnetic moments of clusters of 4d elements in different configurations. While free dimers of the 4d atoms show large magnetic moments, upon deposition on Ag (001) surface, the magnetism decreases [52]. The moment per adatom usually decreases (see Figs. 18 and 19) when the cluster size increases. This decrease happens more quickly for the 5d clusters. The general decrease of the magnetic moment occurs due to the strong hybridization of the 4d/5d wave functions with the valence electrons of Ag which causes a broadening of the local density of states. Calculations [58] have shown that 4d dimers embedded in bulk Ag show negligible magnetic moment. At the surface, the hybridization weakens sufficiently and then the clusters can exhibit magnetism. It is further observed that the moments for the 9-atom islands are almost nonexistent for most clusters except for Ru, Rh, Os, and Ir. For these cases the average moment/atom is comparable to that in the case of monolayers. The chain structures are observed to exhibit enhanced magnetic moments.

The dependence of magnetic moments on the size of supported clusters is illustrated in Fig. 20 for $Rh_n$ deposited on Ag (001) substrate [53]. The dependence is non-monotonic. It is interesting to note that while $Rh_4$ on Ag (001) is magnetic [52], it loses its magnetic character on Rh (001) surface [59]. The magnetic moments of $Ni_n$ ($n < 6$) clusters deposited on Ag (001) substrate, on the other hand, are insensitive to their size and are close to the bulk value of 0.6 $\mu_B$. In the gas phase, the magnetic moment of Ni clusters in this size range change by almost a factor of two.

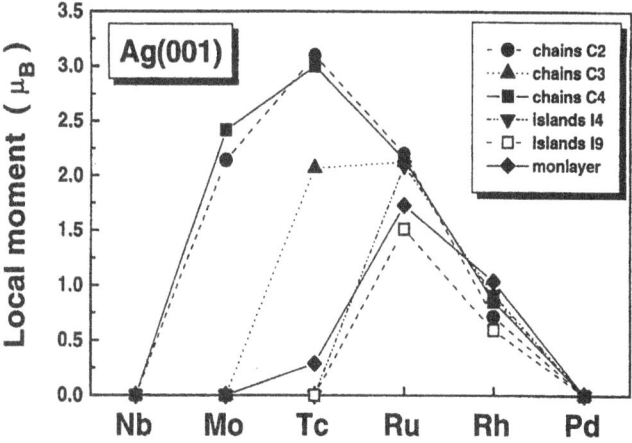

**Fig. 18.** Magnetic moment/atom for small clusters of 4d transition metals on Ag (001) surface.

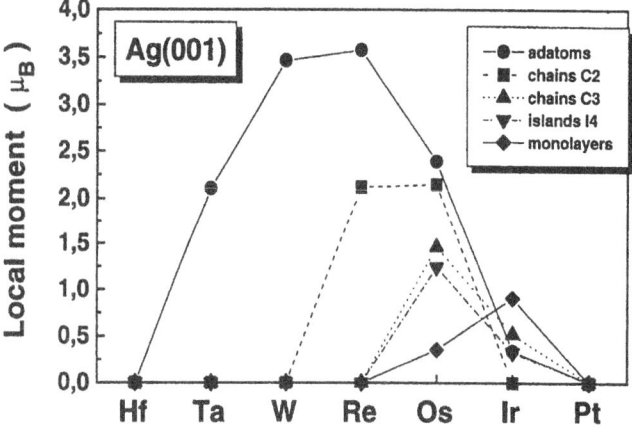

**Fig. 19.** Magnetic moment/atom for small clusters of 5d transition metals on Ag (001) surface.

We have also carried out similar calculations for Mn clusters consisting of 2 and 3 atoms on a Ag (001) substrate [60]. The energetics and magnetic moments at atomic sites for different magnetic configurations (ferromagnetic and anti-ferromagnetic) are given in Table 2. Note that the two magnetic solutions are nearly degenerate. In addition, the moment/atom is about 4.4 $\mu_B$ which is only marginally reduced from the free atom value of 5.0 $\mu_B$. We have repeated similar calculations for $Mn_9$ clusters on Ag (001). It forms a close packed two dimensional island with three different kind of atoms (one central, 4 nearest neighbors, and 4 second nearest neighbors). There are three different anti-ferromagnetic states and one ferromagnetic state. The ground state is the one where the moments are antiparallel with those on the nearest neighbor and lies 71 meV/atom lower than the ferromagnetic state. It is worth noting that the antiferromagnetic configuration of a Mn monolayer on Ag (001) was also found earlier to be 70 meV/atom lower than the ferromagnetic solution. The moment/atom is also about 4.2 $\mu_B$ in all cases. Thus, the supported clusters have rapidly approached the limit set by a monolayer coverage. The fact that the degeneracies of antiferromagnetic and ferromagnetic solutions and the moment/atom are independent of cluster size may make Mn clusters ideal candidates for magnetic storage devices.

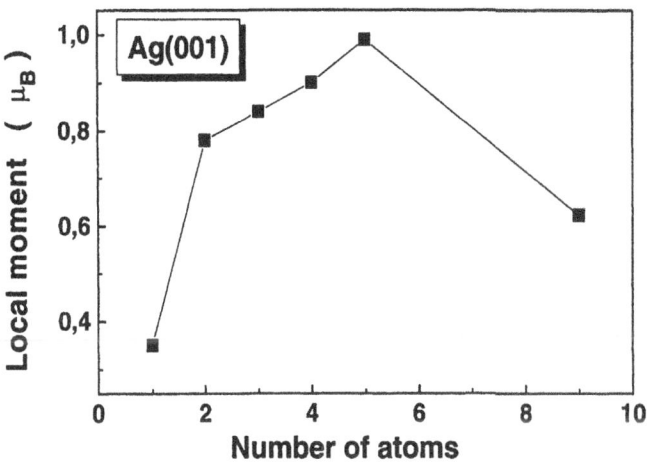

**Fig. 20.** Size dependence of the magnetic moment/atom of small Rh clusters on Ag (001) surface.

**Table 2.** Magnetic moments at atomic sites in $Mn_n$ clusters (n=2,3) on a Ag (001) substrate. The energy of the ground state configuration is given as zero while the energies of the other configurations are given with respect to the ground state configuration.

| Cluster Size | Magnetic Phase | Magnetic Moment/Atom $(\mu_B)$ | $\Delta E$/Atom (meV) |
|---|---|---|---|
| 2 | ↑ ↑ | 4.41, 4.41 | 0 |
|   | ↑ ↓ | 4.46, -4.46 | 0.1 |
| 3 | ↓ ↑ ↓ | -4.41, 4.45, -4.44 | 0 |
|   | ↑ ↑ ↑ | 4.43, 4.32, 4.43 | 29.0 |

50

# 5. Summary

Atomic clusters with a specific size and composition can be designed to mimic the chemistry of atoms in the periodic stable. Using self consistent first principles calculations, we have demonstrated that certain Al-based clusters such as $Al_{13}$, $CAl_{12}$, and $NAl_{12}$ can have the chemistry similar to halogen, rare-gas, and alkali atoms respectively. The stability of the $Al_{13}$ cluster can be further enhanced without changing its chemistry, by replacing the central Al atom by a B Atom. The design of these clusters provides a new way where novel materials with clusters as a building blocks could be synthesized. Changing the building blocks can bring forth new chemistry. For example, the interaction between an Al and K atom is mostly metallic in nature while that between $Al_{13}$ and K is ionic in character. It is interesting to point out that CsAu forms an ionic solid even though Cs and Au are both metallic elements. However, CsAu solid is a semiconductor while the ionically bonded Cs and $BAl_{12}$ clusters form a metastable solid that remains metallic. We have also demonstrated that alkali-halide type crystals could be formed by replacing the halogen atoms with clusters such as $Al_{13}$ and $BAl_{12}$. These crystals, characterized by different length scales and bonding mechanisms may have unique electronic properties that may not exist in crystals of the same constituent atoms, but with atoms and not clusters at lattice sites. It is also possible to synthesize materials with elemental composition that is not allowed by conventional phase diagram. We have also discussed another class of cluster assembled materials where clusters are deposited on a surface. The interaction between the clusters and the substrate cannot only alter the geometry of the clusters they would have in the gas phase, but also their electronic and magnetic properties. In particular, we have shown that supported clusters of non-magnetic transition metal elements may exhibit magnetic bistability. Quantum tunnelling between these magnetic configurations may lead to novel magnetic and magneto-optic device. The ability to alter the properties of materials at an atomic scale and design novel materials with tailored properties will have significant impact on technology and materials science in the coming century.

# 6. Acknowledgment

This work is supported in part by grants from the Army Research Office (DAAH04-95-0158) and the Department of Energy (DEFG05-87E61316).

# References

1. W. D. Knight, K. Clemenger, W. A. deHeer, W. A. Saunders, M. Y. Chou, and M. L. Cohen, Phys. Rev. Lett. **52**, 2141 (1984).
2. M. M. Kappes, M. Schar, U. Rothlischberger, C. Yeretzian, and E. Schumacher, Chem. Phys. Lett. **143**, 251 (1988).
3. *Introduction to Solid State Physics* by C. Kittel (Wiley, New York, 1986), p. 59.

4.    S. N. Khanna, B. K. Rao, P. Jena, and J. L. Martins in *Physics and Chemistry of Small Clusters*, Edited by P. Jena, B. K. Rao, and S. N. Khanna (Plenum, New York, 1987), p. 435; J. L. Martins, J. Buttet, and R. Car, Phys. Rev. B **31**, 1804 (1985).

5.    W. J. Hehre, L. Radom, P. v. R. Schleyer, and J. A. Pople, *Ab initio Molecular Orbital Theory* (John Wiley, New York, 1986).

6.    P. Hohenberg and W. Kohn, Phys. Rev. **136**, B864 (1964); W. Kohn and L. J. Sham, *ibid* **140**, A1133 (1965).

7.    A. D. Becke, Phys. Rev. A **38**, 3098 (1988); J. P. Perdew and Y. Wang, Phys. Rev B **45**, 13244 (1992); A. D. Becke, J. Chem. Phys. **98**, 5648 (1993).

8.    S. N. Khanna and P. Jena, Chem. Phys. Lett. **218**, 383 (1994); S. N. Khanna and P. Jena, Phys. Rev. B **51**, 13716 (1995).

9.    S. N. Khanna and P. Jena, Chem. Phys. Lett. **219**, 479 (1994).

10.   C. Ashman, S. N. Khanna, F. Liu, P. Jena, T. Kaplan, and M. Mostoller, Phys. Rev. B (in press).

11.   R. E. Leuchtner, A. C. Harms, and A. W. Castleman, Jr., J. Chem. Phys. **91**, 2753 (1989).

12.   W. Kratschmer, L. D. Lamb, K. Fostiropoulos, and D. R. Huffman, Nature (London) **347**, 354 (1990); H. W. Kroto, J. R. Heath, S. C. O'Brien, R. F. Curl, and R. E. Smalley, *ibid* **318**, 162 (1985).

13.   A. F. Hebard, M. J. Rosseinsky, R. C. Haddon, D. W. Murphy, S. H. Glarum, T. T. Palstra, A. P. Ramirez, and A. R. Kortan, Nature **350**, 600 (1991).

14.   M. Manninen, J. Mansikka-aho, S. N. Khanna, and P. Jena, Solid State Commun. **85**, 11 (1993).

15.   A. P. Seitsonen, M. J. Puska, M. Alatalo, R. M. Nieminen, V. Milman, and M. C. Payne, Phys. Rev. B **48**, 1981 (1993).

16.   R. Car and M. Parrinello, Phys. Rev. Lett. **45**, 566 (1985).

17.   R. Kawai (Private communication).

18.   F. Liu, M. Mostoller, T. Kaplan, S. N. Khanna, and P. Jena, Chem. Phys. Lett. **248**, 213 (1996).

19.   D. Vanderbilt, Phys. Rev. B **41**, 7892 (1990).

20.   D. M. Ceperley and B. J. Alder, Phys. Rev. Lett. **45**, 566 (1980); J. P. Perdew and A. Zunger, Phys. Rev. B **23**, 5048 (1981).

21.   I. Yamada in *Physics and Chemistry of Finite Systems: From Clusters to Crystals*, Edited by P. Jena, S. N. Khanna, and B. K. Rao (Kluwer, Dordrecht, 1992), p. 1193. See also A. F. Hebard, *ibid*, p. 1213.

22.   K. V. Rao in *Science and Technology of Atomically Engineered Materials*, Edited by P. Jena, S. N. Khanna, and B. K. Rao (World Scientific, Singapore, 1996), p. 619.

23.   J. L. Robins and K. McIsaac, Thin Solid Films **163**, 285 (1988).

24.   M. E. Lin, A. Ramachandra, R. P. Andres, and R. Reifenberger, Z. Phys. D **26**, 59 (1993); see also references therein.

25.   R. Siekmann, E. Holub-Krappe, Bu. Wrenger, Ch. Pettenkofer, and K. H. Meiwes-Broer, Z. Phys. B **90**, 201 (1993).

26. S. D. Berry in *Physics and Chemistry of Small Clusters*, Edited by P. Jena, B. K. Rao, and S. N. Khanna (Plenum, New York, 1987), p. 49.

27. E. Kay in *Metal Clusters*, Edited by F. Träger and G. zu Putlitz (Springer-Verlag, Berlin, 1986), p. 151.

28. C. Furlani, R. Zanoni, C. Dossi, and R. Psaro in *Physics and Chemistry of Small Clusters*, Edited by P. Jena, B. K. Rao, and S. N. Khanna (Plenum, New York, 1987), p. 775.

29. N. E. Bogdanchikova and M. N. Dulin, Z. Phys. D **26**, S48 (1993).

30. K. Sattler in *Physics and Chemistry of Small Clusters*, Edited by P. Jena, B. K. Rao, and S. N. Khanna (Plenum, New York, 1987), p. 713.

31. P. S. Bechtold, U. Kettler, H. R. Schober, and W. Krasser in *Metal Clusters*, Edited by F. Träger and G. zu Putlitz (Springer-Verlag, Berlin, 1986), p. 163.

32. M. Krohn, Thin Solid Films **163**, 291 (1988).

33. E. Choi and R. P. Andres in *Physics and Chemistry of Small Clusters*, Edited by P. Jena, B. K. Rao, and S. N. Khanna (Plenum, New York, 1987), p. 61; T. Castro, R. Reifenberger, E. Choi, and R. P. Andres, Phys. Rev. B **42**, 8548 (1990).

34. H. Shi and K. Jacobi, Surf. Sci. **303**, 67 (1994); H. Tamura, A. Sasahara, and K. Tanaka, Surf. Scci. Lett. **303**, L379 (1994).

35. D. B. Janes, V. R. Kolagunta, J. D. Bielefeld, R. P. Andres, J. I. Henderson, and C. P. Kubiak, Bull. Am. Phys. Soc. **41**, 225 (1996).

36. G. Rosenfeld, A. F. Becker, B. Poelsema, L. K. Verheij, and G. Comsa, Phys. Rev. Lett. **69**, 917 (1992).

37. G. L. Kellogg in *Science and Technology of Atomically Engineered Materials*, Edited by P. Jena, S. N. Khanna, and B. K. Rao (World Scientific, Singapore, 1996), p. 91; G. L. Kellogg in *The Synergy Between Dynamics and Reactivity at Clusters and Surfaces*, Edited by L. J. Farrugia (Kluwer, Dordrecht, 1995), p.21; G. L. Kellogg, Phys. Rev. Lett. **73**, 1833 (1994).

38. C. S. Chang, W. B. Su, and T. T. Tsong, Surf. Sci. Lett. **304**, L456 (1994).

39. W. -D. Schneider, H. -V. Roy, P. Fayet, B. Delley, and C. Massobrio in *Cluster Assembled Materials*, Edited by K. Sattler (Trans Tech, Switzerland, 1996), p. 51; H. -V. Roy, P. Fayet, P. Patthey, W. -D. Schneider, B. Delley, and C. Massobrio, Phys. Rev B **49**, 5611 (1994); G. L. Kellogg, Phys. Rev. Lett. **61**, 578 (1988).

40. A. K. Ray, B. K. Rao, and P. Jena, Phys. Rev. B **48**, 14702 (1993).

41. *Gaussian 94*, Revision B.1, M. J. Frisch, G. W. Trucks, H. B. Schlegel, P. M. W. Gill, B. G. Johnson, M. A. Robb, J. R. Cheeseman, T. Keith, G. A. Petersson, J. A. Montgomery, K. Raghavachari, M. A. Al-Laham, V. G. Zakrzewski, J. V. Ortiz, J. B. Foresman, J. Cioslowski, B. B. Stefanov, A. Nanayakkara, M. Challacombe, C. Y. Peng, P. Y. Ayala, W. Chen, M. W. Wong, J. L. Andres, E. S. Replogle, R. Gomperts, R. L. Martin, D. J. Fox, J. S. Binkley, D. J. Defrees, J. Baker, J. P. Stewart, M. Head-Gordon, C. Gonzalez, and J. A. Pople, Gaussian, Inc., Pittsburgh PA, 1995.

42. B. F. Constance, B. K. Rao, and P. Jena in *Physics and Chemistry of Finite Systems: From Clusters to Crystals*, Edited by P. Jena, S. N. Khanna, and B. K. Rao (Kluwer, Dordrecht, 1992), p. 1065.

43. E. Blaisten-Barojas and S. N. Khanna, Phys. Rev. Lett. **61**, 1477 (1988).

44. A. Antonelli, S. N. Khanna, and P. Jena, Phys. Rev. B **48**, 8263 (1993).

45. H. Häkkinen and M. Manninen, Europhys. Lett. **34**, 177 (1996).

46. H. Häkkinen and M. Manninen, J. Chem. Phys. **105**, 10565 (1996).

47. S. K. Nayak, P. Jena, V. S. Stepanyuk, W. Hergert, and K. Wildberger, Phys. Rev. B **56**, 6952 (1997).

48. A. J. Cox, J. G. Louderback, and L. A. Bloomfield, Phys. Rev. Lett. **71**, 923 (1993); D. C. Douglass, J. P. Bucher, and L. A. Bloomfield, Phys. Rev. B **45**, 6391 (1992).

49. I. M. L. Billas, J. A. Becker, A. Chatelein, and W. A. deHeer, Phys. Rev. Lett. **71**, 4067 (1993).

50. B. V. Reddy, S. N. Khanna, and B. I. Dunlap, Phys. Rev. Lett. **70**, 3323 (1993).

51. J. P. Bucher, D. C. Douglass, and L. A. Bloomfield, Phys. Rev. Lett. **66**, 3052 (1991).

52. K. Wildberger, V. S. Stepanyuk, P. Lang, R. Zeller, and P. H. Dederichs, Phys. Rev. Lett. **75**, 509 (1995).

53. V. S. Stepanyuk, K. Wildberger, R. Zeller, P. H. Dederichs, W. Hergert, and P. Rennert in *Science and Technology of Atomically Engineered Materials*, Edited by P. Jena, S. N. Khanna, and B. K. Rao (World Scientific, Singapore, 1996), p. 361.

54. V. S. Stepanyuk, P. Lang, K. Wildberger, R. Zeller, and P. H. Dederichs, Surf. Sci. Lett. **1**, 477 (1994).

55. P. Lang, V. S. Stepanyuk, K. Wildberger, R. Zeller, and P. H. Dederichs, Solid State Commun. **92**, 755 (1994).

56. K. Wildberger, P. Lang, R. Zeller, and P. H. Dederichs, Phys. Rev. B **52**, 11502 (1995).

57. V. S. Stepanyuk, P. Lang, K. Wildberger, R. Zeller, and P. H. Dederichs, Surf. Rev. Lett. **1**, 477 (1994).

58. K. Willenborg, R. Zeller, and P. H. Dederichs, Europhys. Lett. **18**, 263 (1992).

59. S. K. Nayak, S. E. Weber, P. Jena, K. Wildberger, R. Zeller, P. H. Dederichs, V. S. Stepanyuk, and W. Hergert, Phys. Rev. B **56**, 8849 (1997).

60. V. S. Stepanyuk, W. Hergert, K. Wildberger, S. K. Nayak, and P. Jena, Surf. Sci. Lett. **384**, L892 (1997).

# Structures, Vibrational Frequency Shifts and Photodissociation Dynamics of $Ar_n$HF van der Waals Clusters

Zlatko Bačić*

Department of Chemistry, New York University, New York, New York 10003, USA

**Abstract.** Our theoretical work on the equilibrium structures, HF vibrational frequency shifts and the photodissociation of HF in $Ar_n$HF clusters with $n = 1 - 14, 54$, is reviewed. The main objective of these studies has been to reveal how the frequency shift and the photofragmentation dynamics of the solute HF molecule evolve in the course of formation and closing of the first solvent ($Ar_n$) shell, at $n = 12$ and, in the case of photodissociation, upon addition of the complete second solvent layer, for $n = 54$. All spectroscopic and dynamical properties of $Ar_n$HF clusters that we have studied exhibit strong dependence on the cluster size and isomeric structure.

## 1  Introduction

Solvation of a molecule by either bulk solvent or a finite-size cluster of solvent particles exerts a strong influence on its spectroscopy and photodissociation dynamics. It shifts the vibrational band origins of the solute chromophore to frequencies which are lower ("red shift") or higher ("blue shift") than those of the the isolated gas-phase molecule. The solvent also has a profound effect on the elementary process of laser-induced bond breaking, by modifying the distribution of excess energy among the photofragments, and inducing their recombination through what is known as the "cage effect".

Our microscopic understanding of solvent effects has been greatly enhanced in recent years by the ability to prepare size-selected clusters, which offer a unique opprotunity for exploring fundamental issues of solute-solvent interactions. They are small enough to allow application of sophisticated ultrafast laser spectroscopic techniques and supersonic molecular beams. Yet, these clusters create a controlled environment of sufficient complexity for the spectroscopy and fragmentation dynamics of the solute molecule to exhibit many features characteristic for the condensed matter. What has made clusters particularly attractive is the possibility to vary their size in a stepwise fashion, by adding the solvent particles (atoms or molecules) to the cluster one at a time. This affords an ideal opportunity to observe on atomic scale how various structural, spectroscopic and dynamical properties evolve with cluster size, from those typical for

---

* E-mail address: zlatko.bacic@nyu.edu.

small molecules towards their respective macroscopic, bulk limits. In this sense, clusters truly bridge the gap between isolated molecules and condensed matter.

In addition to control over the system size, clusters generated in supersonic jets offers another significant advantage over bulk condensed matter. Due to very cold temperatures achieved in molecular beams, the solute-solvent hetero-clusters are likely to populate appreciably only a few low-energy isomers whose geometries are well defined and change little on the experimental time scales. Consequently, these clusters provide an exceptionally clear picture of the solute-solvent interactions, which is not obscured by the structural disorder, spatial and temporal inhomogeneities present in bulk solvents. Moreover, it becomes feasible to study how the solvent effects on the solute spectroscopy and photodisociation depend on the isomeric structure of the clusters.

Solvent-induced vibrational frequency shift is caused by the dependence of the solute-solvent interactions on the vibrational state, ground or excited, of the solute chromophore; this alters the energy gap between the initial and the final state of the spectroscopic transition, and therefore its frequency. There is ample experimental evidence that the vibrational frequency shifts constitute a discrim-inating probe of the cluster size and geometry, as well as of the intermolecular potentials [1]. Investigations of size-selected clusters of ethylene ($C_2H_4$) [1, 2], methanol ($CH_3OH$) [1, 3], and acetonitrile ($CH_3CN$) [1, 3], have found a strong link between the frequency shift and cluster structure. The experimental data have been analyzed through calculations of the cluster structures and a perturba-tive treatment of the vibrational frequency shifts developed by Buckingham [4]. Size-specific infrared (IR) spectra have also been recorded for benzene-(water)$_n$ clusters with $n = 1 - 7$ [5, 6]. Their O-H stretch vibrational frequencies show a pronounced dependence on cluster size, and have yielded a great deal of insight into the arrangement of the water molecules in these heteroclusters. Assignments of the observed IR spectra, and their qualitative interpretation in terms of vari-ous types of cluster structures, have been greatly helped by the comparison with harmonic O-H stretching frequency shifts from *ab initio* calculations of small water clusters [7, 8, 9, 10, 11] and benzene-$(H_2O)_n$ ($n = 1 - 3$) clusters [12].

Reaction dynamics in clusters has received a great deal of attention [13]. A particularly well-defined case is the direct photodissociation of a solute molecule within a cluster, upon photoexcitation to a repulsive excited electronic state [14]. The solvent cluster acts as a cage capable of temporarily trapping the so-lute fragments, cooling them and facilitating their recombination. Interestingly, experiments on van der Waals (vdW) dimers $I_2$-M (where M is a rare-gas atom or a diatomic molecule) have shown that the recombination of the dissociating $I_2$ molecule can be induced by a single solvent particle M [15, 16, 17, 18, 19, 20], an example of the so-called one-atom cage effect. Outstanding examples of the experimental work involving larger solvent clusters include the caging dynamics of the photodissociated $I_2^-$ in $(CO_2)_n$ [21, 22, 23] and $Ar_n$ [24] clusters, $ICl^-$ in $(CO_2)_n$ [25], and of $I_2$ in $Ar_n$ clusters [26, 27, 28, 29, 30]. These experiments, of-ten with picosecond to femtosecond time resolution, have provided exceptionally detailed information about the short-time nuclear dynamics of fully or partially solvated dissociating molecules. In particular, the work involving size-selected ionic clusters $(CO_2)_n I_2^-$, $Ar_n I_2^-$, and $(CO_2)_n ICl^-$, has revealed strong variations of the caging efficiency with the size and structure of the clusters, solute-solvent kinematics, and with the nature and relative magnitudes of the solute-solvent and solvent-solvent interactions which determine the configuration and the rigidity of the solvent cage.

The above experimental work on the photodissociation of microsolvated small molecules has prompted a large number of theoretical studies. Many simulations have focused on the one-atom cage effect in $Ar-I_2$ [31, 32, 33, 34, 35, 36], $Ar-HCl$ [37, 38, 39, 40, 41, 42, 43], and $Ar-H_2O$ [44]. The effects of larger solvent clusters on photodissociation and recombination, in particular the size dependence of the fragmentation dynamics, have been investigated theoretically in systems such as $Ar_n Br_2$ [45], $(CO_2)_n I_2^-$ [46, 47, 48], $Ar_n I_2$ [26, 27, 29], $Xe_n HI$ [49], and $Ar_n H_2 O_2$ [50].

In recent years, much attention has been devoted to vdW heteroclusters $Rg_n M$ consisting of $n$ rare-gas atoms Rg bound to an IR-active chromophore M, as finite prototype systems for investigating solvation on a molecular level. $Rg_n M$ clusters are readily produced in supersonic jets and can be probed in detail using high-resolution near- and far-IR spectroscopic techniques. Moreover, abundant information is available about the spectroscopy of many important chromophore molecules in the cryogenic rare-gas matrices. This spectroscopic data define the bulk limit values which various $Rg_n M$ properties are expected to approach with increasing cluster size.

Another reason why $Rg_n M$ clusters are attractive to experimentalists and theorists alike is the fact that the intermolecular potentials for several $Rg-M$ (e.g., $Ar-H_2$ [51, 52], $Ar-HF$ [53], $Ar-HCl$ [54], $Ar-H_2O$ [55]) and $Rg-Rg$ dimers [56, 57] have been characterized far more accurately than those for hydrogen bonded complexes such as $(H_2O)_2$ or benzene–water [14]. In addition, the dominant contributions to the intermolecular potential energy surface (PES) of $Rg_n M$ clusters come from the two-body $Rg-M$ and $Rg-Rg$ interactions. Consequently, pairwise additive PESs constructed from high-quality pair potentials, when combined with accurate dynamical calculations, yield structural and other $Rg_n M$ cluster properties which are typically in very good agreement with experimental results.

Although it may seem paradoxical given the success of pairwise additive PESs, $Rg_n M$ clusters with $n \geq 2$ at the same time offer the best hope for quantitative characterization of nonadditive three-body and higher $n$–body forces. Inclusion of nonadditive intermolecular forces is essential for accurate description of a wide range of phenomena in condensed phases. However, experiments on bulk systems have revealed relatively little about their magnitude and functional form. In contrast, many spectroscopic observables of $Rg_n M$ ($n \geq 2$) clusters, such as vibrational frequency shifts, intermolecular vibrational frequencies, and rotational constants, which can be measured to great precision, are extremely sensitive to the details of the PES. When the potential surfaces of $Rg-M$ and $Rg-Rg$ dimers are known accurately, discrepancies between the spectra of $Rg_n M$

$(n \geq 2)$ calculated using pairwise additive PESs and the observed spectra provide a direct measure of the importance of the nonadditive contributions to the PES [14]. This of course assumes that for the given PES, various spectroscopic observables can be calculated with an error which is negligible compared to that introduced by the PES itself. In fact, as discussed below, the methodology to accomplish this for floppy systems with more than three atoms has been developed only recently [14]. The experimental and theoretical studies of the far-IR spectra of $Ar_2$–HCl by Saykally, Hutson and their co-workers [58, 59, 60], and the work on $Ar_n$HF clusters discussed below, demonstrate the power of this approach.

Scoles et al. [61, 62, 63, 64] have measured vibrational frequency shifts of $SF_6$ and $SiF_4$ in Ar clusters ranging in (average) size from a few tens to about 10000 argon atoms. Their experiments have prompted the the groups of Le Roy [65, 66, 67] and Amar [68] to perform classical Monte Carlo and molecular dynamics simulations of these clusters, which have lead to the interpretation of the observed spectra in terms of isomeric cluster structures, their dynamical state, and preferred location (interior or surface) of the solute ($SF_6$, $SiF_4$) in the solvent ($Ar_n$) cluster. More recently, Nesbitt et al. [69] have employed high-resolution IR spectroscopy to determine the vibrationally averaged structures of $Ar_n CO_2$ ($n = 1, 2$) complexes and their vibrational red shifts; the latter have been found to increase nearly linearly for one and two Ar atoms. The measured red shifts have been extrapolated in a simple manner to the bulk phase and compared with experimental Ar matrix data [69].

This article reviews our theoretical investigations of HF vibrational frequency shift in $Ar_n$HF ($n = 1 - 14$) clusters [70, 71, 72, 73, 74], and of the photodissociation of HF in $Ar_n$HF clusters, $n = 1 - 14, 54$ [75]. The cluster size range considered has permitted us to observe how the frequency shift and the dissociation dynamics evolve in the course of formation and closing of the first solvation shell (for $n = 12$) [70] and, in the case of photodissociation, upon addition of the second complete solvent layer ($n = 54$). In addition, we could study the effects caused by the change, which occurs at $n = 9$ [70], of the energetically optimal location of HF, from a surface site to the interior of a cage. We have found strong dependence of HF frequency shift and dissociation dynamics on the size and structure of $Ar_n$HF heteroclusters, which should be experimentally observable.

These studies were initially prompted by our realization that $Ar_n$HF clusters come close to an ideal model system for studying quantum mechanically the changes in the vibrational frequency of the solute (HF) with the sequential addition of solvent (Ar) particles. The amount and quality of experimental information about small $Ar_n$HF clusters is exceptional; no other rare-gas heteroclusters have been characterized so thoroughly. The vibrationally averaged geometries of $Ar_n$HF clusters for $n = 1 - 4$ have been determined by the microwave studies of Gutowsky et al. [76, 77, 78]. Nesbitt and co-workers have measured the vibrational red shift of HF in $Ar_n$HF ($n = 1 - 4$) clusters [79], and that of DF in $Ar_n$DF, $n = 1 - 3$ [80]. HF and DF frequency shifts in an Ar matrix, which can be viewed as the $n \to \infty$ limit, have also been measured

[81, 82, 83]. The experimental data on the spectroscopy of HF/DF in both very small and very large aggregates of argon atoms have challenged the theory to provide a quantitative account of the vibrational frequency shift between these two limits, and thus fill the gap in our knowledge about HF red shifts in Ar clusters of intermediate size.

The accuracy of the potential energy surfaces for $Rg_n M$ heteroclusters, a key factor determining the reliability of the theoretical predictions about their spectroscopy and fragmentation dynamics, is generally poor. The $Ar_n HF$ clusters are a rare exception in this respect. The anisotropic Ar–HF PES, and its dependence on the HF stretching vibration, is known far more accurately than the intermolecular potentials between rare-gas atoms and other molecule employed so far in the studies of microsolvation ($SF_6$, $SiF_4$, etc.). This is the consequence of the recent work of Hutson [53], whose new anisotropic PES, designated H6(4,3,2), reproduces all existing spectroscopic data for Ar–HF/DF. The H6(4,3,2) PES is actually an effective 2D potential in the vdW stretch ($R$) and bend ($\theta$) coordinates, which depends parametrically on the vibrational quantum number $v$ of HF [53]. This Ar–HF PES can be combined with the high-quality Ar–Ar potential of Aziz and Chen [56] (denoted HFD-C) into pairwise additive PESs for $Ar_n HF$ clusters [70]. They are very close to the limit of what can be achieved by taking only additive interactions into account.

Another distinctive feature of our work on $Ar_n HF$ clusters is that it involved the first coupled multidimensional (in this case 5D) quantum calculation of the large amplitude intermolecular vibrational eigenstates of a molecule (HF) in the clusters. In all previous calculations of $Rg_n M$ clusters, the vibrational cluster dynamics was described by classical molecular dynamics or Monte Carlo simulations, and the frequency shifts were calculated with the help of a first-order perturbation theory model. The combination of exceptionally accurate cluster potentials and nonperturbative quantum calculation of spectral shifts has allowed us to make the most quantitative prediction to date of the evolution of vibrational frequency shift over a wide range of cluster sizes. The multidimensional intermolecular bound states of $Ar_n HF$ clusters were also used in our photodissociation calculations to define accurately the initial quantum state of the system, a critical requirement for reliable modeling of the photofragmentation dynamics of floppy, nonrigid systems.

This article is organized as follows. Section 2 reviews the pairwise additive PESs of $Ar_n HF$ clusters and the low-energy isomeric structures calculated for these PESs. Section 3 outlines the quantum bound state methodology used to calculate the frequency shifts for $Ar_n HF$ clusters. Theoretical predictions regarding the size and isomer dependence of HF red shift are also discussed in Sect. 3. The methodology for, and the results of, the theoretical treatment of the photodissociation of HF in $Ar_n HF$ clusters are presented in Sect. 4. The summary is presented in Sect. 5.

# 2  Minimum-Energy and Low-Lying $\text{Ar}_n\text{HF}$ Isomeric Structures

Accurate $\text{Ar}_n\text{HF}$ cluster structures are the essential prerequisite for the quantitative treatment of both the HF vibrational frequency shift and the photodissociation of rare-gas microsolvated HF. They constitute the key input for the quantum five-dimensional bound state calculations described later. The cluster structures are also indispensable for qualitative understanding of the changes in the HF frequency shift and the fragmentation dynamics as a function of cluster size and structure. We now describe the pairwise additive PESs employed in our work on $\text{Ar}_n\text{HF}$ clusters and the isomeric structures determined for these surfaces.

## 2.1  Pairwise Additive Potential Energy Surfaces for $\text{Ar}_n\text{HF}$ Clusters

One of the two components of our $\text{Ar}_n\text{HF}$ PESs is the 2D anisotropic (effective) Ar–HF potential H6(4,3,2) by Hutson [53] [$V^{\text{eff}}_{\text{Ar}_i\text{HF}(v)}$ in (1) below], which depends parametrically on the HF vibrational quantum number $v$. This potential has two minima: the global minimum at the linear Ar–H–F geometry, $-220.15$ cm$^{-1}$ deep for HF $v = 0$, and a secondary minimum at the linear Ar–F–H configuration, with the well depth of $-107.51$ cm$^{-1}$. Upon vibrational excitation of HF to $v = 1$, the absolute well depth increases by 19 cm$^{-1}$, to $-239.20$ cm$^{-1}$. Increased strength of the Ar–HF interaction when HF is vibrationally excited is responsible for the red shift of the HF vibrational frequency observed in $\text{Ar}_n\text{HF}$ clusters.

The other building blocks for the $\text{Ar}_n\text{HF}$ cluster PES is the Ar–Ar HFD-C potential of Aziz and co-workers [56] [$V_{\text{Ar}_i\text{Ar}_j}$ in (1)], with the well depth of $-99.55$ cm$^{-1}$.

The pairwise additive intermolecular PES of $\text{Ar}_n\text{HF}$ for HF in $v = 0$ or $v = 1$, $V^{tot}_{n,v}$, can be written as a sum of the above Ar–HF and Ar–Ar potentials:

$$V^{tot}_{n,v} = \sum_{i=1}^{n} V^{\text{eff}}_{\text{Ar}_i\text{HF}(v)}(R_i, \theta_i) + \sum_{i<j}^{n} V_{\text{Ar}_i\text{Ar}_j}(r_{ij}) \ . \tag{1}$$

The first term in (1) is a sum of interactions between each of $n$ Ar atoms in the cluster and HF $v = 0, 1$. $R_i$ and $\theta_i$ are Jacobi coordinates of the $i$th argon relative to HF, with $\theta_i$ equal to zero at the linear Ar–H–F geometry. The second term in (1) represents the interactions between all pairs of Ar atoms, with $r_{ij}$ being the distance between Ar atoms $i$ and $j$. $V^{tot}_{n,v}$ depends explicitly on all intermolecular degrees of freedom of the cluster, and also on the vibrational state of HF, through the terms $V^{\text{eff}}_{\text{Ar}_i\text{HF}(v)}$.

## 2.2  Low-Energy Isomers of $Ar_n HF$ ($n = 1 - 14$) Clusters

The number of structurally distinct isomers grows very rapidly with the number of particles, atoms or molecules, in the cluster [84]. Locating all, or most, minima on the potential energy surfaces of larger clusters would involve a major computational effort. Fortunately, such a detailed knowledge of the potential energy landsape is not necessary for adressing many physically significant, and experimentally relevant questions. This should be true in particular for $Ar_n HF$ clusters, which are formed in a supersonic slit jet at low temperatures of about 20 K [79]. One expects that under these circumstances the clusters of any given size mostly populate the lowest-energy structure, with perhaps a small fraction of the population in the local minima with energies only slightly above the global minimum.

With this in mind, for each $Ar_n HF$ cluster size considered, $n = 1 - 14$, we have found, separately for HF $v = 0$ and $v = 1$, the minimum-energy structures corresponding to the global minimum of the pairwise additive PES in (1), and several low-energy isomers energetically closest to it [70]. The search for the global and low-lying local minima was conducted by means of simulated annealing [70, 85, 86] followed by a direct minimization scheme using several Newton-Raphson steps. In the case of larger clusters, the results were verified by the Cerjan-Miller eigenvector-following method [87, 88].

We [70] have used the symbol $V_{n,i}^v$ to designate the $i$th minimum (isomer) of the cluster with $n$ argons, for HF $v = 0$ or $v = 1$. Table 1 gives the energies of the global ($i = 1$) and selected local ($i > 1$) minima of $Ar_n HF$ ($v = 0$) clusters ($V_{n,i}^0$) with $n = 1 - 14$, identified in the course of this work. Also given in Table 1 are our calculated global minima $V_{n,1}^1$ of $Ar_n HF$ ($v = 1$) clusters, $n = 1 - 14$. The lowest-energy structures $V_{n,1}^0$ of these clusters, for HF ($v = 0$), are shown in Fig. 1, while some of the higher-energy isomers are displayed in Fig. 2. For the clusters that we have examined, the geometry of any particular $i$th minimum of $Ar_n HF$ is qualitatively the same for HF $v = 0$ and $v = 1$, apart from minor differences in the atomic coordinates.

There is very close correspondence between the calculated minimum-energy isomers of $Ar_n HF$ with $n = 1 - 4$ Ar atoms in Fig. 1 and the experimentally determined cluster structures [76, 77, 78], giving us confidence in the equilibrium structures predicted for larger $Ar_n HF$ ($n \geq 5$) clusters. They are discussed extensively in [70]. One should note the discontinuity in the size evolution of the energetically optimal cluster structures evident in Fig. 1; it has important consequences for the frequency shifts and the photodissociation dynamics, which are described in later sections. For $n \leq 8$, the lowest-energy configuration of $Ar_n HF$ cluster always has HF bound to the *surface* of the $Ar_n$ subunit, in direct contact with at most three or four Ar atoms. The $Ar_n$ subunits generally appear as the so-called polytetrahedral configurations formed by packing somewhat distorted tetrahedra face to face, well known from theoretical studies of pure $Ar_n$ clusters [89, 90].

In contrast, the global minimum for $Ar_n HF$ clusters with nine or more argons corresponds to HF *inside* a cage formed by the Ar atoms (Fig. 1). Thus, when

**Table 1.** Global and selected low-lying local minima for $Ar_n HF$ clusters, $n = 1 - 14$, for HF $v = 0$ (Ref. 70). $V_{n,i}^0$ (in $cm^{-1}$) represents the $i$th minimum for the cluster of size $n$ and HF $v = 0$; $i = 1$ denotes the global minimum, while $i > 1$ labels the higher-energy local minima. The energies of the isomeric structures $V_{n,i}^0$ for $i > 1$ are relative to the lowest-energy structure $V_{n,1}^0$ of that cluster size. The last column gives the global minima $V_{n,1}^1$ of $Ar_n HF$ clusters, $n = 1 - 14$, for HF $v = 1$ (Ref. 70).

| $n$ | $V_{n,1}^0$ | $V_{n,2}^0$ | $V_{n,3}^0$ | $V_{n,1}^1$ |
|---|---|---|---|---|
| 1 | -220.153 | 112.647 | | -239.167 |
| 2 | -424.08 | 92.72 | 136.09 | -447.17 |
| 3 | -736.60 | 168.54 | 195.50 | -764.75 |
| 4 | -1046.60 | 15.43 | 173.82 | -1075.31 |
| 5 | -1359.84 | 13.42 | 28.49 | -1388.95 |
| 6 | -1756.57 | 29.26 | 98.13 | -1786.43 |
| 7 | -2076.61 | 13.37 | 25.54 | -2106.81 |
| 8 | -2492.70 | 23.88 | 30.96 | -2523.32 |
| 9 | -2883.10 | 66.64 | 83.89 | -2918.13 |
| 10 | -3315.84 | 100.85 | 101.68 | -3352.41 |
| 11 | -3755.09 | 113.76 | 114.42 | -3794.53 |
| 12 | -4430.06 | 361.65 | 368.15 | -4474.76 |
| 13 | -4766.31 | 197.76 | 213.98 | -4811.56 |
| 14 | -5194.24 | 96.36 | | -5240.30 |

$n \geq 9$, it is energetically more favorable for HF to be in the interior of the $Ar_n$ microcluster than on its surface. For $n = 9 - 12$, all Ar atoms form a monolayer cage surrounding HF. The $Ar_{12}HF$ cluster is particularly interesting, for several reasons. Its minimum-energy structure is rather beautiful; the twelve Ar atoms arranged in an icosahedron complete the first solvation shell which fully encloses HF. Moreover, as can be seen in Table 1, the energy gap separating the highly symmetric icosahedral structure from the next higher $n = 12$ isomer, $361.36$ $cm^{-1}$, is much larger than that for any other $Ar_n HF$ cluster that we have investigated, testifying to the extraordinary stability of the icosahedral $Ar_{12}HF$ [70]. With the first solvation shell completed for $Ar_{12}HF$, the subsequent ($n = 13, 14$) Ar atoms occupy sites on the exterior of the $Ar_{12}$ icosahedron, beginning to form the second solvent layer [70].

To allow studying the effect of additional solvent shells on the photodissociation dynamics, the second complete icosahedral shell containing 42 Ar atoms [91] has been added to the exterior of the icosahedral $Ar_{12}HF$ [75]. In this way, we constructed the $Ar_{54}HF$ cluster with the closed-shell icosahedral geometry shown in Fig. 1 of [75]. It is essentially the well known low-energy icosahedral structure of pure $Ar_{55}$ cluster [92, 93, 94], with the central Ar atom replaced by HF.

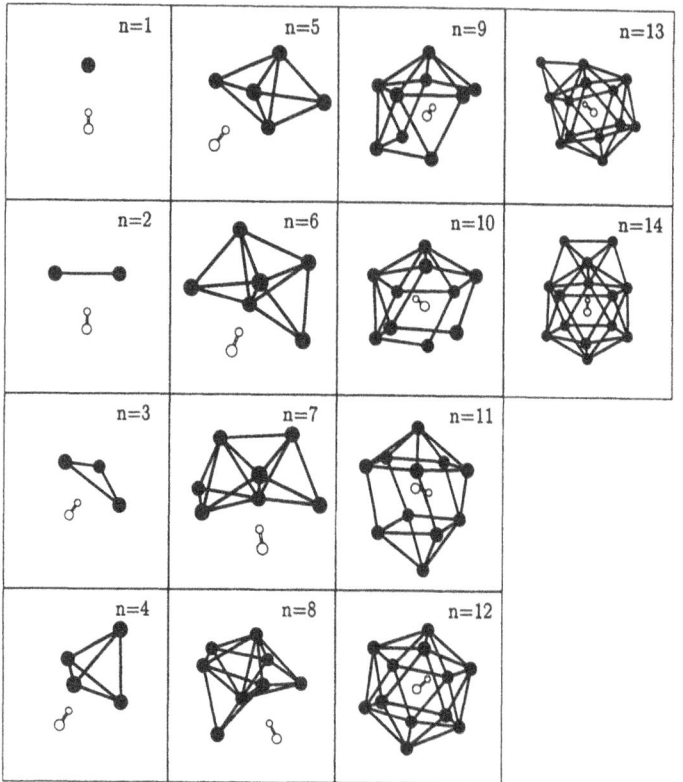

**Fig. 1.** Calculated minimum-energy structures $(V_{n,1}^0)$ of $Ar_n HF$ clusters, $n = 1 - 14$ (Ref. 70). Their energies are given in Table 1.

# 3 Size and Isomer Dependence of HF Vibrational Frequency Shift in $Ar_n HF$ Clusters

## 3.1 Computational Methods

The HF vibrational frequency shift for an $Ar_n HF$ cluster can be obtained as the difference $\mathcal{E}_0(v = 1) - \mathcal{E}_0(v = 0)$, where $\mathcal{E}_0(v = 1)$ and $\mathcal{E}_0(v = 0)$ are the intermolecular vibrational ground state energies of the cluster for HF in $v = 1$ and $v = 0$ vibrational state, respectively. The presence of numerous anharmonic large amplitude motion vibrations, strongly coupled and often delocalized over multiple local minima on the PES, makes accurate quantum mechanical calculation of the intermolecular vibrational levels of weakly bound clusters, even their ground state, an exceedingly difficult problem. Major advances have been made over the past decade in the development of new bound state methods capable of dealing with these challenging features peculiar to floppy systems [95, 96, 97], With the calculations of highly excited rovibrational levels of triatomic floppy

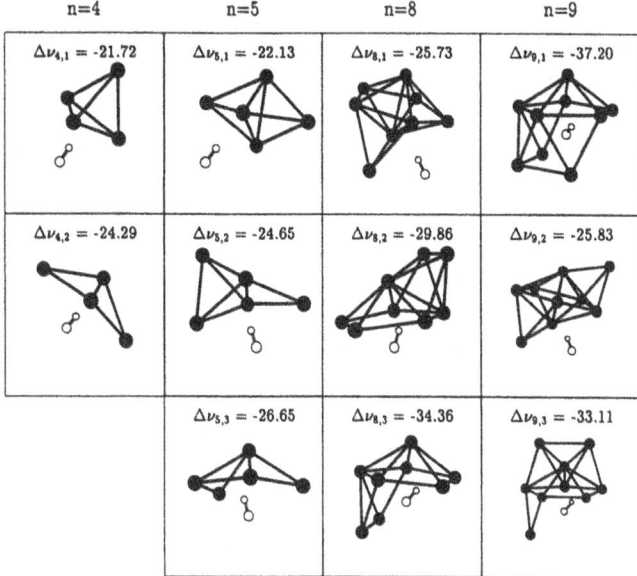

**Fig. 2.** Calculated isomeric structures of $Ar_n HF$ clusters, $n = 4, 5, 8, 9$, from Ref. 70: (top) minimum-energy structure ($V_{n,1}^0$); (middle) next higher isomer ($V_{n,2}^0$); (bottom) second close-lying isomer ($V_{n,3}^0$). Their energies are given in Table 1. Also shown are HF red shifts $\Delta\nu_{n,i}$ (in $cm^{-1}$) of these isomers, from quantum 5D calculations (Ref. 73).

molecules now fairly routine, the computational frontier has moved on to four-atom and larger systems. Variational bound state treatments of tetratomic floppy molecules and clusters have began to appear; they include 5D calculations of $Ar_2 HX$ (X = Cl, F) where the HX bond lenth is fixed [98], and full 6D calculations of excited vibrational levels of HF dimer [99, 100], $H_2O_2$ [101], and HCl dimer [102, 103].

However, clusters of five or more atoms, e.g., $Ar_n HF$ with $n \geq 3$, are beyond reach of these impressive new methodologies at the present time. The only rigorous, full-dimensional quantum mechanical approach applicable to the coupled vibrations of systems of this size, at least in their ground state, is the quantum Monte Carlo (QMC) method [104, 105], in particular its popular variant the diffusion quantum Monte Carlo (DQMC) [104, 106, 107, 108]. The DQMC has been applied to a variety of floppy systems, such as water dimer [104, 109], HF dimer [110, 111], helium clusters, pure or impurity-doped [112], and $Ar_n HF$ ($n = 1 - 4$) clusters [113].

In the clusters like $Ar_n HF$, the high-frequency intramolecular HF vibration and the low-frequency intermolecular motions occur on very different time scales, making their accurate treatment using the standard QMC methods prohibitively costly in computational terms. A recent novel formulation of the DQMC treats the molecules constituting the weakly bound cluster as rigid bodies [114]. Elimination of the high-frequency intramolecular vibrations permits the use of a much larger time step, increasing the speed and accuracy of the DQMC calculations. The rigid-body DQMC has been used to study $H_2$–$H_2O$ [114], $N_2$–$H_2O$ [115], and $(H_2O)_n$ $(n = 2 - 6)$ clusters [109, 116, 117, 118, 119].

We have used a rigid-body DQMC, combined with importance sampling [105], to characterize the vibrationally averaged geometries of $Ar_n HF$ clusters with $n = 1 - 4$ in their intermolecular vibrational ground states, and to obtain the ground-state energies $\mathcal{E}_0(v = 1)$ and $\mathcal{E}_0(v = 0)$ whose difference gives the HF frequency shift [74]. The details of our implementation of the rigid-body DQMC are described in [74]. More recently, such calculations were extended to $Ar_n HF$ clusters up to $n = 12$ [120]. Since all intermolecular degrees of freedom of the clusters are included in the DQMC calculations, their results discussed below are essentially numerically exact for the PES employed. But, for the larger $Ar_n HF$ clusters of interest to us, with up to 12-14 Ar atoms, even the rigid-body DQMC calculations are much too time consuming to allow an exhaustive exploration of the size and isomer variations of the HF frequency shift. It became imperative to develop a reduced-dimensionality approximation, one which would greatly accelerate the calculations without impairing significantly the accuracy of the calculated frequency shifts.

In the ground vibrational state of $Ar_n HF$ clusters, the heavy Ar atoms execute motions which have much smaller amplitudes than those of the light HF molecule. Hence, it is natural to assume that the component of the cluster dynamics which is critical for obtaining accurate frequency shifts are the large amplitude motion (LAM) intermolecular vibrations of HF in the cluster. Based on these considerations, we introduced an approximation in which all vibrational modes of the $Ar_n$ subunit are frozen, while the quantum dynamics of the five remaining coupled LAM intermolecular vibrations of HF relative to the (rigid) $Ar_n$ subunit is treated in its full dimensionality [71, 72, 73]. Detailed description of our quantum 5D bound state methodology employed to calculate HF vibrational frequency shifts of $Ar_n HF$ clusters has been published [73], and only the key steps are summarized here:

(1) The $Ar_n$ subunit is *frozen* in the geometries of the $i$th $Ar_n HF$ minima $V_{n,i}^0$ and $V_{n,i}^1$ (for the same $i$, the two are very similar), either global ($i = 1$) or local ($i > 1$). This effectively reduces $Ar_n HF$ cluster to a weakly bound *dimer*, with the rigid $Ar_n$ microcluster as one monomer and HF molecule as the other.

(2) The 5D intermolecular vibrational energy levels of the floppy $Ar_n$–HF dimer are calculated separately for HF $v = 0$ and $v = 1$, by the quantum 5D bound state method [71, 72, 73] based on the methodology that we have developed for the 3D intermolecular vibrational eigenstates of atom-large molecule complexes [121, 122]. The quantum 5D intermolecular vibrational ground state energies of the $Ar_n$–HF dimer for HF $v = 0, 1$ (and $Ar_n$ geometries defined by the minima $V_{n,i}^v$) are designated $E_{n,i}^0$ and $E_{n,i}^1$, respectively.

(3) HF vibrational frequency shift for the $i$th isomer of $Ar_n HF$, $\Delta\nu_{n,i}$, in this 5D quantum approximation is given by the difference between $E_{n,i}^1$ and $E_{n,i}^0$:

$$\Delta\nu_{n,i} = E_{n,i}^1 - E_{n,i}^0 \ . \tag{2}$$

## 3.2 Size Dependence of HF Vibrational Frequency Shift

This section discusses the HF red shifts for the *minimum-energy* $(i = 1)$ isomers of $Ar_n HF$ clusters, $n = 1 - 14$, whose equilibrium structures can be seen in Fig. 1. The results of our rigid-body DQMC [74] and quantum 5D calculations [72] are presented in Table 2. The quantum 5D red shifts in Table 2 are calculated as $\Delta \nu_{n,1} = E^1_{n,1} - E^0_{n,1}$, following (2). In addition, Table 2 contains the HF red shifts for the $n = 1, 2$ clusters from the variational calculations by Hutson and co-workers [53, 98], and the experimental results for $Ar_n HF$ $(n = 1 - 4)$ clusters obtained by Nesbitt and co-workers [79].

**Table 2.** HF vibrational red shift for $Ar_n HF$ clusters from our rigid-body DQMC calculations for $n = 1 - 4$ (Ref. 74), and from the quantum 5D (rigid $Ar_n$) calculations, $n = 1 - 14$ (Ref. 72). They are compared with the variational results for $n = 1, 2$ of Hutson and co-workers (Refs. 53, 98), and the HF vibrational red shifts measured by Nesbitt and co-workers for the $n = 1 - 4$ clusters (Ref. 79). All red shifts are in $cm^{-1}$.

| $n$ | DQMC | Variational | Quantum 5D (rigid $Ar_n$) | Experiment |
|---|---|---|---|---|
| 1 | $-9.67 \pm 0.08$ | -9.655 | -9.655 | -9.654 |
| 2 | $-15.38 \pm 0.13$ | -15.354 | -15.60 | -14.827 |
| 3 | $-20.57 \pm 0.15$ | | -21.11 | -19.260 |
| 4 | $-21.08 \pm 0.16$ | | -21.72 | -19.697 |
| 5 | | | -22.13 | |
| 6 | | | -25.25 | |
| 7 | | | -25.55 | |
| 8 | | | -25.73 | |
| 9 | | | -37.20 | |
| 10 | | | -38.57 | |
| 11 | | | -39.57 | |
| 12 | | | -42.46 | |
| 13 | | | -42.73 | |
| 14 | | | -42.96 | |

The data in Table 2 can be used to make several comparisons. First, the DQMC red shifts for ArHF and $Ar_2 HF$ agree extremely well with the variational results of Hutson and co-workers [53, 98] on the same PESs. The DQMC values given in Table 2 constitute the definitive, benchmark *pairwise additive* HF vibrational red shifts of $Ar_n HF$ clusters, $n = 1 - 4$, for the PESs employed. Our rigid-body DQMC results are in generally good agreement with the recent DQMC calculations for $Ar_n HF$ $(n = 1-4)$ by Lewerenz [113], which are of lower numerical accuracy.

Our DQMC results in Table 2 are very close to the HF red shifts measured for $Ar_n HF$ clusters, $n = 1-4$, by Nesbitt and co-workers [79]. Evidently, the experimental red shifts can be reproduced nearly quantitatively with the pairwise additive $Ar_n HF$ PESs defined by (1). However, the DQMC red shifts of the clusters with more than one argon are consistently slightly larger than the respective experimental values, the differences being 0.55, 1.31, and 1.38 cm$^{-1}$ for $n = 2-4$, respectively. Since the two-body Ar–Ar and Ar–HF potentials used to construct the pairwise additive Ar$_n$HF potential surfaces have outstanding accuracy, the residual discrepancies between theory and experiment for the $n > 1$ clusters are attributed to the missing *nonadditive*, three-body and higher-order terms. In the case of Ar$_2$HF, the inclusion of some three-body interactions in the PES has led to better, although not complete, agreement of the calculated and experimental red shifts [98]. On the other hand, in the work of Lewerenz [113], addition of the three-body Axilrod-Teller triple-dipole term [123] resulted in DQMC-calculated $n = 2-4$ red shifts in worse agreement with experiment than those based on the pairwise additive PESs. Likewise, the DQMC calculations of Dykstra [120] for a model potential which included a three-body dispersion term yielded $n = 2 - 4$ red shifts whose agreement with the experimental values was comparable to that of our pairwise additive DQMC results.

Our rigid-body DQMC results for the $n = 1 - 4$ clusters also allow us to assess the accuracy of the HF frequency shifts obtained using the quantum 5D bound state methodology outlined in Sect. 3.1. Table 2 shows that the DQMC and quantum 5D red shifts differ by only 1.4%, 2.6%, and 3.0% for $n = 2 - 4$, respectively. From this we conclude that the error which the assumption of rigid Ar$_n$ introduces in the quantum 5D frequency shift calculations is small. Therefore, one can have confidence in the quantitative accuracy of the predictions from such calculation, regarding the size and isomer dependence of the red shift for larger Ar$_n$HF clusters.

The quantum 5D red shifts listed in Table 2 appear in Fig. 3 as well. Actually, Fig. 3 shows the absolute values of the frequency shifts $|\Delta\nu_n|$ (the index $i = 1$ has been dropped), so that the shifts are given as positive numbers. It should be kept in mind that Ar$_n$HF frequency shifts are always to the red, and have negative values. The quantal red shifts agree very well with the available experimental data [79].

Figure 3 also shows the red shifts (their absolute values) approximated by $|V_{n,1}^1 - V_{n,1}^0|$, i.e., by the energy difference between the global minima of Ar$_n$HF cluster calculated for HF $v = 1$ and $v = 0$, respectively [70], given in Table 1. This extremely simple approximation, which neglects completely the vibrations of the cluster, gives red shifts which overestimate the experimental results (and the quantal red shifts) by as much as a factor of two. This demonstrates vividly that accurate calculation of the vibrational frequency shifts is not possible without solving the dynamics of the coupled 5D LAM vibrations of HF in the cluster. That the large amplitude motions of HF must be treated *quantum mechanically* has been illustrated by the recent classical trajectory simulations of the red shifts in Ar$_n$HF clusters [124], which failed to reproduce the experimental results for $n = 2 - 4$ by a wide margin, and disagree seriously with our quantum 5D red shifts for the clusters with $n \leq 8$. Interestingly, Fig. 3 suggests that the quantum vibrational dynamics of HF, which dominates the frequency shifts of the smaller $(n \leq 8)$ Ar$_n$HF clusters where HF is at a surface site, becomes less crucial for

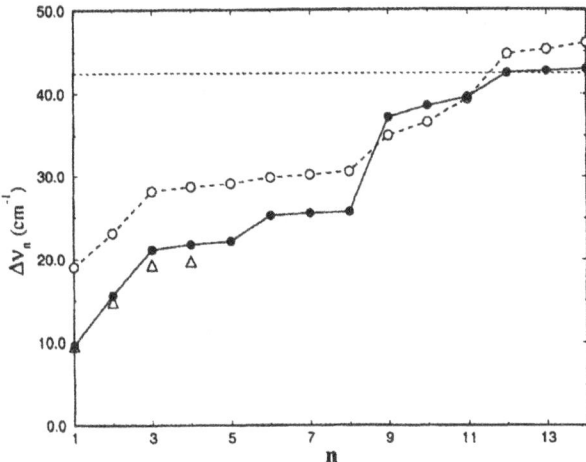

**Fig. 3.** HF red shifts (absolute values) $\Delta\nu_n$ for the minimum-energy isomers of $Ar_nHF$ clusters with $n = 1 - 14$ (Fig. 1), obtained via quantum 5D bound state calculations in Ref. 72 ($\bullet$), and by the classical approximation in Ref. 70 (o). Also shown are experimental $Ar_nHF$ red shifts (Ref. 79), for $n = 1 - 4$ ($\triangle$). The straight dashed line marks the red shift measured for HF in an Ar matrix (Refs. 81-83).

the red shifts of $n \geq 9$ clusters with HF inside partly or fully formed first solvation shell. One explanation for this is that the intermolecular potential which HF experiences inside $Ar_n$ cages, especially for $n \geq 12$ when the first solvent shell is closed, is much less anisotropic than at surface sites. When HF rotates almost freely in such nearly isotropic environment, its coupled large amplitude motions are likely to contribute less to the red shift [70].

Figure 3 shows very nonmonotonic increase in HF red shift (from quantum 5D calculations) with $Ar_nHF$ cluster size until $n = 12$. The red shift grows almost linearly with $n$ for $n = 2, 3$, as these Ar atoms enter the first solvation shell and induce a nearly equivalent red shift. Then, it nearly levels off for clusters with four to eight Ar atoms. Because HF is located on the exterior of the $Ar_n$ subunit in these clusters (Fig. 1), increasing the cluster size changes only slightly the number of Ar atoms in direct contact with HF, from 3 for $n = 4, 5$ to 4 for $n = 6 - 8$, which accounts for a small increase of the red shift at $n = 6$, visible in Fig. 3.

Steep rise of the quantal red shift which occurs for $n = 9$ is the most conspicuous feature of Fig. 3. It coincides with the discontinuity in the size evolution of the minimum-energy cluster structures discussed in Sect. 2.2, which also takes place at $n = 9$ (Fig. 1). While for clusters with $n \leq 8$ HF is preferentially bound to a surface site, starting with $n = 9$ HF is invariably found inside a cage formed by $n = 9 - 12$ Ar atoms. As a result, in going from $n = 8$ to $n = 9$, the number of Ar atoms in the first coordination shell more than doubles from 4 to 9, causing the red shift to rise sharply. For $n = 9 - 12$, all Ar atoms enter the first solvent layer enclosing HF and the red shift increases with their number until $n = 12$, when the first solvation shell is completed. In larger ($n > 12$) clusters, Ar atoms beyond the first twelve begin to form the second solvation shell which affects the frequency shift much less.

One may conclude at this point that the step-like size dependence of the frequency shift arises from the changes in the number of Ar atoms in the *first* solvation shell around HF, which make the dominant contribution to the red shift, with the size of the cluster [70].

It is evident from Fig. 3 and Table 2 that the quantum 5D red shift for icosahedral $Ar_{12}HF$ [71], -42.46 $cm^{-1}$, is virtually identical to the frequency shift measured for HF in an Ar matrix, -42.4 $cm^{-1}$ [81, 82, 83, 125]. In addition, quantum 5D bound state calculations for the $Ar_{12}DF$ cluster [72] have given a red shift of -30.72 $cm^{-1}$, very close to the experimental result for the Ar matrix red shift of DF [83], -29.79 $cm^{-1}$. Taken together, these results indicate that it takes only 12 Ar atoms in the completed first solvation shell around HF for the cluster frequency shift to approach to the Ar matrix value.

It must be pointed out, however, that the embarrassingly good agreement between the red shift calculated for $Ar_{12}HF$ and the experimental Ar matrix result is a bit of a coincidence, and should not be overinterpreted. Our preliminary quantum 5D calculations of $Ar_{12}HF$ using a *three-body* PES [126] have yielded a somewhat smaller red shift, -39.20 $cm^{-1}$. Comparison of the $Ar_{12}HF$ red shift with that for Ar matrix is complicated further by the fact that the influence of additional Ar shells on the HF red shift in the matrix, although small, is not completely negligible. According to the simulations of Schmidt and Jungwirth [127], three (icosahedral) Ar shells are needed for the red shift to converge to about -50 $cm^{-1}$. Moreover, the site occupied by HF in an Ar matrix has octahedral $(O_h)$ symmetry, not the icosahedral symmetry of $Ar_{12}HF$. The frequency shifts calculated for these two environments differ by a couple of wave numbers [127]. Calculations of the HF red shift for two to seven solvation (Ar) shells were also reported by Dykstra [120], showing an appreciable contribution of the second shell to the frequency shift.

## 3.3   Isomer Dependence of HF Vibrational Frequency Shift

In the previous section, we have discussed the size dependence of the HF frequency shift $\Delta\nu_{n,1}$, calculated for the lowest-energy isomer $(i = 1)$ of $Ar_nHF$. But, $Ar_nHF$ clusters have many isomers $(i > 1)$ with drastically different geometries [70]. How large are the variations of the frequency shift among different isomeric structures of a given cluster size? Are they large enough to allow, at least in principle, experimental differentiation of various $Ar_nHF$ isomers?

To answer these questions, we have calculated, by means of the quantum 5D method of Sect. 3.1, the HF vibrational frequency shifts $\Delta\nu_{n,2}$ and $\Delta\nu_{n,3}$ for the low-energy isomers $(i = 2, 3)$ of the $Ar_nHF$ clusters with $n = 4 - 14$ [73]. For a few cluster sizes, the low-lying minima are shown in Fig. 2; a more complete information about the isomeric structures can be found in our initial study of $Ar_nHF$ clusters [70]. Their absolute values $|\Delta\nu_{n,i}|$ $(i = 2, 3)$ are shown in Fig. 4; the quantum 5D frequency shifts $|\Delta\nu_{n,1}|$ for the minimum-energy isomers (Fig. 3) are included for completeness. The calculated HF red shifts also appear in

Figure 2, to facilitate the discussion below about the relationship between the cluster structure and the frequency shift.

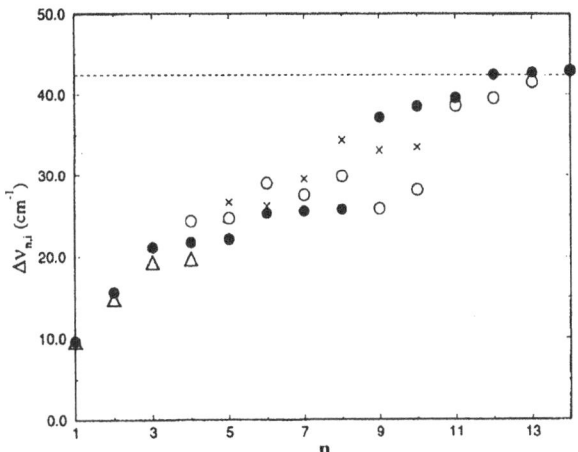

**Fig. 4.** Calculated HF red shifts (absolute values) $\Delta\nu_{n,i}$, for the minimum-energy isomer $V_{n,1}^0$ ($\bullet$), next higher isomer $V_{n,2}^0$ ($\circ$), and the second close-lying isomer $V_{n,3}^0$ ($\times$) of $Ar_nHF$ clusters with $n = 4 - 14$ (Ref. 73). The red shifts (absolute values) $\Delta\nu_{n,1}$ ($\bullet$), for $n = 1 - 14$, are from Ref. 72. The energies of these isomeric structures are given in Table I (Ref. 70). Also shown are experimental $Ar_nHF$ vibrational frequency shifts (Ref. 79), for $n = 1 - 4$ ($\triangle$). The straight dashed line marks the red shift measured for HF in an Ar matrix (Refs. 81-83).

Fig. 4 reveals strong isomer-specificity of HF red shift in $Ar_nHF$ clusters, i.e., different isomers for a cluster of size $n$ have markedly different red shifts. The differences are typically 2-4 cm$^{-1}$, but can be as large as 10-12 cm$^{-1}$. Consequently, HF red shift does provide a distinct spectroscopic signature of the particular cluster isomer, which can be used to identify it experimentally [73]. The isomer variations of the red shift for several $Ar_nHF$ clusters are discussed now, in order to identify those features of the cluster structures to which HF red shift is the most sensitive.

The largest $Ar_nHF$ cluster for which the frequency shift has been measured is $Ar_4HF$ [79]. Our calculations predict that the HF red shift $\Delta\nu_{4,2}$ of the $C_{2v}$ isomer (Fig. 2), -24.29 cm$^{-1}$, is 2.57 cm$^{-1}$ larger (in absolute value) than the red shift $\Delta\nu_{4,1}$ of the energetically optimal, $C_{3v}$ isomer, -21.72 cm$^{-1}$. Subsequent DQMC calculations by Lewerenz [113] gave a frequency shift difference of about 2.8 cm$^{-1}$ between the $C_{2v}$ and $C_{3v}$ isomers, confirming our quantum 5D results. The greater frequency shift of the $C_{2v}$ isomer is expected in view of the fact that it has four Ar atoms in direct contact with HF, compared to only three for the $C_{3v}$ isomer (Fig. 2). Until now only the $C_{3v}$ isomer (Fig. 1) has been observed [78, 79], despite the fact that the quantum 5D vibrational ground state energy of the $C_{2v}$ isomer is only 5.9 cm$^{-1}$ larger than that of the $C_{3v}$ isomer [73]. As discussed above, the experimental red shift of the $C_{3v}$ isomer, -19.697 cm$^{-1}$, agrees well with our quantum 5D and rigid-body DQMC values given in Table 2, -21.72 and -21.08 cm$^{-1}$, respectively.

Figure 4 reveals a substantial spread of HF red shifts $\Delta\nu_{n,i}$ for the three lowest-energy isomers ($i = 1 - 3$) of larger Ar$_n$HF clusters, $n = 5 - 12$. Careful examination shows that in all cases, the magnitudes of the red shifts for various Ar$_n$HF isomers of a given cluster size $n$ correlate directly with the number of Ar atoms they have in close proximity to HF [73]. The Ar$_5$HF cluster (Fig. 2) provides an excellent example. Its three lowest-energy isomers have frequency shifts $\Delta\nu_{5,i}$ ($i = 1 - 3$) of -22.13, -24.65, and -26.65 cm$^{-1}$, respectively; their ordering mirrors the number of Ar atoms, 3, 4, and 5, respectively (Fig. 2), which these isomers have in direct contact with HF (i.e., in the first solvation shell). The same considerations apply to the isomers of other Ar$_n$HF clusters and their corresponding frequency shifts [73].

These findings imply that the isomeric Ar$_n$HF structures differing in the number of Ar atoms occupying the first solvation shell can be readily distinguished by their HF red shifts. The most striking differences in the HF frequency shifts are between the monolayer isomers with fully solvated HF, and the stacked (or nonwetting) isomers where HF is on the surface of the Ar$_n$ subunit [73]. A nice illustration of this is the Ar$_9$HF cluster (Fig. 2). The red shift $\Delta\nu_{9,1}$ of the lowest-energy, fully solvated isomer, -37.20 cm$^{-1}$, and $\Delta\nu_{9,2}$ of the next higher, stacked isomer, -25.83 cm$^{-1}$, differ by 11.4 cm$^{-1}$. The difference between the red shifts $\Delta\nu_{8,1}$ (-25.73 cm$^{-1}$) and $\Delta\nu_{8,3}$ (-34.36 cm$^{-1}$) of the stacked minimum-energy isomer and the solvated second close-lying isomer of Ar$_8$HF (Fig. 2), respectively, is similar, 8.6 cm$^{-1}$. Qualitatively similar conclusions have been reached in the study of Ar$_n$SF$_6$ heteroclusters [65], which employed classical molecular dynamics simulations and a perturbative method for calculating the vibrational frequency shifts [67].

# 4 Photodissociation of HF in Ar$_n$HF clusters

## 4.1 Theoretical Methodology

**Potential Energy Surfaces.** Our treatment of the photodissociation of HF in Ar$_n$HF clusters is restricted to only two potential energy surfaces (PESs), i.e., it assumes that HF is electronically excited, by means of ultrashort $\delta(t)$-pulse excitation, from its ground electronic state $^1\Sigma$ to the repulsive excited state $^1\Pi$ only. Both the ground and excited Ar$_n$HF cluster PESs are modeled in a pairwise additive fashion. Similar assumptions have been made in theoretical studies of the photolysis of HCl in Ar-HCl, by Garcia-Vela *et al.* [37, 38, 39, 40, 41, 42] and Schröder *et al.* [43, 128], and UV photodissociation of Ar$_2$HCl [129]. The total ground-state PES for the cluster of size $n$, $V_n^g$, is written as

$$V_n^g = V_{vdW,n}^g + V_{HF}^g(r). \tag{3}$$

The first term in (3), $V_{vdW,n}^g$, is the pairwise additive intermolecular PES of Ar$_n$HF (for fixed HF bond length $r$) defined in (1). $V_{HF}^g(r)$, the second term of

(3), is the ground-state potential curve for isolated HF molecule, where $r$ is the H–F distance. In this work, $V_{HF}^g(r)$ is approximated by a Morse function with parameters $D = 6.11$ eV, $r_e = 0.93$ Å, and $\alpha = 2.37$ Å$^{-1}$, obtained by fitting the *ab initio* data of Dunning [130]. The separable form of the total ground-state PES $V_n^g$ in (3) implies very weak coupling between the intramolecular HF vibrational coordinate $r$ and the intermolecular degrees of freedom of the cluster.

The excited electronic state PES for Ar$_n$HF, $V_n^e$, is represented as a sum of pairwise interactions between all atoms of the cluster:

$$V_n^e = V_{HF}^e(r) + \sum_{i=1}^n V_{Ar_iH}(r_{Ar_iH}) + \sum_{i=1}^n V_{Ar_iF}(r_{Ar_iF}) + \sum_{i<j}^n V_{Ar_iAr_j}(r_{ij}). \quad (4)$$

$V_{HF}^e(r)$ is the HF potential curve in the $^1\Pi$ excited state, obtained by fitting the *ab initio* calculations of Dunning [130] to the form

$$V_{HF}^e(r) = Ae^{-\alpha r} + \frac{C}{r^{12}}, \quad (5)$$

with $A = 44.95$ eV, $\alpha = 2.38$ Å$^{-1}$, and $C = 15.70$ meVÅ$^{12}$. The Ar–H potential $V_{Ar,H}$ is taken from the Ar–HCl photodissociation calculations [37, 40, 41, 43, 128]. Due to the open-shell character of the fluorine atome the situation is more complicated for the Ar-F interaction. Strictly speaking, when the $p$-orbital of F($^2P$) is properly taken into account, the interaction between Ar and F($^2P$) leads to both $\Sigma$- and $\Pi$-type potential curves ($V_\Sigma$ and $V_\Pi$). In order to simplify the present calculations we represent $V_{Ar,F}$ by an isotropic potential ($V_0$), which depends only on the separation and not on the orientation of the $\pi$ lobe of the F($^2P$) orbital; it represents the average of $V_\Sigma$ and $V_\Pi$. We use the $V_0$ term determined by the molecular-beam scattering experiments of Aquilanti *et al.* [131]. A more rigorous treatment can be found in [132]. In the case of HF dissociation, the H atom gains most of the energy and the F atom plays more the role of a spectator. Therefore we doubt that the energy or the angular distributions of the hydrogen are noticeably affected. If anything, a more realistic description of the Ar-F interaction might modify the fragmentation pattern of the Ar cluster. For computational simplicity, the Ar–Ar interaction $V_{Ar_iAr_j}$ in the excited-state PES of the cluster is described by the Lennard-Jones 12-6 potential

$$V_{Ar_iAr_j}(r_{ij}) = \epsilon\left[\left(\frac{r_m^{Ar-Ar}}{r_{ij}}\right)^{12} - 2\left(\frac{r_m^{Ar-Ar}}{r_{ij}}\right)^6\right], \quad (6)$$

where $\epsilon = 12.34$ meV and $r_m^{Ar-Ar} = 3.759$ Å.

**Classical Trajectory Calculations.** The photodissociation dynamics of Ar$_n$HF cluster on the excited electronic state PES $V_n^e$ in (4) is treated by means of classical trajectories, which is justified by the high-energy (several eV) nature of the process. Following a $\delta(t)$-pulse excitation, Ar$_n$HF makes an instantaneous, vertical Franck-Condon transition from the ground-state PES $V_n^g$ in (3) to the excited-state PES $V_n^e$ of (4), where it evolves subject to the classical equations of motion.

Accurate quantum description of the initial state of the system is absolutely critical for a quantitative treatment of the photofragmentation of a light molecule, such as HF, in a rare-gas heterocluster. This is due to the LAM intermolecular vibrations of these clusters, which are highly nonclassical and whose eigenstates are delocalized over the ground-state PES. They define the initial conditions for the high-energy motions of the H and F atoms in the classical trajectories run on the excited-state PES. Use of inaccurate initial state can seriously distort even the qualitative features of the photodissociation dynamics, as demonstrated by the comparison of our exact quantum 3D wave packet calculations of the UV photodissociation of HCl in the Ar-HCl vdW complex [128] with approximate treatments of this problem [41].

The LAM intermolecular vibrations of $Ar_n$HF clusters mainly involve the HF molecule, because it is considerably lighter (especially the H atom) than Ar atoms. Therefore, in the classical trajectory simulations on the excited-state PES, the argons were initially set at their equilibrium positions defined by the minimum-energy $Ar_n$HF cluster structures shown in Fig. 1, with zero kinetic energy. This choice of the initial configuration and momenta of Ar atoms is reasonable for clusters in their ground vibrational state; moreover, the energy released by breaking of the H–F bond (several eV) vastly exceeds the vdW vibrational zero-point energy of the clusters.

The problem which remained was that of generating the probability distributions of the initial coordinates and momenta of the H and F atom. Quantum 5D bound-state calculations (rigid HF) of $Ar_n$HF clusters (for frozen $Ar_n$) [71, 72, 73] described earlier give vdW vibrational eigenstates of the coupled LAM vibrations on the ground-state intermolecular cluster PES $V_{n,v}^{tot}$ in (1) [denoted $V_{vdW,n}^g$ in (3)]. In the following, the vdW vibrational ground state wave function of $Ar_n$HF (minimum-energy isomer) is denoted $\Psi_{vdW,n}(x, y, z, \theta, \phi)$. The first three coordinates $x, y, z$ are the Cartesian coordinates of the center of mass (c.m.) of HF, while the two polar angles $(\theta, \phi)$ specify the orientation of HF relative to $Ar_n$. The $z$ axis is aligned with HF in its equilibrium geometry in the $Ar_n$HF cluster. Contour plots of the probability amplitude $|\Psi_{vdW,3}(x, y, z, \theta, \phi)|^2 \sin \theta$, of $Ar_3$HF, displayed in Fig. 2 of [75], show substantial wave function delocalization already in the ground state of the cluster. The full 6D vibrational ground state wave function, which includes dependence on the HF stretching coordinate $r$, can be reasonably represented in the product form as

$$\Psi_n(r, x, y, z, \theta, \phi) = \Phi_{HF}(r)\Psi_{vdW,n}(x, y, z, \theta, \phi). \qquad (7)$$

In (7), $\Phi_{HF}(r)$ is the vibrational ground state wave function of the HF Morse potential $V_{HF}^g(r)$ appearing in (3).

It is well known that there is no unique way of generating phase-space distributions from general quantum mechanical wave functions. When the wave function is approximated by a product of 1D Gaussians in the coordinate space, the (separable) Wigner distribution function of cooordinates and momenta is easily obtained [133]. However, this straightforward recipe is not appropriate for systems like $Ar_n HF$, whose multidimensional ground-state wave function is extensively delocalized. We had to resort to a lengthy procedure outlined below; the details can be found in [75].

The probability distribution of the Cartesian coordinates and momenta for the H and F atom, $P(x_1, \ldots, x_6, p_1, \ldots, p_6)$, is assumed to have a factorized form:

$$P(x_1, \ldots, x_6, p_1, \ldots, p_6) = P_x(x_1, \ldots, x_6) P_p(p_1, \ldots, p_6), \qquad (8)$$

$P_x$ and $P_p$ in (8) are the probability distributions of (H,F) coordinates and momenta; $(x_1, x_2, x_3, p_1, p_2, p_3)$ and $(x_4, x_5, x_6, p_4, p_5, p_6)$ are associated with H and F, respectively. It is further assumed that the 6D wave function $\Psi_n(\mathbf{q})$ $(q_1 = r, q_2 = x, \ldots, q_5 = \theta, q_6 = \phi)$ in Eq. (7) could be approximated by a product of 1D wave functions in each coordinate $q_i$:

$$\Psi_n(\mathbf{q}) = \prod_{i=1}^{6} \psi_n(q_i). \qquad (9)$$

The 1D wave functions $\psi_n(q_i)$ are obtained from $\Psi_n(\mathbf{q})$ by setting the five coordinates $q_j \neq q_i$ to the values for which $\Psi_n(\mathbf{q})$ has maximum amplitude. The probability distribution $P_{q_i}(q_i)$ of coordinate $q_i$ $(i = 1 - 6)$ is obtained by fitting the modulus squared of the 1D wave function $\psi_n(q_i)$ by a sum of Gaussians in $q_i$:

$$P_{q_i}(q_i) = [1 - \delta_{i5} + \delta_{i5} \sin q_5] \sum_k a_k e^{-b_k^2 (q_i - q_i^o)^2}, 1 \leq k \leq 3, \qquad (10)$$

where $a_k, b_k, q_i^o$ are the fitting parameters, $q_5 = \theta$, and $\delta_{i5}$ is the Kronecker delta. The probability distribution of the momenta $\Gamma_p(p_1, \ldots, p_6)$ is obtained in a similar fashion, although the procedure is complicated by the fact that the 6D wave function $\Psi_n$ is in the coordinates $r, x, y, z, \theta, \phi$, while the momenta in $P_p(p_1, \ldots, p_6)$ are Cartesian [75].

## 4.2 Results and Discussion

For each $Ar_n HF$ cluster size $n = 1 - 14, 54$, 20000 trajectories were run, which was sufficient to produce converged resultes discussed in this section. In order to discern the influence of $Ar_n$ cages of different sizes and geometry on the HF photodissociation dynamics, we have carefully analyzed the H-atom kinetic energy and angular distributions, the survival probability, and cluster fragmentation patterns [75].

**Kinetic Energy Distributions of Hydrogen Atom.** The kinetic energy distributions (KEDs) of the H photofragment for $Ar_n HF$ $(n = 1-14, 54)$ clusters are displayed in Fig. 5. The "noise"-like structures visible in each KED result from the limited number of trajectories and have no physical meaning. The KEDs for ArHF and $Ar_2 HF$ show little evidence of energy transfer due to H/Ar and

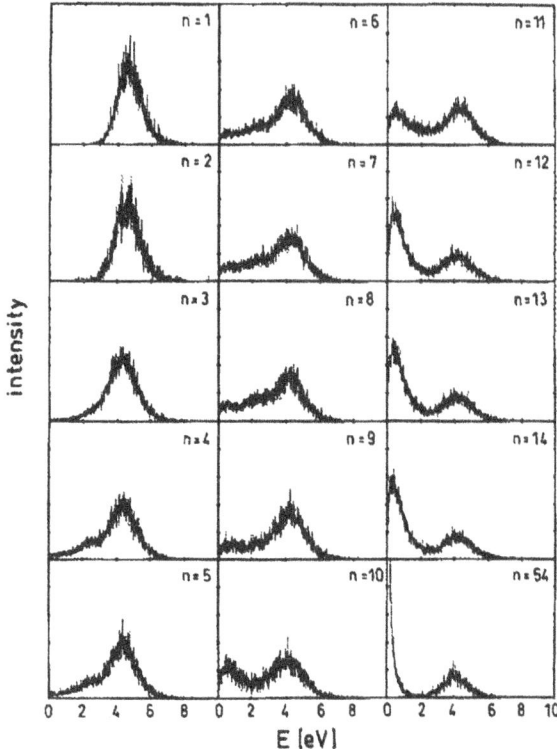

**Fig. 5.** The H-atom kinetic energy distributions for $Ar_nHF$ ($n = 1 - 14, 54$) clusters.

H/F collisions. Their widths merely reflect the broad range of energies with which the clusters emerge on the excited-state PES, as a result of $\delta(t)$-pulse excitation.

Collisional energy transfer from the H atom to argons becomes clearly evident in the H-atom KED for $Ar_3HF$, as a tail extending towards the low-energy region of the distribution. The three Ar atoms of $Ar_3HF$ arranged in an equilateral triangle (Fig. 1) present a more sizable "wall" to the H fragment than one or two argons, leading to more effective energy transfer from the hydrogen. The low-energy tail is considerably more intense in the H-atom KED for $Ar_4HF$, which can be understood from the comparison of the equilibrium structures of $Ar_4HF$ and $Ar_3HF$ (Fig. 1). The fourth argon of $Ar_4HF$ caps the $Ar_3$ base, thus creating a significantly thicker barrier to the dissociating H atom than a single $Ar_3$ layer.

The H-atom KED remains qualitatively the same as the number of Ar atoms in the cluster grows from 4 to 8. The low-energy tail continues to gain in intensity while the high-energy peak, whose position does not shift with $n$, diminishes. Again, the explanation for this trend lies in the minimum-energy structures of $Ar_nHF$ clusters shown in Fig. 1. For $n = 4 - 8$, HF is bound to a surface site of $Ar_n$, and the number of Ar atoms in direct contact with it changes very little, from three ($n = 4, 5$) to four ($n = 6-8$)[70]. Instead of bulding the first solvation shell around HF, the additional argons enter into the second layer, making the solvent cage thicker rather than more complete. Growing thickness of the $Ar_n$ ($n = 4 - 8$) microcluster means that the H atom has to spend an increasing fraction of its kinetic energy to break through the obstacle.

These simple, qualitative pictures based solely on consideration of the equilibrium cluster structures are supported by the more quantitative information presented in Fig. 6, which gives the probability of the H photofragment undergoing zero, one, two, three, or more collisions with the Ar atoms before dissociating completely from the cluster, for $Ar_n$HF with $n = 2, 4, 6, 8, 10, 12$. It is clear from Fig. 6 that trajectories in which the H atom undergoes multiple collisions with the Ar atoms, causing it to lose an appreciable fraction of its kinetic energy, are common only for clusters with four or more argons, explaining why the low-energy trail is absent in the KEDs for ArHF and $Ar_2$HF, and is barely visible in the $Ar_3$HF KED. For larger clusters ($n = 6, 8$), a substantial fraction of H atoms escapes after numerous (3-10) collisions with the Ar atoms and a large loss of kinetic energy; they form the intense low-energy tail of the KED. The probability of trajectories with many H/Ar collisions increases with cluster size, accounting for the growing tail of the distribution.

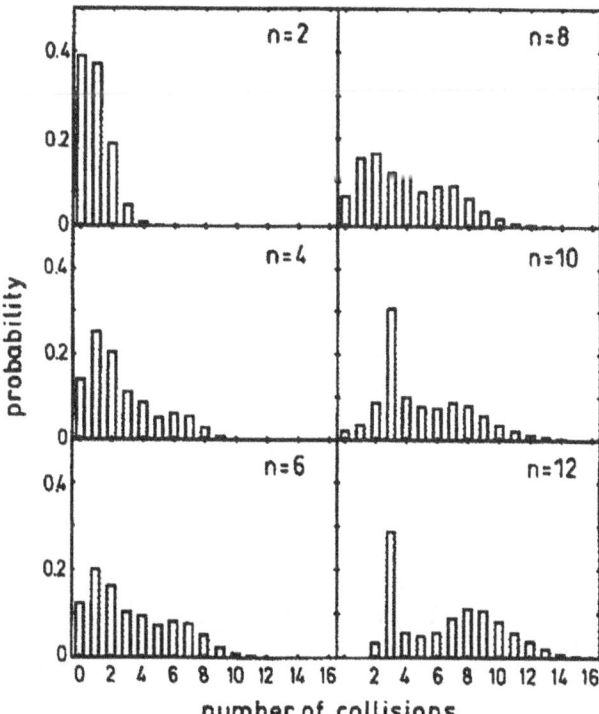

Fig. 6. The probability that the H atom escaping the cluster undergoes a given number of collisions with the Ar atoms, for selected $Ar_n$HF cluster sizes.

The KEDs of $n \geq 10$ $Ar_n HF$ clusters are qualitatively different from those of the smaller clusters; they are distinctly *bimodal*, unlike their $n < 9$ counterparts which display a single high-energy peak. This qualitative change of the H-atom KED is a reflection of the discontinuity in the size evolution of $Ar_n HF$ minimum-energy cluster structures described in Sect. 2.2 and evident in Fig. 1. The clusters with $n < 9$ have HF at an exterior, surface site, while for $n \geq 9$ HF is inside a cage formed by all available Ar atoms (up to $n = 12$). Fig. 5 demonstrates that the Ar cage around HF leads to much more extensive energy transfer from hydrogen to the Ar atoms than is the case for surface-bound HF in $n < 9$ clusters. Fig. 6 shows that this is due to the fact that formation of a more complete solvent shell increases significantly the probability of trajectories in which many (6-10) H/Ar collisions occur, accompanied by large energy transfer.

Fig. 5 reveals another interesting feature of the bimodal KEDs of $Ar_n HF$ clusters for $n \geq 10$. The relative magnitudes of the two KED peaks change with increasing $n$; the high-energy peak diminishes, while the low-energy peak grows and becomes dominant for $n = 12$. This trend is readily understood with the help of Fig. 6, which shows (for $n = 10, 12$) the existence of two distinct classes of H photofragments when HF is in the interior of $Ar_n$ cage. The first class is comprised of the H atoms which break through the Ar shell (or escapes through a "hole" in it, for $n < 12$) fairly directly, after only a few collisions with the Ar atoms. The most probable trajectories of this type, $\sim30\%$ of *all* trajectories, are those with three H/Ar collisions which produce still very hot H atoms, and contribute to the high-energy peak of the KED. In the second class are the H atoms trapped inside the cage, undergoing multiple collisions with the Ar atoms prior to their escape. For $Ar_{10}HF$, the trajectories with 6-8 H/Ar collisions are the most probable in this class; for $Ar_{12}HF$, the probability maximum shifts to trajectories with 7-10 H/Ar collisions. This shift is understandable since, unlike the $n = 12$ icosahedral cage, the $n = 10$ cage is incomplete and allows the H atom to leave after fewer collisions. In any case, after 6-10 collisions with the Ar atoms the H atom loses most of its kinetic energy; these H atoms form the low-energy peak in the KEDs of the $n \geq 10$ clusters.

The the KED for $Ar_{54}HF$ cluster, with *two* complete icosahedral shells, is very different from that of any other cluster considered here, including $Ar_{12}HF$ which has only the first solvation shell completed. The low-energy peak is exceptionally narrow and the high-energy peak is very small, indicating almost complete cooling of the H atom. Clearly, the second solvation shell enhances dramatically the collisional energy transfer. The H atom which penetrates the first solvent layer is reflected back by the second shell. It rattles back and forth much longer than in the $Ar_{12}$ cage, until it has transferred virtually all its kinetic energy to the solvent cluster. Thus, the $Ar_{54}HF$ cluster with only two complete solvent layers mimics very closely the efficiency of collisional energy transfer in bulk Ar matrix.

**Angular Distributions of Hydrogen Atom.** Fig. 7 displays the final angular distributions (ADs) of the H photofragment which we have calculated for $Ar_n HF$

$(n = 1 - 14, 54)$ clusters. The scattering angle $\Theta$ is the angle enclosed by the position vector of the escaping H atom and the $z$ axis of the cluster, aligned with the equilibrium orientation of HF relative to the $Ar_n$ subunit in $Ar_n HF$ (see the minimum-energy geometries in Fig. 1). When $\Theta$ equals zero, the H atom moves along the $z$ axis towards $Ar_n$, and for $\Theta = 180°$ it escapes in the opposite direction.

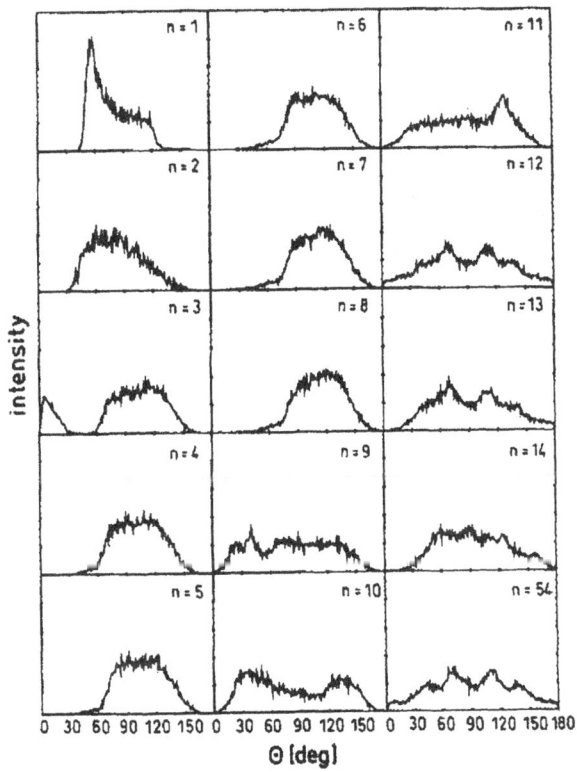

**Fig. 7.** The H-atom angular distributions for $Ar_n HF$ $(n = 1 - 14, 54)$ clusters. For additional information, see the text.

It is clear from Fig. 7 that the H-atom AD is strongly dependent on the size of $Ar_n HF$ cluster. The variation of the AD with cluster size is the most pronounced for small $Ar_n HF$ clusters, $n = 1 - 4$, which should facilitate its experimental observation; the spectroscopy of these clusters has been thoroughly investigated [76, 77, 78, 79]. This may be contrasted with the size dependence of the H-atom KED, where the most dramatic changes take place for larger clusters having nine or more Ar atoms. Detailed analysis of the ADs for $Ar_n HF$ $(n = 1 - 14, 54)$ clusters has been made in [75]. In this review, we discuss only the ADs for the $n = 1, 3, 4$ clusters, which we consider to be the most interesting.

The AD for ArHF differs qualitatively from those for the larger clusters. It is essentially zero for angles $\Theta$ less than 45°; it then rises steeply and peaks around 50°. Following a gradual descrease for larger scattering angles, around 120° the AD abruptly decays practically to zero. The unusual shape of the ArHF AD is best understood in terms of the deflection function $\Theta(\Theta_i)$, i.e., the final angle of the escaping H atom as a function of the initial angle $\Theta_i$, which is shown in Fig. 8. Also shown in Fig. 8 is $P(\Theta_i)$, the quantum mechanical distribution function of the initial angles, obtained from the bound state calculations. $\Theta(\Theta_i)$ in Fig. 8 shows that no classical trajectories lead to angles smaller than 45°, which explains why the AD is zero below $\sim$ 45°. Without going into details, which can be found in [75], the sharp rise of the AD around 50° and its sudden decay around 120° are manifestations of so-called rainbow structures, which stem from the extrema of the deflection function. They are very similar to the the rainbows known in elastic scattering [134] or rotational rainbows [135, 136] observed in atom-molecule inelastic scattering and photodissociation of triatomic molecules. The negligible intensity of the AD for final scattering angles $\Theta$ larger than $\sim$ 120°, although they are in principle accessible for large initial angles $\Theta_i$ (see $\Theta(\Theta_i)$ in Fig. 8), comes from the near-zero values of $P(\Theta_i)$ for $\Theta_i \geq 120°$. In conclusion, the presence of a single solvent (Ar) atom is manifested much more clearly in the H-atom AD for ArHF than in its KED.

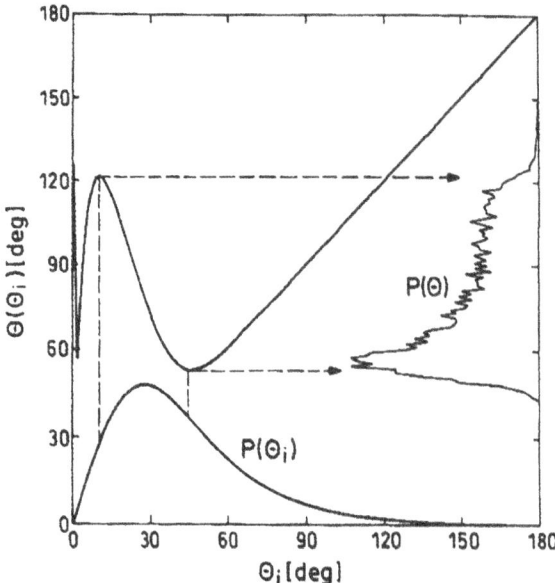

**Fig. 8.** Plot of the deflection function $\Theta(\Theta_i)$ for ArHF. Also shown are $P(\Theta_i)$, the quantum probability amplitude of the initial angles $\Theta_i$, and the H-atom angular distribution $P(\Theta)$ of ArHF. For additional information, see the text.

The ADs for $Ar_3HF$ and $Ar_4HF$ provide a beautiful example of the strong impact which one additional Ar atom can have on the photofragmentation dynamics. Two maxima are present in the $Ar_3HF$ AD; a rather narrow one at smaller scattering angles ($0° < \Theta < 30°$), which is not found in the AD of any other $Ar_nHF$ cluster, and a broad maximum at large scattering angles ($60° < \Theta < 150°$). The AD for $Ar_4HF$ shows only the latter broad maximum at larger scattering angles, but there is no trace of the small-angle peak. What has caused its disappearance? Inspection of the trajectories for $Ar_3HF$ showed that the peak at smaller angles comes from the H atoms which initially move nearly parallel to the cluster $C_3$ axis, and pass through the "hole" in the $Ar_3$ triangle relatively undisturbed, with only minor deflections. The broad large-angle peak is due to the H atoms which hit the $Ar_3$ triangle at angles less favorable for direct penetration, and are scattered to larger $\Theta$'s. In $Ar_4HF$, the fourth Ar atom on the $C_3$ axis (Fig. 1) quite literally plugs the "hole" in the $Ar_3$ base so that the H atom, even if aimed straight towards the center of $Ar_3$, finds its path blocked and is reflected to larger scattering angles, thus eliminating the small-angle peak from the AD.

**Survival Probabilities of Hydrogen Atom.** The solvent ($Ar_n$) cluster in close proximity to the solute (HF) undergoing fragmentation can act as a cage, delaying for some time, or prevent entirely, the separation of the H and F photofragments. The $Ar_nHF$ clusters are well suited for investigating how the length of the confinement of the fragments varies with the size and the geometry of the solvent cluster. In order to characterize the caging efficiency, for each cluster size considered ($n = 1 - 14, 54$), we have calculated the H-atom survival probability (SP) as a function of time, which is the probability that at time $t$ the H atom has still not separated completely (in this work, by more than 12 Bohr, a distance at which the H–Ar potential is essentially zero) from all other atoms in the cluster. The time dependence of the H-atom SP for selected $Ar_nHF$ clusters is displayed in Fig. 9.

First, we discuss the SPs for the $Ar_nHF$ clusters with $n = 1 - 8$, in which HF is on the surface of $Ar_n$; the SPs for $n = 1 - 4, 8$ as a function of time are shown in Fig. 9(a). For ArHF and $Ar_2HF$, the SPs decay to zero after only 30-40 fs, demonstrating virtually no caging with one or two Ar atoms, in agreement with our findings regarding the UV photodissociation of ArHCl [43, 128]. The SP develops a barely visible tail towards slightly longer times for $n = 3$. With increasing cluster size, this tail extends to longer times and gains in intensity, showing that, as expected, larger ($n = 3 - 8$) clusters cage the dissociating HF more effectively. It should be noted that the growth of the longer-time tail of the SPs with $n$ mirrors the trend of increasing prominence of the low-energy tail of the H-atom KEDs in the same cluster size range (Fig. 5), discussed earlier. This is understandable, since both tails have their origins in multiple H/Ar collisions, the probability of which increases with $n$ (Fig. 6).

Fig. 9(b) shows the time dependence of the SPs for the $n = 9 - 12, 54$ clusters where, in contrast to the clusters with $n < 9$, HF is inside $Ar_n$ microcluster.

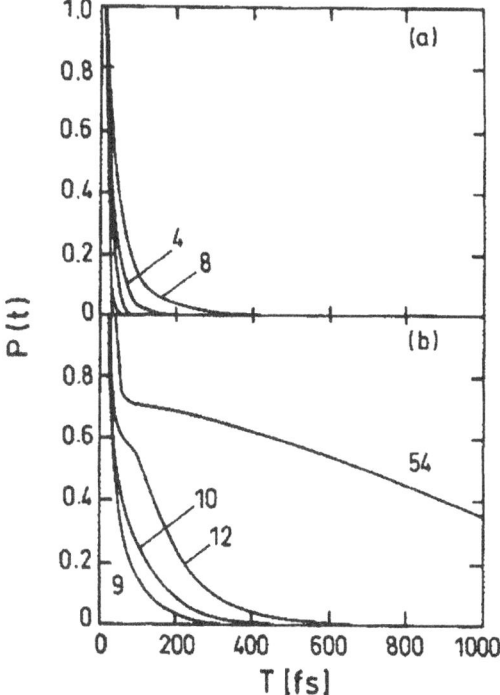

**Fig. 9.** The survival probability of selected $Ar_nHF$ clusters as a function of time. The first three curves, which are not marked, are for the $n = 1, 2, 3$ clusters, respectively.

These SPs take significantly more time to decay, evidence that a partial ($n = 9 - 11$) or complete ($n = 12$) solvent shell around HF provides a much better cage than $Ar_n$ with HF on its exterior. Closed first solvation shell, for $n = 12$, is particularly effective in delaying the escape of the H atom. Unlike the SPs of the smaller clusters which decay monotonically, the SP for $Ar_{12}HF$ exhibits a shoulder. This feature can be traced to the existence of two classes of dissociating trajectories, discussed earlier in connection with the bimodal H-atom KED for this cluster (Fig. 5). The trajectories in which the H atom undergoes just a few collisions with the Ar atoms and leave the cluster quickly, give rise to the initial, rapidly decaying segment of the SP. The second class of trajectories for $Ar_{12}HF$ involves numerous H/Ar collisions (Fig. 6) which delay the escape of the H atom; they are responsible for the more slowly decreasing portion of the $n = 12$ SP.

Completion of the second solvent shell in $Ar_{54}HF$ affects profoundly the H-atom SP, just as it did the KED. The time scale for decay of the SP is dramatically extended to picoseconds, far beyond that of other clusters considered. At $t = 1$ ps, which is about twice the time it takes the SP for $Ar_{12}HF$ to decay to zero, the SP for $Ar_{54}HF$ still has a substantial value of about 0.4. The H-atom SP confirms our conclusion based on the consideration of the KED for $Ar_{54}HF$ above, that the dissociation dynamics of the solute molecule surrounded by two solvation shells is qualitatively different from that when only one solvent layer is present, and displays much more of the behaviour expected in bulk solvent.

# 5 Conclusions

This article reviews our calculations of the equilibrium structures [70], HF vibrational frequency shift [71, 72, 73, 74], and HF photodissociation [75] of $Ar_nHF$ clusters. We have achieved for the first time a comprehensive and highly accurate quantum mechanical description of the evolution of HF vibrational frequency shift over a wide range of cluster sizes and isomeric structures. In the cluster size range considered, $n = 1 - 14$, the calculated HF red shift grows from -9.66 cm$^{-1}$ for the smallest cluster, ArHF, to the value for the icosahedral $Ar_{12}HF$ with completed first solvation shell, -42.46 cm$^{-1}$, which is comparable to that measured for HF in bulk Ar matrix. For the $n = 1 - 4$ clusters, the red shifts which we have calculated using both a quantum five-dimensional approach and rigid-body diffusion quantum Monte Carlo method agree very well with the available experimental data. The size dependence of $Ar_nHF$ red shift predicted by our quantum calculations is distinctly nonmonotonic, and correlates perfectly with the changes in the number of Ar atoms in the first solvation shell around HF. Our quantum 5D bound state calculations also predict a strong isomer dependence of HF red shift for the $n = 4 - 14$ clusters. The red shifts of different $Ar_nHF$ cluster isomers typically differ by several wave numbers, which in principle can be used to differentiate them experimentally.

Our theoretical study of the photodissociation of HF in $Ar_nHF$ clusters with $n = 1 - 14, 54$, for ultrashort $\delta(t)$-pulse excitation, has elucidated the influence of the size and structure of the solvent $(Ar_n)$–solute (HF) heterocluster on the photofragmentation dynamics of the solute molecule. This range of cluster sizes spans the formation and closing of the first solvation shell, for $n = 12$, and the addition of the complete second solvent layer, $n = 54$. The dissociation was treated using classical trajectories; the distributions of the initial coordinates and momenta of the H and F atom were generated from the quantum 5D intermolecular vibrational ground state wave functions of $Ar_nHF$ clusters, obtained in the course of the HF frequency shift calculations. This was essential for quantitative description of the photofragmentation, in view of the highly quantum nature of the initial bound state of the very floppy $Ar_nHF$ clusters. Our investigation has revealed strong size dependence of all aspects of the dissociation dynamics studied, namely the H-atom kinetic energy and angular distributions, the survival probability, and cluster fragmentation patterns, some of which should be experimentally observable.

# Acknowledgments

The research described in this article has been supported in part by the National Science Foundation. I am most grateful to my co-workers, Dr. Suyan Liu, Prof. Jules M. Moskowitz (NYU), Prof. Kevin E. Schmidt (Arizona State U.),

Prof. Reinhard Schinke and Dr. Thomas Schröder (MPI für Strömungsforschung, Göttingen), and Dr. Parhat Niyaz (U. of California-Berkeley). This project would have never been realized without their participation. The help of Mr. Atul Bahel (NYU) with some of the figures is gratefully acknowledged.

# References

1. U. Buck, J. Phys. Chem. **98**, 5190 (1994).
2. U. Buck and B. Schmidt, J. Chem. Phys. **101**, 6365 (1994).
3. U. Buck and I. Ettischer, Fraday Discuss. **97**, 215 (1994).
4. A. D. Buckingham, J. Chem. Soc., Faraday Trans. **56**, 753 (1960).
5. R. N. Pribble and T. S. Zwier, Science **265**, 75 (1994).
6. R. N. Pribble and T. S. Zwier, Faraday Discuss. **97**, 229 (1994).
7. E. Honneger and S. Leutwyler, J. Chem. Phys. **88**, 2582 (1988).
8. A. Furlan, T. Troxler and S. Leutwyler, J. Phys. Chem. **97**, 13527 (1993).
9. S. S. Xantheas and T. H. Dunning, Jr., J. Chem. Phys. **98**, 8037 (1993).
10. S. S. Xantheas and T. H. Dunning, Jr., J. Chem. Phys. **99**, 8774 (1993).
11. K. Kim, K. D. Jordan and T. S. Zwier, J. Am. Chem. Soc. **116**, 11568 (1994).
12. S. Y. Fredericks, K. D. Jordan and T. S. Zwier, J. Phys. Chem. **100**, 7810 (1966).
13. A. W. Castleman, Jr. and K. H. Bowen, Jr., J. Phys. Chem. **100**, 12911 (1996).
14. Z. Bačić and R. E. Miller, J. Phys. Chem. **100**, 12945 (1996).
15. K. L. Saenger, G. M. McClelland and D. R. Herschbach, J. Phys. Chem. **85**, 3333 (1981).
16. J. J. Valentini and J. B. Cross, J. Chem. Phys. **77**, 572 (1982).
17. J. M. Philippoz, P. Melinon, R. Monot and H. van den Bergh, Chem. Phys. Lett. **138**, 579 (1987).
18. J. M. Philippoz, H. van den Bergh and R. Monot, J. Phys. Chem. **91**, 2545 (1987).
19. J. M. Philippoz, R. Monot and H. van den Bergh, J. Chem. Phys. **93**, 8676 (1990).
20. M. L. Burke and W. M. Klemperer, J. Chem. Phys. **98**, 1797 (1993).
21. J. M. Papanikolas, J. R. Gord, N. E. Levinger, D. Ray, V. Vorsa and W. C. Lineberger, J. Phys. Chem. **95**, 8028 (1991).
22. J. M. Papanikolas, V. Vorsa, M. E. Nadal, P. J. Campagnola, J. R. Gord and W. C. Lineberger, J. Chem. Phys. **97**, 7002 (1992).
23. J. M. Papanikolas, V. Vorsa, M. E. Nadal, P. J. Campagnola, H. K. Buchenau and W. C. Lineberger, J. Chem. Phys. **99**, 8733 (1993).
24. V. Vorsa, P. J. Campagnola, S. Nandi, M. Larsson and W. C. Lineberger, J. Chem. Phys. **105**, 2298 (1996).
25. M. E. Nadal, P. D. Kleiber and W. C. Lineberger, J. Chem. Phys. **105**, 504 (1996).
26. A. H. Zewail, Ber. Bunsenges. Phys. Chem. **99**, 474 (1995).
27. A. H. Zewail. In *Femtosecond Chemistry*, edited by J. Manz and L. Wöste, page 15. (VCH, Weinheim, New York, 1995).
28. J. K. Wang, Q. Liu and A. H. Zewail, J. Phys. Chem. **99**, 11309 (1995).
29. E. D. Potter, Q. Liu and A. H. Zewail, Chem. Phys. Lett. **200**, 605 (1992).
30. Q. Liu, J. K. Wang and A. H. Zewail, Nature **364**, 427 (1993).
31. I. NoorBatcha, L. M. Raff and D. L. Thompson, J. Chem. Phys. **81**, 5658 (1984).
32. J. A. Beswick, R. Monot, J. M. Philippoz and H. van den Bergh, J. Chem. Phys. **86**, 3965 (1987).

33. M. P. de Miranda, J. A. Beswick and N. Halberstadt, Chem. Phys. **187**, 185 (1994).
34. O. Roncero, J. A. Beswick and N. Halberstadt, Chem. Phys. Lett. **226**, 82 (1994).
35. O. Roncero, J. A. Beswick and N. Halberstadt, J. Chem. Phys. **104**, 7554 (1996).
36. J. Y. Fang and C. C. Martens, J. Chem. Phys. **105**, 9072 (1966).
37. A. Garcia-Vela, R. B. Gerber and J. J. Valentini, Chem. Phys. Lett. **186**, 223 (1991).
38. A. Garcia-Vela, R. B. Gerber and J. J. Valentini, J. Chem. Phys. **97**, 3297 (1992).
39. A. Garcia-Vela, R. B. Gerber and D. G. Imre, J. Chem. Phys. **97**, 7242 (1992).
40. A. Garcia-Vela, R. B. Gerber, D. G. Imre and J. J. Valentini, Chem. Phys. Lett. **202**, 473 (1993).
41. A. Garcia-Vela and R. B. Gerber, J. Chem. Phys. **98**, 427 (1993).
42. A. Garcia-Vela, R. B. Gerber, D. G. Imre and J. J. Valentini, Phys. Rev. Lett. **71**, 931 (1993).
43. T. Schröder, R. Schinke, M. Mandziuk and Z. Bačić, J. Chem. Phys. **100**, 7239 (1994).
44. K. M. Christoffel and J. M. Bowman, J. Chem. Phys. **104**, 8348 (1996).
45. F. G. Amar and B. J. Berne, J. Phys. Chem. **88**, 6720 (1984).
46. L. Perera and F. G. Amar, J. Chem. Phys. **90**, 7354 (1989).
47. F. G. Amar and L. Perera, Z. Phys. D. **20**, 173 (1991).
48. J. M. Papanikolas, P. E. Maslen and R. Parson, J. Chem. Phys. **102**, 2452 (1995).
49. R. Alimi and R. B. Gerber, Phys. Rev. Lett. **64**, 1453 (1990).
50. L. M. Finney and C. C. Martens, J. Phys. Chem. **97**, 13477 (1993).
51. R. J. Le Roy, J. S. Carley and J. E. Grabenstetter, Faraday Discuss. Chem. Soc. **62**, 169 (1977).
52. R. J. Le Roy and J. M. Hutson, J. Chem. Phys. **86**, 837 (1987).
53. J. M. Hutson, J. Chem. Phys. **96**, 6752 (1992).
54. J. M. Hutson, J. Phys. Chem. **96**, 4237 (1992).
55. R. C. Cohen and R. J. Saykally, J. Chem. Phys. **98**, 6007 (1993).
56. R. A. Aziz and H. H. Chen, J. Chem. Phys. **67**, 5719 (1977).
57. R. A. Aziz, J. Chem. Phys. **99**, 4518 (1993).
58. M. J. Elrod, D. W. Steyert and R. J. Saykally, J. Chem. Phys. **94**, 58 (1991).
59. M. J. Elrod, J. G. Loeser and R. J. Saykally, J. Chem. Phys. **98**, 5352 (1993).
60. A. R. Cooper and J. M. Hutson, J. Chem. Phys. **98**, 5337 (1993).
61. S. Goyal, D. L. Schutt and G. Scoles, Acc. Chem. Res. **26**, 123 (1993).
62. S. Goyal, D. L. Schutt and G. Scoles, J. Chem. Phys. **102**, 2302 (1995).
63. X. J. Gu, D. J. Levandier, G. Scoles and D. Zhang, J. Chem. Phys. **93**, 4898 (1990).
64. S. Goyal, G. N. Robinson, D. L. Schutt and G. Scoles, J. Phys. Chem. **95**, 4186 (1991).
65. M. A. Kmetic and R. J. Le Roy, J. Chem. Phys. **95**, 6271 (1991).
66. D. J. Chartrand, J. C. Shelley and R. J. Le Roy, J. Phys. Chem. **95**, 8310 (1991).
67. D. Eichenauer and R. J. LeRoy, J. Chem. Phys. **88**, 2898 (1988).
68. L. Perera and F. G. Amar, J. Chem. Phys. **93**, 4884 (1990).
69. J. M. Sperhac, M. J. Weida and D. J. Nesbitt, J. Chem. Phys. **104**, 2202 (1996).
70. S. Liu, Z. Bačić, J. W. Moskowitz and K. E. Schmidt, J. Chem. Phys. **100**, 7166 (1994).
71. S. Liu, Z. Bačić, J. W. Moskowitz and K. E. Schmidt, J. Chem. Phys. **101**, 6359 (1994).

84

72. S. Liu, Z. Bačić, J. W. Moskowitz and K. E. Schmidt, J. Chem. Phys. **101**, 10181 (1994).
73. S. Liu, Z. Bačić, J. W. Moskowitz and K. E. Schmidt, J. Chem. Phys. **103**, 1829 (1995).
74. P. Niyaz, Z. Bačić, J. W. Moskowitz and K. E. Schmidt, Chem. Phys. Lett. **252**, 23 (1996).
75. T. Schröder, R. Schinke, S. Liu, Z. Bačić and J. W. Moskowitz, J. Chem. Phys. **103**, 9228 (1995).
76. H. S. Gutowsky, T. D. Klots, C. Chuang, C. A. Schmuttenmaer and T. Emilsson, J. Chem. Phys. **86**, 569 (1987).
77. H. S. Gutowsky, T. D. Klots, C. Chuang, J. D. Keen, C. A. Schmuttenmaer and T. Emilsson, J. Am. Chem. Soc. **109**, 5653 (1987).
78. H. S. Gutowsky, T. D. Klots, C. Chuang, T. Emilsson, R. S. Ruoff and K. R. Krause, J. Chem. Phys. **88**, 2919 (1988).
79. A. McIlroy, R. Lascola, C. M. Lovejoy and D. J. Nesbitt, J. Phys. Chem. **95**, 2636 (1991).
80. J. T. Farrell, Jr., S. Davis and D. J. Nesbitt, J. Chem. Phys. **103**, 2395 (1995).
81. M. T. Bowers, G. I. Kerley and W. H. Flygare, J. Chem. Phys. **45**, 3399 (1966).
82. M. G. Mason, W. G. Von Holle and D. W. Robinson, J. Chem. Phys. **54**, 3491 (1971).
83. D. T. Anderson. *High resolution infrared spectroscopy and photolysis of solutes in cryogenic solids*. Ph.D. thesis, Dartmouth College, (1993).
84. D. J. Wales, Science **271**, 925 (1966).
85. W. H. Press, S. A. Teukolsky, W. T. Vetterling and B. P. Flannery. *Numerical Recipes in FORTRAN. The Art of Scientific Computing*. Cambridge University Press, Cambridge, (1992).
86. S. Kirkpatrick, S. D. Gelatt, Jr. and M. P. Vecchi, Science **220**, 4598 (1983).
87. C. J. Cerjan and W. H. Miller, J. Chem. Phys. **75**, 2800 (1981).
88. J. Nichols, H. Taylor, P. Schmidt and J. Simons, J. Chem. Phys. **92**, 340 (1990).
89. M. R. Hoare and P. Pal, Advan. Phys **20**, 161 (1971).
90. M. R. Hoare and P. Pal, J. Cryst. Growth **17**, 77 (1972).
91. A. L. Mackay, Acta Crystallogr. **15**, 916 (1962).
92. J. Farges, M. F. de Feraudy, B. Raoult and G. Torchet, J. Chem. Phys. **78**, 5067 (1983).
93. J. Farges, M. F. de Feraudy, B. Raoult and G. Torchet, J. Chem. Phys. **84**, 3491 (1986).
94. J. A. Northby, J. Chem. Phys. **87**, 6166 (1987).
95. Z. Bačić and J. C. Light, Annu. Rev. Phys. Chem. **40**, 469 (1989).
96. Z. Bačić. In *Domain-Based Parallelism and Problem Decomposition Methods in Computational Science and Engineering*, edited by D. E. Keyes, Y. Saad and D. G. Truhlar, page 263. (SIAM, Philadelphia, 1995).
97. A. van der Avoird, P. E. S. Wormer and R. Moszynski, Chem. Rev. **94**, 1931 (1994).
98. A. Ernesti and J. M. Hutson, Phys. Rev. A **51**, 239 (1995).
99. D. H. Zhang, Q. Wu, J. Z. H. Zhang, M. von Dirke and Z. Bačić, J. Chem. Phys. **102**, 2315 (1995).
100. W. C. Necoechea and D. G. Truhlar, Chem. Phys. Lett. **231**, 125 (1994).
101. J. Antikainen, R. Friesner and C. Leforestier, J. Chem. Phys. **102**, 1270 (1995).
102. Y. Qiu and Z. Bačić, J. Chem. Phys. **106**, 2158 (1997).
103. Y. Qiu, J. Z. H. Zhang and Z. Bačić, J. Chem. Phys. **108**, 4804 (1998).

104. M. A. Suhm and R. O. Watts, Phys. Rep. **204**, 293 (1991).
105. Z. Bačić, M. Kennedy-Mandziuk, J. W. Moskowitz and K. E. Schmidt, J. Chem. Phys. **97**, 6472 (1992).
106. J. B. Anderson, J. Chem. Phys. **73**, 3897 (1980).
107. J. W. Moskowitz, K. E. Schmidt, M. A. Lee and M. H. Kalos, J. Chem. Phys. **77**, 349 (1982).
108. P. J. Reynolds, D. M. Ceperley, B. J. Alder and W. A. Lester, Jr., J. Chem. Phys. **77**, 5593 (1982).
109. J. K. Gregory and D. C. Clary, Chem. Phys. Lett. **228**, 547 (1994).
110. H. Sun and R. O. Watts, J. Chem. Phys. **92**, 603 (1990).
111. M. Quack and M. A. Suhm, Chem. Phys. Lett. **234**, 71 (1995).
112. K. B. Whaley, Int. Rev. Phys. Chem. **13**, 41 (1994).
113. M. Lewerenz, J. Chem. Phys. **104**, 1028 (1996).
114. V. Buch, J. Chem. Phys. **97**, 726 (1992).
115. P. Sandler, J. O. Jung, M. M. Szczesniak and V. Buch, J. Chem. Phys. **101**, 1378 (1994).
116. K. A. Franken and C. E. Dykstra, J. Chem. Phys. **100**, 2865 (1994).
117. J. K. Gregory and D. C. Clary, J. Chem. Phys. **102**, 7817 (1995).
118. J. K. Gregory and D. C. Clary, J. Chem. Phys. **105**, 6626 (1996).
119. K. Liu, M. G. Brown, C. Carter, R. J. Saykally, J. K. Gregory and D. C. Clary, Nature **381**, 501 (1996).
120. C. E. Dykstra, J. Chem. Phys. **108**, 6619 (1998).
121. M. Mandziuk and Z. Bačić, J. Chem. Phys. **98**, 7165 (1993).
122. M. Mandziuk and Z. Bačić, Faraday Discuss. **97**, 265 (1994).
123. B. M. Axilrod and E. Teller, J. Chem. Phys. **11**, 299 (1943).
124. B. L. Grigorenko, A. V. Nemukhin and V. A. Apkarian, J. Chem. Phys. **104**, 5510 (1996).
125. D. T. Anderson and J. S. Winn, Chem. Phys. **189**, 171 (1994).
126. S. Liu, Z. Bačić, J. W. Moskowitz and J. M. Hutson (manuscript in preparation).
127. B. Schmidt and P. Jungwirth, Chem. Phys. Lett. **259**, 62 (1996).
128. T. Schröder, R. Schinke and Z. Bačić, Chem. Phys. Lett. **235**, 316 (1995).
129. A. Garcia-Vela, R. B. Gerber and U. Buck, J. Phys. Chem. **98**, 3518 (1994).
130. T. H. Dunning, J. Chem. Phys. **65**, 3854 (1976).
131. V. Aquilanti, E. Luzzatti, F. Pirani and G. G. Volpi, J. Chem. Phys. **89**, 6165 (1988).
132. A. I. Krylov, R. B. Gerber and V. A. Apkarian, Chem. Phys. **189**, 261 (1994).
133. R. Schinke. *Photodissociation Dynamics.* Cambridge University Press, Cambridge, (1993).
134. M. S. Child. *Molecular Collision Theory.* Academic Press, London, (1974).
135. R. Schinke and J. M. Bowman. In *Molecular Collision Dynamics*, edited by J. M. Bowman. (Springer-Verlag, Heidelberg, 1983).
136. H. J. Korsch and F. Wolf, Comments At. Mol. Phys. **15**, 139 (1984).

# Rearrangements of Water Dimer and Hexamer

David J. Wales

University Chemical Laboratories, Lensfield Road, Cambridge, CB2 1EW, United Kingdom

**Abstract.** Rearrangement mechanisms of the water dimer and the cage form of the water hexamer are examined theoretically with particular reference to tunneling splittings and spectroscopy. The three lowest barrier rearrangements of the water dimer are characterized by *ab initio* methods and compared with the results of previous constrained calculations. The acceptor-tunneling pathway does not proceed via a direct rotation around the $C_2$ axis of the acceptor, but rather via relatively asynchronous rotation of the donor about the hydrogen bond and an associated 'wag' of the acceptor. Rearrangements between different cage isomers of the water hexamer are studied for two empirical potentials. The experimentally observed triplet splittings may be the result of flip and bifurcation rearrangements of the two single-donor, single-acceptor monomers. Two-dimensional quantum calculations of the nuclear dynamics suggest that delocalization over more than one cage isomer may occur, especially in excited states.

## 1 Introduction

Water clusters have proved to be attractive systems for study to both theory and experiment, particularly in the last two decades. This popularity may be ascribed in part to the role of water as an almost universal solvent in chemistry and biochemistry, but also to the importance of developing a fundamental understanding of intermolecular forces [1-4]. A recent flurry of activity has been sparked by the advent of far-infrared vibration-rotation tunneling (FIR-VRT) spectroscopy [5-8], where resolutions of up to 1 MHz have been achieved. These results provide a new challenge to theory because the tunneling splittings that can now be resolved necessitate a global view of the potential energy surface (PES) if they are to be explained. The large amplitude motions which are typical of such weakly bound Van der Waals complexes sample regions of the PES that are far removed from the bottom of potential wells, and provide new information about the nature of the intermolecular forces. Furthermore, to explain or predict the tunneling splittings theory must characterize not only minima but transition states and rearrangement mechanisms.

The interplay between theory and experiment is perhaps best illustrated by our recently improved understanding of the dynamics exhibited by water trimer, which began with the FIR-VRT experiment of Pugliano and Saykally [9]. These authors reported spectra for $(D_2O)_3$ characteristic of an oblate symmetric rotor, with each line split into a regularly spaced quartet and a spacing of roughly

6 MHz, i.e. $2 \times 10^{-4}$ cm$^{-1}$. Accurate values for the vibrationally averaged rotational constants revealed a large negative inertial defect, indicative of extensive out-of-plane motion of the non-hydrogen-bonded hydrogens. However, a cyclic, asymmetric global minimum structure for the trimer was established some years ago in *ab initio* calculations [10], in agreement with earlier experimental results [11].

The oblate symmetric top spectrum of the trimer has now been explained by vibrational averaging over large amplitude torsional motions of the free hydrogens on the timescale of the FIR-VRT experiment. These large amplitude motions are associated with a facile 'flip' rearrangement where a free hydrogen moves from one side of the plane defined by the three oxygen atoms to the other. This mechanism was probably first characterized for an empirical potential by Owicki *et al.* [12]. An *ab initio* pathway has been presented by Wales [13] and the corresponding transition state was also characterized by Fowler and Schaefer [14]. Wales identified two other degenerate rearrangement mechanisms for the trimer with rather larger barriers than the flip, and suggested that one of them (christened the 'donor' or 'bifurcation' pathway) might be responsible for the quartet splittings observed experimentally [13]. In the associated transition state one monomer acts as a double donor to a neighbour which acts as a double acceptor in a configuration similar to that of the 'donor tunneling' transition state in the water dimer, discussed in §4.

Subsequent experiments [15-16] assigned new transitions for $(H_2O)_3$ and $(D_2O)_3$ which revealed rigorously symmetric rotor structure, in contrast to the strongly perturbed band that was initially investigated by Pugliano and Saykally [9]. All the transitions reported by Liu *et al.* [15] and Suzuki and Blake [16] show a regular quartet splitting of every rovibrational transition. Liu *et al.* reported that their new spectra were consistent with the $G(48)$ group suggested by the mechanisms discovered by Wales [13]. Walsh and Wales have recently studied the bifurcation mechanism in more detail and discovered that six slightly different alternatives exist, depending upon the level of the calculation employed [17].

Numerous other treatments of the water trimer have focused on the breakdown of the total energy into many-body contributions and the effects of electron correlation [18-22], the development of torsional potential energy surfaces [23-25], the calculation of torsional energy levels using the discrete variable representation (DVR) [26-29] and quantum simulations of the nuclear dynamics using the diffusion Monte Carlo (DMC) method [30-31].

In the present contribution we report new results for the water dimer and water hexamer, which lie at the two limits of the size range for which high resolution data have been obtained. Unconstrained pathways do not seem to have been described before for the three rearrangements of the water dimer which lead to observable splittings or shifts in the pattern of rovibrational energy levels. In particular, we find that the acceptor tunneling mechanism is not simply a rotation about the local $C_2$ axis of the acceptor monomer. For water hexamer we employ two empirical intermolecular potentials to examine the cage isomers and show that these structures are connected by facile, non-degenerate single flips of the two single-donor, single-acceptor monomers. Two-dimensional DVR calculations are performed to investigate these motions. First we give a brief overview of the effective molecular symmetry group which must be employed to classify the energy levels of floppy molecules.

## 2 The Effective Molecular Symmetry Group

The observation of quantum tunneling effects in high resolution spectra provides indirect information about the corresponding rearrangement mechanisms. To make use of this information we need to employ group theory to classify the energy levels of such systems. Point groups prove satisfactory when the molecule under consideration is rigid on the appropriate experimental timescale. However, molecular potential energy surfaces generally contain local minima corresponding to permutational isomers of any given stationary point. We will adopt the nomenclature of Bone *et al.* [32] where a 'structure' is understood to mean a particular molecular geometry and a 'version' is a particular labelled permutational isomer of a given structure. Minima which are directly connected by a given rearrangement are said to be 'adjacent'.

Tunneling splittings may be observed when rovibronic wavefunctions localized in the potential wells corresponding to different versions interfere with each other. For example, the ammonia molecule displays doublet splittings because pairs of permutational isomers are interconverted by the inversion mechanism in which the molecule passes through a planar transition state [33-34]. If the barrier corresponding to a certain rearrangement mechanism is sufficiently low, the path length sufficiently short, and the associated effective mass sufficiently small then tunneling may occur. Hence tunneling splittings can usually be associated with a low energy transition state corresponding to a degenerate rearrangement mechanism [35] which links different versions of the same structure, as opposed to different structures where the energy levels would not be in resonance. Degenerate rearrangements can be either symmetric or asymmetric, depending upon whether the two sides of the corresponding path are related by symmetry [36]. We follow Murrell and Laidler's definition of a transition state as a stationary point with a single negative Hessian eigenvalue [37].

The energy levels of non-rigid molecules can be classified using the Complete Nuclear Permutation-Inversion (CNPI) group which is the direct product of the group containing all possible permutations of identical nuclei and the inversion group. The latter group contains only the identity operation, $E$, and the operation of inversion of all particle coordinates through the space-fixed origin, $E^*$, which commutes with all the permutations. The CNPI group is a true symmetry group of the full molecular Hamiltonian in the absence of external fields, and its elements are generally referred to as permutation-inversion operations. However, the order of this group increases factorially for systems containing increasing numbers of atoms of the same element, and rapidly becomes difficult to use. Fortunately, Longuet-Higgins showed that it not necessary to consider the whole of the CNPI group, but rather only the permutation-inversions which correspond to tunneling splittings that are resolvable for a given experiment [38]. The corresponding rearrangement mechanisms are said to be 'feasible'. The resulting subgroup of the CNPI group is known as the effective molecular symmetry (MS) group. For rigid molecules the appropriate MS group is isomorphic to the usual rigid molecule point group [39-42]. The corresponding permutation-inversions simply correspond to overall rotation of the system, and are therefore always feasible.

If there exists a non-trivial feasible rearrangement with a finite barrier then the MS group is enlarged and the wavefunctions which transform according to irreducible representations of this group are linear combinations of the functions localized in each well. In general, a given mechanism will not link all the possible versions of a given structure but rather the versions will be partitioned into a number of closed sets with equivalent reaction graphs for each set. If we consider a representative set of versions all of which can be interconverted by repeated application of the feasible rearrangement, then the corresponding wavefunction must be a linear combination of the localized functions from members of this set. The delocalized wavefunctions can be found by solving a secular problem, just as in the linear combination of molecular orbitals approach to electronic structure. If the wavefunctions decay rapidly in the classically forbidden regions of the PES between minima then it may be sufficient to consider only nearest-neighbour interactions and the resulting splitting pattern is then determined largely by symmetry. The connectivity of the reaction graph, the associated MS group and the splitting pattern can all be found automatically by a computer program once a minimal set of generator permutation-inversions is known [43].

## 3  Geometry Optimizations and Potentials

The geometry optimizations and calculations of rearrangement pathways described in the following sections were all performed by eigenvector-following [44]. Details of the precise implementation have been given elsewhere [45-46]. Analytic first and second derivatives of the energy were calculated at every step. In the *ab initio* calculations these derivatives were all generated by the CAD-PAC program [47], and Cartesian coordinates were used throughout. Pathways were calculated by taking small displacements of $0.03\,a_0$ away from a transition state both parallel and antiparallel to the transition vector, and then employing eigenvector-following energy minimization to find the associated minimum. The pathways obtained by this procedure have been compared to steepest-descent paths and pathways that incorporate a kinetic metric [48] in previous work—the mechanism is generally found to be represented correctly [17].

Calculations employing rigid body intermolecular potentials were performed using the ORIENT3 program [49-51], which contains the same optimization package adapted for centre-of-mass/orientational coordinates. This program can treat intermolecular potentials based upon Stone's distributed multipoles [52-53] and distributed polarizabilities [54-55]; simpler models based upon point charges and Lennard-Jones interactions fall within this framework. Calculations for the water hexamer in §5 were performed using the relatively sophisticated ASP-W2 potential of Millot and Stone [56] (somewhat modified from the published version) and the much simpler but widely-used TIP4P form [57-58].

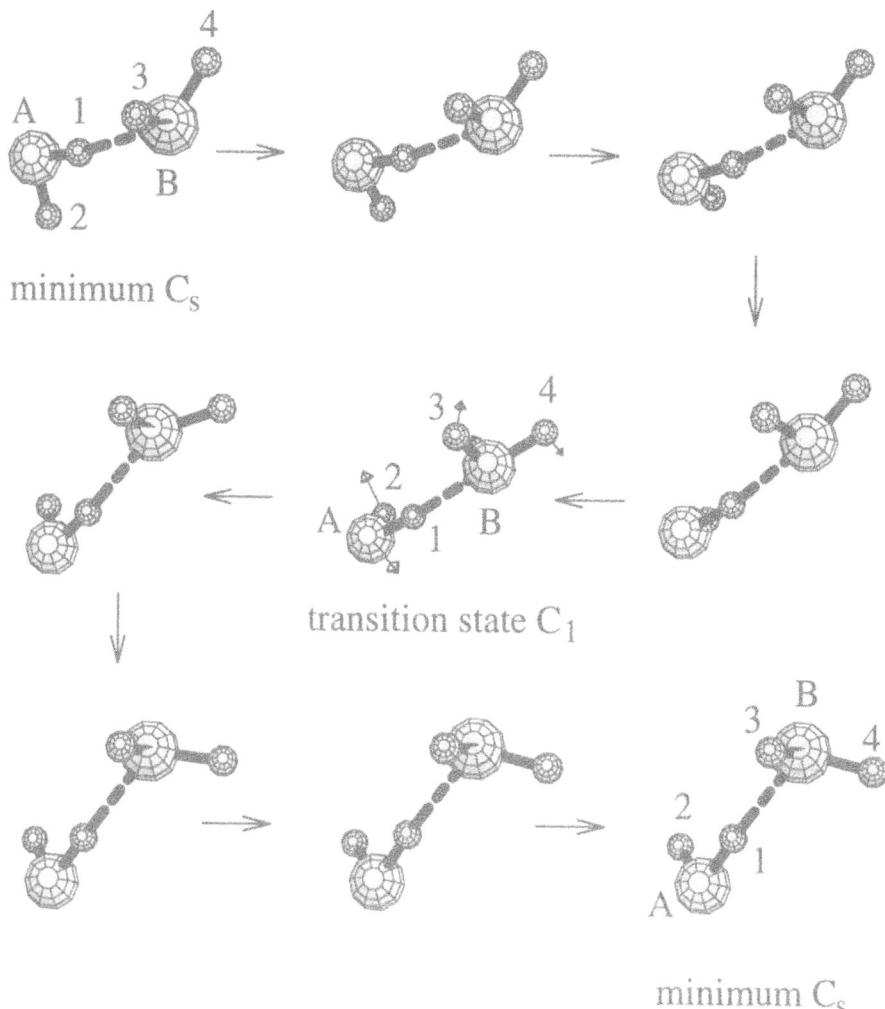

**Fig. 1.** Acceptor-tunneling path for $(H_2O)_2$ calculated at the DZP+diff/BLYP level. For this path $S = 4.2\,a_0$, $D = 2.8\,a_0$ and $\gamma = 1.3$ (see §3).

In the *ab initio* Hartree-Fock (HF) calculations for $(H_2O)_2$ two basis sets were considered. The smaller double-$\zeta$ [59-60] plus polarization (DZP) basis employed polarization functions consisting of a single set of $p$ functions on each hydrogen atom (exponent 1.0) and a set of six $d$ functions on each oxygen atom (exponent 0.9) to give a total of 26 basis functions per monomer. The larger basis set, denoted DZP+diff, includes the above DZP functions with an additional diffuse $s$ function on each hydrogen atom (exponent 0.0441) and diffuse sets of $s$ and $p$ functions on each oxygen atom (exponents 0.0823 and 0.0651 for $s$ and $p$ respectively) [14], to give 32 basis functions per monomer. Correlation corrections were obtained through both second order Møller-Plesset (MP2) theory [61] and density functional theory (DFT). In the DFT calculations we employed the Becke nonlocal exchange functional [62] and the Lee-Yang-Parr correlation functional [63] (together referred to as BLYP); derivatives of the grid weights were not included and the core electrons were not frozen. Numerical integration of the BLYP functionals was performed using grids between the CADPAC 'MEDIUM' and 'HIGH' options. The 'MEDIUM' grids were not accurate enough to give the right number of negative Hessian eigenvalues, whereas the 'HIGH' grids contained more points than necessary. CADPAC actually uses different sized grids for different parts of the calculation [47]; in the present work these grids contained 14,386 and 97,008 points after removal of those with densities below the preset tolerances. Calculations were deemed to be converged when the root-mean-square gradient fell below $10^{-6}$ atomic units. This is sufficient to reduce the six 'zero' normal mode frequencies to less than $1\,\mathrm{cm}^{-1}$ in the HF and MP2 calculations. Because derivatives of the grid weights were not included, the largest of the six 'zeros' can be as big as $20\,\mathrm{cm}^{-1}$ for the DFT stationary points.

Three additional parameters are useful in describing the rearrangement mechanisms. The first is the integrated path length, $S$, which was calculated as a sum over eigenvector-following steps. The second is the distance between the two minima in nuclear configuration space, $D$. The third is the moment ratio of displacement [64], $\gamma$, which gives a measure of the cooperativity of the rearrangement:

$$\gamma = \frac{N \sum_i [\mathbf{Q}_i(s) - \mathbf{Q}_i(t)]^4}{\left( \sum_i [\mathbf{Q}_i(s) - \mathbf{Q}_i(t)]^2 \right)^2}, \tag{1}$$

where $\mathbf{Q}_i(s)$ is the position vector in Cartesian coordinates for atom $i$ in minimum $s$, etc., and $N$ is the number of atoms. If every atom undergoes the same displacement in one Cartesian component then $\gamma = 1$, while if only one atom has one non-zero component then $\gamma = N$, i.e. 18 for $(H_2O)_6$.

## 4 Rearrangements of Water Dimer

There have been many experimental investigations of the water dimer [65] following the original work of Dyke, Mack and Muenter [66]. Most recently, a complete characterization of the tunneling dynamics in a vibrationally excited state of $(D_2O)_2$ has been presented [67]. The dimer rovibronic energy levels were first classified in terms of permutation-inversion group theory by Dyke [68], and Coudert and Hougen applied their internal axis approach to the intermolecular dynamics using an empirical potential [69-71]. Smith et al. [72] performed ab

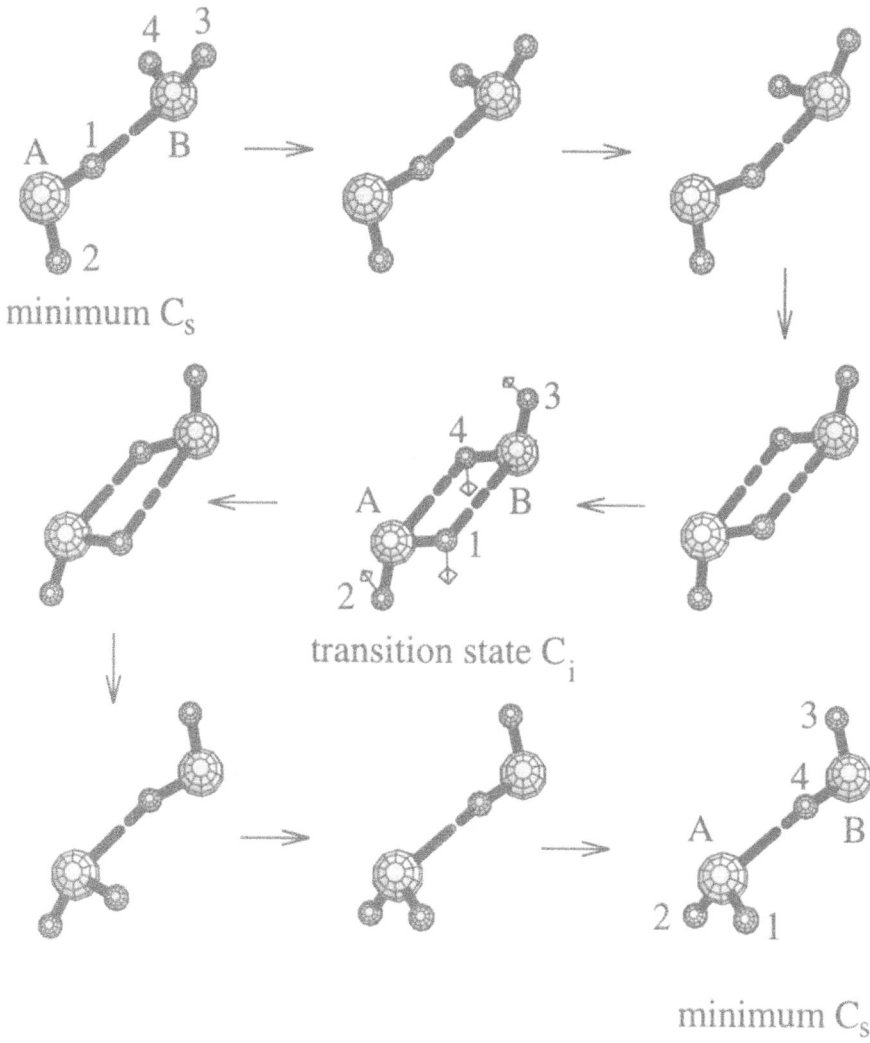

**Fig. 2.** Donor-acceptor-interchange tunneling mechanism for $(H_2O)_2$ calculated at the DZP+diff/BLYP level. For this path $S = 5.4\,a_0$, $D = 4.4\,a_0$ and $\gamma = 1.4$ (see §3).

*initio* calculations and identified three true transition states for $(H_2O)_2$; they also performed constrained calculations of the donor-tunneling pathway. In this section we will describe unconstrained pathway calculations for all three rearrangements; the energetics of the various stationary points at different levels of theory are summarized in Table 1 and counterpoise-corrected [73] binding energies (including monomer relaxation [74]) are given in Table 2.

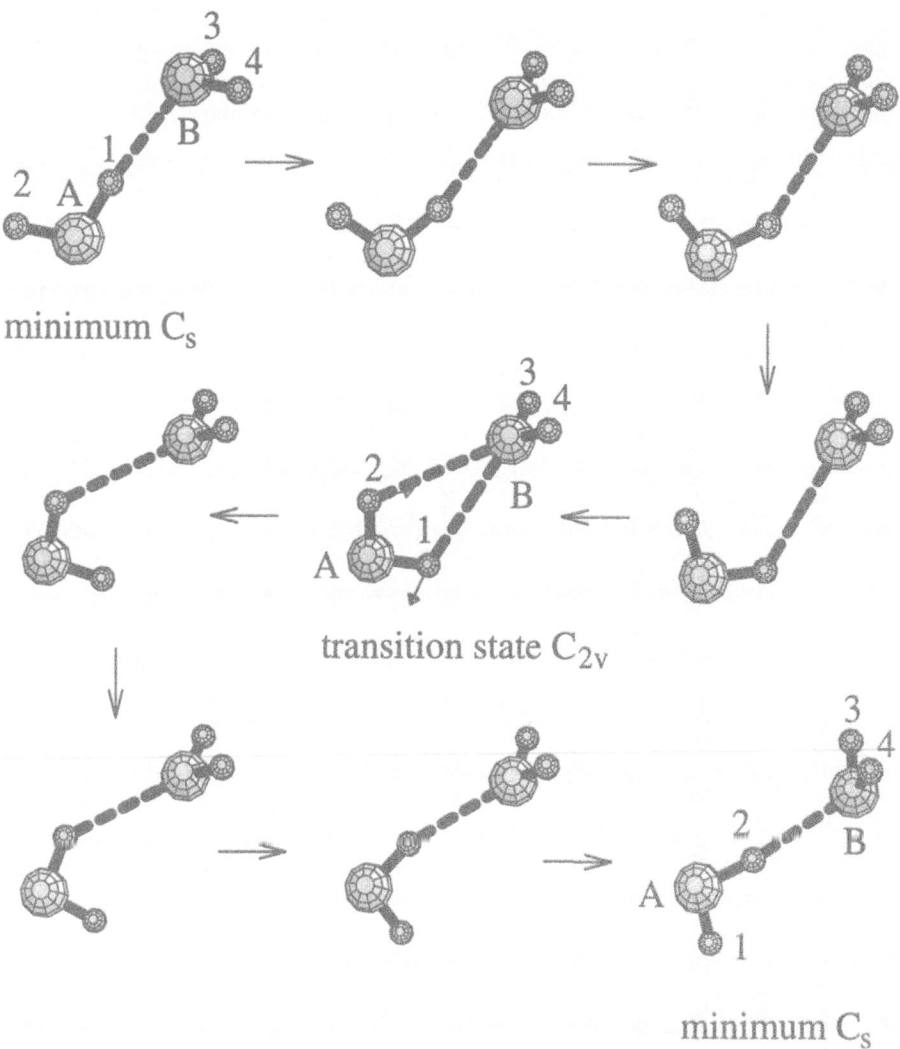

**Fig. 3.** Donor-tunneling path for $(H_2O)_2$ calculated at the DZP+diff/BLYP level. For this path $S = 5.8\,a_0$, $D = 4.5\,a_0$ and $\gamma = 2.2$ (see §3).

There are a total of $2 \times 2! \times 4!/2 = 48$ distinct versions of the water dimer global minimum on the PES (we divide by two because the rigid molecule point group has order two [32]). However, mechanisms which involve the making and breaking of covalent bonds lie much too high in energy to give rise to observable tunneling splittings. The largest possible MS group which can pertain when covalent bond-breaking is not feasible has order $2 \times 2! \times (2!)^2 = 16$, where the first factor accounts for the inversion mechanism, the second factor accounts for permutation of the two oxygen nuclei, and the last term accounts for the permutation of the two hydrogen (or deuterium) atoms within each monomer. This group may be denoted $G(16)$ and is isomorphic to the point group $D_{4h}$ [68]. Since the equilibrium geometry has $C_s$ symmetry the maximum number of distinct versions that can be interconverted without breaking covalent bonds is $16/2 = 8$.

Table 1. Energies in hartree and point groups at various levels of theory for the water dimer global minimum and the transition states for acceptor tunneling, donor-acceptor interchange and donor tunneling.

|         | DZP/HF | DZP+diff/HF | DZP+diff/BLYP | DZP+diff/MP2 |
|---------|--------|-------------|---------------|--------------|
| minimum | $-152.102057(C_s)$ | $-152.107971(C_s)$ | $-152.87817(C_s)$ | $-152.540770(C_s)$ |
| acceptor | $-152.101274(C_s)$ | $-152.107282(C_s)$ | $-152.87701(C_1)$ | $-152.539673(C_1)$ |
| don-acc | $-152.100688(C_i)$ | $-152.106287(C_i)$ | $-152.87591(C_i)$ | $-152.538743(C_i)$ |
| donor | $-152.099832(C_{2v})$ | $-152.105946(C_{2v})$ | $-152.87535(C_{2v})$ | $-152.538149(C_{2v})$ |

In the equilibrium $C_s$ geometry one 'donor' monomer acts as a single hydrogen-bond donor to the other 'acceptor' molecule [75]. The largest tunneling splitting is due to a mechanism in which the two hydrogen atoms of the acceptor monomer are effectively interchanged, for which Smith et al. [72] reported a barrier of $206\,\mathrm{cm}^{-1}$. In the present work a planar transition state of $C_s$ symmetry was found for both basis sets in the HF calculations, but this changed to an out-of-plane $C_1$ structure when correlation corrections were applied, in agreement with Smith et al. [72]. The pathway is shown in Fig. 1, and corresponds to a 'methylamine-type' process [76] rather than a direct rotation about the local $C_2$ axis of the acceptor monomer. The path is represented by nine snapshots where the first and last frames are the two minima, the middle frame is the transition state and three additional frames on each side of the path were selected to best illustrate the mechanism. All the pathways in the present work were visualized using Mathematica [77].

The above mechanism is in agreement with the analysis of Pugliano et al. [67] for the ground state acceptor tunneling path. The generator permutation-inversion corresponding to the labelling scheme of Fig. 1 is (34), and it connects the versions in pairs. Hence each rigid-dimer rovibrational level is split into two by this process. Experimentally the ground state splitting due to acceptor tunneling is [78] $2.47\,\mathrm{cm}^{-1}$, in good agreement with a five-dimensional treatment of the nuclear dynamics by Althorpe and Clary [79], which yielded a value of $2.34\,\mathrm{cm}^{-1}$. The MS group for the rigid dimer labelled according to Fig. 1 contains only the identity, $E$, and the permutation-inversion (34)*, where hydrogens three and four change places and all coordinates are inverted through the space-fixed origin. The appropriate MS group when the generator (34) operation is feasible contains the elements $E$, $E^*$, (34) and (34)$^*$.

Table 2. Couterpoise-corrected [73] binding energies in millihartree at various levels of theory for the water dimer global minimum and the transition states for acceptor tunneling, donor-acceptor interchange and donor tunneling.

| | DZP/HF | DZP+diff/HF | DZP+diff/BLYP | DZP+diff/MP2 |
|---|---|---|---|---|
| minimum | 7.42 | 6.92 | 7.75 | 7.45 |
| acceptor | 6.88 | 6.38 | 6.64 | 6.65 |
| don-acc | 5.73 | 5.82 | 6.03 | 6.52 |
| donor | 5.13 | 5.07 | 4.79 | 5.28 |

The next largest splitting is caused by donor-acceptor interchange. For this process Smith et al. [72] found a cyclic transition state with $C_i$ symmetry and an associated barrier of 304 cm$^{-1}$. In this rearrangement, for which the calculated pathway is shown in Fig. 2, the roles of the donor and acceptor monomers are interchanged. An appropriate generator for the labelling scheme of Fig. 2 is (AB)(1423), i.e. oxygen A is replaced by oxygen B, hydrogen 1 is replaced by hydrogen 4, hydrogen 4 is replaced by hydrogen 2 etc. This generator is not unique—one could also choose (AB)(14)(23)$^*$ and obtain the same MS group.

If donor-acceptor interchange is the only feasible mechanism then the versions are connected in sets of four and the MS group has eight members: class 1 contains $E$, class 2 contains (12)(34), class 3 contains (AB)(1423) and (AB)(1324), class 4 contains (34)$^*$ and (12)$^*$, and class 5 contains (AB)(14)(23)$^*$ and (AB)(13)(24)$^*$. The splitting pattern in the simplest Hückel-type approximation [80] is:

$$2\beta_{da}(A_1), \qquad 0(E), \qquad -2\beta_{da}(B_1), \tag{2}$$

where $\beta_{da}$ is the donor-acceptor-interchange tunneling matrix element and we have labelled the levels according to appropriate irreducible representations of the $G(8)$ group which is a subgroup of $G(16)$ [68]. When this process and acceptor tunneling are both feasible the MS group has order 16, i.e. the largest MS group possible without breaking covalent bonds, and is isomorphic to $D_{4h}$ [68]. The versions are then connected in sets of eight and the splitting pattern is:

$$\beta_a + 2\beta_{da}(A_1^+), \qquad \beta_a(E^+), \qquad \beta_a - 2\beta_{da}(B_1^+),$$
$$-\beta_a + 2\beta_{da}(A_2^-), \qquad -\beta_a(E^-), \qquad -\beta_a - 2\beta_{da}(B_2^-), \tag{3}$$

where $\beta_a$ is the acceptor-tunneling matrix element. Experimentally, the tunneling splittings due to donor-acceptor-interchange are about a factor of five smaller than those associated with acceptor tunneling.

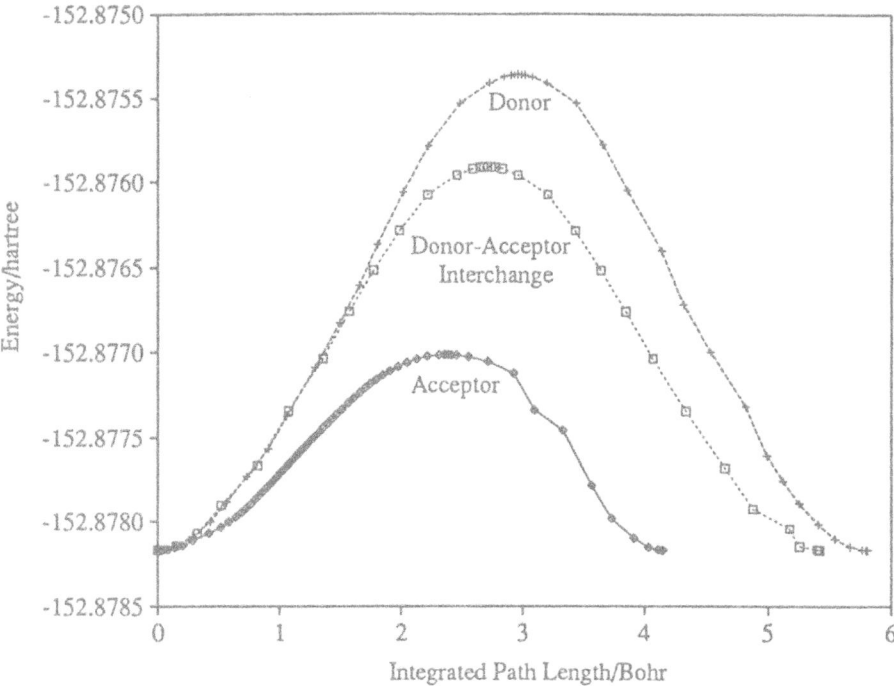

**Fig. 4.** Energy versus the integrated path length, $S$, for the three degenerate rearrangement mechanisms of the water dimer described in §4.

The third process which is generally presumed to have an observable effect on the energy level diagram is donor tunneling, but this can only lead to energy level shifts rather than further splittings because $G(16)$ is the largest MS group possible without breaking covalent bonds. Smith *et al.* [72] calculated a barrier of $658\,cm^{-1}$ for this mechanism, and the corresponding pathway found in the present work is shown in Fig. 3. An appropriate generator permutation-inversion for the labelling scheme of Fig. 3 is $(12)(34)$, and on its own this process would simply lead to doublet splittings with versions linked in pairs and an MS group of order 4. However, when combined with the other two mechanisms the eigenvalues of the corresponding secular problem (assuming a Hückel-type approximation [80]) are:

$$\beta_a + 2\beta_{da} + \beta_d\,(A_1^+), \qquad \beta_a - \beta_d\,(E^+), \qquad \beta_a - 2\beta_{da} + \beta_d\,(B_1^+),$$
$$-\beta_a + 2\beta_{da} + \beta_d\,(A_2^-), \qquad -\beta_a - \beta_d\,(E^-), \qquad -\beta_a - 2\beta_{da} + \beta_d\,(B_2^-). \quad (4)$$

This pattern is in complete agreement with that obtained by Althorpe and Clary [79] and with the model calculations of Coudert and Hougen [69-71]. Since the tunneling levels are no longer in plus-minus pairs we conclude that the presence of donor tunneling introduces odd-membered rings into the reaction graph [81].

The DZP+diff/BLYP energy profiles for all three rearrangements are shown in Fig. 4. A maximum step size of $0.1\,a_0$ was used for the left-hand side of the acceptor-tunneling path. This value was increased to $0.15\,a_0$ for all the other paths, resulting in slightly less smooth profiles but much faster execution time.

## 5   Water Hexamer

Liu *et al.* [82] have recently identified a VRT band of $(H_2O)_6$ at $83\,cm^{-1}$ on the basis of an isotope mixture test. In contrast to smaller water clusters the lowest energy isomer of the hexamer is probably not cyclic, and the four lowest energy structures found by Tsai and Jordan [83] lie within an energy range of only $100\,cm^{-1}$. The most accurate calculations conducted so far suggest that a 'cage' structure lies lowest, followed closely by 'prism' and 'book' forms (Fig. 5) [84]. DMC calculations were used to find the vibrationally averaged rotational constants for each structure using an empirical potential, and the best match was found for the cage structure. On this basis the spectrum was assigned to the cage [82].

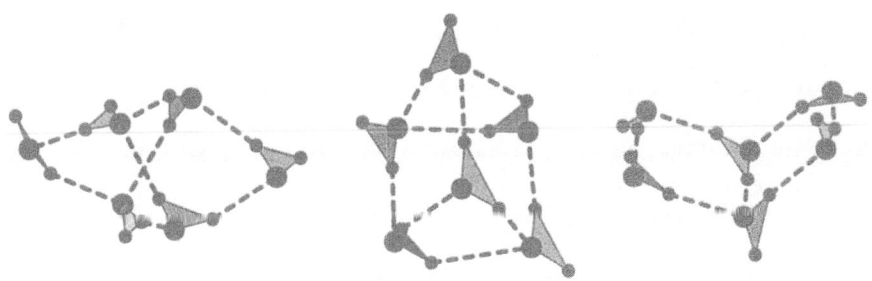

**Fig. 5.** Cage, prism and book forms of $(H_2O)_6$.

There are several unanswered questions concerning the interpretation of this experiment. First, if a number of isomers lie so close together in energy, how is it that only one of them seems to be observed? One possibility might be that the other isomers are also present in the beam, but do not have a spectral feature in the range scanned experimentally. In fact, simulations reveal numerous isomers of all three morphologies illustrated in Fig. 5 with different arrangements of the hydrogen bonds. The remaining discussion will concentrate on the cage and the results of calculations using the ASP-W2 and TIP4P potentials described in §3. The ASP-W2 form should be very similar to that employed in the DMC calculations of nuclear dynamics by Liu *et al.* [82]. However, we note that the cage isomer is also not the lowest in energy for either of these potentials, although inclusion of zero-point energy can alter the ordering [82]. In fact, for the ASP-W2 potential the cyclic structure of the water pentamer is also not the global

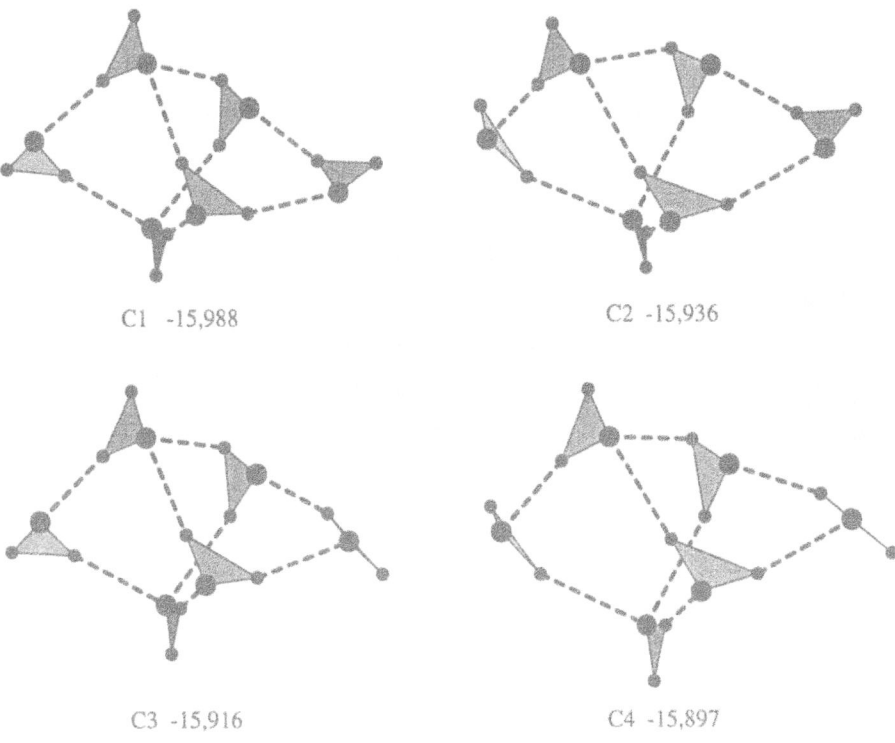

C1  -15,988

C2  -15,936

C3  -15,916

C4  -15,897

**Fig. 6.** Isomers of the cage structure for $(H_2O)_6$ calculated using the ASP-W2 potential with binding energies in $cm^{-1}$.

minimum, but nevertheless, the lowest energy rearrangements of this isomer are quite well reproduced [46].

For the ASP-W2 potential only the first order induction energy was considered, since iterating this term to convergence is time-consuming and was found to make no qualitative difference to results for the water pentamer in previous work [46]. There are then four isomers of the cage structure shown in Fig. 5 differing only in the position of the free hydrogen atoms of the two terminal, single-donor, single-acceptor monomers (Fig. 6). No low energy degenerate rearrangements of these isomers were found, but transition states were located for non-degenerate single flip and bifurcation mechanisms. For every single flip process there is an analogous bifurcation which links different permutational isomers of the same structures. Details of the paths are given in Table 3 and the single flip and bifurcation pathways linking cage isomers C1 and C2 are illustrated in Fig. 7 and Fig. 8. The other paths are omitted for brevity.

The experimental VRT transition has been assigned to a torsional motion of one of the single-donor, single-acceptor monomers, and the form of the spectrum has been described as 'near-prolate' [82]. Note that even with full vibrational av-

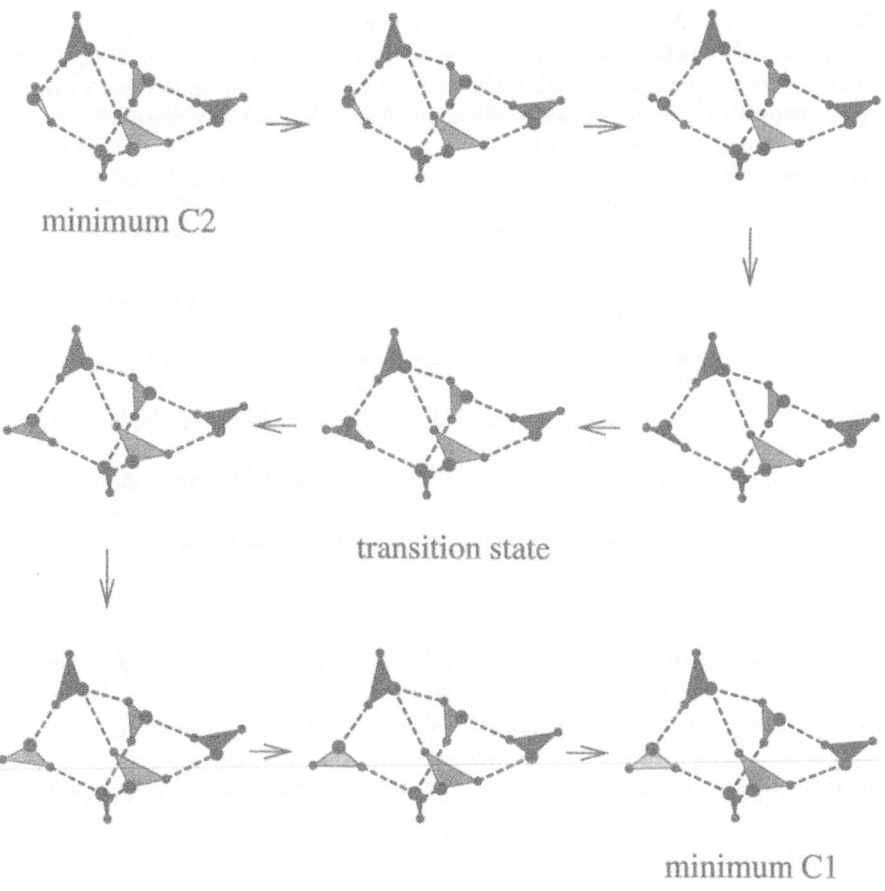

**Fig. 7.** Single flip mechanism which interconverts C1 and C2 for the ASP-W2 potential.

eraging over the four cage isomers in Fig. 6 a symmetric top spectrum would not be expected. Experimentally, every line was found to be split into a triplet with intensities in the ratio 9:6:1 and equal spacings of 1.92 MHz [82]. Liu *et al.* have explained how this pattern might emerge in terms of hypothetical degenerate rearrangements of the cage which they assumed must interchange the hydrogens of two monomers in almost equivalent environments. This could lead to a doublet of doublets where the middle lines are essentially superimposed. The resulting nuclear spin weights can then reproduce the observed intensity pattern [82].

Liu *et al.* additionally suggested that the two monomers in question might be the two terminal molecules which we have found to undergo flip and bifurcation rearrangements above. However, no direct degenerate rearrangements of the cage isomers have been found to date. Since a combination of sequential bifurcation and flip rearrangements would achieve the desired effect we will now

**Table 3.** Rearrangement mechanisms which interconvert cage isomers of $(H_2O)_6$ calculated with the ASP-W2 potential. The energies are in $cm^{-1}$. $Min_1$ is the lower minimum, $\Delta_1$ is the higher barrier, TS is the transition state and $\Delta_2$ is the smaller barrier corresponding to the higher minimum $Min_2$. $S$ is the integrated path length in Å, $D$ is the displacement between minima in Å and $\gamma$ is the cooperativity index. All these quantities are defined in §3.

| $Min_1$ | $\Delta_1$ | TS | $\Delta_2$ | $Min_2$ | $S$ | $D$ | $\gamma$ | description |
|---|---|---|---|---|---|---|---|---|
| C1 | 560 | -15,428 | 508 | C2 | 2.8 | 2.1 | 13.7 | single flip |
| C1 | 641 | -15,347 | 589 | C2 | 2.9 | 2.2 | 7.6 | bifurcation |
| C1 | 323 | -15,664 | 251 | C3 | 1.7 | 1.3 | 15.4 | single flip |
| C1 | 1,135 | -14,853 | 1,063 | C3 | 3.2 | 2.1 | 8.5 | bifurcation |
| C2 | 325 | -15,612 | 285 | C4 | 1.7 | 1.3 | 16.2 | single flip |
| C2 | 1,058 | -14,878 | 1,018 | C4 | 3.4 | 2.1 | 8.6 | bifurcation |
| C3 | 524 | -15,391 | 506 | C4 | 2.7 | 2.0 | 14.1 | single flip |
| C3 | 650 | -15,266 | 631 | C4 | 2.7 | 2.2 | 7.9 | bifurcation |

consider the splittings that might result in more detail. The eight single flip and bifurcation processes described above for the cage isomers link 16 versions: four of each isomer. To distinguish between versions of each isomer we need only specify which of the two hydrogens is free, and so we label the hydrogens on the terminal monomer with two double-acceptor neighbours a and b and those on the monomer with two double-donor neighbours c and d. The four relevant versions of the C1 isomer may then be written C1(ac), C1(bc), C1(ad) and C1(bd). The interconnectivity of all 16 versions is shown in Fig. 9. If we make a Hückel-type approximation then the resulting secular determinant is:

$$\begin{pmatrix}
0 & 0 & 0 & 0 & f_{12} & b_{12} & 0 & 0 & f_{13} & 0 & b_{13} & 0 & 0 & 0 & 0 & 0 \\
0 & 0 & 0 & 0 & b_{12} & f_{12} & 0 & 0 & 0 & f_{13} & 0 & b_{13} & 0 & 0 & 0 & 0 \\
0 & 0 & 0 & 0 & 0 & 0 & f_{12} & b_{12} & b_{13} & 0 & f_{13} & 0 & 0 & 0 & 0 & 0 \\
0 & 0 & 0 & 0 & 0 & 0 & b_{12} & f_{12} & 0 & b_{13} & 0 & f_{13} & 0 & 0 & 0 & 0 \\
f_{12} & b_{12} & 0 & 0 & 52 & 0 & 0 & 0 & 0 & 0 & 0 & 0 & f_{24} & 0 & b_{24} & 0 \\
b_{12} & f_{12} & 0 & 0 & 0 & 52 & 0 & 0 & 0 & 0 & 0 & 0 & 0 & f_{24} & 0 & b_{24} \\
0 & 0 & f_{12} & b_{12} & 0 & 0 & 52 & 0 & 0 & 0 & 0 & 0 & b_{24} & 0 & f_{24} & 0 \\
0 & 0 & b_{12} & f_{12} & 0 & 0 & 0 & 52 & 0 & 0 & 0 & 0 & 0 & b_{24} & 0 & f_{24} \\
f_{13} & 0 & b_{13} & 0 & 0 & 0 & 0 & 0 & 72 & 0 & 0 & 0 & f_{34} & b_{34} & 0 & 0 \\
0 & f_{13} & 0 & b_{13} & 0 & 0 & 0 & 0 & 0 & 72 & 0 & 0 & b_{34} & f_{34} & 0 & 0 \\
b_{13} & 0 & f_{13} & 0 & 0 & 0 & 0 & 0 & 0 & 0 & 72 & 0 & 0 & 0 & f_{34} & b_{34} \\
0 & b_{13} & 0 & f_{13} & 0 & 0 & 0 & 0 & 0 & 0 & 0 & 72 & 0 & 0 & b_{34} & f_{34} \\
f_{24} & 0 & b_{24} & 0 & f_{24} & 0 & b_{24} & 0 & f_{34} & b_{34} & 0 & 0 & 91 & 0 & 0 & 0 \\
0 & f_{24} & 0 & b_{24} & 0 & f_{24} & 0 & b_{24} & b_{34} & f_{34} & 0 & 0 & 0 & 91 & 0 & 0 \\
b_{24} & 0 & f_{24} & 0 & b_{24} & 0 & f_{24} & 0 & 0 & 0 & f_{34} & b_{34} & 0 & 0 & 91 & 0 \\
0 & b_{24} & 0 & f_{24} & 0 & b_{24} & 0 & f_{24} & 0 & 0 & b_{34} & f_{34} & 0 & 0 & 0 & 91
\end{pmatrix}$$

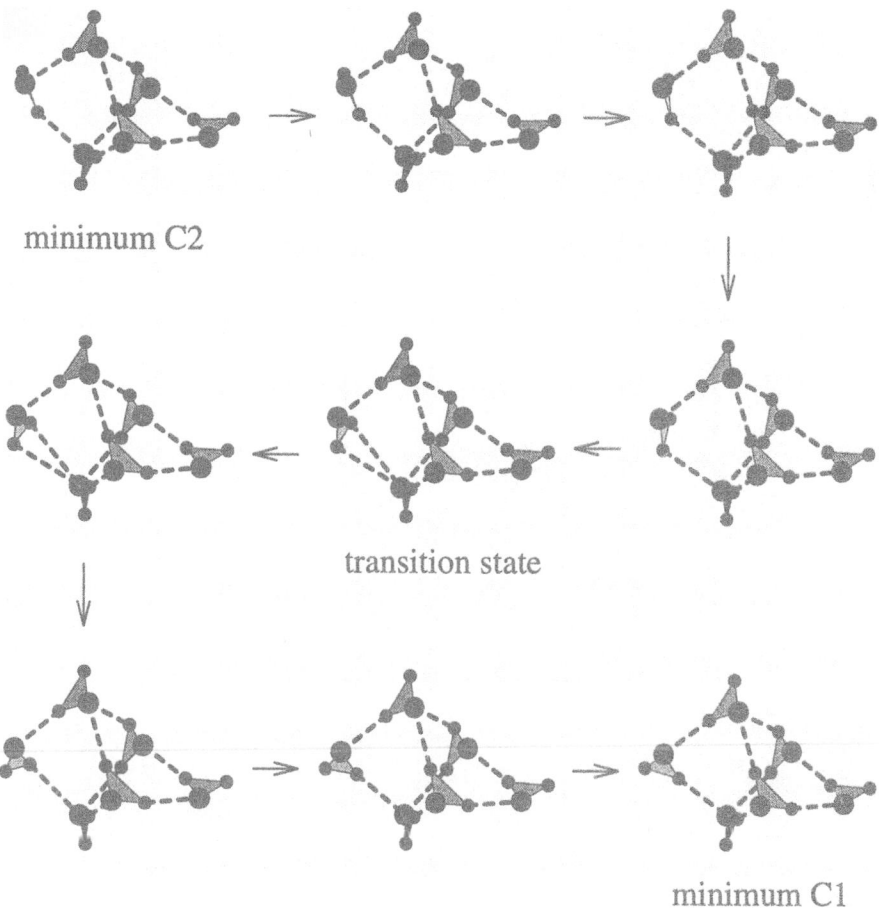

minimum C2

transition state

minimum C1

**Fig. 8.** Bifurcation mechanism which interconverts C1 and C2 for the ASP-W2 potential.

where $f_{ij}$ and $b_{ij}$ are the appropriate tunneling matrix elements between C$i$ and C$j$. We can simplify this problem by focusing on one of the isomers, say C1, and considering an effective generator [85] corresponding to a flip and a bifurcation (in either order). The reaction graph can then be represented as shown in Fig. 10, where the double lines indicate that there are two routes between each pair of versions depending upon whether the flip or bifurcation comes first. The effective MS group contains the elements $E$, (ab), (bc) and (ab)(bc) and is isomorphic to $C_{2v}$. If the four connections are each represented by the same tunneling matrix element $\beta$ then the energy level pattern will be identical to that obtained in the Hückel treatment of the $\pi$ system of butadiene, i.e.

$$2\beta(A_1), \qquad 0(A_2, B_1), \qquad -2\beta(B_2), \tag{5}$$

C4(bd)
C3(bd)
C3(ad)
C4(ad)
C2(bc)
C1(bc)
C1(ac)
C2(ac)
C3(ac)
C4(ac)
C4(bc)
C3(bc)
C1(ad)
C2(ad)
C2(bd)
C1(bd)

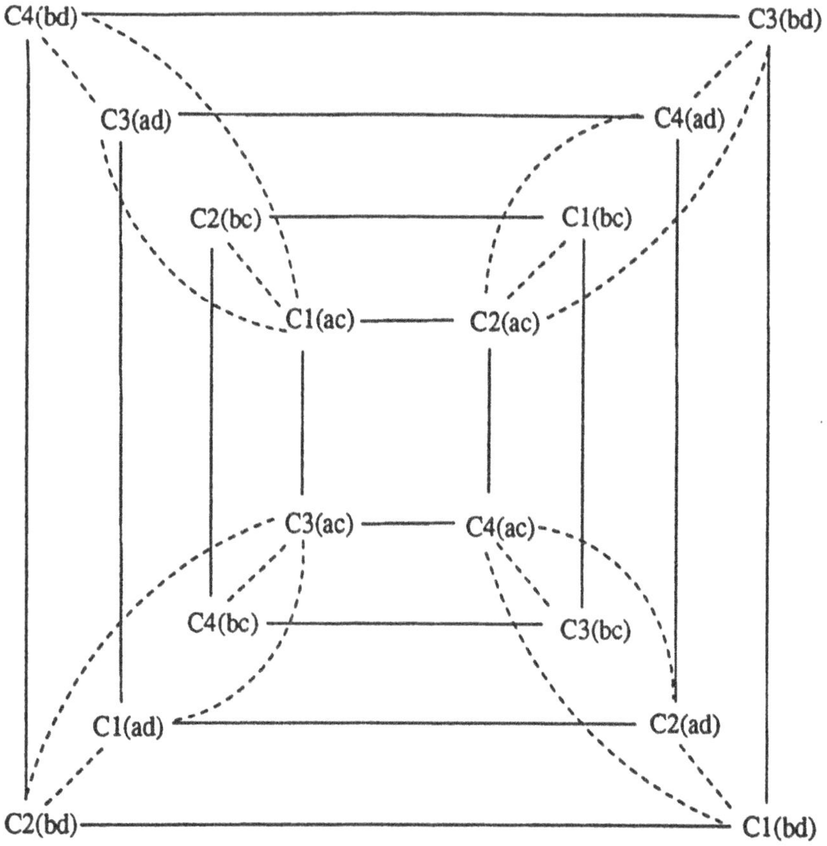

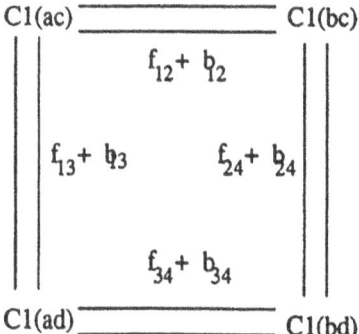

**Fig. 9.** Connectivity of versions of the cage isomers C1, C2, C3 and C4 according to the rearrangements listed in Table 3. Solid lines represent the four single flip processes and dashed lines the four bifurcations.

C1(ac)
C1(bc)
$f_{12}+ b_{12}$
$f_{13}+ b_{13}$
$f_{24}+ b_{24}$
$f_{34}+ b_{34}$
C1(ad)
C1(bd)

**Fig. 10.** Effective reaction graph for the four versions of cage isomer C1 that are interconnected indirectly by flip and bifurcation rearrangements.

where the correspondence between the MS group elements and the operations of $C_{2v}$ has been chosen as (cd) $\equiv C_2$, (ab) $\equiv \sigma_v(xz)$ and (ab)(cd) $\equiv \sigma_v'(yz)$. The accidental degeneracy of the $A_2$ and $B_1$ states would be broken at higher resolution. The relative nuclear spin weights for rovibronic states are easily shown to be 9:3:3:1 for $(H_2O)_6$ and 4:2:2:1 for $(D_2O)_6$ corresponding to $A_1:A_2:B_1:B_2$. If the accidental degeneracy is unresolved then the relative intensities of the three triplet components would be 9:6:1 for $(H_2O)_6$ and 4:4:1 for $(D_2O)_6$.

The above analysis leads to the same result as that obtained by Liu $et$ $al.$ [82], who considered the consequences of hypothetical direct permutations (ab) and (cd). However, in the present picture the tunneling between different versions of the four cage isomers does not occur through a single barrier, which might explain why the splittings are rather small. For this framework to be consistent we would expect to find similar splitting patterns for all four cage isomers, although the magnitude of the splitting could be different for each one.

The same analysis would also hold for a second family of cage isomers where the double acceptor monomers are adjacent. However, the four corresponding isomers in this case lie several hundred wavenumbers higher in energy than those considered above. Several mechanisms interconverting the two cage families were found in the set of 900 pathways calculated for the hexamer in this study. One further point worthy of consideration is that the experiment measures the difference in tunneling splittings between different rovibrational states. If the observed vibrational transition does indeed correspond to the torsional motion of a single-donor, single-acceptor monomer then the splitting in the excited state could be significantly different from the ground state.

For the TIP4P potential the situation is slightly different because there are only two cage isomers rather than four. The single-donor, single acceptor that has two double-donor neighbours exhibits only one torsional minimum intermediate between the two states found in C1-C4 above. The other single-donor, single-acceptor has the same two torsional states as in C1-C4, and undergoes single flip and bifurcation rearrangements as before. For the single-donor, single-acceptor monomer in the intermediate torsional state a direct degenerate rearrangement corresponding to a bifurcated transition state was found. The pathways are summarized in Table 4 where we label the two isomers C1' and C2'. Illustrations of these rearrangements are omitted for brevity. A degenerate bifurcation rearrangement of the C2' isomer could not be located, despite starting a number of transition state searches from points around the expected geometry.

Although the topology of the PES is different for the TIP4P potential the above rearrangements could produce the same tunneling splittings and intensity pattern as before, since there are still pathways linking all four versions of each isomer. Unfortunately there is insufficient experimental information to distinguish the two possibilities, and it is quite possible that neither scenario is correct. Due to the relatively large number of degrees of freedom and the uncertainties associated with the choice of empirical potential more accurate theoretical treatments of the dynamics will not be easy. However, it may not be a bad approximation to neglect relaxations of the rest of the cage in considering the dynamics

Table 4. Rearrangement mechanisms which interconvert cage isomers of $(H_2O)_6$ calculated with the TIP4P potential. The energies are in $cm^{-1}$. $Min_1$ is the lower minimum, $\Delta_1$ is the higher barrier, TS is the transition state, and $\Delta_2$ is the smaller barrier corresponding to the higher minimum $Min_2$. $S$ is the integrated path length in Å, $D$ is the displacement between minima in Å and $\gamma$ is the cooperativity index. All these quantities are defined in §3.

| $Min_1$ | $\Delta_1$ | TS | $\Delta_2$ | $Min_2$ | $S$ | $D$ | $\gamma$ | description |
|---------|------------|------|------------|---------|-----|-----|----------|-------------|
| C1' (-16,533) | 14 | -16,519 | 2 | C2' (-16,521) | 1.3 | 1.2 | 14.1 | single flip |
| C1' (-16,533) | 825 | -15,708 | 813 | C2' (-16,521) | 3.6 | 2.2 | 8.6 | bifurcation |
| C1' (-16,533) | 1,361 | -15,172 | 1,361 | C1' (-16,533) | 4.5 | 2.2 | 8.7 | bifurcation |

of the two single-donor, single-acceptor monomers. Two-dimensional quantum calculations of the torsional dynamics of these two molecules were performed under this approximation using the C1 geometry and the discrete variable representation [86] (DVR). These calculations follow the three-dimensional DVR calculations of Sabo et al. [29] for $(H_2O)_3$ in employing a model Hamiltonian in which only rotation about the hydrogen-bonded O−H bond is permitted:

$$\hat{\mathcal{H}} = -B_{\text{eff}} \left( \frac{\partial^2}{\partial\phi_1^2} + \frac{\partial^2}{\partial\phi_2^2} \right) + V(\phi_1, \phi_2), \tag{6}$$

where $\phi_1$ and $\phi_2$ are the two torsional angles in question and the value of $B_{\text{eff}}$ was taken to be $19.63\,cm^{-1}$ for $H_2O$ and $9.82\,cm^{-1}$ for $D_2O$. For each torsional degree of freedom the basis functions were chosen as:

$$\psi(\phi) = \frac{1}{\sqrt{2\pi}} e^{im\phi}, \qquad m = 0, \pm1, \ldots, \pm N, \tag{7}$$

giving a total of $(2N+1)^2$ basis functions for a given value of $N$. The DVR grid points are then uniformly spaced in each coordinate:

$$\phi^j = \frac{2\pi j}{2N+1}, \qquad j = 1, 2, \ldots, 2N+1. \tag{8}$$

Hence we obtain a direct product DVR with grid points $(\phi_1^j, \phi_2^k)$, and using the known analytic form for the kinetic energy operator [87] we obtain the Hamiltonian matrix elements:

$$H_{cd}^{ab} = T_{ca}\delta_{db} + T_{db}\delta_{ca} + V(\phi_1^c, \phi_2^d)\delta_{ac}\delta_{db}, \tag{9}$$

where

$$T_{ca} = B_{\text{eff}}(-1)^{c-a} \begin{cases} N(N+1)/3, & a = c, \\[2mm] \dfrac{\cos\left[\pi(c-a)/(2N+1)\right]}{2\sin^2\left[\pi(c-a)/(2N+1)\right]}, & a \neq c. \end{cases} \tag{10}$$

The lowest eigenvalues were converged to an accuracy better than $0.1\,\mathrm{cm}^{-1}$ for both $(H_2O)_6$ and $(D_2O)_6$ at a value $N = 14$; an iterative Lanczos matrix diagonalization procedure was employed.

The lowest eight eigenvectors for $(H_2O)_6$ along with assignments are shown in Fig. 11 for ASP-W2 and Fig. 12 for TIP4P. In both cases the ranges of the torsional angles have been restricted to exclude regions where the amplitude is essentially always zero. The centre of each surface corresponds to the geometry with both free hydrogens in intermediate positions. Not surprisingly, the wavefunctions for the TIP4P potential are all delocalized over torsional space, and can be classified according to the number of nodes in the $\phi_1$ and $\phi_2$ directions. The three lowest energy wavefunctions for the ASP-W2 potential are localized in the wells corresponding to C1, C2 and C3. However, the fourth, seventh and eighth functions appear to be delocalized over two isomers in each case. For this potential the wavefunctions are described in terms of localized functions in the four different wells, e.g. C1(1,0) is the function localized in the C1 well with one node in the $\phi_1$ direction and none in the $\phi_2$ direction. The results for $(D_2O)_2$ are omitted for brevity, and can be obtained from the author on request.

Clearly we cannot assign the experimental transition on the basis of these model calculations. However, the fact that both potentials exhibit some delocalization between different cage isomers suggests that vibrational averaging of the structure is likely, especially in an excited torsional state. Unfortunately, it does not seem to be possible to admit both the flip and bifurcation rearrangements without including at least three angular degrees of freedom for each of the two monomers in question. Such calculations are considerably more difficult, and are left for future work.

# References

1. D. D. Nelson, G. T. Fraser and W. Klemperer, Science **238**, 1670 (1987).
2. G. C. Maitland, M. Rigby, E. B. Smith and W. A. Wakeham, *Intermolecular Forces* (Clarendon Press; Oxford, 1981).
3. R. C. Cohen and R. J. Saykally, Annu. Rev. Phys. Chem. **42**, 369 (1991).
4. J. M. Hutson, Annu. Rev. Phys. Chem. **41**, 123 (1990).
5. R. C. Cohen and R. J. Saykally, J. Phys. Chem. **94**, 7991 (1990).
6. N. Pugliano and R. J. Saykally, J. Chem. Phys. **96**, 1832 (1992).
7. R. J. Saykally and G. A. Blake, Science **259**, 1570 (1993).
8. K. Liu, J. D. Cruzan and R. J. Saykally, Science **271**, 929 (1996).
9. N. Pugliano and R. J. Saykally, Science **257**, 1937 (1992).
10. J. Del Bene and J. A. Pople, J. Chem. Phys. **58**, 3605 (1973).
11. M. F. Vernon, D. J. Krajnovich, H. S. Kwok, J. M. Lisy, Y. R. Shen and Y. T. Lee, J. Chem. Phys. **77**, 47 (1982).
12. Owicki, J. C.; Shipman, L. L.; Scheraga, H. A., J. Phys. Chem. **79**, 1794 (1975).
13. D. J. Wales, J. Amer. Chem. Soc. **115**, 11180 (1993).
14. J. E. Fowler and H. F. Schaefer, J. Amer. Chem. Soc. **117**, 446 (1995).

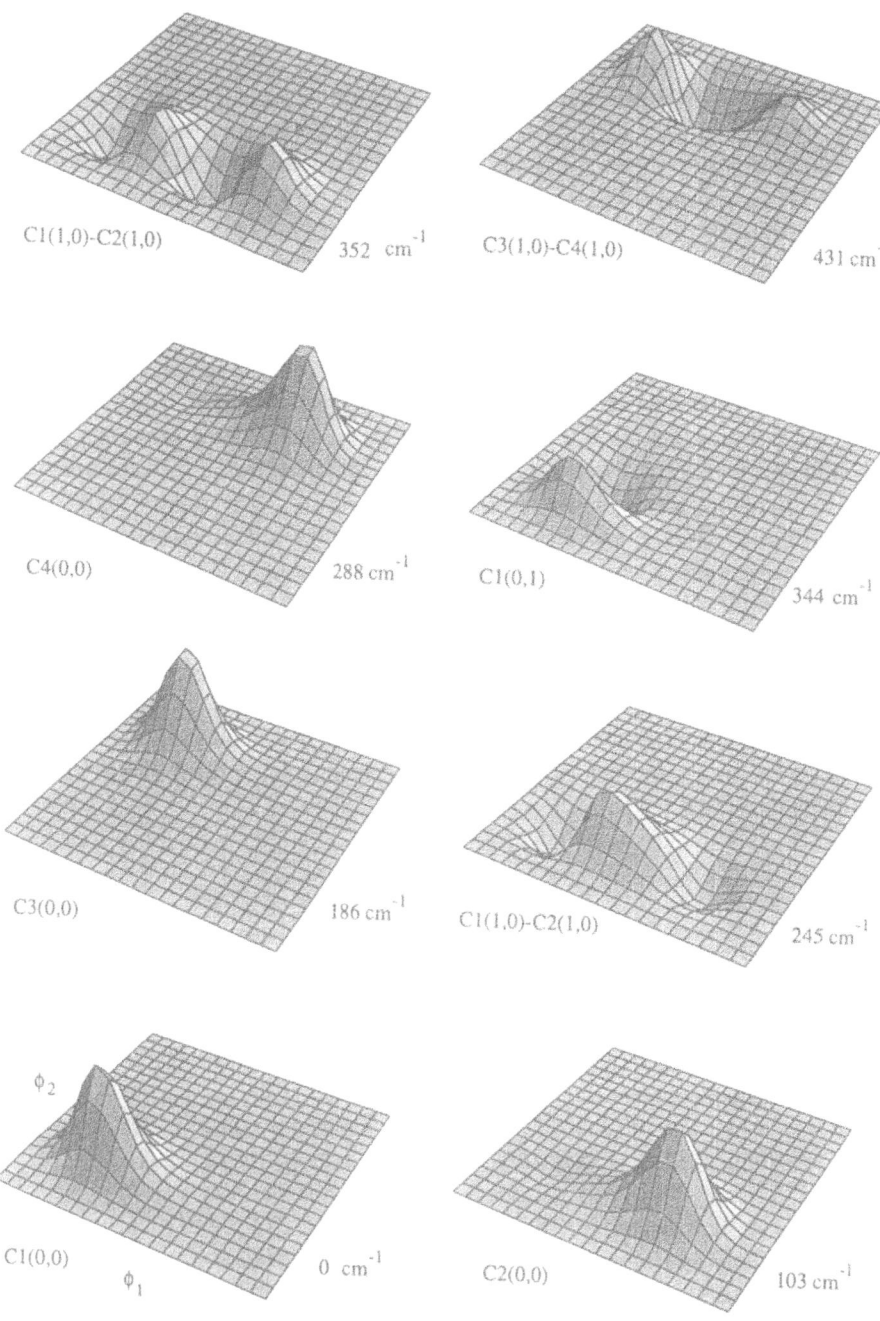

**Fig. 11.** Amplitudes of the lowest eight eigenvectors in torsional space found by dimensional DVR calculations for $(H_2O)_6$ with the ASP-W2 potential. The int between grid lines is $12.41°$ in both directions.

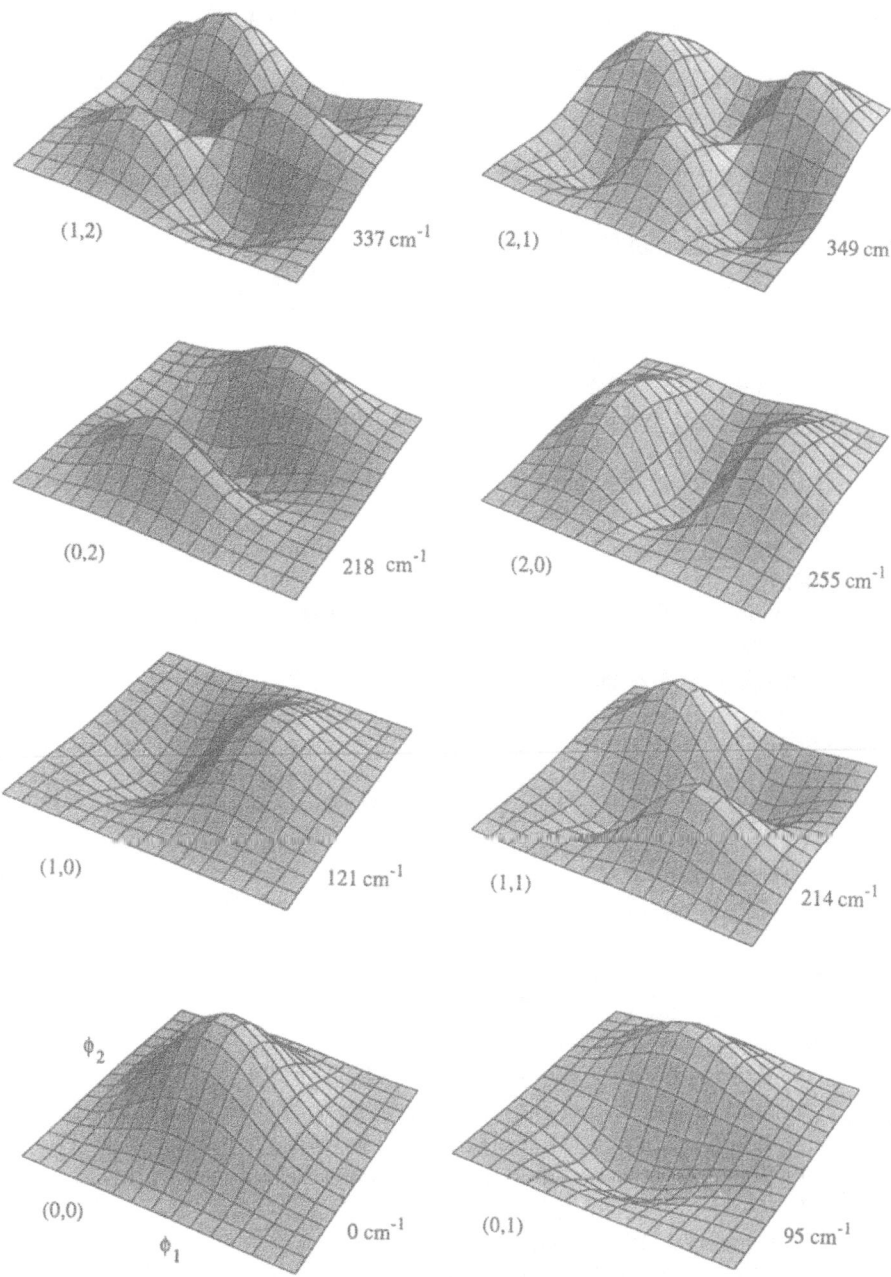

**Fig. 12.** Amplitudes of the lowest eight eigenvectors in torsional space found by two-dimensional DVR calculations for $(H_2O)_6$ with the TIP4P potential. The interval between grid lines is $12.41°$ in both directions.

15. K. Liu, J. G. Loeser, M. J. Elrod, B. C. Host, J. A. Rzepiela, N. Pugliano and R. J. Saykally, J. Amer.'Chem. Soc. **116**, 3507 (1994).
16. S. Suzuki and G. A. Blake, Chem. Phys. Lett. **229**, 499 (1994).
17. T. R. Walsh and D. J. Wales, J. Chem. Soc., Faraday Trans. **92**, 2505 (1996).
18. J. G. C. M. van Duijneveldt-van de Rijdt and F. B. van Duijneveldt, J. Chem. Phys. **97**, 5019 (1982).
19. G. Chalasiński, M. M. Szczęśniak, P. Cieplak and S. Scheiner, J. Chem. Phys. **94**, 2873 (1991).
20. S. S. Xantheas and T. H. Dunning, J. Chem. Phys. **98**, 8037 (1993).
21. S. S. Xantheas and T. H. Dunning, J. Chem. Phys. **99**, 8774 (1993).
22. S. S. Xantheas, J. Chem. Phys. **100**, 7523 (1994).
23. van Duijneveldt-van de Rijdt, J. G. C. M.; van Duijneveldt, F. B., Chem. Phys. Lett. **237**, 560-567 (1995).
24. W. Klopper, M. Schütz, H. P. Lüthi and Leutwyler, S., J. Chem. Phys. **103**, 1085 (1995).
25. T. Bürgi, S. Graf, S. Leutwyler and W. Klopper, J. Chem. Phys. **103**, 1077 (1995).
26. M. Schütz, T. Bürgi, S. Leutwyler and H. B. Bürgi, J. Chem. Phys. **99**, 5228 (1993).
27. M. Schütz, T. Bürgi, S. Leutwyler and H. B. Bürgi, J. Chem. Phys. **100**, 1780 (1994).
28. W. Klopper and M. Schütz, Chem. Phys. Lett. **237**, 536 (1995).
29. D. Sabo, Z. Bačić, T. Bürgi and S. Leutwyler, Chem. Phys. Lett. **244**, 283 (1995).
30. J. K. Gregory and D. C. Clary, J. Chem. Phys. **102**, 7817 (1995).
31. J. K. Gregory and D. C. Clary, J. Chem. Phys. **103**, 8924 (1995).
32. R. G. A. Bone, T. W. Rowlands, N. C. Handy and A. J. Stone, Molec. Phys. **72**, 33-73 (1991).
33. D. M. Dennison and J. D. Hardy, Phys Rev. **39**, 938-947 (1932).
34. R. P. Bell, *The Tunnel Effect in Chemistry* (Chapman and Hall; New York, 1980).
35. R. E. Leone and P. v. R. Schleyer, Angew. Chem. Int. Ed. Engl. **9**, 860 (1970).
36. J. G. Nourse, J. Amer. Chem. Soc. **102**, 4883 (1980).
37. J. N. Murrell and K. J. Laidler, J. Chem. Soc., Faraday II **64**, 371 (1968).
38. H. C. Longuet-Higgins, Molec. Phys. **6**, 445 (1963).
39. J. T. Hougen, J. Chem. Phys. **37**, 1433 (1962).
40. J. T. Hougen, J. Chem. Phys. **39**, 358 (1962).
41. J. T. Hougen, J. Phys. Chem. **90**, 562 (1986).
42. P. R. Bunker, *Molecular Symmetry and Spectroscopy* (Academic Press; New York, 1970).
43. D. J. Wales, J. Amer. Chem. Soc. **115**, 11191 (1993).
44. C. J. Cerjan and W. H. Miller, J. Chem. Phys. **75**, 2800 (1981); J. Simons, P. Jørgenson, H. Taylor and J. Ozment, J. Phys. Chem. **87**, 2745 (1983); D. O'Neal, H. Taylor and J. Simons, J. Phys. Chem. **88**, 1510 (1984);

A. Banerjee, N. Adams, J. Simons and R. Shepard, J. Phys. Chem. **89**, 52 (1985); J. Baker, J. Comput. Chem. **7**, 385 (1986); J. Baker, J. Comput. Chem. **8**, 563 (1987).

45. D. J. Wales, J. Chem. Phys. **101**, 3750 (1994).

46. D. J. Wales and T. R. Walsh, J. Chem. Phys. **105**, 6957 (1996).

47. R. D. Amos and J. E. Rice, *CADPAC: the Cambridge Analytic Derivatives Package, Issue 4.0*; Cambridge, 1987.

48. A. Banerjee and N. P. Adams, Int. J. Quant. Chem. **43**, 855 (1992).

49. P. L. A. Popelier, A. J. Stone and D. J. Wales, J. Chem. Soc., Faraday Discuss. **97**, 243 (1994).

50. D. J. Wales, P. L. A. Popelier and A. J. Stone, J. Chem. Phys. **102**, 5556 (1995).

51. D. J. Wales, A. J. Stone and P. L. A. Popelier, Chem. Phys. Lett. **240**, 89 (1995).

52. A. J. Stone, Chem. Phys. Lett. **83**, 233 (1981).

53. A. J. Stone and M. Alderton, Molec. Phys. **56**, 1047 (1985).

54. A. J. Stone, Molec. Phys. **56**, 1065 (1985).

55. C. R. Le Sueur and A. J. Stone, Molec. Phys. **78**, 1267 (1993).

56. C. Millot and A. J. Stone, Molec. Phys. **77**, 439 (1992).

57. W. L. Jorgensen, J. Amer. Chem. Soc. **103**, 335 (1981).

58. W. L. Jorgensen, J. Chandraesekhar, J. W. Madura, R. W. Impey and M. L. Klein, J. Chem. Phys. **79**, 926 (1983).

59. T. H. Dunning Jr., J. Chem. Phys. **53**, 2823 (1970).

60. S. J. Huzinaga, J. Chem. Phys. **47**, 1293 (1965).

61. C. Møller and M. S. Plesset, Phys. Rev. **46**, 618 (1934).

62. A. D. Becke, Phys. Rev. A **38**, 3098 (1988).

63. C. Lee, W. Yang and R. G. Parr, Phys. Rev. B **37**, 785 (1988).

64. F. H. Stillinger and T. A. Weber, Phys. Rev. A **28**, 2408 (1983).

65. G. T. Fraser, Int. Rev. Phys. Chem. **10**, 189 (1991).

66. T. R. Dyke, K. M. Mack and J. S. Muenter, J. Chem. Phys. **66**, 498 (1977).

67. N. Pugliano, J. D. Cruzan, J. G. Loeser and R. J. Saykally, J. Chem. Phys. **98**, 6600 (1993).

68. T. R. Dyke, J. Chem. Phys. **66**, 492 (1977).

69. J. T. Hougen, J. Mol. Spectr. **114**, 395 (1985).

70. L. H. Coudert and J. T. Hougen, J. Mol. Spectr. **130**, 86 (1988).

71. L. H. Coudert and J. T. Hougen, J. Mol. Spectr. **139**, 259 (1990).

72. B. J. Smith, D. J. Swanton, J. A. Pople, H. F. Schaefer and L. Radom, J. Chem. Phys. **92**, 1240 (1990).

73. S. F. Boys and F. Bernardi, Molec. Phys. **19**, 553 (1970).

74. S. S. Xantheas, J. Chem. Phys. **104**, 8821 (1996).

75. J. A. Odutola and T. R. Dyke, J. Chem. Phys. **72**, 5062 (1980).

76. M. Tsuboi, A. Y. Hirakawa, T. Ino, T. Sasaki and K. Tamagake, J. Chem. Phys. **41**, 2721 (1964).

77. Mathematica 2.0; Wolfram Research Inc., Champaign, IL, 1989.

78. E. Zwart, J. J. ter Muelen, W. L. Meerts and L. H. Coudert, J. Mol. Spectrosc. **147**, 27 (1991).

79. S. C. Althorpe and D. C. Clary, J. Chem. Phys. **101**, 3603 (1995).

80. D. J. Wales, J. Amer. Chem. Soc. **115**, 11191 (1993).

81. C. A. Coulson and S. Rushbrooke, Proc. Camb. Phil. Soc. **36**, 193 (1940).

82. K. Liu, M. G. Brown, C. Carter, R. J. Saykally, J. K. Gregory and D. C. Clary, Nature **381**, 501 (1996).

83. C. J. Tsai and K. D. Jordan, Chem. Phys. Lett. **213**, 181-188 (1993).

84. K. Kim, K. D. Jordan and T. S. Zwier, J. Amer. Chem. Soc. **116**, 11568-11569 (1994).

85. B. J. Dalton and P. D. Nicholson, Int. J. Quantum Chem. **9**, 325 (1975).

86. Z. Bačić and J. C. Light, Annu. Rev. Phys. Chem. **40**, 469 (1989).

87. D. T. Colbert and W. H. Miller, J. Chem. Phys. **96**, 1982 (1992).

# Fullerene Collisions

Olaf Knospe and Rüdiger Schmidt

Technische Universität Dresden, Institut für Theoretische Physik, D-01062 Dresden, Germany

*Dedicated to Hans O. Lutz on his 60th birthday.*

**Abstract.** After an introduction and overview on the field of fullerene collisions three types of collisions are discussed in detail: *First*, results from a theoretical study of collisions of noble-gas atoms with $C_{60}$ using classical molecular dynamics are shown. The possible reaction channels are described. The microscopically calculated differential cross section for $C_{60}$ (5 keV) + He collisions reproduces the experimentally observed second maximum in the lowest tail of the kinetic-energy spectrum. *Second*, a theoretical investigation on collisions between fullerenes for the systems $C_{60}^+$ + $C_{60}$, $C_{70}^+$ + $C_{60}$ and $C_{70}^+$ + $C_{70}$ is presented covering a wide range of collision energies. Structural as well as energetic aspects of the reaction mechanism are discussed from a microscopic point of view using quantum molecular dynamics. The fusion barriers for the investigated systems are determined and compared with experimental data. A simple phenomenological fusion model is described and compared to the experimental fusion cross section as a function of the collision energy. For $C_{60}^+$ (1 keV) + $C_{60}$ collisions it is shown that multifragmentation and collective-flow effects can be probed in fullerene–fullerene collisions, and these phenomena are discussed in terms of nuclear heavy-ion collisions. Mass and angular distributions of the fragments have been calculated microscopically. *Third*, it is demonstrated with molecular dynamics simulations that bound states consisting of two oppositely charged atomic clusters ("Rydberg clusters") can be formed in three-body collisions occurring in the plasma of a $C_{60}$-laser desorption source. The promising properties of this new class of Rydberg systems are sketched.

## 1 Introduction

A wealth of collision experiments with fullerenes has been performed in the past few years enabled by the easy production of intense *and* mass-selected fullerene ion beams or fullerene targets. In the majority of these experiments, collisions of fullerenes (mostly $C_{60}$ or $C_{70}$) with atoms or small molecules have been investigated favoring *fragmentation* and *endohedral-complex formation* as the topics of interest. The formation of endohedral complexes in *collisions*, where an atom is captured within the fullerene cage, represents one of the most exciting observations in fullerene research. These complexes have been found in collisions with noble-gas atoms (Weiske et al. 1991, Ross and Callahan 1991, Caldwell et al. 1992, Campbell et al. 1992, Wan et al. 1992a, Mowrey et al. 1992, Callahan et al. 1993a, Christian et al. 1993, Kleiser et al. 1993) as well as metal ions (Wan et al. 1992b and 1993). Other atoms (and ions) are "candidates" in ongoing investigations (Basir et al. 1994, Basir and Anderson 1995). The interesting reaction $C^+ + C_{60}$ has been studied in detail in Christian et al. 1992.

The observed fragmentation mass spectra exhibit a typical bimodal distribution for moderate collision energies around 100 eV in the center-of-mass frame (Young et al. 1991, Doyle and Ross 1991, Weiske et al. 1991, Caldwell et al. 1992, Hvelplund et al. 1992, Wan et al. 1992b and 1993, Takayama and Shinohara 1993, Christian et al. 1993, Brink et al. 1994, Lorents et al. 1995, McHale et al. 1995, Ehlich et al. 1996a). The larger fragments appear as a sequence of peaks differing by a $C_2$ unit (e.g. $C_{60}^+$, $C_{58}^+$, $C_{56}^+$, ...$C_{30}^+$), whereas the peak sequence for the small fragments is single spaced with pronounced maxima for the "magic numbers" ($C_{11}^+$, $C_{15}^+$, $C_{19}^+$, $C_{23}^+$). Surprisingly, this fragmentation pattern is very similar to that resulting from interactions of fullerenes with electrons (Völpel et al. 1993, Kolodney et al. 1995) as well as photons (O'Brien et al. 1988, Jones et al. 1993, Wurz and Lykke 1994, Gaber et al. 1992, Hohmann et al. 1994 and 1995). At high collision energies (MeV), a typical multifragmentation mass spectrum with a power-law distribution for the small fragments has been found (LeBrun et al. 1994).

In contrast to the large amount of experimental material, the number of theoretical investigations, which are molecular dynamics (MD) simulations in most cases, is limited. The difficulties in a microscopic description arise in general from the large number of degrees of freedom in the system, and in particular from the covalent character of the carbon–carbon interaction which requires (at least) the incorporation of three-body forces or the quantum-mechanical treatment of the binding forces. The formation of endohedral complexes could be reproduced directly in MD simulations (Mowrey et al. 1992, Ehlich et al. 1993, Kaplan et al. 1993, Seifert and Schulte 1994, Cui et al. 1994, Ohno et al. 1996), because the process develops on a picosecond time scale. Although the mechanism has been qualitatively understood, particular problems are still open, e.g. those connected with the penetration barriers and the "windowing mechanism" (Murry and Scuseria 1994, Cui et al. 1994). In contrast, the mechanism leading to the *final* fragmentation pattern is still not understood because of the importance of secondary processes on the microsecond time scale which cannot be followed in MD simulations (Ehlich et al. 1993, Kaplan et al. 1993, Yueyuan Xia et al. 1995). Here, the application of statistical models (Ehlich et al. 1996a, Campbell 1996) can help to understand the fragmentation dynamics of fullerenes.

As a "theoretical introduction" to atom–fullerene collisions, the possible reaction channels in noble-gas atom – $C_{60}$ collisions are discussed in Sect. 2.2. In the experimental investigations on endohedral complexes, kinetic-energy spectra of the scattered fullerenes have been obtained with the characteristic "foot" due to inelastic collisions (Caldwell et al. 1992, Mowrey et al. 1992, Callahan et al. 1993a). In Sect. 2.3, a kinetic-energy spectrum for $C_{60}^+(5 \text{ keV})$ + He collisions calculated in a fully microscopic framework based on MD simulations is presented, in which the "foot" is perfectly reproduced and its origin is explained.

The interest in atom–fullerene collision experiments inducing electronic processes like excitation (Finch et al. 1995), charge transfer (Hvelplund et al. 1994) and ionization (Walch et al. 1994) is widely increasing. Ionization (and neutralization) processes have also been studied experimentally bombarding fullerenes with electrons (Wörgötter et al. 1994, Scheier et al. 1994, Rauth et al. 1995). Charge transfer has been observed in fullerene–fullerene collisions as well (Javahery et al. 1992, Rohmund and Campbell 1995, Shen et al. 1995). First model calculations of the charge transfer in collisions of a highly charged ion with fullerenes have been carried out (Thumm 1994 and 1995, Thumm et al. 1995, Barany and Setterlind 1995).

Considering fullerene–surface collisions a lot of experimental results have been accumulated (Beck et al. 1991, Busmann et al. 1991 and 1993, Lill et al. 1992 and 1993a, Lill et al. 1993b, Yeretzian et al. 1993 and 1994, Toglhofer et al. 1993, Callahan et al. 1993b, Lill et al. 1994, 1995 and 1996, Weis et al. 1996, Beck et al. 1996), where the stability of fullerenes (i.e. inelastic scattering or fragmentation) in reflections from or the deposition on a variety of solid surfaces have been studied. The MD simulation of these collisions, however, requires considerable computational efforts. That's why their theoretical investigation is still at the beginning (Mowrey et al. 1991, Blaudeck et al. 1994, Galli and Mauri 1994).

In the main part of this review (Sect. 3), fullerene–fullerene collisions as particular realizations of cluster–cluster collisions (CCC) are investigated with emphasis on *fusion* and *fragmentation*. Already in the first experiment on CCC, performed on the system $C_{60}^+ + C_{60}$ (Campbell et al. 1993), the fusion of clusters was observed. Similar fusion (or coalescence) reactions have been found in collisions of $C_{60}$ with fullerite surfaces (Lill et al. 1993b) and in connection with laser desorption of fullerite surfaces (Yeretzian et al. 1992, Hunter et al. 1994, Mitzner et al. 1995). In a recent experiment (Rohmund et al. 1996a), fusion in $C_{60}^+ + C_{60}$ collisions has been investigated systematically, i.e. measuring the fusion cross section as a function of the collision energy. Moreover, the experimental setup allowed the extension of the investigations to $C_{70}^+ + C_{60}$ and $C_{70}^+ + C_{70}$ collisions (Rohmund et al. 1996b). The fragmentation patterns obtained in $C_{60}^{q+}, C_{70}^{q+} (100 \text{ keV}) + C_{60}$ collisions (Shen et al. 1995) are very similar to that for atom–$C_{60}$ collisions described above. A further CCC experiment has been carried out (Scheidemann et al. 1994) in which the scattering of sodium clusters in $C_{60}$ vapor has been studied.

The first systematic theoretical study of CCC and in particular the prediction of fusion was given for metallic clusters (Schmidt et al. 1991 and 1992, Schmidt and Lutz 1992, Schmidt and Lutz 1993; for a review see Schmidt and Lutz 1995). The experimental verification, however, is still not feasible at present due to the difficulties of producing mass-selected metal cluster beams. The success in the $C_{60}^+ + C_{60}$ experiment (Campbell et al. 1993) drew the attention of the theoreticians to this collisional system (Seifert and Schmidt 1992a, Strout et al. 1993, Zhang et al. 1993, Schmidt et al. 1994, Kim and Tomanek 1994, Long at al. 1994, Schulte et al. 1995, Robertson et al. 1995, Yueyuan Xia et al. 1996). Investigations of fusion with the help of molecular dynamics simulations have been performed on different theoretical levels concerning the calculation of the interatomic forces using empirical potentials (Ballone and Milani 1990, Robertson et al. 1995, Yueyuan Xia et al. 1996), a semiempirical MNDO approach (Long at al. 1994), a tight-binding scheme (Zhang et al. 1993) or the density functional theory in the local density approximation (called quantum molecular dynamics – QMD – see Sect. 3.1; Seifert and Schmidt 1992a, Rohmund et al. 1996a, Knospe et al. 1996a).

In Sect. 3.2, fusion and deep inelastic scattering in $C_{60}^+ + C_{60}$, $C_{70}^+ + C_{60}$ and $C_{70}^+ + C_{70}$ collisions are investigated using QMD simulations in the energy range from 50 to 250 eV in the center-of-mass frame. The resulting fusion barriers including the effect of a finite cluster temperature are described. A qualitative understanding of the energy dependence of the experimental fusion cross section can be obtained with a simple phenomenological fusion model based on transparent physical assumptions. At higher collision energies (above 400 eV in the center-of-mass frame), multifragmentation, i.e. the multiple break-up of a highly excited piece of matter within a short time interval, becomes the dominating reaction channel in near central fullerene–fullerene collisions. In Sect. 3.3, multifragmentation and collective-flow effects in $C_{60}^+(1 \text{ keV}) + C_{60}$ collisions are discussed in terms of heavy-ion collisions. The mass spectrum and the angular distribution of the primary fragments are calculated in a fully microscopic framework based on QMD simulations.

Recently a new class of Rydberg systems has been proposed consisting of two oppositely charged atomic clusters in a bound state with high angular momentum ("Rydberg clusters"; Knospe and Schmidt 1996a). In Sect. 4 it is demonstrated with MD simulations that in a plasma containing $C_{60}^+$ and $C_{60}^-$ (e.g. in a laser-desorption source), effective three-body collisions may occur which lead to bound states. Such exotic fullerene collisions could open a new field of research touching the fundamental question of the transition from quantum to classical mechanics (Schrödinger 1926).

# 2 Collisions of $C_{60}$ with Noble-Gas Atoms

## 2.1 Molecular Dynamics Model

In this section, the ingredients and some details of the MD model (Ehlich et al. 1993) are described in which Newton's equations with empirical interatomic forces for the atomic motion are solved. To simulate high-energy collisions with $C_{60}$ and related fragmentation processes the most serious demands must be put on the empirical potential. For the atom–$C_{60}$ collisions investigated in this section, the cage-like structure of $C_{60}$ is more or less preserved, although large deformations and direct fragmentation of $C_{60}$ occur (cf. Fig. 1). Therefore the empirical potential should simultaneously reproduce the ground-state properties of $C_{60}$, that of highly excited and deformed $C_{60}$ as well as that of small carbon clusters.

In our classical MD model used for the simulation of noble-gas atom + $C_{60}$ collisions, the forces between carbon atoms are calculated from a potential of the type (Tersoff 1988, Brenner 1990)

$$V^{C-C} = \frac{1}{2} \sum_i \sum_{i \neq j} [V_R(r_{ij}) - B_{ij} V_A(r_{ij})] \qquad (1)$$

with a repulsive $(V_R)$ and an attractive $(V_A)$ two-body interaction of the form $V(r) \sim \exp[-\lambda r]$ and a bond-order function $B_{ij}$ that contains a sum over three-body interaction terms. This three-body interaction is determined by the angle between two "bonds" from one atom. At a certain equilibrium angle, which represents one of the potential parameters, the three-body force vanishes which essentially influences the resulting geometrical equilibrium structures. Different available parametrizations of the carbon potential (1) have been tested by calculating ground-state and excited-state properties of $C_{60}$ and geometrical structures of small carbon clusters (Ehlich et al. 1993) and comparing the predictions with available experimental data and ab initio calculations. For the MD simulations in this section, the Brenner parametrization (first parameter set in Brenner 1990) was used.

The interaction between the noble-gas atom and the carbon atoms is described (as in Mowrey et al. 1992) by a screened Coulomb potential

$$V^{A-C} = \sum_{j=1}^{60} \frac{Z_1 Z_2 e^2}{r_{Aj}} \chi(r_{Aj}/a) \qquad (2)$$

with $Z_1$, $Z_2$ the charge numbers of the atoms and the screening function $\chi$ of Moliere (Moliere 1947). It could be shown (Ehlich et al. 1993) that a consistent interpretation of experimental threshold energies for endohedral-compound formation in $C_{60}^+$ + He and $C_{60}^+$ + Ne collisions cannot be obtained from the standard expression for the screening radius $a$ given in O'Connor and MacDonald 1977 and used in Mowrey et al. 1992. Instead, $a$ is fixed by a fit to experimental threshold energies for capture taking into account the finite temperature of the $C_{60}$ (for more details see Ehlich et al. 1993).

## 2.2 Reaction Channels

As one of the main results of our MD simulations of Ne + $C_{60}$ collisions we recognized that the reaction channels can be classified according to four categories: deep inelastic scattering, fragmentation, capture and inelastic scattering. Their preferential appearance depend strongly on the impact parameter,

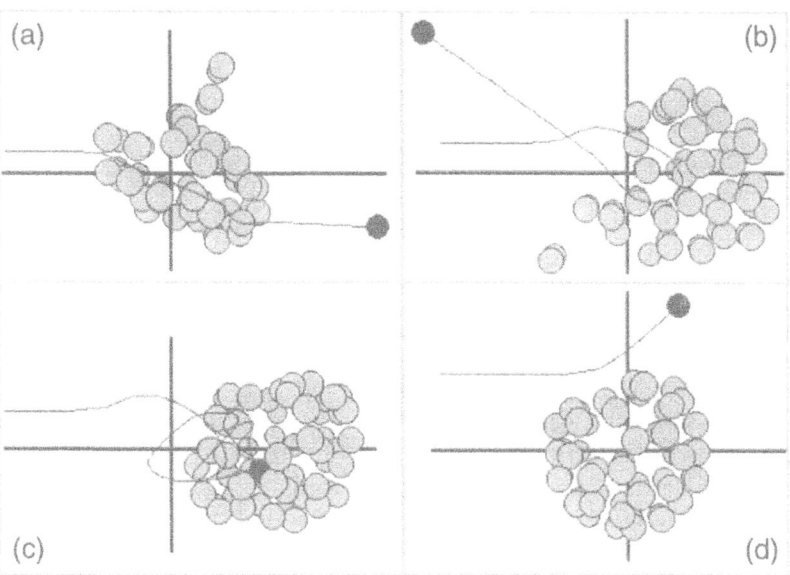

**Fig. 1.** Reaction channels in Ne(80 eV) + $C_{60}$ collisions: (**a**) deep inelastic scattering ($b/R = 0.3$, (**b**) fragmentation ($b/R = 0.4$), (**c**) capture ($b/R = 0.5$), (**d**) scattering ($b/R = 1.0$). The same initial orientation of the $C_{60}$ cage was used in all cases. (The thin lines mark the trajectories of the Ne atom.)

the incident energy and the orientation of the $C_{60}$ cage with respect to the collisional axis. Snapshots of typical events for $E_i = 80$ eV lab incident energy (i.e. $E_{cm} = 78$ eV), zero initial internal energy and the same initial orientation of the $C_{60}$ are shown in Fig. 1.

In deep inelastic collisions (Fig. 1a for $b/R = 0.3$; $R$ is the cage radius) the Ne atom penetrates through the $C_{60}$ cage and loses nearly all available kinetic energy. The energy loss of the projectile is stored, nearly equally partitioned, into deformation energy and intrinsic kinetic energy (Ehlich et al. 1993). During and just after the interaction the $C_{60}$ cage is significantly deformed, however, it remains stable during the time period studied. Of course, the large excitation energy will lead to secondary evaporation processes of (mainly) dimers within time scales that cannot be followed in MD studies.

In Fig. 1b, a typical fragmentation event is shown, where a small fragment (in this case a dimer) is emitted. The energy needed to break the bonds decreases the total kinetic energy. Small fragments $C_N$ up to $N = 6$ have been observed at $E_i = 80$ eV. As a rule, fragmentation appears in the intermediate range of impact parameters. In the event presented in Fig. 1b, the Ne atom loses (and gains) kinetic energy in a multi-step process in four collisions. Again, approximately one half of the total kinetic-energy loss of the projectile is stored into deformation energy (Ehlich et al. 1993).

In Fig. 1c, a capture event is presented where an endohedral complex Ne@C$_{60}$ is formed. The common motion of the compound can be followed by the trajectory of the Ne atom (thin line in Fig. 1c). The Ne atom moves within the cage but with very low kinetic energy. Full energy as well as angular-momentum transfer occur. For this exotic reaction channel, the trajectory has been followed over a long time period. At this high incident energy the actual "lifetime" of the endohedral compound is very short and the Ne leaves the C$_{60}$ after colliding many times with the cage after about 1.5 ps (Ehlich et al. 1993). Therefore, a significant amount of stable endohedral compounds Ne@C$_{60}$ are expected to be formed only at lower energies, in accord with experimental data (Campbell et al. 1992).

At large impact parameters $b/R \geq 1$ the Ne atom is inelastically scattered at the outer front of the C$_{60}$ cage (see Fig. 1d). It loses a fraction of its kinetic energy in a single collision. However, even at this grazing collision the total kinetic-energy loss is not small ($\approx 28$ eV). In tangential collisions the deformation energy of the C$_{60}$ cage is small compared to that part stored into vibrational energy. As an interesting feature of nearly grazing collisions the breathing mode is excited (Ehlich et al. 1993).

## 2.3 Kinetic-Energy Spectra

One of the most striking features concerns the extreme *inelasticity* of C$_{60}$ as collisional partner. This property leads to characteristic structures in the kinetic-energy spectra, which will be discussed in this section. Experimentally, a maximum in the lowest kinetic-energy tail of C$_{60}^+$ + He collisions has been found although no explanation could be given (the "foot", see Mowrey et al. 1992, Caldwell et al. 1992 and Callahan et al. 1993a).

In Ehlich et al. 1993, the total kinetic-energy loss of the Ne atom in Ne + C$_{60}$ collisions for different bombarding energies as function of the impact parameter was discussed. It could be shown that the energy loss does not depend on the impact parameter $b$ in a large range of partial waves ($b/R < 0.4$) for incident energies $E_i < 50$ eV. Therefore a pronounced maximum was expected in the lowest tail of the kinetic-energy distribution.

For a quantitative comparison with experimental spectra of C$_{60}^+$ + He collisions, an enormous amount of collision events was simulated ($\sim 10^4$ events for a given energy, Ehlich et al. 1996b), from which *absolute* values of differential cross sections have been obtained. The kinetic-energy spectrum $d\sigma/dE$ is calculated from the probability $P(E, b)$ to find clusters with kinetic energy $E$ in the simulated events with an impact parameter $b$ by

$$\frac{d\sigma}{dE} = \frac{2\pi}{\Delta E} \int\limits_0^{b_{max}} P(E, b)\, b \, db \tag{3}$$

with the energy bin $\Delta E$. The probability $P(E, b)$ is approximated by the relative abundance

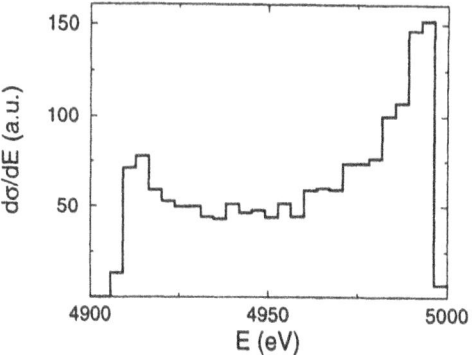

**Fig. 2.** The calculated kinetic-energy spectrum $d\sigma/dE$ for $C_{60}$ (5 keV) + He collisions (without capture events).

$$P(E, b) \approx \frac{n(E, b)}{n_{tot}(b)} \quad , \tag{4}$$

where $n(E, b)$ and $n_{tot}(b)$ are the number of fragments with energy $E$ and the total number of fragments found in the simulated events with impact parameter $b$, respectively (Schulte et al. 1995).

The result for $C_{60}$(5 keV) + He collisions with zero initial temperature of the projectile is shown in Fig. 2. Besides the trivial maximum in the range of the beam energy (which arises from peripheral collisions due to a finite $b_{max}$), a second maximum corresponding to those events with maximum energy loss can be seen at the lower end of the spectrum. The difference between both maxima ($\approx$ 80 eV) agrees perfectly with the experimental observation (cf. Fig. 3 in Mowrey et al. 1992). Thus, our MD simulations provide a straightforward explanation for the existence of this second maximum in the kinetic-energy spectra in $C_{60}^+$ + He collisions (Ehlich et al. 1996b).

# 3 Fullerene–Fullerene Collisions

## 3.1 Quantum Molecular Dynamics

In order to obtain a microscopic insight into the dynamics of fullerene–fullerene collisions, large-scale QMD simulations have been performed the basics of which are described in this section. Newton's equations for the atomic motion are solved simultaneously with approximated Kohn–Sham (KS) equations of the electronic structure using an LCAO ansatz for the KS orbitals. This approach has been successfully applied to a variety of collision processes (e.g. Seifert and Schmidt 1992b, Schmidt et al. 1994, Seifert and Schulte 1994).

The electronic KS orbitals $\psi_n(\mathbf{r})$ are written as linear combination of atomic orbitals $\phi_\nu(\mathbf{r} - \mathbf{R}_j)$ centered at the nuclei $j$:

$$\psi_n(\mathbf{r}) = \sum_{\nu} C_{\nu n} \phi_\nu(\mathbf{r} - \mathbf{R}_j) \quad . \tag{5}$$

Note that the KS orbitals $\psi_n$ as well as the coefficients $C_{\nu n}$ depend parametrically on the atomic coordinates $\mathbf{R}_j$. Inserting this ansatz into the KS equations one obtains the secular equations

$$\sum_{\nu} (H_{\mu\nu} - \epsilon_n S_{\mu\nu}) C_{\nu n} = 0 \tag{6}$$

with the KS Hamiltonian matrix $H_{\mu\nu}$, the overlap matrix $S_{\mu\nu}$ and the KS eigenvalues $\epsilon_n$.

The effective potential $V_{\text{eff}}$ in the KS Hamiltonian is approximated by a sum of potentials of neutral atoms

$$V_{\text{eff}} \approx \sum_{j} V_j^{(0)} \quad , \tag{7}$$

and for consistency one has to assemble the Hamiltonian matrix containing only one- and two-center terms, which are obtained in an atomic-structure calculation (Seifert et al. 1986, Seifert and Schmidt 1992b). The calculation of the forces is considerably simplified if only the valence electrons of the atoms are taken into account. Furthermore, the ionic Coulomb forces and the orbital terms from the electron–electron interaction (i.e. Hartree plus exchange–correlation) cancel each other to a large extent (Seifert and Schmidt 1992b). The remaining repulsive force is modeled as the gradient of a short-range potential $U(R)$ which can be parametrized to reproduce structural properties of molecules or obtained from ab initio calculations. The equations of motion for the atoms (mass $M_i$) can be written as

$$M_i \ddot{\mathbf{R}}_i = -\sum_{j} \frac{\partial U(R_{ij})}{\partial \mathbf{R}_i} - \sum_{n} \sum_{\mu\nu} \left[ C_{n\mu} \left( \frac{\partial}{\partial \mathbf{R}_i} H_{\mu\nu} - \epsilon_n \frac{\partial}{\partial \mathbf{R}_i} S_{\mu\nu} \right) C_{\nu n} \right] \quad . \tag{8}$$

The algebraic KS equations (6) have to be solved simultaneously with the equations of motion (8). This requires much less computational effort compared to ab initio MD schemes (e.g. Car and Parinello 1985) and enables the systematic study of the dynamics in large systems like fullerenes. It is worth to note that the structure of (8) resembles that of tight-binding MD (e.g. Zhang et al. 1993). However, the Hamiltonian matrix $H_{\mu\nu}$ in (8) does not contain free parameters in contrast to tight-binding parametrizations.

## 3.2  Fusion and Deep Inelastic Scattering

**Fusion versus scattering.** Our systematic investigation of fullerene–fullerene collisions using QMD simulations includes at present $C_{60}^+ + C_{60}$ (Schmidt et al. 1994, Schulte et al. 1995, Rohmund et al. 1996a), $C_{70}^+ + C_{60}$ and $C_{70}^+ + C_{70}$ collisions (Knospe et al. 1996a). In this section, the fusion-relevant range

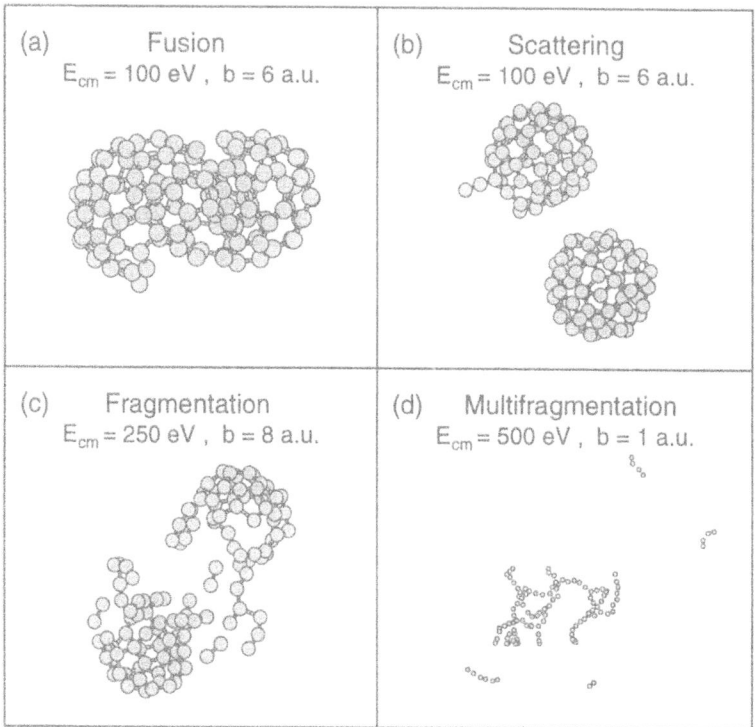

**Fig. 3.** Snapshots from the time evolution in QMD simulations of $C_{60}^+ + C_{60}$ collisions for typical events characterizing the reaction channels – fusion (**a**), deep inelastic scattering (**b**), fragmentation (**c**) and multifragmentation (**d**). Different (randomly chosen) initial orientations of the $C_{60}$ cages are used for the two events with $E_{cm} = 100$ eV and $b = 6$ a.u. (a and b).

of collision energies $50 < E_{cm} < 250$ eV (in the center-of-mass frame) is considered. A large number of collision events was simulated for zero cluster temperature as well as for a temperature of $T = 2000$ K in both clusters.

Snapshots of typical events for $C_{60}^+ + C_{60}$ are shown in Fig. 3. For all three collisional systems considered, fusion and deep inelastic scattering were found to be competing reaction channels in the energy range above the fusion·barrier (see next section) before fragmentation becomes the dominating reaction channel at higher collision energies (see Sect. 3.3).

The discriminating mechanism between fusion and scattering in these collisions can be recognized by investigating the kinetic energy as a function of time (Rohmund et al. 1996a), which is shown for two $C_{70}^+ + C_{70}$ events in Fig. 4. The general increase of the center-of-mass (and also the total) kinetic energy after the system has reached the distance of closest approach

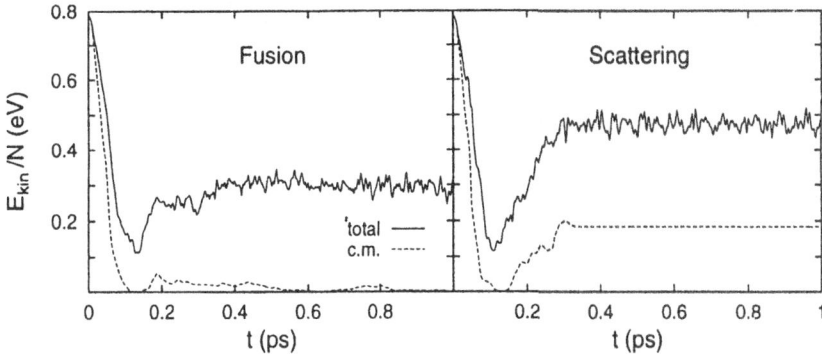

**Fig. 4.** Calculated total kinetic energy per atom (full lines) and center-of-mass kinetic energy per atom of the relative motion between the two colliding fullerenes (dashed lines) for QMD simulations of two central $C_{70}^+ + C_{70}$ collisions at a collision energy $E_{cm} = 110$ eV, zero initial temperature and the same principal orientation (Knospe et al. 1996a). The different random orientations around the symmetry axes lead to different reaction channels – fusion (left part) and deep inelastic scattering (right part).

(corresponding to zero center-of-mass energy in Fig. 4) indicates that a part of the stored potential energy can be converted back into kinetic energy of relative motion. If this "bouncing-off" is suppressed (left part of Fig. 4) due to the rearrangement of atoms in the contact zone a fusion event will be detected, whereas if the increase of kinetic energy proceeds to a large enough extent (right part of Fig. 4) the effective repulsion of the (more or less) intact fullerene structure leads to a scattering event (Knospe et al. 1996a).

In fusion events, strongly deformed (and highly excited) compound clusters are formed which resemble a "peanut"-like structure (see Fig. 3a). The "memory" on the entrance channel (i.e. on the orientation) is lost in most cases. The large excitation energy in these compounds can be reduced by successive dimer evaporation (Rohmund et al. 1996a, Rohmund et al. 1996b). However, even *complete* fusion, i.e. $C_{120}$, $C_{130}$ and $C_{140}$, respectively, is observed experimentally (Rohmund et al. 1996a, Rohmund et al. 1996b). The stabilization of the fused compound against evaporation is a direct consequence of the formation of such "peanut" isomers, which drastically reduces the final vibrational energy (Rohmund et al. 1996a). Experimental hints on stable "$C_{60}$ dimer" structures has been found in coalescence reactions induced by laser ablation of fullerene films (Hunter et al. 1994).

**Fusion barriers.** The fusion barriers for the three considered systems at zero cluster temperature have been determined by varying the collision energy and (for each energy) the random rotation of both fullerenes (for $C_{70}$ only around the symmetry axis). A surprising result for $C_{70}^+ + C_{60}$ and $C_{70}^+ + C_{70}$ collisions was the *independence* of the fusion barrier on the orientation of the $C_{70}$ symmetry axis (see Knospe et al. 1996a, figs. 1 and 2). This finding is in striking contrast to nuclear fusion in heavy-ion collisions of deformed nuclei, where a marked difference in the calculated fusion barriers for parallel or perpendicular orientation of the symmetry axes was noticed (e.g. in the case U + U, Bock 1980).

The values obtained for the fusion barriers are compared with the experimental data (Rohmund et al. 1996b) in Table 1. Comparing the theoretical values at zero temperature for the three systems, a remarkable increase of the fusion barrier with increasing particle number of the colliding fullerenes can be seen. The relatively large energy thresholds for fullerene fusion (compared to thermal energies) are a consequence of the fast and effective energy transfer from collision energy (relative motion of the two clusters) into internal energy (vibrational modes) (Rohmund et al. 1996a). Large deformations have to be induced, i.e. the closed fullerene structure has to be broken, before a rearrangement of atoms in the overlap region can lead to stable "intercluster" bonds. The larger the number of internal degrees of freedom in which energy can be transferred the smaller is the amount of remaining energy for the rearrangement of atoms, which gives a qualitative explanation of the increase of the fusion barrier with increasing number of atoms in the system (Knospe et al. 1996a).

The theoretical barriers for $T = 0$ are significantly larger than the experimental values (cf. second and fourth column in Table 1), which has been understood for $C_{60}^+ + C_{60}$ to be an effect of the finite cluster temperature in the experiment (Rohmund et al. 1996a, Zhang et al. 1993, Robertson et al. 1995). The available phase space is considerably enlarged due to the additional energy and the softening of the tight fullerene structure. Unfortunately, the actual cluster temperature in the experiment cannot be measured but only roughly estimated by indirect methods (Rohmund et al. 1996b). To be comparable with the earlier theoretical studies on $C_{60}^+ + C_{60}$ (Zhang et al.

**Table 1.** Fusion barriers for $C_{60}^+ + C_{60}$ , $C_{70}^+ + C_{60}$ and $C_{70}^+ + C_{70}$ collisions obtained from QMD simulations (Knospe et al. 1996a) at zero temperature (second column) and at $T = 2000$ K for projectile and target (third column) compared with the experimental data (fourth column, taken from Rohmund et al. 1996b).

| | QMD | | Experiment |
|---|---|---|---|
| | $V_B(T = 0)$/eV | $V_B(2000$ K$)$/eV | $V_B/$(eV) |
| $C_{60}^+ + C_{60}$ | 80 | 60 | $60 \pm 1$ |
| $C_{70}^+ + C_{60}$ | 94 | 70 | $70 \pm 6.5$ |
| $C_{70}^+ + C_{70}$ | 104 | 75 | $76 \pm 4$ |

1993, Robertson et al. 1995) we have chosen for our simulations a temperature of $T \approx 2000$ K for projectile and target, which fits into the experimental estimation (1800...2000 K, Rohmund et al. 1996b). The temperature was simulated in a simple manner by random, Gaussian distributed velocity components for the carbon atoms, which leads to slightly different vibrational energies in each MD event (with an average corresponding to 2000 K). The resulting fusion barriers for $C_{60}^+ + C_{60}$, $C_{70}^+ + C_{60}$ and $C_{70}^+ + C_{70}$ agree perfectly with the experimental values (cf. third and fourth column in Table 1). The temperature effect for $C_{60}^+ + C_{60}$ $\Delta V_B = V_B(0) - V_B(2000\,\text{K}) = 20$ eV agrees well with that obtained in the tight-binding MD simulations of $\Delta V_B = 18$ eV (Zhang et al. 1993), whereas from the classical MD with a phenomenological potential (Robertson et al. 1995) an unreasonably large effect $\Delta V_B \sim 50$ eV was predicted. Comparing the QMD results for the three systems (third column in Table 1) an almost linear increase of the temperature effect $\Delta V_B$ with increasing number of atoms in the system is apparent. This compensates partly for the growing fusion barrier for $T = 0$ (second column in Table 1, Knospe et al. 1996a).

**Fusion model.** With a simple model for $C_{60}^+ + C_{60}$ collisions (Rohmund et al. 1996a), a qualitative understanding of the fusion cross section as a function of the collision energy could be achieved. This model has been refined (Knospe et al. 1996a) taking into account more realistic assumptions about the fusion probability.

In the (semi-classical) partial-wave expansion, the fusion cross section $\sigma_{CF}(E)$ as a function of the collision energy $E$ is given by

$$\sigma_{CF}(E) = \frac{\pi \hbar^2}{2\mu E} \sum_{l=0}^{\infty} (2l + 1) P_l(E) \tag{9}$$

where $\mu$ is the reduced mass of the colliding system. $P_l(E)$ represents the probability that the partial wave $l$ contributes to fusion at the energy $E$. In the "sharp cut-off" approximation, this probability is written as

$$P_l(E) = \begin{cases} P(E) & ; \quad l \le l_{CF}(E) \\ 0 & ; \quad l > l_{CF}(E) \end{cases} \tag{10}$$

with $P(E)$ the mean fusion probability (assumed to be angular-momentum independent) and $l_{CF}(E)$ the maximum angular momentum which contributes to fusion. This limiting angular momentum can be estimated from the centrifugal barrier

$$E \approx V_B + \frac{\hbar^2 l_{CF}^2(E)}{2\mu R_{12}^2} \tag{11}$$

where it is assumed that the barrier radius does not depend on $l$ and is located in the vicinity of the contact radius $R_{12}$. If the angular momentum exceeds a critical value $l_{cr}$ the fused compound is not stable against centrifugal fragmentation (Schmidt and Lutz 1992). Thus, for energies larger than

$$E_{cr} = V_B + \frac{\hbar^2 \, l_{cr}^2}{2 \, \mu \, R_{12}^2} \tag{12}$$

the critical angular momentum $l_{cr}$ limits the contributing $l$ range. In both energy regions, the sum (9) can be carried out due to the "cut-off'" leading for $l_{CF} \gg 1$ to

$$\sigma_{CF}(E) = \begin{cases} \pi R_{12}^2 \left(1 - \dfrac{V_B}{E}\right) P_<(E) \quad ; & V_B \leq E \leq E_{cr} \\[3mm] \pi R_{12}^2 \, \dfrac{E_{cr} - V_B}{E} \, P_>(E) \quad ; & E > E_{cr} \end{cases} \tag{13}$$

where a different fusion probability $P_>(E)$ has been introduced for the "high-energy" region $(E > E_{cr})$, because it was shown for $C_{60}^+ + C_{60}$ collisions (Rohmund et al. 1996a) that a further reaction channel – fragmentation (as opposed to evaporation, see Rohmund et al. 1996b) – comes into play. With increasing collision energy $E$, more and more fused compounds decay into (more than two) fragments on a picosecond time scale. This observation from the QMD simulations can be modeled by a linear decrease of the fusion probability $P_>(E)$ with the collision energy $E$

$$P_>(E) = P_1 - P_2 \cdot (E - E_{cr}) \tag{14}$$

with the parameters $P_1$ and $P_2$. In the "low-energy" region $(V_B \leq E \leq E_{cr})$, a "steric effect" (Levine and Bernstein 1987) is taken into account, i.e. every initial orientation of the colliding fullerenes can have a different fusion barrier. At the measured fusion barrier, only a few optimal relative orientations of the fullerenes lead to fusion. With increasing collision energy, the number of initial orientations which are above the barrier increases, which may be expressed by a linear increase of the fusion probability

$$P_<(E) = P_0 \cdot (E - V_B) \tag{15}$$

with a further parameter $P_0$. Inserting (14) and (15) into (13) one obtains

$$\sigma_{CF}(E) = \begin{cases} \sigma_0 \, \dfrac{(E - V_B)^2}{E} \quad ; & V_B \leq E \leq E_{cr} \\[3mm] \dfrac{\sigma_1}{E} - \sigma_2 \quad ; & E > E_{cr} \end{cases} \tag{16}$$

with the model parameters $\sigma_0 = \pi R_{12}^2 P_0$, $\sigma_2 = \pi R_{12}^2 (E_{cr} - V_B) P_2$, $V_B$ and $E_{cr}$ ($\sigma_1 = \sigma_0 (E_{cr} - V_B)^2 + \sigma_2 E_{cr}$ is fixed to guarantee continuity at $E = E_{cr}$).

As a first test of the fusion model, eq. (16) is fitted to the experimental fusion cross section for the three considered collisional systems. The results

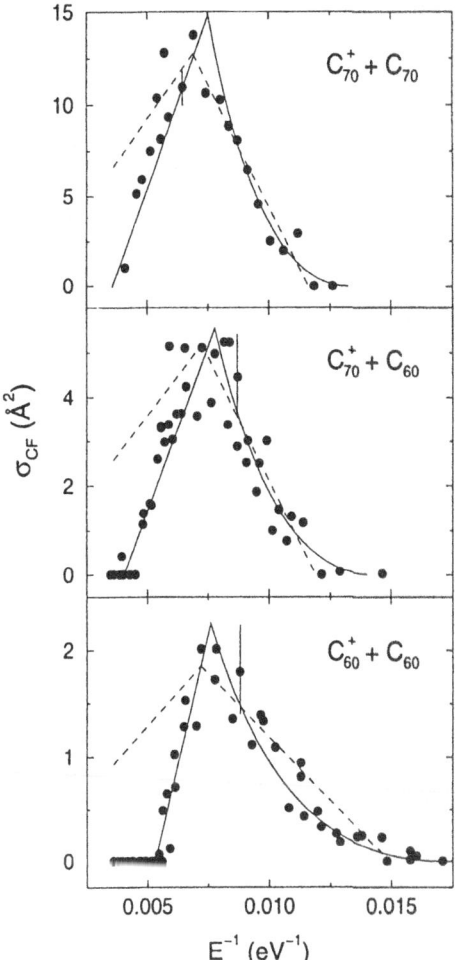

**Fig. 5.** Absolute fusion cross section as a function of the inverse collision energy for $C_{60}^+ + C_{60}$ (lower part), $C_{70}^+ + C_{60}$ (middle) and $C_{70}^+ + C_{70}$ (upper part). The full lines represent the results of the phenomenological fusion model according to (16) in comparison with the experimental data (dots, Rohmund et al. 1996b) and with a model with energy-independent fusion probability (dashed lines, Rohmund et al. 1996b). The model parameters in (16) are adjusted to give the best overall agreement with the data (Knospe et al. 1996a).

as a function of $1/E$ are shown in Fig. 5, where the experimental values for $V_B$ were used, and $E_{cr}$ was located (approximately) at the maximum of the experimental cross section. (The experimental values of $V_B$ given in Table 1 have been obtained by a fit with the low-energy part of (16), see Rohmund et al. 1996b.) The energy dependence of the fusion cross section is in very good agreement with the experimental data, which demonstrates that the assumptions concerning the fusion probability in (14) and (15)) reflect the essential physical aspects properly (Knospe et al. 1996a).

## 3.3 Multifragmentation and Collective-Flow Effects

**Collisional dynamics.** Spontaneous multifragmentation and collective-flow effects are macroscopic phenomena like shattering of glass, or shock wave propagation. Their existence in nuclear heavy-ion collisions with a large but *finite* number of degrees of freedom is nontrivial, and their specific nuclear properties enlarged our knowledge about nuclear dynamics far from equilibrium (e.g. the validity range of macroscopic concepts, like fluid dynamics; Scheid et al. 1974, Stöcker and Greiner 1986). From a more general point of view, this holds for *any* many-body system with a *finite* number of particles, like fullerenes. In this section, $C_{60}^+ + C_{60}$ collisions at a bombarding energy of $E_{cm} = 500$ eV are considered with an excitation energy of 30 eV for the projectile ion to obtain realistic experimental conditions. The whole range of (fragmentation-) relevant impact parameters ($b \leq 14$ a.u.) was investigated (Schmidt et al. 1994, Schulte et al. 1995).

Most important for the scenario of near central collisions ($b \leq 7$ a.u.) is the formation of a super-dense state during the approach phase. After a very short lifetime ($< 0.1$ ps) the compressed matter explodes leading to a nonisotropic expansion accompanied by cluster emission (Schmidt et al. 1994). A snapshot from a typical multifragmentation event is shown in Fig. 3d in which 10 fragments are formed finally. For more peripheral collisions ($7 < b < 12$ a.u.) fragmentation processes similar to Fig. 3c can be observed, and for $b > 12$ a.u. only scattering events have been found.

In Fig. 6, the mean translational kinetic energy per atom $\langle E_{kin} \rangle / N$ of the emitted clusters with $N \leq 30$ is presented (histogram). It is compared to that of a thermal source which simply yields $\langle E_{kin} \rangle / N = 3kT/2N$ (dashed curve). The temperature $kT = 1$ eV has been estimated from the mean internal excitation energy per atom of the clusters (Schulte et al. 1995). The nearly N-independent shift of the calculated $\langle E_{kin} \rangle / N$ compared to that of a thermal source is a first clear signal of an additional mechanism that contributes to the final translational cluster velocities in terms of collective-flow effects (Schmidt et al. 1994).

One of the universal features of multifragmentation concerns the power law in the mass distribution of the fragments, which is observed in very different fields of physics, e.g., in the mass distribution of the fragments in high-energy nuclear reactions, for the collision debris of macroscopic stones or in the size distribution of asteroids in the planetary system (Hüfner and Mukhopadhyay 1986). The calculated mass distribution of the multifragmentation events ($b \leq 7$ a.u.) for our collisional system is compatible with a simple power-law distribution $\sim N^{-\tau}$ up to the largest clusters formed ($N \leq 94$) with an exponent between $\tau \approx 1$ and $\tau \approx 1.5$ (Schulte et al. 1995).

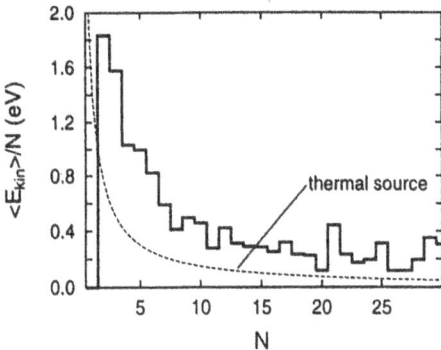

**Fig. 6.** Mean translational kinetic energy per atom $\langle E_{\mathrm{kin}}\rangle/N$ as a function of the cluster size $N$ in near central $C_{60}^{+}(1\ \mathrm{keV}) + C_{60}$ collisions (histogram). The dashed curve corresponds to the expectations of a thermal source (see text, Schmidt et al. 1994).

**Mass- and angular distributions.** The cross section of the mass distribution $d\sigma/dN$ is calculated from the probability $P(N,b)$ to find clusters of size $N$ in the simulated events with an impact parameter $b$ according to (3) and (4) with the size bin $\Delta N = 1$.

In contrast to experiment, in our MD simulation the mass distribution can be followed as a function of time providing a transparent insight into different mechanisms contributing to the final mass spectrum. After $t \approx 0.5$ ps only three narrow mass ranges are populated: compound-like clusters ($N > 110$) resulting from near central events, $C_{60}$-like clusters ($50 < N < 70$) from peripheral collisions, and complementary small fragments ($N < 10$) originating from both regions of $b$ values. The compound-like clusters decay very rapidly by multifragmentation processes filling up the intermediate mass range ($10 \leq N \leq 50$), at $t \approx 1$ ps. During that time interval the distribution of the $C_{60}$-like clusters remains practically unchanged. After about 2 ps almost all heavy fragments ($N > 60$) have decayed. The resulting distribution for the intermediate-mass fragments (left part in Fig. 7) reminds of a U-shape, which is also found in heavy-ion collisions in certain energy regions (Stöcker and Greiner 1986). The further time evolution of the mass spectrum is dominated by statistical evaporation processes in time scales of microseconds, which cannot be followed by a MD simulation.

In the right part of Fig. 7, the calculated angular distribution $d\sigma/d\theta$ of all fragments is presented. The center-of-mass scattering angle $\theta$ is taken in the plane of the beam axis and the impact parameter and is counted as absolute value between $0°$ and $180°$ (negative angles are included), where $0°$ and $180°$ correspond to the initial direction of projectile and target movement, respectively. (Negative angles can simply be "measured" in the simulations but hardly in an experiment.) The calculation of $d\sigma/d\theta$ is carried out ac-

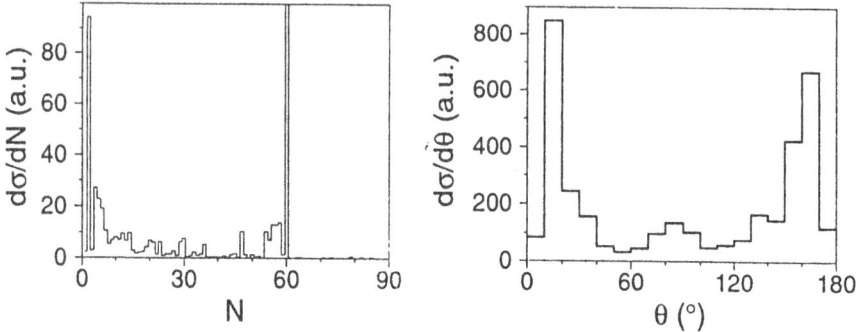

**Fig. 7.** The calculated mass distribution $d\sigma/dN$ (left part, cutting the dominating peak at $N = 60$) and the calculated angular distribution $d\sigma/d\theta$ of the fragments in $C_{60}^+(1\text{ keV}) + C_{60}$ collisions (Schulte et al. 1995).

cording to (3) and (4) counting the number of fragments $n(\theta, b)$ which are "detected" within one angle bin ($\Delta\theta \sim 10°$). The peaks of $d\sigma/d\theta$ at $\theta \approx 15°$ and $\theta \approx 165°$ arise due to the "cut-off" with respect to the impact parameter ($b_{\max} = 14$ a.u.). As an interesting feature of the angular distribution one observes a bump at angles of about $\theta = 90°$. This enhancement of $d\sigma/d\theta$ around $\theta = 90°$ can be interpreted as a further signal of a collective transverse flow in high-energetic $C_{60}^+ + C_{60}$ collisions (Schulte et al. 1995).

## 4 Exotic Fullerene Collisions and Rydberg Clusters

The interest in the fundamental question of the transition between quantum and classical physics, already discussed in Schrödingers pioneering work (Schrödinger 1926), has been renewed considerably (in particular, the Kepler dynamics of electronic wave packets formed by a superposition of Rydberg states). Despite of the intensive efforts in the last years, the ultimate goal of these investigations – the preparation of a Rydberg wave packet localized with respect to radial *and* angular coordinates moving along a Kepler orbit – was not yet reached experimentally (cf. Alber and Zoller 1991, Knospe and Schmidt 1996b and refs. therein). Apparently, it is difficult to excite a "classical" state (a spatially localized wave packet) starting from the "quantum-mechanical limit" (the electronic ground state of an atom). Alternatively, we have recently proposed to form Coulomb systems by objects, the sizes of which are in the transition region from microscopic to macroscopic dimensions, e.g. two oppositely charged atomic clusters strongly bound in states with arbitrarily high quantum numbers ("Rydberg clusters", Knospe and Schmidt 1996a) In this case, mass and charge can be changed independently leading to an infinite variety of systems. The motion of such massive objects is expected to be in essence classical, i.e. characterized by spatially localized wave packets. In addition, qualitatively new phenomena are expected, be-

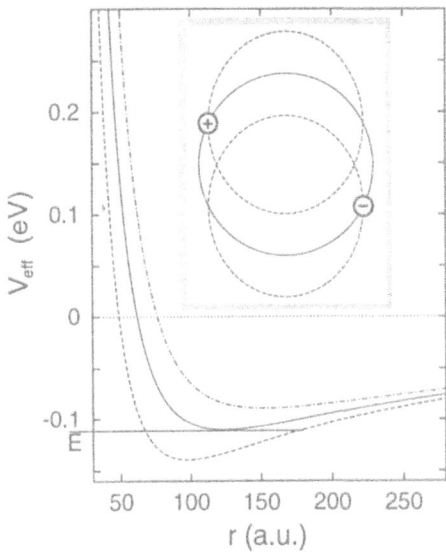

**Fig. 8.** Rydberg clusters $C_{60}^+ + C_{60}^-$: effective potential $V_{\text{eff}}$ as a function of the intercluster separation $r$ for three different angular momenta $L = 8$, 9 and $10 \cdot 10^3\, \hbar$ (dashed, solid and dashed-dotted line, respectively) and the corresponding classical trajectories (insert) to a given total energy $E = -0.1102$ eV (Knospe and Schmidt 1996a). The same length scale is used in both drawings.

cause an atomic cluster represents a system with a large number of internal degrees of freedom, and their influence on the wave-packet dynamics opens an interesting new field of research (Knospe and Schmidt 1996b).

One of the most challenging questions concerns the possible formation process of these states and, thus, their experimental verification. In contrast to electronic Rydberg states, Rydberg clusters cannot (or hardly) be formed by electronic transitions. Instead, *preformed* and *unbound* positively and negatively charged cluster ions (as e.g. existing in the plasma of a $C_{60}$-laser desorption source) can be used as starting point. Assuming a large intercluster distance and a high orbital angular momentum between both cluster ions, their relative kinetic energy may dissipate during the collision with a third particle leading to a bound state, e.g. as presented in Fig. 8.

To investigate this kind of formation processes such ("effective" three-body) collisions have been simulated microscopically using molecular dynamics (Knospe and Schmidt 1996a). As "third particles", noble-gas atoms as well as neutral $C_{60}$ clusters have been considered. The ingredients of the molecular dynamics model are the same as described in Sect. 2.1 taking into account additional Coulomb forces. Thereby, the total charge of an individual $C_{60}$-cluster ion (+1 or -1) is equally distributed over 60 atoms corresponding to the delocalization of the binding electrons. The initial conditions for the sim-

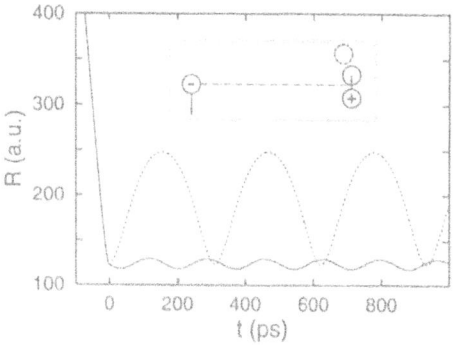

**Fig. 9.** Calculated distance $R$ between the centers of $C_{60}^+$ and $C_{60}^-$ as a function of time for two formation events of Rydberg clusters in molecular dynamics simulations of $C_{60} + C_{60}^+ (+ C_{60}^-)$. In the first event (full line), the $C_{60}^+$ collides centrally at $t \gtrsim 0$ with the resting $C_{60}$, whereas in the second event (dashed line) the collision occurs at $t \approx 6$ ps using the same initial conditions for the ions illustrated by the insert (Knospe and Schmidt 1996a).

ulations were chosen in such a way that the subsystem $C_{60}^+ + C_{60}^-$ represents an unbound state with kinetic energies up to 0.5 eV.

A neutral $C_{60}$ as third collision partner offers suitable prerequisites for the considered formation process. In Fig. 9, the distance between two oppositely charged $C_{60}$ ions is plotted as a function of time for two simulated events, in which the initial kinematics guarantee a sufficient loss of kinetic energy, i.e. the $C_{60}^+$ hits centrally on the resting $C_{60}$ (see insert in Fig. 9). The rare case of a circular orbit is realized if, just after the collision with the third particle, the distance vector and the relative velocity of the ions are perpendicular to each other, and if the relative kinetic energy is equal to half of the absolute value of the Coulomb energy. In the first event in Fig. 9 (full line), these two conditions are approximately fulfilled after the collision at $t \gtrsim 0$ resulting in an almost circular orbit with $L \approx 9000\,\hbar$ and $E \approx -0.11$ eV very close to the example shown in Fig. 8 (full lines). In the second case (dashed line in Fig. 9), the same initial conditions are used for the ions (the "history" $t < 0$ is the same), but the collision of the $C_{60}^+$ with the neutral $C_{60}$ takes place at $t \approx 6$ ps, and an elliptic orbit results. Thus, it has been shown that the formation of Rydberg-cluster states starting from preformed and unbound $C_{60}$ ions is in principle possible (Knospe and Schmidt 1996a).

In contrast to electronic Rydberg systems, the large binding energy of Rydberg clusters (Knospe and Schmidt 1996a) guarantees stability against external perturbations. Therefore, their lifetime is expected to be determined mainly by two processes, i.e. electromagnetic radiation and mutual neutralization. However, macroscopically large lifetimes were estimated provided that the states are formed with sufficiently large angular momenta or, equivalently, at large intercluster separations (Knospe and Schmidt 1996a).


# 5 Summary and Outlook

After an overview on the rapidly growing field of fullerene collisions (Sect. 1) we have presented some results from our systematic study of collisions of noble-gas atoms with $C_{60}$ (Sect. 2) using MD with empirical two- and three-body forces. The reaction channels can be classified according to four categories: deep inelastic scattering, fragmentation, capture and inelastic scattering. An important result of the calculations is the extreme inelasticity of $C_{60}$ in the collisions leading to a nearly impact-parameter independent energy loss. This specific behavior leads to the appearance of a second maximum in the lowest tail of the kinetic-energy spectra of the projectile in $C_{60}^+$ + He collisions at low collision energy. The microscopically calculated differential cross section of the kinetic-energy spectrum in $C_{60}$ (5 keV) + He collisions agrees perfectly with the experimental observation.

In the main part of this review (Sect. 3), we have presented theoretical investigations on $C_{60}^+ + C_{60}$, $C_{70}^+ + C_{60}$ and $C_{70}^+ + C_{70}$ collisions using QMD. The competing reaction channels in the energy range above the fusion barrier – fusion and deep inelastic scattering – exhibit the same general features for all three considered systems. The "bouncing-off" of the kinetic energy as a function of time, where a part of the stored potential energy is converted back into kinetic energy of relative motion, can be viewed as the discriminating mechanism between fusion and scattering. It has been found that the fusion process is determined mainly by the *local* geometry in the contact region of the colliding fullerenes during the early stages of the collision. The fusion barriers for the three considered systems $C_{60}^+ + C_{60}$, $C_{70}^+ + C_{60}$ and $C_{70}^+ + C_{70}$ have been determined. The growing barrier for $T = 0$ with increasing number of atoms in the system is partly compensated for by an increasing temperature effect leading to a perfect agreement between calculated barriers for $T = 2000$ K and the experimental data.

In order to understand the energy dependence of the experimental fusion cross section a simple phenomenological fusion model has been derived using reasonable assumptions about the fusion probability. A fit of the model parameters to the experimental data yields a very good agreement for all three considered systems, which demonstrates that the model reflects the essential physical aspects properly.

At higher collision energies, fragmentation processes become dominant. For $C_{60}^+$ (1 keV) + $C_{60}$ collisions we have shown that multifragmentation and collective-flow effects can be probed in fullerene–fullerene collisions and discussed these phenomena in terms of nuclear heavy-ion collisions. The simultaneous investigation of multifragmentation in different fields of physics may reveal some universal features. The mass- and the angular distribution of the primary fragments have been calculated microscopically from the results of the QMD simulations.

Exotic (three-body) collisions involving oppositely charged $C_{60}$ may lead to bound states ("Rydberg clusters", Sect. 4). In the plasma of a $C_{60}$-laser desorption source, such formation events may occur. Cluster-beam techniques or the storage of clusters in ion traps represent experimental alternatives. The existence of Rydberg-cluster states may be verified by field ionization of these neutral systems with succeeding detection of the charged clusters. The dynamics in such a two-cluster system could be studied (in the case of an elliptic orbit) by resonant time-resolved absorption spectroscopy using a laser in resonance with a sensitive state within one of the clusters. Then, the classical-like motion of a wave packet would be indicated by regular oscillations in the absorption signal.

QMD simulations provide a microscopic insight into the mechanism of fullerene collisions concerning the *atomic* degrees of freedom. The atomic dynamics is governed by forces resulting from the electronic ground state of the considered system. For *electronic* processes like charge transfer or ionization, however, such an adiabatic dynamics on the Born–Oppenheimer surface of the ground state cannot be adequate. Recently, a general formalism based on time-dependent density functional theory has been presented (Saalmann and Schmidt 1996) which treats simultaneously classical atomic motion and quantum electronic excitations in atomic many-body systems ("non-adiabatic QMD"). First applications of this powerful tool to fullerene collisions are already in progress.

## Acknowledgment

Financial support by the Deutsche Forschungsgemeinschaft through Schwerpunkt "Zeitabhängige Phänomene und Methoden in Quantensystemen der Physik und Chemie" and by the EU through HCM networks "Formation, stability and photophysics of fullerenes" and "Collision induced cluster dynamics" is gratefully acknowledged.

# References

Alber G., Zoller P. (1991), Phys. Rep. **199**, 231

Ballone P., Milani P. (1990), Phys. Rev. B **42**, 4201

Barany A., Setterlind C.J. (1995), Nucl. Instr. Meth. B **98**, 184

Basir Y., Zhimin Wan, Christian J.F., Anderson S.L. (1994), Int. J. Mass Spectr. Ion Proc. **138**, 173

Basir Y., Anderson S.L. (1995), Chem. Phys. Lett. **243**, 45

Beck R.D., St. John P.M., Alvarez M.M., Diederich F., Whetten R.L. (1991), J. Phys. Chem. **95**, 8402

Beck R.D., Rockenberger J., Weis P., Kappes M.M. (1996), J. Chem. Phys. **104**, 3658

Blaudeck P., Frauenheim Th., Busmann H.-G., Lill Th. (1994), Phys Rev. B **49**, 11409

Bock R. (1980): *Heavy Ion Collisions*, vols. 1-3 (North-Holland, Amsterdam)

Brenner D.W. (1990), Phys. Rev. B **42**, 9458

Brink C., Andersen L.H., Hvelplund P., Yu D.H. (1994), Z. Phys. D **29**, 45

Busmann H.-G., Lill T., Hertel I.V. (1991), Chem. Phys. Lett. **187**, 459; Busmann H.-G., Lill Th., Reif B., Hertel I.V., Maguire H.G. (1993), J. Chem. Phys. **98**, 7574

Caldwell K.A., Giblin D.E., Gross M.L. (1992), J. Am. Chem. Soc. **114**, 3743

Callahan J.H., Ross M.M., Weiske T., Schwarz H. (1993a), J. Phys. Chem. **97**, 20

Callahan J.H., Somogyi A., Wysocki V.H. (1993b), Rapid Comm. Mass Spectr. **7**, 693

Campbell E.E.B., Ehlich R., Hielscher A., Frazao J.M.A, Hertel I.V. (1992), Z. Physik D **23**, 1

Campbell E.E.B., Schyja V., Ehlich R., Hertel I.V. (1993), Phys. Rev. Lett. **70**, 263

Campbell E.E.B. (1996), Chem. Phys. Lett. **253**, 261

Car R., Parinello M. (1985), Phys. Rev. Lett. **55**, 2471

Christian J.F., Wan Z., Anderson S.L. (1992), J. Phys. Chem. **96**, 3574

Christian J.F., Wan Z., Anderson S.L. (1993), J. Chem. Phys. **99**, 3468

Cui F.Z., Liao D.X., Li H.D. (1994), Phys. Lett. A **195**, 156

Doyle R.J., Ross M.M. (1991), J. Phys. Chem. **95**, 4954

Ehlich R., Campbell E.E.B., Knospe O., Schmidt R. (1993), Z. Phys. D **28**, 153

Ehlich R., Westerburg A.M., Campbell E.E.B. (1996a), J. Chem. Phys. **104**, 1900

Ehlich R., Knospe O., Schmidt R. (1996b), J. Phys. B **30**, 5429

Finch C.D., Popple R.A., Nordlander P., Dunning F.B. (1995), Chem. Phys. Lett. **244**, 345

Gaber H., Hiss R., Busmann H.-G., Hertel I.V. (1992), Z. Phys. D **24**, 307

Galli G., Mauri F. (1994), Phys. Rev. Lett. **73**, 3471

Hohmann H., Callegari C., Furrer S., Grosenick D., Campbell E.E.B., Hertel I.V. (1994), Phys. Rev. Lett. **73**, 1919; Hohmann H., Ehlich R., Furrer S., Kittelmann O., Ringling J., Campbell E.E.B. (1995), Z. Phys. D **33**, 143

Hüfner J., Mukhopadhyay D. (1986), Phys. Lett. B **173**, 373

Hunter J.M., Fye J.L., Boivin N.M., Jarrold M.F. (1994), J. Phys. Chem. **98**, 7440

Hvelplund P., Andersen L.H., Haugen H.K., Lindhard J., Lorents D.C., Malhotra R., Ruoff R. (1992), Phys. Rev. Lett. **69**, 1915

Hvelplund P., Andersen L.H., Brink C., Yu D.H., Lorents D.C., Ruoff R. (1994), Z. Phys. D **30**, 323

Javahery G., Petrie S., Wang J., Bohme D.K. (1992), Int. J. Mass Spectr. Ion Proc. **120**, R5

Jones A.C., Dale M.J., Banks M.R., Gosney I., Langridge–Smith (1993), Mol. Phys. **80**, 583

Kaplan T., Rasolt M., Karimi M., Mostoller M. (1993), J. Phys. Chem. **97**, 6124

Kim S.G., Tomanek D. (1994), Phys. Rev. Lett. **72**, 2418

Kleiser R., Sprang H., Furrer S., Campbell E.E.B (1993), Z. Phys. D **28**, 89

Knospe O., Glotov A.V., Seifert G., Schmidt R. (1996a), J. Phys. B **29**, 5163

134

Knospe O., Schmidt R. (1996a), Z. Phys. D **37**, 85

Knospe O., Schmidt R. (1996b), Phys. Rev. A **54**, 1154

Kolodney E., Tsipinyuk B., Budrevich A. (1995), J. Chem. Phys. **102**, 9263

LeBrun T., Berry H.G., Cheng S., Dunford R.W., Esbensen H., Gemmell D.S., Kanter E.P., Bauer W. (1994), Phys. Rev. Lett. **72**, 3965

Levine R.D., Bernstein R.B. (1987): *Molecular Reaction Dynamics and Chemical Reactivity*, (2nd ed., Oxford University Press, New York)

Lill Th., Busmann H.-G., Reif B., Hertel I.V. (1992), Appl. Phys. A **55**, 461; Lill Th., Busmann H.-G., Hertel I.V. (1993a), Z. Phys. B **91**, 267

Lill Th., Lacher F., Busmann H.G., Hertel I.V. (1993b), Phys. Rev. Lett. **71**, 3383

Lill Th., Busmann H.-G., Reif B., Hertel I.V. (1994), Surf. Sc. **312**, 124; Lill Th., Busmann H.-G., Lacher F., Hertel I.V. (1995), Chem. Phys. **193**, 199; Lill Th., Busmann H.-G., Lacher F., Hertel I.V. (1996), Int J. Mod. Phys. **10**, 11

Long X., Graham R.L., Lee C., Smithline S. (1994), J. Chem. Phys. **100**, 7223

Lorents D.C., Yu, D.H., Brink C., Jensen N., Hvelplund P. (1995), Chem. Phys. Lett. **236**, 141

McHale K.J., Polce M.J., Wesdemiotis C. (1995), J. Mass Spectrosc. **30**, 33

Mitzner R., Winter B., Kusch Ch., Campbell E.E.B., Hertel I.V. (1995), Z. Phys. D **37**, 89

Moliere G. (1947), Z. Naturforschung **2a**, 133

Mowrey R.C., Brenner D.W., Dunlap B.I., Mintmire J.W., White C.T. (1991), J. Phys. Chem. **95**, 7138

Mowrey R.C., Ross M.M., Callahan J.H. (1992), J. Phys. Chem. **96**, 4755

Murry R.L., Scuseria G.E. (1994), Science **263**, 791

O'Brien, Heath J.R., Curl R.F., Smalley R.E. (1988), J. Chem. Phys. **88**, 220

O'Connor D.J., MacDonald R.J. (1977), Radiation Effects **34**, 247

Ohno K., Maruyama Y., Esfarjani K., Kawazoe Y., Sato N., Hatakeyama R., Hirata T., Niwano M. (1996), Phys. Rev. Lett. **76**, 3590

Rauth T., Echt O., Scheier P., Märk T.D. (1995), Chem. Phys. Lett. **247**, 515

Robertson D.H., Brenner D.W., White C.T. (1995), J. Phys. Chem. **99**, 15721

Rohmund F., Campbell E.E.B. (1995), Chem. Phys. Lett. **245**, 237

Rohmund F., Campbell E.E.B., Knospe O., Seifert G., Schmidt R. (1996a), Phys. Rev. Lett. **76**, 3289

Rohmund F., Glotov A.V., Hansen K., Campbell E.E.B. (1996b), J. Phys. B **29**, 5143

Ross M.M., Callahan J.M. (1991), J. Phys. Chem. **95**, 5720

Saalmann U., Schmidt R. (1996), Z. Phys. D **38**, 153

Scheid W., Müller H., Greiner W. (1974), Phys. Rev. Lett. **32**, 741

Scheidemann A.A., Kresin V.V., Knight W.D. (1994), Phys. Rev. A **49**, R4293

Scheier P., Dünser B., Wörgötter R., Lezius M., Robl R., Märk T.D. (1994), Int. J. Mass Spectr. Ion Proc. **138**, 77

Schmidt R., Lutz H.O. (1992), Phys. Rev. A **45**, 7981

Schmidt R., Lutz H.O. (1993), Phys. Lett. A **183**, 338

Schmidt R., Lutz H.O. (1995), Comments on Atomic and Molecular Physics **31**, 461

Schmidt R., Seifert G., Lutz H.O. (1991), Phys. Lett. A **158**, 231; Schmidt R., Seifert G., Lutz H.O (1992), in *Nuclear Physics Concepts in the Study of Atomic*

*Cluster Physics*, ed. by R. Schmidt, H.O. Lutz and R.M. Dreizler, Lecture Notes in Physics, Vol. **404** (Springer, Berlin 1992), p. 128

Schmidt R., Schulte J., Knospe O., Seifert G. (1994), Phys. Lett. A **194**, 101

Schrödinger E. (1926), Naturwiss. **14**, 137

Schulte J., Knospe O., Seifert G., Schmidt R. (1995), Phys. Lett. A **198**, 51

Seifert G., Eschrig H., Bieger W. (1986), Z. Phys. Chem. (Leipzig) **267**, 529

Seifert G., Schmidt R. (1992a), J. of Modern Physics B **96**, 3845

Seifert G., Schmidt R. (1992b), New J. Chem. **16**, 1145

Seifert G., Schulte J. (1994), Phys. Lett. A **188**, 365

Shen H., Hvelplund P., Mathur D., Barany A., Cederquist H., Selberg N., Lorents D.C. (1995), Phys. Rev. A **52**, 3847

Stöcker H., Greiner W. (1986), Phys. Rep. **137**, 277

Strout D.L., Murry R.L., Xu C., Eckhoff W.C., Odom K., Scuseria G.E. (1993), Chem. Phys. Lett. **214**, 576

Takayama M., Shinohara H. (1993), Int. J. Mass Spectr. Ion Proc. **123**, R7

Tersoff J. (1988), Phys. Rev. B **37**, 6991; Phys. Rev. Lett. **61**, 2879

Thumm U., J. Phys. B **27**, 3515 (1994); J. Phys. B **28**, 91 (1995)

Thumm, U., Bastug T., Fricke B. (1995), Phys. Rev. A **52**, 2955

Toglhofer K., Aumayr F., Kurz H., Winter H.P., Scheier P., Märk T.D. (1993), J. Chem. Phys. **99**, 8254

Völpel R., Hofmann G., Steidl M., Stenke M., Schlapp M., Trassl R., Salzborn E. (1993), Phys. Rev. Lett. **71**, 3439

Walch B., Cocke C.L., Völpel R., Salzborn E. (1994), Phys. Rev. Lett. **72**, 1439

Wan Z., Christian J.F., Anderson S.L. (1992a), J. Chem. Phys. **96**, 3344

Wan Z., Christian J.F., Anderson S.L. (1992b), Phys. Rev. Lett. **69**, 1352; Wan Z., Christian J.F., Basir Y., Anderson S.L. (1993), J. Chem. Phys. **99**, 5858

Weis P., Rockenberger J., Beck R.D., Kappes M.M. (1996), J. Chem. Phys. **104**, 3629

Weiske T., Böhme D.K., Hrusak J., Krätschmer W., Schwarz H. (1991), Angew. Chem. Int. Ed. Engl. **30**, 884; Weiske T., Hrusak J., Böhme D.K., Schwarz H. (1991), Chem. Phys. Lett. **186**, 459; Weiske T., Böhme D.K., Schwarz H. (1991), J. Phys. Chem. **95**, 8451

Wörgötter R., Dünser B., Scheier P., Märk T.D. (1994), J. Chem. Phys. **101**, 8674

Wurz P., Lykke K.R. (1994), Chem. Phys. **184**, 335

Yeretzian C., Hansen K., Diederich F., Whetten R.L. (1992), Nature **359**, 44

Yeretzian C., Hansen K., Beck R.D., Whetten R.L. (1993), J. Chem. Phys. **98**, 7480; Yeretzian C., Beck R.D., Whetten R.L. (1994), Int. J. Mass Spectr. Ion Proc. **135**, 79

Young A.B., Cousins L.M., Harrison A.G. (1991), Rap. Comm. Mass Spectr. **5**, 226

Yueyuan Xia, Yuelin Xing, Chunyu Tan, Liangmo Mei (1995), Phys. Rev. B **52**, 110

Yueyuan Xia, Yuelin Xing, Chunyu Tan, Liangmo Mei (1996), Phys. Rev. B **53**, 13871; Nucl. Instr. Meth. B **111**, 41

Zhang B.L., Wang C.Z., Chan C.T., Ho K.M. (1993), J. Phys. Chem. **97**, 3134

# Predicting the Properties
# of Semiconductor Clusters

James R. Chelikowsky[1], Serdar Öğüt[1], Igor Vasiliev[1], Andreas Stathopoulos[2], and Yousef Saad[2]

[1] Department of Chemical Engineering and Materials Science, Minnesota Supercomputer Institute, University of Minnesota, Minneapolis, MN 55455, USA
[2] Department of Computer Science, Minnesota Supercomputer Institute, University of Minnesota, Minneapolis, MN 55455, USA

**Abstract.** The electronic and structural properties of semiconductor clusters are examined using pseudopotentials and a *real space* method. Examples of this method are applied to predict the photoemission spectra, the vibrational modes, and the polarizabilities of Si and Ge clusters.

## 1 Introduction

Suppose we perform the following thought experiment. Let us make a perfect crystalline solid by starting from one atom and adding one atom at a time. Initially, our "crystal" would be composed of a few interacting atoms and would be far removed from a macroscopic crystal. With a few hundreds of atoms, the crystal, or better, the *cluster* of atoms, would still not resemble a macroscopic crystal. The surface of such a system would dominate its properties. The structural properties and electronic properties would be very different from a "real" crystal owing to unsaturated bonds at the surface, and the development of energy bands would be far from complete. The spacing between the electronic levels might be much larger than $kT$, and removed from the quasi-continuous limit of a true energy band. Also, the thermodynamic properties of the cluster would be quite different, *e.g.*, the cluster might melt well below the bulk melting point of the crystal. This thought experiment illustrates an important point: *Size* is a variable which can be used to alter or tailor the properties of matter.

The role of *size* in modifying the properties of a material has not been exploited until recently (Bloomfield et al., 1985, Jena et al., 1987, Alford et al., 1990, Jarrold, 1991, Chelikowsky and Louie, 1996). The task of nalyzing and preparing small assemblages of atoms, or clusters, presents a number of theoretical and experimental challenges. By definition, clusters of atoms are stable only in isolation. Maintaining an isolated system, and simultaneously probing the system, requires a highly sophisticated experimental set-up. One example of such an experimental technique allows clusters to be deposited on an inert substrate, and then probed with photons (Honea et al., 1993). Another experimental procedure is to produce a cluster beam, and to perform to the structural energy of the configuration. As the cluster is cooled, only energetically favorable structures are sampled. Provided the cluster is cooled slowly, the resulting structure at zero temperature should be close to the true ground state. In practice, once a cluster exceeds a dozen atoms or so, one can never cool the system at a rate which insures the final structure is the ground state structure as opposed to a local, or metastable structure (Ballone et al., 1988, Rötlisberg et al., 1991). Simulated annealing times are many orders of magnitude shorter than experimental annealing times, and this situation is not likely to change in the near future.

The accuracy of theoretical methods and dynamical effects is another issue as the cluster size and the number of competing structures increases. Even for some small clusters, such as $Si_6$, these effects are already of the same order of magnitude as the energy difference between the lowest energy isomers (Raghavachari, 1986). As such, comparison between theory and experiment is often essential to identify the relevant isomers.

In this Chapter, we will illustrate some new theoretical procedures to determine the structure of semiconductor clusters, and we make comparisons with a variety of experiments: photoemission, Raman spectroscopy, and polarizability measurements. We focus on semiconductor clusters because of their technological importance and the difficulty these clusters present in terms of predicting their structural properties. Specifically, the covalent nature of the bond in semiconductors precludes the use of simple interatomic forces.

## 2 Theoretical Methods

### 2.1 Langevin Dynamics: Isothermal Simulations and Simulated Annealing

In Langevin dynamics, the ionic positions, $R_j$, evolve according to

$$M_j \, \ddot{R}_j = F(\{R_j\}) - \gamma M_j \, \dot{R}_j + G_j \tag{1}$$

where $F(\{R_j\})$ is the interatomic force on the $j$-th particle, and $\{M_j\}$ are the ionic masses. The last two terms on the right hand side of Eq. ( 1) are the dissipation and fluctuation forces, respectively. The dissipative forces are defined by the friction coefficient, $\gamma$. The fluctuation forces are defined by random Gaussian variables, $\{G_i\}$, with a white noise spectrum:

$$\langle G_i^\alpha(t) \rangle = 0 \quad \text{and} \quad \langle G_i^\alpha(t) G_j^\alpha(t') \rangle = 2\gamma \, M_i \, k_B \, T \, \delta_{ij} \, \delta(t - t') \tag{2}$$

The angular brackets denote ensemble or time averages, and $\alpha$ stands for the Cartesian component. The coefficient of $T$ on the right hand side of Eq. (2) insures that the fluctuation-dissipation theorem is obeyed, i.e., the work done on the system is dissipated by the viscous medium (Kubo, 1966, Risken, 1984).

Langevin molecular dynamics coupled to the simulated annealing procedure can provide a general tool for complex structural optimization (Adelman and Garrison, 1976, Doll and Dion, 1976, Tully et al., 1979, Biswas and Hamann, 1986). The temperature can be controlled without rescaling the velocities as is often done in Newtonian molecular dynamics. Energy can exchange into and out of the system as required by the temperature of the heat bath. Simulated annealing need not follow each time step of the "natural evolution" of the physical system. Annealing rates can be significantly faster if the dynamics lead to acceptable "shortcuts" relative to the true evolution of a cluster anneal.

Langevin dynamics can also be used for *isothermal simulations*. The heat bath can play the role of a buffer gas in experimental situations, although the time frame is quite different. The use of Langevin dynamics as a thermostat can be rigorously justified (Binggeli and Chelikowsky, 1994) in the same sense as a Nosé-Hoover thermostat (Nosé, 1984, Hoover, 1985). Although Langevin dynamics is not appropriate for following the "physical" dynamics of the system such as vibrational modes, it is an appropriate isothermal simulation procedure for *thermodynamic properties*.

Molecular dynamics simulations sample the configuration space by collectively moving the particles, in contrast to Monte Carlo simulations. Also, molecular dynamics simulations can locate minima in a potential energy surface by exploiting the interatomic forces in contrast to Monte Carlo methods. On the other hand, the stochastic nature of the random forces present in Langevin molecular dynamics helps the system to escape from metastable states in a manner reminiscent of "uphill" moves in Monte Carlo simulations (Kirkpatrick et al., 1983).

Determining the interatomic forces, $\mathbf{F}$, in Eq. (1) is the most demanding issue in implementing Langevin dynamics. There are two general approaches used in the literature. One approach is to compute the forces from interatomic potentials which have been fit to some data base. The data base can be constructed from experimental data or *ab initio* calculations. In either case, the particular form for the potential is usually *ad hoc*. Effects such as charge transfer, rehybridization, and Jahn-Teller distortions are very difficult to incorporate into such classical potentials. The other approach is to use "quantum forces." This approach is more accurate, but it is more computationally intensive when compared to interatomic potential approaches. However, the computational effort in determining the quantum forces is not nearly as great as a few years ago. A number of new algorithms have been developed to expedite the evaluation of the quantum forces as outlined in the next two sections.

## 2.2 Computing Quantum Interatomic Forces

Within the local density approximation (LDA) (Hohenberg and Kohn, 1964, Kohn and Sham, 1965), the total ground state energy may be expressed as photoemission measurements on the clusters within the beam (Cheshnovsky et al., 1987). In both cases, extracting a sufficient "signal to noise" ratio is a real problem. However, as semiconductor devices approach nanostructural limits, e.g., quantum dots, questions of "size" and the role it plays in controlling electronic properties becomes more than just an academic issue.

Before any accurate theoretical calculations can be performed for a cluster, the atomic structure must be known. However, determining the atomic structure of clusters is a formidable exercise as it is generally believed that semiconductor clusters undergo important surface reconstructions relative to bulk crystalline fragments. Methods for structural predictions are confronted with major hurdles when applied to clusters. Serious problems arise from the existence of multiple local minima in the potential-energy-surface of these systems.

While it is appealing to consider empirical force fields, or interatomic potentials to compute the structural properties of clusters, these approaches require careful construction and deep insight into the nature of the chemical bond. Since clusters often contain atoms in "unusual" configurations, it may be incorrect to transfer interatomic interactions from known crystalline to cluster environments. For example, silicon atoms are four fold coordinated in a crystalline environment at ambient conditions, but may exhibit two, three, and possibly five fold coordinated states in clusters (Raghavachari, 1986, Raghavachari, 1990, Raghavachari and Rohlfing 1991, Binggeli et al., 1992, Binggeli and Chelikowsky, 1994). For this reason, it is useful to concentrate on *ab initio* methods. Such methods will allow one to determine structural energies in an accurate, albeit computationally intensive manner.

One promising procedure for calculating structural energies is based on *ab initio* pseudopotentials constructed within the local density approximation (Chelikowsky and Cohen, 1992). Within the local density approximation, if we are given the spatial and energetic distributions of the valence electrons, then we can compute the electronic energy for a given structure (Hohl et al., 1988, Ballone et al., 1988, Kawai and Weare, 1990, Yi and Bernholc, 1991, Kumar and Car, 1991, Rötlisberg et al., 1991, Chelikowsky and Cohen, 1992). The pseudopotential approximation effectively removes the chemically inert core electrons from the problem. The resulting wave functions are smoothly varying since the core states have been excluded. Such wave functions permit the efficient application of simple approaches such as a plane wave basis or a real space grid.

Given a formalism to compute structural energies, there remains a serious issue in determining which structures are energetically and kinetically viable. For cluster sizes exceeding a few atoms, one generally relies on simulated annealing procedures for global geometry optimization (see e.g., Car et al., 1987, Binggeli and Chelikowsky, 1994). In this method, the cluster is heated to a high temperature and gradually cooled. Initially, the "hot" cluster samples numerous configurations in the dynamical simulation with little regard follows:

$$E_{tot} = T[\rho] + E_{e-i}(\mathbf{R}_a, [\rho]) + E_{hart}[\rho] + E_{xc}[\rho] + E_{i-i}(\mathbf{R}_a) \qquad (3)$$

where $T[\rho]$ is the kinetic energy, $E_{e-i}$ $(\mathbf{R}_a, [\rho])$ is the ionic potential energy, $E_{hart}[\rho]$ is the Hartree potential energy, $E_{xc}[\rho]$ is the exchange-correlation energy (see e.g., Perdew and Zunger, 1981), $E_{i-i}(\mathbf{R}_a)$ is the inter-ionic core interaction energy, $\rho(\mathbf{r}) = \sum_n |\psi_n(\mathbf{r})|^2$ is the ground state valence charge density where the sum is over occupied states, and $\psi_n(\mathbf{r})$ are the ground state wave functions (Ihm et al., 1979).

In Eq. (3), the contributions from the electron-ion and ion-ion interactions are the only two parts which have explicit dependence on the nuclear coordinates. Since the Hellmann-Feynman (Feynman, 1939) theorem asserts that the first-order change in the wave functions does not contribute to the forces, only the $E_{e-i}$ and $E_{i-i}$ terms are relevant to the interatomic forces. The total force, $F_a^\alpha$, on an atom located at $\mathbf{R}_a$ in the $\alpha$ direction for a finite system is,

$$F_a^\alpha = -\frac{dE_{tot}}{dR_a^\alpha} = -\frac{\partial E_{e-i}}{\partial R_a^\alpha} - \frac{\partial E_{i-i}}{\partial R_a^\alpha} \qquad (4)$$

The inter-ionic core interaction is simply the point-charge point-charge interaction under the frozen core approximation. It is the direct pair summation of Coulomb interactions for an isolated system, and an Ewald summation for a periodic system.

A complicating issue is that the ionic term is described by a non-local ionic pseudopotential (Troullier and Martins, 1991). The interactions between valence electrons and pseudo-ionic cores may be separated into a local potential and a Kleinman and Bylander (Kleinman and Bylander, 1982) form of a non-local pseudopotential in *real space* (Troullier and Martins, 1991, Chelikowsky et al., 1994, Chelikowsky et al., 1994),

$$V_{ion}(\mathbf{r})\psi_n(\mathbf{r}) = \sum_a V_{loc}(|\mathbf{r}_a|)\psi_n(\mathbf{r}) + \sum_{a,\,lm} K_{n,lm}^a u_{lm}(\mathbf{r}_a)\Delta V_l(r_a) \qquad (5)$$

$$K_{n,lm}^a = \frac{1}{<\Delta V_{lm}^a>} \int u_{lm}(\mathbf{r}_a)\Delta V_l(r_a)\psi_n(\mathbf{r})d^3r, \qquad (6)$$

and $<\Delta V_{lm}^a>$ is the normalization factor,

$$<\Delta V_{lm}^a> = \int u_{lm}(\mathbf{r}_a)\Delta V_l(r_a)u_{lm}(\mathbf{r}_a)d^3r, \qquad (7)$$

where $\mathbf{r}_a = \mathbf{r} - \mathbf{R}_a$, and the $u_{lm}$ are the atomic pseudopotential wave functions of angular momentum and azimuthal quantum numbers, $(lm)$, from which the $l$ dependent ionic pseudopotential $V_l(r)$ are generated. $\Delta V_l(r) = V_l(r) - V_{loc}(r)$ is the difference between the $l$ component of the ionic pseudopotential and the local ionic potential.

The energy from the electron-ion interaction, $E_{e-i}$ can be obtained by using Eq. (5) as,

$$E_{e-i} = \sum_a \int \rho(\mathbf{r}) V_{loc}(r_a) d^3 r + \sum_{a,n,lm} <\Delta V_{lm}^a > [K_{n,lm}^a]^2 \qquad (8)$$

where the sum on $n$, is over the occupied states. Combining Eq. (4) and Eq. (6), one can get an expression for the force,

$$F_a^\alpha = \int \rho(\mathbf{r}) \frac{\partial V_{loc}(r_a)}{\partial r_a^\alpha} d^3 r + 2 \sum_{n,lm} <\Delta V_{lm}^a > K_{n,lm}^a \frac{\partial K_{n,lm}^a}{\partial r_a^\alpha} - \frac{\partial E_{i-i}}{\partial R_a^\alpha} \qquad (9)$$

The force from the electronic contribution comprises two parts. The first term at the right hand side of Eq. (9) is the contribution from the local ionic potential, and the second term is from the non-local potential.

## 2.3 Real Space Grid Methods for Solving the Eigenvalue Problem: Higher Order Finite Difference

To obtain quantum forces, we need to compute eigenvalues and eigenvectors of a one-electron Schrödinger equation or the Kohn-Sham equation (Kohn and Sham, 1965). Unlike the traditional approach of expressing the wave functions with a basis, $e.g.$, plane waves, we employ a higher-order finite difference approach. A key aspect of our work is the availability of $higher$-$order$ $finite$ $difference$ expansions for the kinetic energy operator, $i.e.$, expansions of the Laplacian. We impose a simple uniform grid on our system where the points are described in a finite domain by $(x_i, y_j, z_k)$. We approximate $\frac{\partial^2 \psi}{\partial x^2}$ at $(x_i, y_j, z_k)$ by

$$\frac{\partial^2 \psi}{\partial x^2} = \sum_{n=-N}^{N} C_n \psi(x_i + nh, y_j, z_k) + O(h^{2N+2}) \qquad (10)$$

where $h$ is the grid spacing and $N$ is a positive integer. This approximation is accurate to $O(h^{2N+2})$ upon the assumption that $\psi$ can be approximated accurately by a power series in $h$ around a grid point. Algorithms are available to compute the coefficients $C_n$ for arbitrary order in $h$ (Fornberg and Sloan, 1994).

With the kinetic energy operator expanded as in Eq. (10), one can set up the Schrödinger equation over a grid. A uniform grid over the three dimensions is employed for this purpose, but this is not a necessary assumption. [1]

---

[1] Non-uniform grids can be employed with real space methods, e.g., Gygi and Galli, 1995, Zumbach et al. 1996, have implemented nonuniform grids. While the use of nonuniform grids allows adaptation to different length scales, it complicates the problem considerably. This is especially true for dynamical simulations where the grids must be updated as the atoms move.

One can obtain $\psi(x_i, y_j, z_k)$ on the grid by solving the secular equation:

$$-\frac{\hbar^2}{2m} \left[ \sum_{n_1=-N}^{N} C_{n_1} \psi_n(x_i + n_1 h, y_j, z_k) + \sum_{n_2=-N}^{N} C_{n_2} \psi_n(x_i, y_j + n_2 h, z_k) \right.$$
$$\left. + \sum_{n_3=-N}^{N} C_{n_3} \psi_n(x_i, y_j, z_k + n_3 h) \right] + [\, V_{ion}(x_i, y_j, z_k) + V_H(x_i, y_j, z_k)$$
$$+ V_{xc}(x_i, y_j, z_k)\,]\, \psi_n(x_i, y_j, z_k) = E_n\, \psi_n(x_i, y_j, z_k). \tag{11}$$

The above equation is a matrix eigenvalue problem. If there are $M$ grid points, the size of the full matrix resulting from the above eigenvalue problem is $M \times M$. Here, $V_{ion}$ is the nonlocal ionic pseudopotential, $V_H$ is the Hartree potential, and $V_{xc}$ is the local density expression for the exchange and correlation potential. Two parameters used in setting up the matrix are the grid spacing $h$ and the order $N$.

The full matrix $H$ for these isolated systems is real, symmetric, and sparse. These attributes can be exploited in expediting the diagonalization procedure. The sparsity of the matrix is a function of the order $N$ to which the kinetic energy is expanded. To solve this eigenvalue problem, we can utilize one of several iterative procedures developed in the literature for sparse matrices, see Parlett and Saad, 1987.

Two popular such procedures are the accelerated subspace iteration and the Lanczos algorithm. These methods consist of projecting the original problem into a small subspace in which standard techniques can be used. For example, in the simplest version of the subspace iteration algorithm, an initial basis $X = [x_0, x_1, \ldots, x_m]$ is chosen and the power $X_k = H_0^k$ is formed for a certain power $k$. Then, a Ritz procedure is applied to $H$ with this matrix, i.e., $X_k$ is orthonormalized into a matrix $Y_k$ and the eigenvalues $\lambda_i$ and eigenvectors $\phi_i$ of the small $m \times m$ matrix $Y_k^T H Y_k$ are computed by a standard method such as the QR algorithm (Parlett and Saad, 1987). The eigenvalues $\lambda_i$ are then used as approximations to the eigenvalues of $H$, and the vectors $Y_k \phi_i$ are used as approximations to the eigenvectors of $H$. The procedure is repeated with $X$ replaced by the set of approximate eigenvectors until convergence is reached. In realistic implementations of this procedure, it is common to use a Chebyshev polynomial $C_k(H)$ instead of the powers $H^k$ to obtain the next basis $X_k$ from $X$. The Lanczos algorithm also utilizes a Ritz procedure, but the basis used consists of the successive powers $H^k v_0, k = 1, \ldots, m$ where $v_0$ is an initial vector. Thus, in this case, the dimension $m$ increases at each step. Both procedures can and should be "preconditioned" to improve convergence rates (Saad, 1992). Preconditioning consists of enhancing a given vector introduced in the new basis by a process which amplifies the desired eigenvector components and dampens the others.

One important observation is that these diagonalization algorithms use the coefficient matrix $H$ only to perform matrix-by-vector products. The sparse matrix $H$ can be stored in one of several sparse formats available (Saad, 1992) which avoid storing the zero elements. Performing matrix-vector products with these formats is inexpensive. However, because of the special structure of the matrix, an even more appealing alternative is to perform these matrix vector products in "stencil" or "operator" form (Ortega, 1992). Indeed, there is no need to store the matrix in any sparse form since the coefficients $C_{n_i}, i = 1, 2, 3; n = -N, N$ in Eq. (11) are constant. As a result, the matrix-by-vector kinetic operations required by the diagonalization routine can be performed by only accessing the desired components of the current vector $\Psi$ and forming a small linear combination using these coefficients $C_{n_i}$. Similarly, the non-local operations are accomplished by performing vector-by-vector operations. This strategy not only saves storage, but also leads to an efficient implementation on most high-performance vector and parallel computers.

Several other issues must be addressed to solve Eq. (11). The first concerns the procedure by which the self-consistent field is constructed. The exchange-correlation potential, $V_{xc}$, is constructed trivially once the charge density has been constructed over the grid. The Hartree potential can be determined by setting up a matrix equation and solving via a conjugate gradient method. In this method, we determine the boundary conditions by exploiting a multipole expansion of the Hartree potential outside the domain which encompasses the cluster. A mixing scheme is often used to expedite the convergence of the self-consistent field (Broyden, 1965).

Another technical issue concerns the nonlocality of the ionic pseudopotential. Usually, a specific component is taken as the local component. For example, one may take $V_{loc} = V_s$, where $V_s$ is the s-component. It is safe to ignore contributions to the potential higher than $l = 1$ for semiconductor clusters such as silicon or germanium. This can been verified by direct comparisons to calculations using a plane wave basis. The integral $K^a_{n,lm}$ involving $\psi_n(x, y, z)$ is performed over a grid:

$$
\int u_{lm}(x, y, z) \, \Delta V_l(x, y, z) \, \dot{\psi}_n(x, y, z) dx dy dz =
$$
$$
\sum_{ijk} u_{lm}(x_i, y_j, z_k) \Delta V_l(x_i, y_j, z_k) \psi_n(x_i, y_j, z_k) h^3. \tag{12}
$$

The local potential resides only on the diagonal of the matrix; as a result only the diagonal part of the matrix needs to be updated during the self-consistency iterations.

In performing calculations for the electronic structure of localized systems, three parameters must be fixed: the size of domain to contain the cluster, the order $N$ for the kinetic energy expansion, and the grid spacing $h$.

One commonly starts the calculations by solving for the electronic structure of the *isolated* atom by direct integration. This allows one to estimate the size and density of the grid. A spherical domain is often chosen to enclose the cluster such that no atom is within ~ 5-10 a.u. [2] of the domain boundary. In the standard finite difference method, the order $N$ is fixed (at $N = 1$) and the mesh-size $h$ is varied to obtain a desired accuracy. To determine the accuracy, the results of the two meshes are compared ($h$ and $h/2$), and an estimate of the error is then determined. A more appropriate mesh-spacing $h$ can then be derived, if necessary. However, since we have some prior knowledge of the eigenvalues and the pseudo-wavefunctions of the atomic pseudopotentials, we can use this information to fix an initial $h$ and vary to the grid spacing to establish the convergence of the system. Finding an appropriate value for $N$ is more complex. Roughly speaking, the value of $N$ and $h$ are coupled. A coarse grid with a large value of $N$ and a fine grid with a small value for $N$ can often yield comparable results. The advantage of keeping $N$ small is that the sparsity of the $H$ matrix is greater than for large $N$. Also, in terms of computational issues, small $N$ requires less communication between grid points than a large value of $N$. The disadvantage of a small value of $N$ is that more grid points are required. As general guideline, we found a value of $N = 4 - 6$ to be optimal.

## 3  Simulated Annealing Simulations for the Structural Properties

Langevin simulated annealing is a convenient method to determine the structure of small clusters. To illustrate the procedure, we consider a germanium cluster of seven atoms.

With respect to the technical details for this example, the initial temperature of the simulation was taken to be 2800 K; the final temperature was taken to be 300 K. The annealing schedule lowered the temperature 500 K each 50 time steps. The time step was taken to be 7 fs. The friction coefficient in the Langevin equation was taken to be $6 \times 10^{-4}$ a.u. After the cluster reached a temperature of 300 K, the clusters were quenched to 0 K. The ground state structure was found through a direct minimization by a steepest descent procedure.

Choosing an initial atomic configuration takes some care. If the atoms are too far apart, they will exhibit Brownian motion and may not form a stable cluster as the simulation proceeds. If the atoms are too close together, they may form a metastable cluster from which the ground state may be kinetically inaccessible. Typically, the initial cluster is formed by a random placement of the atoms with a constraint that any given atom must reside within 1.05 and 1.3 times the dimer bond length of at least one atom.

With respect to computing the quantum forces, the cluster in question is placed in a spherical domain. Outside of this domain, the wave function is required to vanish. The radius of the sphere is such that the outmost atom is

---

[2] We use atomic units, $a.u.$, where $\hbar = e = m = 1$

at least 6 a.u. from the boundary. Initially, the grid spacing was 0.8 a.u. For the final quench to a ground state structure, the grid spacing was reduced to 0.5 a.u. As a rough estimate, one can compare this grid spacing with a plane wave cutoff of $(\pi/h)^2$ or about 30 Ry for h=0.5 a.u.

In Figure 1, we illustrate the simulated anneal for this $Ge_7$ cluster. While the initial cluster contains several of bonds, the structure is still somewhat removed from the ground state. After ~200 time steps, the ground state structure is essentially formed. The ground state of $Ge_7$ is a bicapped pentagon, as is the corresponding structure for the $Si_7$ cluster. The binding energy shown is relative to the isolated Ge atom. We have not included gradient corrections, or spin polarization in our work (Kutzler and Painter, 1992). Therefore, the values indicated are likely to overestimate the binding energies by ~ 20% or so.

In Figure 2, we present the ground state structures for $Ge_n$ for $n \leq 10$. The structures for $Ge_n$ are very similar to $Si_n$. The primary difference resides in the bond lengths. The Si bond length in the crystal is 2.35 Å, whereas in Ge the bond length is 2.44 Å. This difference is reflected in the bond lengths for the corresponding clusters. $Ge_n$ bond lengths are typically a few percent larger than the corresponding $Si_n$ clusters.

It should be emphasized that this annealing simulation is an optimization procedure. As such, other optimization procedures may be used to extract the minimum energy structures. Recently, a genetic algorithm has been used to examine carbon clusters (Deaven and Ho, 1995). In this algorithm, an initial set of clusters is "mated" with the lowest energy offspring "surviving". By examining several thousand generations, it is often possible to extract a reasonable structure for the ground state. The genetic algorithm has some advantages over a simulated anneal, especially for clusters which contain more than ~20 atoms. One of these advantages is that kinetic barriers are more easily overcome. However, the implementation of the genetic algorithm is more involved than an annealing simulation, e.g., in some cases "mutations," or *ad hoc* structural rearrangements, must be introduced to obtain the correct ground state.

# 4  Photoemission Spectra

One of the earliest experiments performed to examine the electronic structures of small semiconductor clusters is a photoemission study on negatively charged $Si_n$ and $Ge_n$ ($n \leq 12$) clusters (Cheshnovsky et al., 1987). The photoemission spectra obtained in this work were used to gauge the energy gap between the highest occupied state and the lowest unoccupied state. Large gaps were assigned to the "magic number" clusters, while other clusters appeared to have vanishing gaps. Unfortunately, the first theoretical estimates (Tomanek and Schlüter, 1986) for these gaps showed substantial disagreements with the measured values. It was proposed by Cheshnovsky et al., 1987,

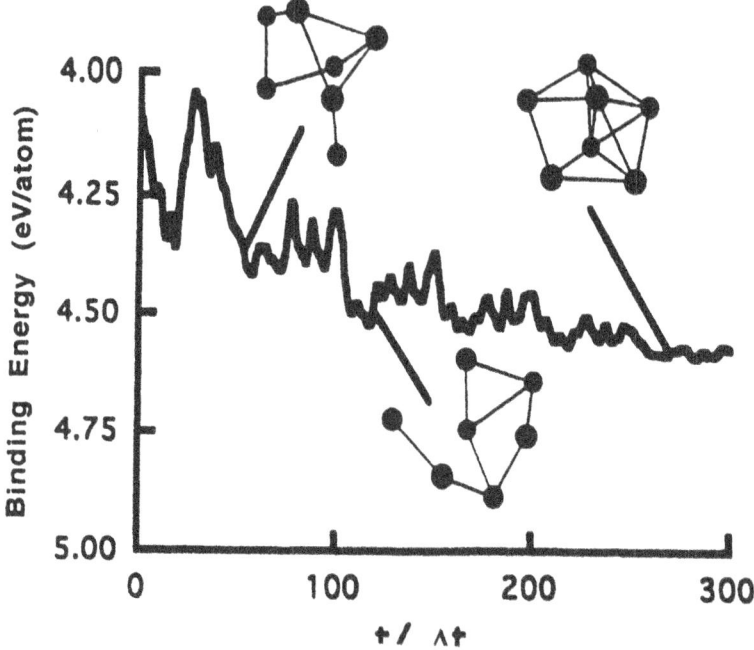

**Fig. 1.** Binding energy of Ge$_7$ during a Langevin simulation. The initial temperature is 2800 K; the final temperature is 300 K. Bonds are drawn for interatomic distances of less than 2.5Å. The time step is 7 fs.

that elaborate calculations including transition cross sections and final states were necessary to identify the cluster geometry from the photoemission data. However, these data were first interpreted in terms of the gaps obtained for *neutral* clusters. It was later demonstrated that *atomic relaxations* within the *charged* cluster are important in analyzing the photoemission data (Binggeli and Chelikowsky, 1994). In particular, atomic relaxations as a result of charging may change dramatically the electronic spectra of certain clusters. These charge induced changes in the gap were found to yield very good agreement with the experiment.

We illustrate this situation for the calculated density of states for Ge$_5^-$ in Figure 3, and compare to the photoemission experiment. The neutral cluster yields a gap of 2.13 eV; however, when charged this gap closes in the simulated spectra in agreement with the photoemission spectra. The calculated spectrum was generated in the constant matrix approximation by using an average density of states over a 3 ps isothermal Langevin molecular dynamics simulation. One does not expect transition cross sections to modify qualitatively the spectrum, since the electronic levels involved here mostly derive from the same type of Ge atomic 4$p$ states.

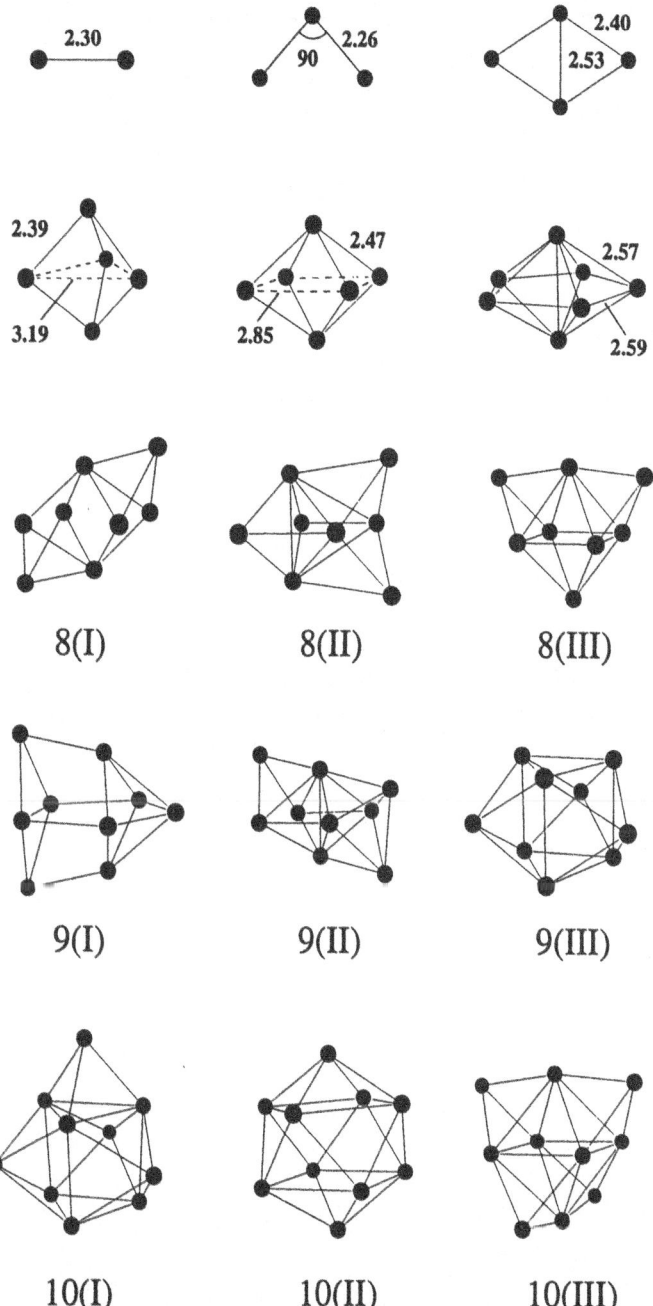

**Fig. 2.** Ground state geometries and some low-energy isomers of $Ge_n$ ($n \leq 10$) clusters. Interatomic distances (in Å) are given for clusters with $n \leq 7$. For $n > 8$, the lowest energy isomer is given by (I).

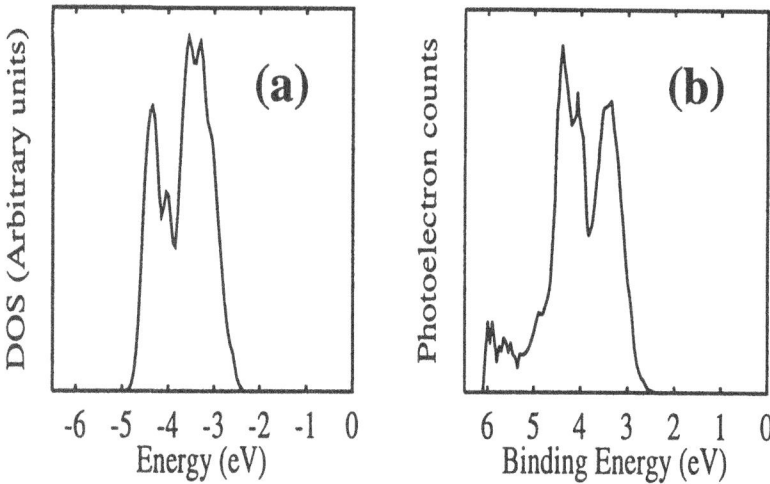

**Fig. 3.** (a) Calculated density of states for $Ge_5^-$. (b) Experimental photoemission spectra (Cheshnovsky et al., 1987).

In some cases, it is possible to distinguish between structures by comparing the computed and calculated density of states. For example, in the case of $Si_6^-$ (or $Ge_6^-$), two structures, the bicapped tetrahedron and a distorted octahedral structure, are nearly identical in energy (*i.e.*, they differ by $\sim$0.01 eV/atom). However, the photoemission spectrum agrees with the density of states for the bicapped tetrahedron, and is strongly at variance with the density of states for the distorted octahedral (Binggeli and Chelikowsky, 1994). structure.

## 5 Polarizabilities

Recently polarizability measurements (Schäfer et al., 1996) have been performed for small semiconductor clusters. These measurements allow us to compare our computed values with experiment, although it should be noted that the measurements are done on clusters somewhat larger than those which can easily be handled at present by theory.

The polarizability tensor, $\alpha_{ij}$, is defined as the second derivative of the energy with respect to electric field components. For a noninteracting quantum mechanical system, the expression for the polarizability can be easily obtained by using second order perturbation theory where the external electric field, $\mathcal{E}$, is treated as a weak perturbation.

Within the density functional theory, since the total energy is not the sum of individual eigenvalues, the calculation of polarizability becomes a nontrivial task. One approach is to use density functional perturbation theory which has been developed recently in Green's function and variational formulations (Baroni et al., 1987, Gonze et al., 1992).

Another approach, which is very convenient for handling the problem for *confined* systems, like clusters, is to solve the full problem exactly within the one electron approximation. In this approach, the external ionic potential $V_{ion}(\mathbf{r})$ experienced by the electrons is modified to have an additional term given by $-e\mathcal{E} \cdot \mathbf{r}$. The Kohn-Sham equations are solved with the full external potential $V_{ion}(\mathbf{r}) - e\mathcal{E} \cdot \mathbf{r}$. For quantities like polarizability, which are derivatives of the total energy, one can compute the energy at a few field values, and differentiate numerically. Real space methods are very suitable for such calculations on confined systems, since the position operator $\mathbf{r}$ is not ill-defined, as is the case for supercell geometries in plane wave calculations. In Table 1,

**Table 1.** Static dipole moments and average polarizabilities of small silicon and germanium clusters.

| Silicon | | | Germanium | | |
|---|---|---|---|---|---|
| cluster | $|\mu|$ (D) | $\langle\alpha\rangle$ ($\text{Å}^3/atom$) | cluster | $|\mu|$ (D) | $\langle\alpha\rangle$ ($\text{Å}^3/atom$) |
| $Si_2$ | 0 | 6.29 | $Ge_2$ | 0 | 6.67 |
| $Si_3$ | 0.33 | 5.22 | $Ge_3$ | 0.43 | 5.89 |
| $Si_4$ | 0 | 5.07 | $Ge_4$ | 0 | 5.45 |
| $Si_5$ | 0 | 4.81 | $Ge_5$ | 0 | 5.15 |
| $Si_6$ (I) | 0 | 4.46 | $Ge_6$ (I) | 0 | 4.87 |
| $Si_6$ (II) | 0.19 | 4.48 | $Ge_6$ (II) | 0.14 | 4.88 |
| $Si_7$ | 0 | 4.37 | $Ge_7$ | 0 | 4.70 |

we present some recent calculations for the polarizability of small Si and Ge clusters. (This procedure has recently been extended to heteropolar clusters such as $Ga_mAs_n$, see Vasiliev et al., 1997) It is interesting to note that some of these clusters have permanent dipoles. For example, $Si_6$ and $Ge_6$ both have nearly degenerate isomers. One of these isomers possesses a permanent dipole, the other does not. Hence, in principle, one might be able to separate the one isomer from the other via an inhomogeneous electric field.

# 6    Vibrational Modes

Experiments on the vibrational spectra of clusters can provide us with very important information about their physical properties. Recently, Raman experiments have been performed on clusters which have been deposited on inert substrates (Honea et al., 1993). Since different structural configurations of a given cluster can possess different vibrational spectra, it is possible to compare the vibrational modes calculated for a particular structure with experiment. If the agreement between experiment and theory is good, this is a necessary condition for the validity of the theoretically predicted structure.

There are two common approaches for determining the vibrational spectra of clusters. One approach is to calculate the dynamical matrix for the ground state structure of the cluster:

$$M_{i\alpha,j\beta} = -\frac{1}{m}\frac{\partial^2 E}{\partial R_i^\alpha \partial R_j^\alpha} = \frac{1}{m}\frac{\partial F_i^\alpha}{\partial R_j^\alpha} \qquad (13)$$

where $m$ is the mass of the atom, $E$ is the total energy of the system, $F_i^\alpha$ is the force on atom $i$ in the direction $\alpha$, $R_i^\alpha$ is the $\alpha$ component of coordinate for atom $i$. One can calculate the dynamical matrix elements by calculating the first order derivative of force versus atom displacement numerically. From the eigenvalues and eigenmodes of the dynamical matrix, one can obtain the vibrational frequencies and modes for the cluster of interest (Jing et al., 1995).

The other approach to determine the vibrational modes is to perform a molecular dynamics simulation. The cluster in question is excited by small random displacements. By recording the kinetic (or binding) energy of the cluster as a function of the simulation time, it is possible to extract the power spectrum of the cluster and determine the vibrational modes. This approach has an advantage for large clusters in that one never has to do a mode analysis explicitly. Another advantage is that anaharmonic modes and mode coupling can be examined. It has the disadvantage in that the simulation must be performed over a long time to extract all the modes.

As a specific example, let us consider the vibrational modes for a small silicon cluster: $Si_4$. The starting geometry was taken to be a planar structure for this cluster as established from a higher order finite difference calculation (Jing et al., 1995).

Table 2. Calculated and experimental vibrational frequencies in a $Si_4$ cluster. See Figure 4 for an illustration of the normal modes. The frequencies are given in $cm^{-1}$.

|  | $B_{3u}$ | $B_{2u}$ | $A_g$ | $B_{3g}$ | $A_g$ | $B_{1u}$ |
|---|---|---|---|---|---|---|
| Experiment (Honea et al., 1993) |  |  | 345 |  | 470 |  |
| Dynamical Matrix (This work) | 160 | 280 | 340 | 460 | 480 | 500 |
| MD simulation (This work) | 150 | 250 | 340 | 440 | 490 | 500 |
| HF (Rohlfing and Raghavachari, 1992) | 117 | 305 | 357 | 465 | 489 | 529 |
| LCAO (Fournier et al., 1992) | 55 | 248 | 348 | 436 | 464 | 495 |

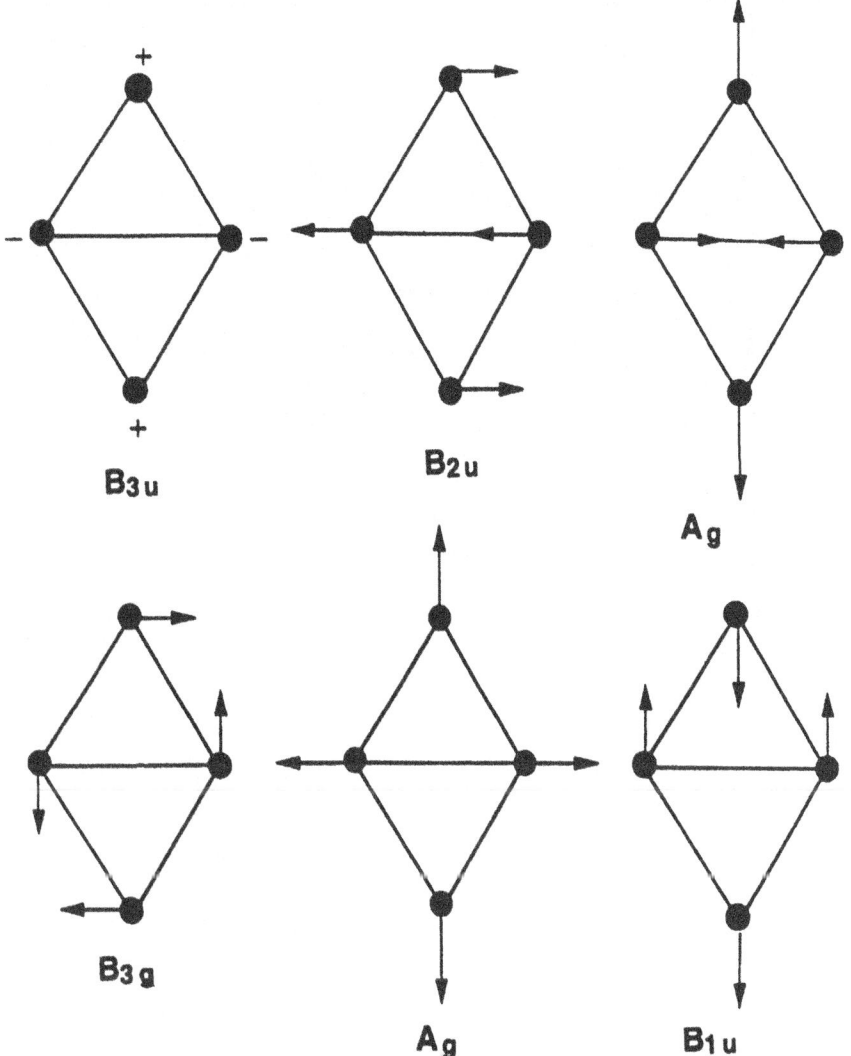

**Fig. 4.** Normal modes for a Si₄ cluster. The + and − signs indicate motion in and out of the plane, respectively.

It is straightforward to determine the dynamical matrix and eigenmodes for this cluster. In Figure 4, the fundamental vibrational modes are illustrated. In Table 2, the frequency of these modes are presented. One can also determine the modes via a simulation. To initiate the simulation, one can perform a Langevin simulation (Binggeli and Chelikowsky, 1994) with a fixed temperature at 300K. After a few dozen time steps, the Langevin simulation is turned off, and the simulation proceeds following Newtonian dynamics with "quantum" forces. This procedure allows a stochastic element to be introduced and establish initial conditions for the simulation without bias toward

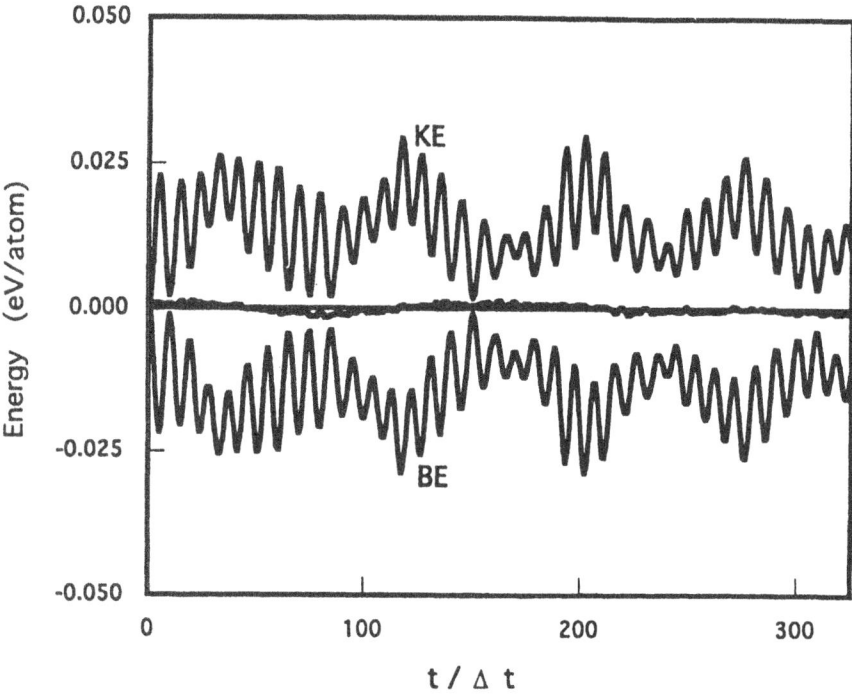

**Fig. 5.** Simulation for a Si$_4$ cluster. The kinetic energy (KE) and binding energy (BE) are shown as a function of simulation time. The total energy (KE+BE) is also shown with the zero of energy taken as the average of the total energy. The time step, $\Delta t$, is 7.38fs.

a particular mode. For this example, time step in the MD simulation was taken to be 3.7 fs, or approximately 150 a.u. The simulation was allowed to proceed for 1000 time steps or roughly 4 ps. The variation of the kinetic and binding energies is given in Figure 5 as a function of the simulation time. Although some fluctuations of the total energy occurs, these fluctuations are relatively small, *i.e.*, less than $\sim 1$ meV, and there is no noticeable drift of the total energy. Such fluctuations arise, in part, because of discretization errors As the grid size is reduced such errors are minimized (Jing et al., 1995). Similar errors can occur in plane wave descriptions using supercells, *i.e.*, the artificial periodicity of the supercell can introduce erroneous forces on the cluster. By taking the power spectrum of either the KE or BE over this simulation time, the vibrational modes can be determined. These modes can be identified with the observed peaks in the power spectrum as illustrated in Figure 6.

A comparison of the calculated vibrational modes from the MD simulation and from a dynamical matrix calculation are listed in Table 2. Overall, the agreement between the two simulations and the dynamical matrix analysis

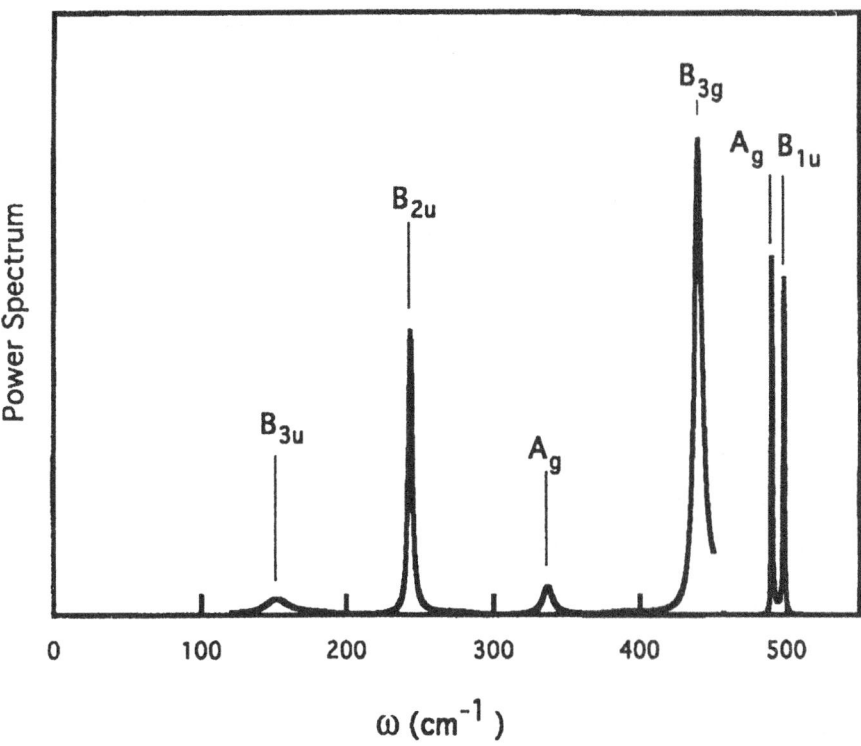

**Fig. 6.** Power spectrum of the vibrational modes of the $Si_4$ cluster. The simulation time was taken to be 2.4 ps. The intensity of the $B_{3g}$ and $(A_g, B_{1u})$ peaks has been scaled by $10^{-2}$.

is quite satisfactory. In particular, the softest mode, *i.e.*, the $B_{3u}$ mode, and the splitting between the $(A_g, B_{1u})$ modes are well replicated in the power spectrum. The splitting of the $(A_g, B_{1u})$ modes is less than 10 cm$^{-1}$, or about 1 meV, which is probably at the resolution limit of any *ab initio* method.

The theoretical values are also compared to experiment. The predicted frequencies for the two $A_g$ modes are surprisingly close to Raman experiments on silicon clusters (Honea et al., 1993). The other allowed Raman line of mode $B_{3g}$ is expected to have a lower intensity and has not been observed experimentally.

The theoretical modes using the formalism outlined here are in good accord (except the lowest mode) with other theoretical calculations given in the Table: an LCAO calculation (Fournier et al., 1992) and a Hartree-Fock (HF) calculation (Rohlfing and Raghavachari, 1992). The calculated frequency of the lowest mode, *i.e.*, the $B_{3u}$ mode, is problematic. The general agreement of the $B_{3u}$ mode as calculated by the simulation and from the dynamical matrix is reassuring. Moreover, the real space calculations agree with the HF value to within $\sim$ 20-30 cm$^{-1}$. On the other hand, the LCAO method yields a value which is $50 - 70\%$ smaller than either the real space or HF

calculations. The origin of this difference is not apparent. For a poorly converged basis, vibrational frequencies are often overestimated as opposed to
the LCAO result which underestimates the value, at least when compared
to other theoretical techniques. Setting aside the issue of the $B_{3u}$ mode, the
agreement between the measured Raman modes and theory for $Si_4$ suggests
that Raman spectroscopy can provide a key test for the structures predicted
by theory.

# 7  Conclusions

We have presented in this review some contemporary calculational methods
for describing the structural and electronic properties of semiconductor clusters. In particular, we illustrated how simulated annealing methods utilizing
quantum forces can be used to predict the ground state structures of clusters
with up to a dozen atoms or so. Moreover, we found the electronic and vibrational properties predicted for the resulting structures to be in very good
accord with experiment.

We also demonstrated the use of *real space* methods for determining electronic states for clusters. These methods have a number of advantages over
the conventional momentum space methods, *e.g.*, polarizabilities of clusters
can determined in a simple and straightforward manner.

At present, the theoretical techniques outlined within this Chapter, and
related methods (Gygi and Galli, 1995, Briggs et al., 1995, Zumbach et al.
1996) are being extended and implemented on highly parallel platforms. This
combination of new algorithms which take advantage of new computer platforms will in the very near future allow one to consider much large clusters,
*e.g.*, clusters of up to 1,000 atoms have been recently examined using these
techniques to examine quantum dots (Öğüt et al., 1997).

# 8  Acknowledgments

We wish to thank the National Science Foundation and the Minnesota Supercomputer Institute for supporting this work. One of us (JRC) would like
to thank the John Simon Guggenheim Foundation for its support.

# References

Adelman S.A., Garrison B.J. (1976): J. Chem. Phys. **65**, 3751.
Alford J.M., Laaksonen R.T., Smalley R.E. (1990): J. Chem. Phys. **94**, 2618.
Ballone P., Andreoni W., Car R., Parrinello M. (1988): Phys. Rev. Lett. **60**, 271.
Baroni S., Gianozzi P., Testa A. (1987): Phys. Rev. Lett. **58**, 1861 (1987).
Binggeli N., Martins J.L., Chelikowsky J.R. (1992): Phys. Rev. Lett. **68**, 2956.
Binggeli N., Chelikowsky J.R. (1994): Phys. Rev. B **50**, 11764.

Biswas R., Hamann D.R. (1986): Phys. Rev. B **34**, 895.

Bloomfield L.A., Freeman R.R., Brown W.L. (1985): Phys. Rev. Lett. **54**, 2246 (1985).

Briggs E.L., Sullivan D.J., Bernholc J. (1995): Phys. Rev. B **52**, R5471.

Broyden C.G. (1965): Math. Comp. **19**, 577.

Car R., Parrinello M., Andreoni W. (1987): *Microclusters*, Sugano S., Nishina Y., Ohnishi S., editors, Springer Series in Materials Science (Springer-Verlag, Berlin), Volume 4, p. 134.

Chelikowsky J.R., Cohen M.L. (1992): "Ab initio Pseudopotentials for Semiconductors," *Handbook on Semiconductors*, Peter Landsburg, editor, (Elsevier, Amsterdam), Volume 1, p.59.

Chelikowsky J.R., Louie S.G., editors (1996): *Quantum Theory of Real Materials*, (Kluwer Press, NY).

Chelikowsky J.R., Troullier N., Saad Y. (1994): Phys. Rev. B **50**, 11355.

Chelikowsky J.R., Troullier N., Wu K., Saad Y. (1994): Phys. Rev. Lett. **72**, 1240.

Cheshnovsky O., Yang S.H., Pettiett C.L., Craycraft M.J., Liu Y., Smalley R.E. (1987): Chem. Phys. Lett. **138**, 119.

Deaven D., Ho K.M. (1995): Phys. Rev. Lett. **75**, 288.

Doll J.D., Dion D.R. (1976): J. Chem. Phys. **65**, 3762.

Feynman R.P. (1939): Phys. Rev. **56**, 340.

Fornberg B., Sloan D. (1994): "A review of pseudospectral methods for solving partial differential equations," Acta Numerica, Iserles A., editor, (Cambridge Press, Cambridge), p. 203.

Fournier R., Sinnott S.B., DePristo A.E. (1992): J. Chem. Phys. **97**, 4149.

Gonze X., Allan D.C., Teter M.P. (1992): Phys. Rev. Lett. **68**, 3603 (1992).

Gygi F., Galli G. (1995): Phys. Rev. B **52**, R2229.

Hohenberg P., Kohn W. (1964): Phys. Rev. **136**, B864.

Hohl D., Jones R.O., Car R., Parrinello M. (1988): J. Chem. Phys. **89**, 6823.

Honea E.C., Ogura A., Muarry C.A., Raghavachari K., Sprenger O., Jarrold M.F., Brown W.L. (1993): Nature **366**, 42.

Hoover W.G. (1985): Phys. Rev. A **31**, 1695.

Ihm J., Zunger A., Cohen M.L. (1979): J. Phys. C **12**, 4409.

Jarrold M.F., (1991): Science **252**, 1085.

Jena P., Rao B.K., Khanna S.N., editors (1987): *Physics and Chemistry of Small Clusters*, (Plenum, NY), 1987.

Jing X., Troullier N., Chelikowsky J.R., Wu K., Saad Y. (1995): Solid State Comm. **96**, 231.

Kawai R., Weare J.H. (1990): Phys. Rev. Lett. **65**, 80.

Kirkpatrick S., Gelatt C.D., Vecchi M.P. (1983): Science **220**, 671.

Kleinman L., Bylander D.M. (1982): Phys. Rev. Lett. **48**, 1425.

Kohn W., Sham L. (1965): Phys. Rev. **140**, A1133.

Kubo R. (1966): Rep. Prog. Theor. Phys. **29**, 255.

Kumar V., Car R. (1991): Phys. Rev. B **44**, 8243.

Kutzler F.W., Painter, G.S. (1992): Phys. Rev. B **45**, 3236.

Nosé, S. (1984): Mol. Phys. **52**, 255.

Öğüt S., Chelikowsky J.R. , Louie S.G. (1997): Phys. Rev. Lett. **79**, 1770, and to be published.

Ortega J.M. (1992): *Introduction to Parallel and Vector Solutions of Linear Systems* (Manchester University Press).

Parlett B.N., Saad Y. (1987): Linear Algebra and Its Applications **88/89**, 575.

Perdew J.P., Zunger A. (1981): Phys. Rev. B **23**, 5048 (1981).

Raghavachari K. (1986): J. Chem. Phys. **84**, 5672.

Raghavachari K. (1990): Phase Transitions **24-26**, 61.

Raghavachari K., Rohlfing C.M. (1991): J. Chem. Phys. **94**, 3670.

Risken H. (1984): *The Fokker-Planck Equation* (Springer-Verlag, Berlin).

Rohlfing C., Raghavachari K., (1992): J. Chem. Phys. **96**, 2114.

Rötlisberg U., Andreoni W., Car R. (1991): J. Chem. Phys. **96**, 1248.

Saad Y. (1992): *Numerical Methods for Large Eigenvalue Problems*, (Halstead Press).

Schäfer R., Schlect S., Woenckhaus J., Becker J.A. (1996): Phys. Rev. Lett. **76**, 471.

Tomanek D., Schlüter, M. (1986): Phys. Rev. Lett. **56**, 1055.

Troullier N., Martins J.L. (1991): Phys. Rev. B **43**, 8861.

Tully J.C., Gilmer G.H., Shugard M. (1979): J. Chem. Phys. **71**, 1630.

Vasiliev I., Öğüt S., Chelikowsky J.R. (1997) : Phys. Rev. Lett. **78**, 4805.

Yi J.Y., Bernholc J. (1991): Phys. Rev. Lett. **67**, 1594.

Zumbach G., Modine N.A., Kaxiras E. (1996): Solid State Comm. bf 99, 57; Modine N.A., Zumbach G., Kaxiras E. (1997): Phys. ReV. B **55**, 10289.

# Structure, Reactivity and Dynamics of Atomic and Molecular Clusters Using Density Functional Theory (DFT) and Other Tools

Dennis R. Salahub[1,3], Ana Martinez[2], Dongqing Wei[3]

[1] Département de Chimie, Université de Montréal, C.P. 6128, Succursale Centre-ville, Montréal, Québec, Canada H3C 3J7
[2] Departamento de Quimica, Division de Ciencias Basicas e Ingenieria, Universidad Autonoma Metropolitana-Iztapalapa, A.P. 55-534, Mexico, D.F. C.P 09340. Mexico
[3] Centre de Recherche en Calcul Appliqué, 5160 Boulevard Décarie, Bureau 400, Montréal, Québec, Canada H3X 2H9

**Abstract.** DFT calculations were carried out to study the structure and spectroscopic properties of transition-metal and hydrated proton clusters. For the latter, the thermodynamic and dynamic properties are also described. These two systems present challenges of computational chemistry, one has a significant electron exchange-correlation problem to deal with, another explores ways to treat weak interactions, namely, hydrogen bonds. Good accord with experimental results has been achieved. These examples reflect the current status of DFT applications to clusters and point to future perspectives.

## 1 Introduction

Interest in the physical and chemical properties of metal clusters [1, 2] has grown rapidly. Developments in both theory and experiments will lead to improved understanding of many properties af atomic aggregates, particularly those that reflect the transition from molecular to bulk behavior.

Much work has been dedicated to the study of transition-metal clusters in an effort to understand the contribution of the d electrons to the bonding. Investigations into the bonding of small transition-metal clusters have been carried out for many years now. Understanding the physical and chemical properties of these aggregates has important consequences for both fundamental and applied areas.

From a fundamental point of view, studying transition-metal clusters allows us to characterize better metal-metal bonds and the associated electronic structure. Knowledge of the electronic structure of transition-metal clusters yields an important component of a general understanding of chemical bonding between transition-metal atoms in various environments. For instance, metal clusters may be viewed as representing a natural bridge between molecules or atoms in the gas phase and solids. Hence, the evolution in the onset of metallic behavior from the gas phase to the condensed phase can be inferred from investigations of the size-dependent electronic structure of clusters.

Concerning the applied areas, the knowledge gained can contribute to a better understanding of chemisorption and catalysis and be helpful in the development of new materials. The reactivity of small metal clusters is a topic of major interest because these clusters can be used as models in the study of the reactivity of metal centers in catalytic processes. A molecular-level understanding of the catalytic activity of a metal is expected to be of considerable value in the design of new generations of industrial catalysts such as supported transition-metal particles, which are required to be cheap but efficient and selective. There is also a fundamental desire to understand the nature of the bonding between various transition metals and different types of ligands. The investigation of physisorption and chemisorption on clusters, as well as the incorporation of atoms and molecules into clusters is very important for catalytic processes.

Transition metal clusters provide a considerable challenge to state-of-the-art experimental and theoretical techniques [3]. Much progress in this area has been made with the aid of beam techniques, which allow the synthesis and characterization of metallic clusters of well defined size. There are relatively few methods currently available for determining the geometrical structure of metal clusters. Theoretical calculations of cluster energetics do not yet predict the most stable structures with complete reliability, since energy differences between various isomers are often less than, or at least comparable to, the estimated errors in the calculated energies. However, in some of the work reviewed here we show that a combination of theory and experiment has allowed, for the first time, unequivocal determination of the geometry and vibrational structure of a gas phase transition-metal cluster.

The study of molecular clusters is also of great interest. Particularly, hydrated proton clusters are of importance in understanding the role of protons in chemically and biologically interesting systems [4, 5, 6, 7, 8]. Advances in theory and the availability of increased computer power have allowed substantial progress to be made towards understanding proton transfers in solution [9, 10, 11] and in clusters [12, 13]. Study of the structural, spectroscopic and other properties of hydrated proton clusters has gained significant attention in recent years not only due to the availability of computational tools for small systems that allow accurate studies, but also because of the unique solvent environment the small system provides, where the reaction dynamics can be very different from that in aqueous solution. We [14, 15] have recently optimized structures for the hydrated proton clusters up to $(H_2O)_8H^+$. For small clusters (less than 5 waters), it is believed that the ground state has been achieved, while, for large clusters, several structures with small energy differences are obtained. Comparison has been made with experimental results of Lee et al. [16, 17]. Generally, the theoretical spectroscopic data is in very good agreement with experiment. For large clusters, the theoretical calculations could give valuable information on contributions of low energy structures so that experimental observation can be properly interpreted.

As the energy states get closer, thermal fluctuation plays a dynamic role to allow various structures to be visited on the time scale of experimental studies. Then real time dynamics takes on great importance for weakly interacting molecular cluster systems, where bond breaking and making may be viewed on the same time scale as the electronic structure change [15]. Molecular dynamics simulation is very useful in generating anharmonic vibrational frequencies for small clusters. We have shown for a small hydrated proton cluster that the frequencies obtained are quite close to the frequencies calculated by traditional quantum chemical methods. The role of such "on-the-fly" dynamics studies of clusters will surely increase in the near future as more and more complex systems and questions are addressed.

As one can see by a glance at the chapters of this book, there is great activity in cluster science on both experimental and theoretical fronts for a wide variety of questions of structure and dynamics. In this chapter, we will treat just a few examples in two classes, covalent (metal) and molecular (H-bonded) clusters. We hope that they will contribute, along with other works described in this volume, to accurately reflecting the state of the art.

## 2 Methodology

Density Functional Theory (DFT) and other ab-initio techniques are able to calculate structures for transition-metal clusters containing up to eight atoms or so [3], or water/hydrated proton clusters with 8 water molecules [18, 14, 15, 19]. The program package deMon-KS [20, 21] is a quantum chemistry program that, within the framework of DFT [21, 22, 23], performs accurate calculations of the properties of organic and organometallic compounds, biologically interesting systems and metal clusters.

Like most conventional ab-initio quantum chemistry packages based on the Hartree-Fock approximation, the molecular orbitals used in deMon-KS are expressed as a Linear Combination of Gaussian-Type Orbitals (LCGTO) and solved by an iterative Self-Consistent Field (SCF) approach. The Gaussian basis sets are similar to those developed for traditional ab-initio methods such as those of Huzinaga et al. [24]. Typically, the orbital basis sets in deMon-KS are obtained by re-optimizing an LCGTO-HF basis for DF calculations [25]. As proposed by Sambe and Felton [26] and Dunlap et al. [27], the charge density and the exchange-correlation potentials are fitted with an LCGTO expansion in an auxiliary basis.

The exchange-correlation energy in deMon-KS is given by a functional that depends only on the electronic density for each spin (a local functional or Local Spin Density (LSD)) or a functional that depends also on the norm of the gradient of the electronic density for each spin (the Generalized Gradient Approximation (GGA)). For an LSD calculation, deMon-KS uses the functional of Vosko, Wilk and Nusair [28]. For GGA calculations, deMon-KS calls upon the correlation functional of Perdew [29] and either the exchange functional of Perdew and Wang [30] or that of Becke [31]. The availability of the gradient-corrected DFT Hamiltonian in deMon-KS allows for accurate calculations of energies for the study of chemical reactions, and of the structure and dynamics of clusters.

Geometries are optimized using analytical energy gradients. To identify each final geometry as a true minimum or a transition state, a vibrational analysis was done using numerical differentiation (two-point differences with a step of 0.02 bohrs) of the gradients. Harmonic frequencies can be obtained by diagonalizing the mass-weighted Cartesian force constant matrix. The force constants are evaluated by numerical differentiation of the analytical gradient using a displacement of 0.02 a.u. from the optimized geometry for all 3N coordinates.

Often geometry optimization for transition-metal clusters are performed at the LSD level, because the results of bond distances and angles for known systems are close to the experimental values. Indeed, in some cases they are somewhat better than the results obtained with gradient-corrected functionals. For some systems, the tendency of the gradient-corrected functional is to overestimate the distances and the angles slightly. However, for relative energies the gradient-corrected functionals are very important because the LSD functional overestimates many energy differences seriously. A methodology that is useful to obtain the transition-metal cluster structures and properties is to perform the geometry optimization and the vibrational analysis at the LSD level, and then to make a single point calculation for the optimized geometry using the gradient-corrected functional. For clusters involving hydrogen bonds [14], the gradient-corrected functional yields better energies, bond lengths and angles as compared with LSD calculations. The LSD approximation is not useful in this arena.

The DFT methods typically yield bond lengths, including those for transition metal-ligand bonds, transition metal-transition metal bonds and hydrogen bonds to an accuracy of a few hundredths of an Angstrom. Relative energies are more difficult to predict and are less completely validated, but with the gradient-corrected functionals, accuracy of several tenths of an eV (5-10 kcal/mol), and often better, can usually be obtained for transition-metal systems. Calculations for the second and third transition series are complicated by appreciable relativistic effects. Some of these difficulties have been avoided by using relativistic model core potentials (MCP) which are available in deMon-KS [20, 32]. Where experimental data are available, the calculated properties agree well with experiment. For systems with weak interactions, such as hydrogen bonds, DFT calculations [18, 14, 15] are typically reliable to within 0.5-1.0 kcal/mol. Better accuracy, in the 0.1-0.5 kcal/mol range, remains a challenge to those developing new functionals and new computational techniques. Vibrational frequencies are expected to be accurate to a few percent.

Another aspect of the current study is to obtain real time dynamics of cluster systems where bond breaking and making are on the same time scale as the electronic structure change. To this end, we developed an *ab initio* molecular dynamics simulation method implemented in the deMon-KS package [15] which uses the Linear Combination of Gaussian Type Orbital (LCGTO) density functional formalism.

*Ab initio* molecular dynamics (AIMD) simulation was pioneered by Car and Parrinello (CP) [33, 34]. In the original formulation one solves a coupled dynamic equation, moving classical atoms according to the correct self-consistent forces, moving the electronic variables according to the energy gradients and constraints. The AIMD simulations can also be carried out by simply integrating the classical motion of the nuclei on the Born-Oppenheimer surface [35, 36] where at each MD step, the forces on the nuclei are given by the quantum energy gradients and the molecular orbitals are updated by solving a Schrödinger type equation in the Born-Oppenheimer approximation. This approach is often referred to as Born-Oppenheimer MD (BOMD). It has been found recently [37] that trajectories obtained utilizing the Born-Oppenheimer approach can be both more accurate and less costly than their Car-Parrinello counterparts.

Our density functional MD simulation is carried out at the Born-Oppenheimer level [38], where nuclei and electrons are treated as explicit species. The motion of electrons is described by quantum mechanics. It is assumed that the motion of the nuclei is much slower than that of electrons, and is always in equilibrium with the electronic structure. Therefore the classical dynamics equations can be used to give complete trajectories if the potential surface is known.

# 3   Results for Transition Metal Clusters

The question of the geometrical structure of transition-metal clusters remains an unsolved, yet very important, problem in cluster science. This is because one of the major themes in cluster research uses small transition-metal clusters as models for reactive processes in heterogeneous catalysis [39, 40]. In order to take full advantage of the cluster-surface analogy, it is necessary to establish the structure of metal clusters containing several transition-metal atoms, both in the presence and absence of ligands.

In order to find the most stable structure of a specific cluster, it is necessary to perform full geometry optimizations without symmetry constraints, starting from several initial geometries to locate different minima on the potential energy surface (PES), and considering different spin-multiplicities for each geometry in order to find the lowest spin state. It is difficult to establish the geometry of lowest energy because it is possible to have several local minima on the PES, and even if one tried several initial geometries, there is always a possibility of missing another, more stable, one. Even though we cannot be sure that the most stable geometries that we found correspond to the absolute minimum configuration, the total number of initial geometries that usually were considered is sufficiently large to feel confident that they are not far away from the true absolute minimum.

It is attractive to consider the application of the PFI-ZEKE technique ( Pulsed Field Ionization (PFI)-Zero Electron Kinetic Energy (ZEKE) photoelectron spectroscopy) to the problem of the structure of small transition-metal clusters. The principle advantage is that threshold photoelectron spectroscopy links the ground electronic state of the neutral cluster to the ground electronic state of the cation. Thus, one obtains the vibrational spectra of the neutral and the

**Table 1.** Geometry, Energy and Vibrational Frequencies For $Nb_3O$

| Spin multiplicity | Geometry (Å) | Energy (kcal/mol) | Vibrational Frequencies $(cm^{-1})$ | Symmetry | 2S+1 |
|---|---|---|---|---|---|
| Doublet | planar | 0.0 | 238 | $b_1$ | 2 |
| | $Nb - Nb$ bond: | | 300 | $b_2$ | |
| | 2.26 | | 334 | $a_1$ | |
| | 2.70 | | 382 | $a_1$ | |
| | $Nb - O$ bond: | | 579 | $b_1$ | |
| | 1.87 | | 753 | $a_1$ | |
| Doublet | pyramid | 23.74 | 161 | e | 2 |
| | $Nb - Nb$ bond: | | 221 | e | |
| | 2.32 | | 266 | e | |
| | 2.42 | | 354 | e | |
| | $Nb - O$ bond: | | 387 | e | |
| | 2.0 | | 671 | e | |
| | 2.27 | | | | |

**Table 2.** Geometry, Energy and Vibrational Frequencies For $Nb_3O^+$

| Spin multiplicity | Geometry (Å) | Energy (kcal/mol) | Vibrational Frequencies $(cm^{-1})$ | Symmetry | 2S+1 |
|---|---|---|---|---|---|
| Singlet | planar | 0.0 | 269 | $b_1$ | 1 |
| | $Nb - Nb$ bond: | | 289 | $b_2$ | |
| | 2.27 | | 337 | $a_1$ | |
| | 2.65 | | 392 | $a_1$ | |
| | $Nb - O$ bond: | | 625 | $b_1$ | |
| | 1.85 | | 790 | $a_1$ | |
| Triplet | planar | 12.82 | 213 | $b_1$ | 3 |
| | $Nb - Nb$ bond: | | 265 | $b_2$ | |
| | 2.28 | | 324 | $a_1$ | |
| | 2.70 | | 372 | $a_1$ | |
| | $Nb - O$ bond: | | 572 | $b_1$ | |
| | 1.84 | | 750 | $a_1$ | |
| Triplet | pyramid | 23.72 | 258 | e | 3 |
| | $Nb - Nb$ bond: | | 259 | e | |
| | 2.35 | | 304 | e | |
| | 2.35 | | 307 | e | |
| | $Nb - O$ bond: | | 386 | $a_1$ | |
| | 2.02 | | 684 | $a_1$ | |

**Table 3.** Geometry, Energy and Vibrational Frequencies For $Nb_3C_2$

| Spin multiplicity | Geometry (Å) | Energy (kcal/mol) | Vibrational Frequencies $(cm^{-1})$ | Symmetry | 2S+1 |
|---|---|---|---|---|---|
| Doublet | bipyramid | 0.0 | 65 | $a''$ | 2 |
| | $Nb-Nb$ bond: | | 214 | $a''$ | |
| | 2.46 | | 256 | $a'$ | |
| | 2.48 | | 289 | $a'$ | |
| | 2.58 | | 387 | $a'$ | |
| | $Nb-C$ bond: | | 518 | $a''$ | |
| | 2.0 | | 539 | $a'$ | |
| | 2.02 | | 693 | $a'$ | |
| | 2.05 | | 826 | $a'$ | |
| Doublet | bipyramid | 3.79 | 225 | $b_1$ | 2 |
| | $Nb-Nb$ bond: | | 257 | $a_1$ | |
| | 2.40 | | 298 | $b_2$ | |
| | 2.82 | | 353 | $a_1$ | |
| | $Nb-C$ bond: | | 395 | $a_1$ | |
| | 1.94 | | 472 | $a_2$ | |
| | 2.31 | | 692 | $b_2$ | |
| | | | 722 | $b_1$ | |
| | | | 824 | $a_1$ | |

**Table 4.** Geometry, Energy and Vibrational Frequencies For $Nb_3C_2^+$

| Spin multiplicity | Geometry (Å) | Energy (kcal/mol) | Vibrational Frequencies $(cm^{-1})$ | Symmetry | 2S+1 |
|---|---|---|---|---|---|
| Singlet | bipyramid | 0.0 | 282 | $e''$ | 1 |
| | $Nb-Nb$ bond: | | 290 | $e'$ | |
| | 2.48 | | 398 | $a_1'$ | |
| | | | 586 | $e'$ | |
| | $Nb-C$ bond: | | 749 | $a_2''$ | |
| | 2.02 | | 840 | $a_1'$ | |
| Doublet | bipyramid | 26.56 | 198 | $b_2$ | 1 |
| | $Nb-Nb$ bond: | | 209 | $b_1$ | |
| | 2.41 | | 251 | $a_1$ | |
| | 2.88 | | 372 | $a_1$ | |
| | $Nb-C$ bond: | | 385 | $a_2$ | |
| | 1.95 | | 459 | $a_1$ | |
| | 2.24 | | 691 | $b_2$ | |
| | | | 694 | $b_1$ | |
| | | | 844 | $a_1$ | |

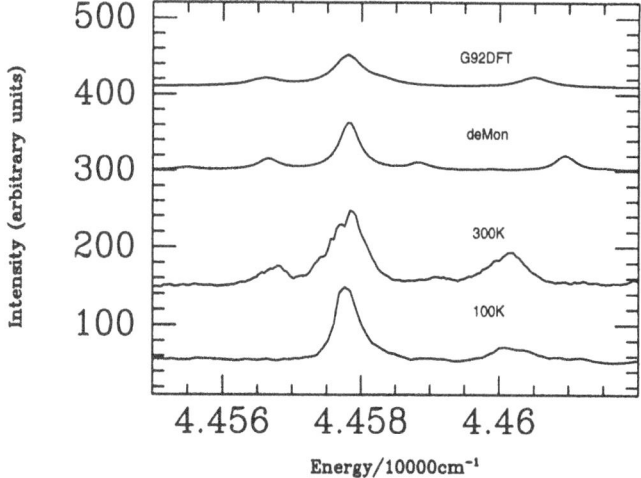

**Fig. 1.** Simulated PFI-ZEKE spectra of $Nb_3O$, at 300K, in the region of the 0-0 band, using GAUSSIAN92 and deMon-KS DFT codes (top two traces). The experimentally determined spectra at 300 and 100 K are also shown (bottom two traces) (a similar figure appears in reference 43).

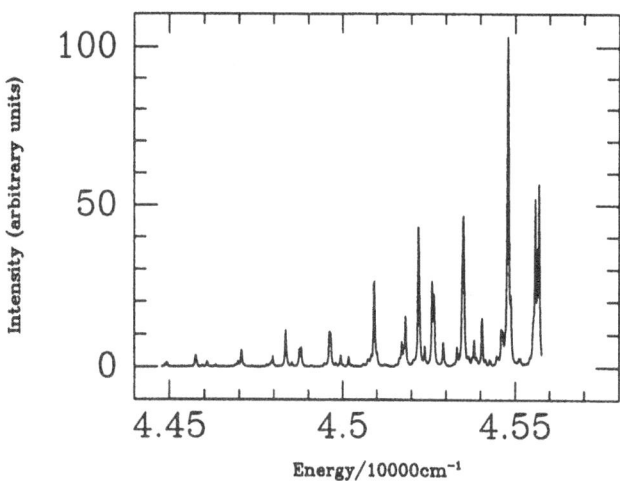

**Fig. 2.** A simulation of the PFI-ZEKE spectrum of the three-dimensional form of $Nb_3O$. The simulation was performed with the geometries and vibrational frequencies from the deMon-KS calculation. Notice the disagreement with the experimental data, Fig. 1 (a similar figure appears in reference 43).

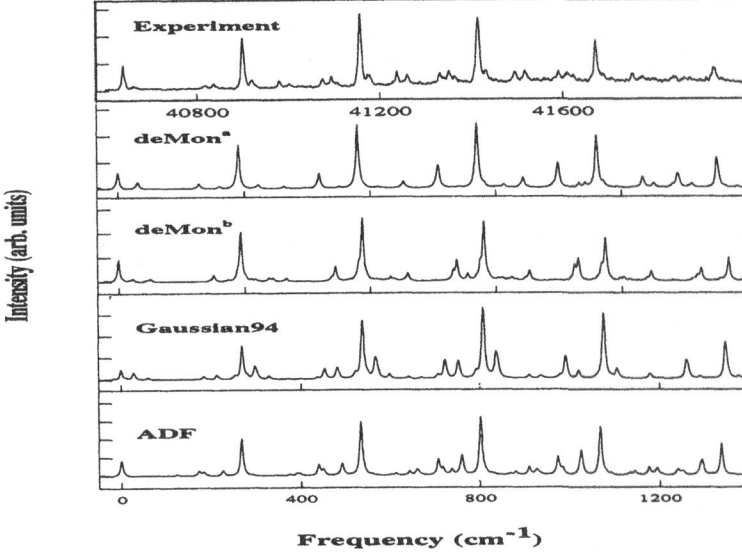

**Fig. 3.** Experimental and simulated PFI-ZEKE spectra of $Nb_3C_2$ at 300K. The spectra were calculated using the trigonal bipyramid geometry. In the $deMon^a$ calculations, model core potentials were used for the niobium atoms and all electrons for the carbon. In the $deMon^b$ calculations, model core potentials were used for the niobium and carbon atoms (a similar figure appears in reference 44).

cation carried on the strong, electronically allowed photoionization transition. This is a major advantage because the ground electronic states of the neutral and the cation will be the most well understood, and most easily calculable, electronic states of any metal clusters and, thereby, the comparison of experimental data with theoretical predictions will be facilitated. A second reason to be optimistic about the potential of PFI-ZEKE techniques is that photoionization thresholds for many transition-metal clusters are to be found in a spectral region easily accessible to tunable dye lasers, making the application of PFI-ZEKE spectroscopy fairly routine and, in principle, universally applicable. The photoionization thresholds for many transition-metal elements as a function of cluster size have recently been determined [1].

PFI-ZEKE spectroscopy was applied to small vanadium clusters [41, 42]. These studies showed that PFI-ZEKE spectra could be obtained for transition-metal clusters but, as yet, they have not led to structural assignments for clusters larger than dimers. However, it is possible to use the PFI-ZEKE approach, coupled with DFT electronic structure and Franck-Condon calculations, to resolve questions concerning the structure of small transition-metal clusters. This methodology was used for the structural determination of two of the more complex transition-metal cluster systems studied to date, triniobium monoxide ($Nb_3O$) [43] and triniobium dicarbide ($Nb_3C_2$) [44].

Niobium is a second row, group 5, transition series element. The ground electronic configuration of the niobium atom is $4d^4 5s^1$. The complexity of the electronic structure in a molecule containing three niobium atoms can be adjudged from the fact that there are 507 states that arise in Hund's case c from the first asymptote of the niobium dimer, and that the niobium atom has 15 electronic terms below 2 eV [45]. Nonetheless, most of this complexity is avoided using the PFI-ZEKE approach and well resolved vibrational bands were obtained. These have been simulated using multidimensional Franck-Condon factors calculated using the stable geometries and harmonic frequencies obtained from density functional theory for the neutral and the cationic systems.

In order to obtain the most stable geometries of $Nb_3O$ (neutral and cation) and $Nb_3C_2$ (neutral and cation) we used the deMon-KS program with model core potentials. In Tables 1 and 2 we present the most stable structures for $Nb_3O$ (neutral and cationic). As outlined before, these geometries were obtained by the minimization of the total energy without symmetry constraints and using model core potentials. The optimization was performed at the local level. After the optimization we calculated the nonlocal energy for the final geometries. The relative energies that we present in Tables 1 and 2 are at the GGA level.

For $Nb_3O$ (neutral and cationic) the most stable structure is planar. The oxygen atom is bonded with equal bond distances to two Nb atoms. Two different Nb-Nb bond distances are present in the cluster. The most stable three-dimensional structure lies higher by 23.74 and 23.72 kcal/mol, for the neutral and for the cationic system respectively. The ionization does not change the symmetry of the planar structures but, for the three-dimensional structures, the cationic system has higher symmetry than the neutral. For $Nb_3O$ we found only one stable spin multiplicity, while for the cationic system two values of the spin multiplicity are close in energy. For $Nb_3O$, the most stable spin-multiplicity is a doublet. For $Nb_3O^+$, the most stable spin-multiplicity is a singlet, with a triplet-singlet splitting value equal to 12.82 kcal/mol.

In Table 3 and 4 we present the vibrational analysis for the most stable structures of $Nb_3O$ (neutral and cationic). The assignment by symmetry of the normal modes of vibration is also reported. All these results are at the local level. The higher vibrational frequencies (579 and 753 $cm^{-1}$ for the neutral, 625 and 790 $cm^{-1}$ for the cation) correspond to the vibration of the oxygen atom.

Fig. 1 compares the simulated spectra using the results for the most stable structures, at 300 K, with the experimental one. The simulation labeled G92DFT used the geometries and frequencies from the GAUSSIAN92/DFT code [46] . It is clear that both methods produce very similar simulations. The slight differences between the two sets of simulated spectra are due to slight differences in the calculated geometry. The comparison with the experimental spectra establishes how well the two implementations of DFT theory are able to calculate these differences.

Fig. 2 shows a simulation for the PFI-ZEKE spectrum of the three-dimensional structure. The three-dimensional structures of $Nb_3O$ (neutral and cation) shown in the tables can be rejected on the basis of a cursory comparison with experimental data. The large differences between the geometries of the neutral and the ion in this structure would lead to strong progressions in the PFI-ZEKE spectrum [43], as we can see in Fig. 2.

The results for $Nb_3C_2$ are shown in Tables 5 to 8. $Nb_3C_2^+$ has $D_3h$ symmetry. Adding an electron to form the neutral system populates a degenerate orbital leading to a Jahn-Teller distorted structure of lower symmetry $(C_s)$. Fig. 3 compares the experimental and theoretical spectra from the trigonal structures. From the PFI-ZEKE spectrum we expect that the geometries of the neutral and ion are rather different since relatively long progressions were observed. Comparing the experimental with the simulated spectra, we see that the main spectral features are in fairly good agreement. The comparison between experiment and theory establishes that the triniobium dicarbide cluster exists in the trigonal bipyramid geometry under the experimental conditions. The trigonal bipyramid and the doubly-bridged geometries are calculated to have similar stability (the relative energy is equal to 3.79 kcal/mol), but the latter have higher ionization energy. This may imply that the photon energies used in the experiment may not be high enough to probe the doubly-bridged structure.

In these works, the quality of the agreement between the experimental spectrum and the theoretical simulations is sufficient to assign the geometry of the (electronically) complex transition-metal clusters, triniobium oxide and triniobium dicarbide. With these studies, it has been possible, for the first time, to establish unequivocally the geometry and vibrational structure of a gas phase transition-metal cluster with more than three atoms.

# 4  Hydrated Proton Clusters

Our calculations are carried out using the DZVP basis set [25], which is of double-zeta quality with polarization functions on heavy atoms. Previous work has indicated that the Perdew nonlocal potential for the electron correlation and exchange [29] yields reasonable results for the geometries and vibrational frequencies for water clusters [18, 19] and hydrated proton clusters [14, 47]. Therefore the Perdew nonlocal potential has been used in all our calculations.

If a positively charged proton is introduced to a water cluster we expect to see a significant structure change. The proton tends to react with a water molecule to form a very stable hydronium ion $H_3O^+$. It thus should significantly alter the ring structures bridged by hydrogen bonds found in water clusters [48, 49, 50]. This is very apparent from the ball-and-stick structures shown in Figs. 4-6. The hydronium ion is, indeed, the central unit of many structures we optimized.

It can be seen from Fig. 4-6 that the length of $O - H$ bonds within the hydronium unit decreases whereas the hydrogen bond lengths between water and the hydronium unit increase as the cluster size increases. This is a good indication that the hydrogen bonds become weaker while the O-H bonds in the hydronium unit become stronger. It is interesting to note that for $H_7O_3^+$ and $H_9O_4^+$ the hydrogen bond length between water and the hydronium unit of the DFT calculation is smaller than that of Remington and Schaefer [51, 16]. This is also observed for water clusters where DFT gives shorter hydrogen bond lengths than MP2 [18, 19]. But the chemical bond length given by DFT is larger than that of Remington and Schaefer [51, 16]. For example, the DFT $O - H$ distance in the hydronium unit of $H_9O_4^+$ is 1.04 Å while Remington and Schaefer find 0.983 Å [51, 16], the hydrogen bond in DFT is 1.58 Å while Remington and Schaefer give 1.61 Å.

The optimized structure for $H_{11}O_5^+$ is essentially a $H_9O_4^+$ cluster hydrogen bonded to an additional water, which is in the second solvation shell. We also tried to optimize the structure of the $H_{11}O_5^+$ cluster by attaching a water to $H_9O_4^+$ with a hydrogen close to the oxygen of the hydronium unit. This gives an opportunity to form a 4-coordinate first solvation shell by electrostatic inter- action or hydrogen bonding. However after many optimization steps essentially the same structure as shown in Fig. 4 was obtained. This supports the argument that the existence of a 4-coordinate gas phase species is very unlikely [52] for $H_{11}O_5^+$.

We have obtained two structures for $H_{13}O_6^+$, one (structure b) with a central $H_5O_2^+$ unit hydrogen bonded to 4 water molecules, another (structure a) with a central hydronium ion ($H_3O^+$), with three water molecules in the first solvation shell and two in the third solvation shell. The $H_3O^+$-centered structure is 1.14 kcal/mol more stable than the $H_5O_2^+$-centered one, a smaller difference than the Hartree-Fock (HF) value of 2.2 kcal/mol reported by Newton [52]. Such a small energy difference between the two structures is already suggestive that a variety of solvent-solute geometries will have to be allowed in any dynamical treatment. A similar structure to $H_{13}O_6^+$(b) was obtained for $H_9O_4^+$, $H_9O_4^+$ (b). The energy difference between $H_9O_4^+$(a) and $H_9O_4^+$(b) is 3.37 kcal/mol, which is much larger than for the case of $H_{13}O_6^+$.

Fig. 5 shows the ball-and-stick representation of the optimized structures for $H_{15}O_7^+$. Fig. 5(a) is achieved by adding a water to the most stable $(H_2O)_6H^+$(a) cluster to form a three-coordinated second solvation shell so that the water molecules in the first solvation shell become equivalent. It is expected that this structure is among those whose energy is close to the global minimum. Indeed, its energy is 1.19 kcal/mol lower than that shown in Fig. 5(b) in which a water molecule has been displaced from the second to the third solvation shell.

As we indicated earlier, the small energy difference between optimized struc- tures is already seen for $H_{13}O_6^+$, the energy of $H_3O^+-$ and $H_5O_2^+$-centered structures that we optimized become quite close, i.e., about 1 kcal/mol. The possibility of other, perhaps 3-dimensional structures for larger clusters with similar energy differences has to be addressed. To be able to generate a large number of possible structures a simulated annealing technique is employed. The simulated annealing was started with an initial geometry and equilibrated at finite temperature for a few hundred steps. Then the temperature was decreased slowly by scaling the velocity. However, we found it to be faster and easier to locate a low energy local minimum by using a traditional quantum chemistry structure optimization method [20]. In the current calculation, we use the MD to allow the system to visit various regions of phase space and we use quantum chemistry optimization methods to find quickly a low energy local minimum.

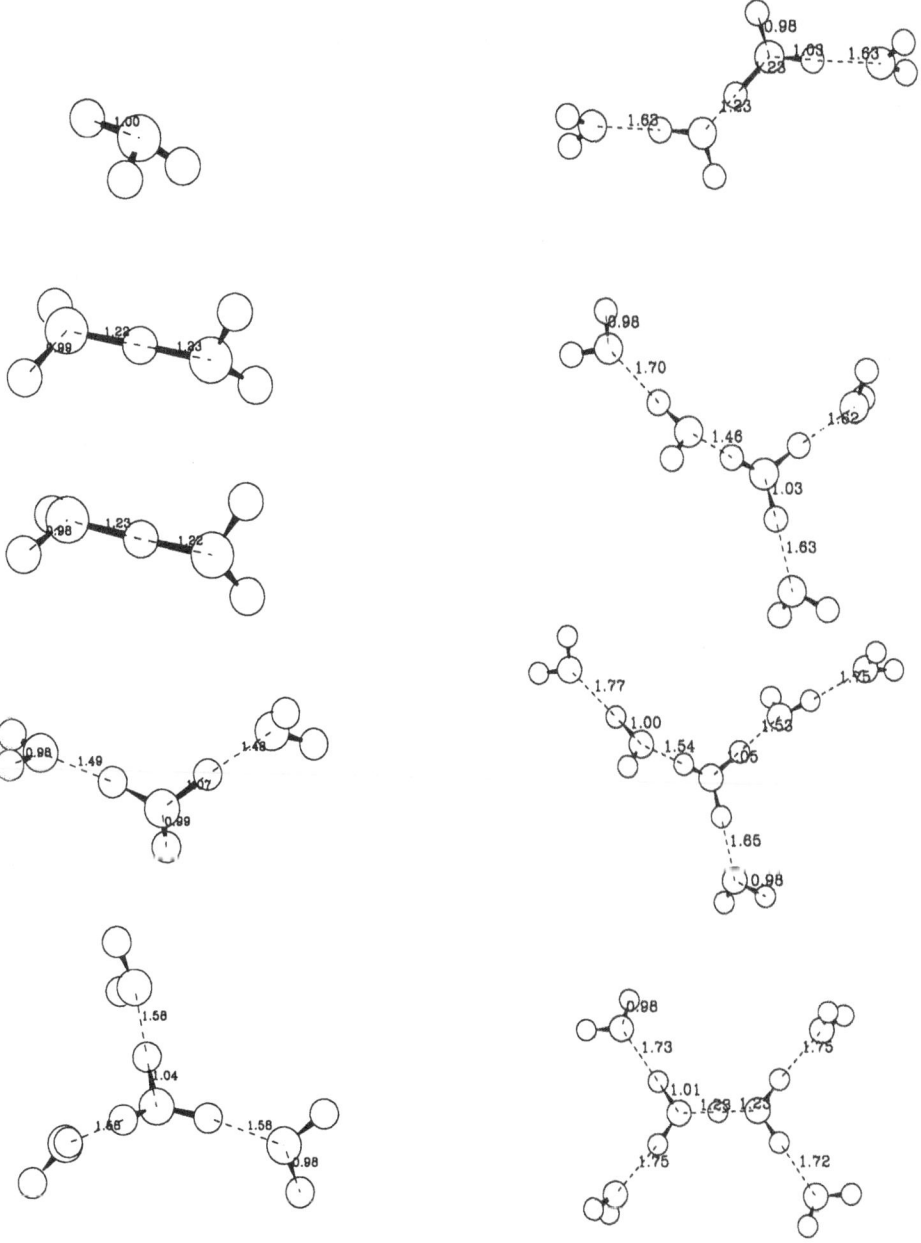

**Fig. 4.** The ball-and-stick representation of the hydrated proton clusters. The larger ball represents oxygen and the smaller one represents hydrogen ( a similar figure appears in reference 14).

If we start from the structure in Fig. 5(a) and scale the velocity of atoms to achieve a finite temperature it is found that a rather high temperature is required to allow the atoms to escape from vibrating around the equilibrium position in the original structure. We can easily create a fast "collapse" of a "spread" structure by scaling the coordinates of each atom. The temperature rises quickly so that we can sample many configurations in a limited number of time steps. Usually a cluster of smaller size compared with the starting position is obtained. We then optimize the structure using a traditional quantum geometry optimization method [20]. Figs. 5(c) and 5(d) show examples of the structures optimized using this method.The energy for the structure in Fig. 5(c) is 1.35 kcal/mol higher than that of Fig. 5(a). On the other hand, the energy of Fig. 5(d) is 0.85 lower than that of Fig. 5(a).

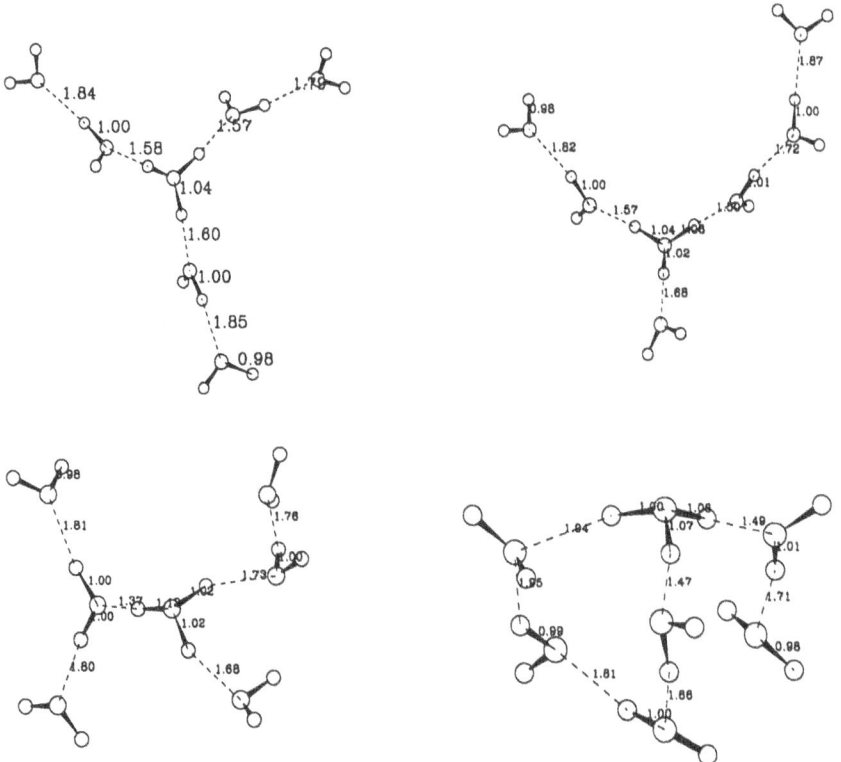

**Fig. 5.** The ball-and-stick representation of $H_{15}O_7^+$ cluster, (a) and (b) refer to the first and second figure in the first row, and (c) and (d) refer to the first and second figure in the second row, respectively. The symbols are as in Fig.4 (a similar figure appears in reference 15).

The structure in Fig. 5(c) deserves special attention. The central unit looks rather like $H_5O_2^+$. One of the $O-H$ bonds in $H_3O^+$ is stretched from its normal value 1.02-1.04 Å to 1.12 Å. The $O-H$ hydrogen bond in the central unit is

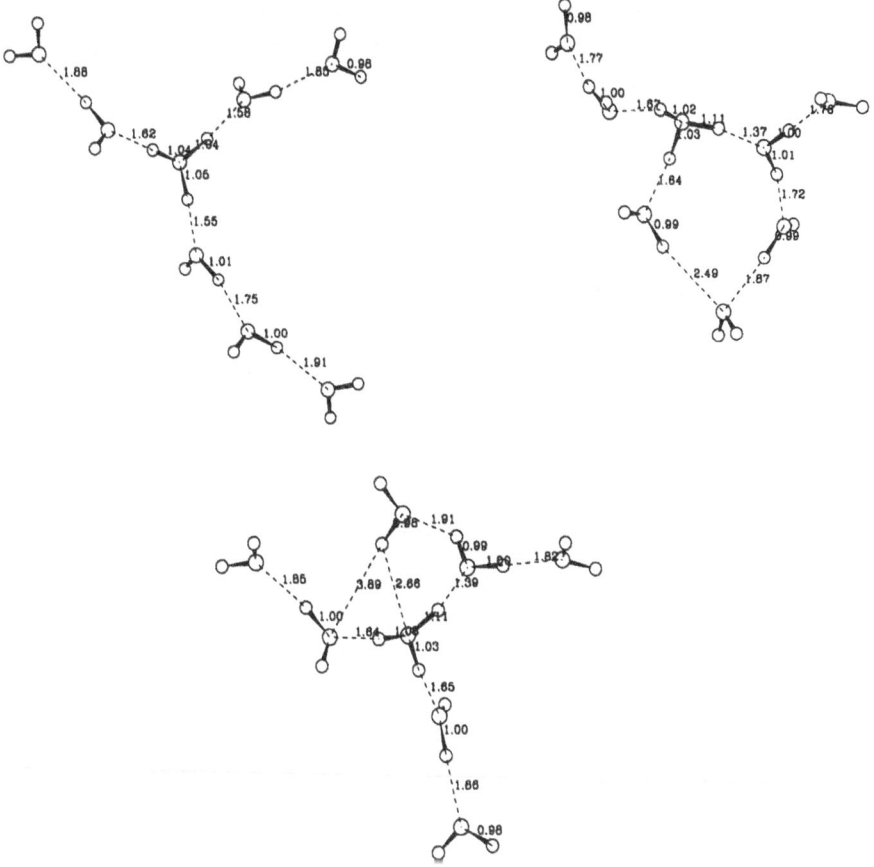

**Fig. 6.** The ball-and-stick representation of $H_{17}O_8^+$ cluster. (a) and (b) refer to the first and second figure in the first row, and (c) refers to the figure in the second row, respectively. The symbols are as in Fig. 4 (a similar figure appears in reference 15).

1.37 Å which is much shorter than other hydrogen bonds (for an isolated $H_5O_2^+$ cluster, the $O - H$ bond length is 1.23 Å for $C_2$ symmetry [14], and 1.12 Å and 1.34 Å respectively for $C_s$ symmetry [54]). Clearly this is due to the cooperative effect of the two water molecules hydrogen bonded to two hydrogen atoms on the left. This structure can be used to demonstrate the idea of solvation-facilitated proton transfer. Starting with a structure with $H_3O^+$ as the central unit, a few water molecules are attached in an unbalanced way to one side of the cluster. The bond between one of the protons in the $OH_3^+$ becomes weaker. Eventually it leaves the original $OH_3^+$ to join the other part of the structure. This completes a proton transfer process. This is consistent with the Grotthuss mechanism [59]. It is obvious that we can not exclude the $H_5O_2^+$-centered structures from the low energy ones for large hydrated proton clusters.

The structure optimization for $H_{17}O_8^+$ is achieved in a similar fashion. The structure shown in Fig. 6(a) is obtained starting from an initial geometry constructed by attaching a water to the third solvation shell of Fig. 6(a). The structure 6(b) is optimized starting from a low energy structure in a simulated annealing run. It is found that the energy for 6(b) is 1.53 kcal/mol lower than that for 6(a). This shows that indeed, low energy structures can be found by simulated annealing. The "designed" structure 6(a) given by Price et al. [17] is less stable than structure 6(b).

As we indicated before, a water molecule can bind to the lone pair of the oxygen in the central $H_3O^+$ unit using its hydrogen to form a four-coordinate solvation shell. We have tried to optimize such a structure for $(H_2O)_5H^+$ without success; the water molecule initially attached to the oxygen eventually moves away to the second solvation shell of $H_3O^+$. For $H_{17}O_8^+$, however, we have achieved a four-coordinate structure 6(c) whose energy is 1.52 kcal/mol higher than 6(a).

Both structures 6(b) and 6(c) have a central unit similar to that which we found in structure 5(c), $H_5O_2^+$ with a structure similar to an isolated $H_5O_2^+$ of $C_s$ symmetry. The $O - proton$ and $proton - O$ distances are quite similar in these structures, i.e, 1.12 Å and 1.37 Å for 2(c), 1.11 Å and 1.38 Å for 6(b), 1.11 Å and 1.39 Å for 6(c), which is similar to the isolated case, 1.12 Å and 1.34 Å. It is also worthwhile to note that 6(b) is the lowest in energy among the three structures we have optimized for $H_{17}O_8^+$. It appears that the $H_5O_2^+$ unit can be clearly identified in the structures of large hydrated proton clusters.

A symmetric $H_5O_2^+$ unit was identified by Tuckerman et al. [34] in their simulation of proton solvation in the liquid environment provided by 32 waters in a cubic box with periodic boundary conditions, ie, an $H_5O_2^+$ unit with $C_2$ geometry where a proton is equally shared by two water molecules, which is similar to the situation illustrated in $H_{13}O_6^+$. This is probably because solvation of the $H_5O_2^+$ ion by water molecules is more balanced in the simulated liquid than in a finite cluster.

It is apparent that, in some of the structures we optimized, the central unit is an $H_3O^+$ ion. Naturally, three waters are bonded to the hydrogens of the $H_3O^+$ unit to form an $H_9O_4^+$ structure which is consistent with the simulation results of Tuckerman et al. [34]. Electron population analysis indicates that a significant portion of the positive charge is localized in the central unit, $H_3O^+$. So the stability is a result of strong ionic hydrogen bonds between a positively charged $H_3O^+$ and water. Indeed, the bond length is around 1.6 Å which is much smaller than the normal hydrogen bond in systems involving neutral water. For example, the hydrogen bond lengths in water clusters are larger than 1.8 Å [48, 49, 50]. As we can see from Figs. 5 and 6, the hydrogen bond length in the second and third solvation shells is comparable with that of water clusters.

Both DFT [14, 47] and post-Hartree Fock [53, 54, 12, 13, 55, 56, 57] calculations yield a ground state of $C_2$ structure for $H_5O_2^+$, i.e., the proton is equally shared by two oxygen atoms. However, there is a very small energy difference between the ground state and a state with $C_s$ symmetry in which the proton is close to one of the oxygen atoms. This shows that the energy surface is very flat along the O-Proton-O axis. The proton undergoes large-amplitude motion. The theoretical model with harmonic parabolic energy surface used in the quantum harmonic frequency calculation is certainly not adequate. The anharmonic vibrational frequency of the proton can be calculated using 1-D and 2-D potential surfaces [53]. However 3-D calculation would be too expensive to do.

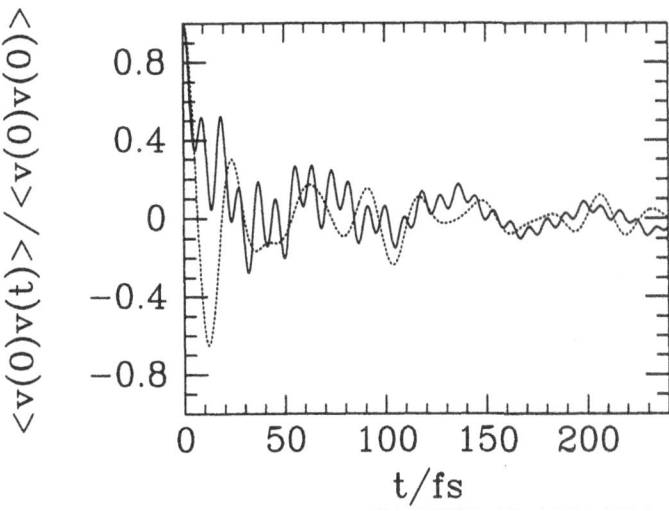

**Fig. 7.** Velocity auto correlation function as a function of time. The solid and dotted curves are for hydrogen and proton respectively (a similar figure appears in reference 15).

For a small system such as $H_5O_2^+$ we can afford a long dynamical study. A constant energy AIMD simulation at about 180 K was carried out. The Newton Equation was integrated using a Verlet algorithm [58] A time step of 0.24 fs was used. Statistical averages were calculated using data collected for 20000 MD steps. The velocity auto correlation functions for the hydrogen atom and the proton are plotted in Fig 7. The difference in period can be a reflection of the amplitude of atomic motion. It shows that the hydrogen correlation function decays faster than the proton function. The initial decay of the auto correlation function is very important in determining the high frequency vibrational modes when it is subjected to the Fourier transformation that yields the vibrational spectrum. The $O - H$ stretching frequency is above 3000 $cm^{-1}$ [15]. As it is shown in Fig 8, the proton vibrational frequencies are lower than 2000 $cm^{-1}$. The signature band for the proton is around 1000 $cm^{-1}$. It is interesting to see that the anharmonic frequency by AIMD is not distinguishable from the harmonic one, i.e., 1034 $cm^{-1}$ within the accuracy of the current simulation. However, the 1-D and 2-D potential surface calculations by Ojamäe et al. [53] give very different values,i.e., 1623 and 1275 $cm^{-1}$, respectively. It would be

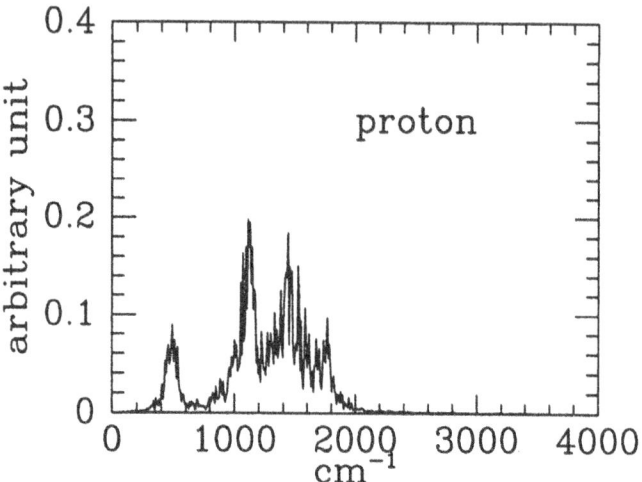

**Fig. 8.** Proton power spectrum of $H_5O_2^+$ cluster (a similar figure appears in reference 15).

necessary to calculate the quantum anharmonic frequency in a 3-D potential surface to verify the prediction of our MD simulation.

Table 9 lists the energy change for the reaction $H_3O^+(H_2O)_{(n-1)} + H_2O = H_3O^+(H_2O)_n$. We have included the vibrational energy contribution at 298K, calculated using the following formula [60]:

$$E_{zero} = \sum_\alpha (h\nu_\alpha/2) + \sum_\alpha \frac{h\nu_\alpha}{exp(h\nu_\alpha/kT) - 1} \qquad (1)$$

where h is the Planck constant, $\nu_\alpha$ is the vibrational frequencies. The deMon results are qualitatively in good agreement with the experimental values. A $6-31**G$ and $DZ+p$ SCF calculation by Yamabe et al. [61] yields a value of 33 kcal/mol for the enthalpy. For reactions involving larger clusters than $H_5O_2^+$, the basis set superposition error (BSSE) has not been calculated. We expect a smaller value will be obtained if BSSE is included. We notice that the value for (4,5) is larger than for (3,4). The contribution from the interaction energy still follows the normal trend, but the zero point energy for (4,5) is significantly larger than for (3,4).

The harmonic vibrational frequencies and IR intensities of various vibrational modes have been generated for all the structures optimized. The vibrational frequencies for $H_2O$ symmetric and asymmetric stretches are plotted for the different cluster species in Figs. 9 and 10, respectively. A very large frequency increase was observed on going from $H_3O^+$ to $H_5O_2^+$ and from $H_5O_2^+$ to $H_7O_3^+$ due to a significant change of O-H bond strength as we pointed out earlier. Such a large increase is not reflected by the scaled frequencies of Remington and Schaefer [51]. The increase becomes much smaller on going to $H_{11}O_5^+$ and $H_{11}O_6^+$ clusters.

**Table 5.** Energy Change for the Reactions: $H_3O^+(H_2O)_{(n-1)} + H_2O = H_3O^+(H_2O)_n$

| (n-1,n) | deMon | HF | $exp_1$ | $exp_2$ |
|---------|-------|-----|---------|---------|
| $(0,1)_1$ | 34.5 | | | |
| $(0,1)_2$ | 34.9 | | | |
| 0,1 | 37.2 | 37 | 31.6 | 36 |
| 1,2 | 24.4 | 26 | 19.5 | 22 |
| 2,3 | 19.6 | 22 | 17.9 | 17 |
| 3,4 | 13.8 | 16 | 12.7 | 15 |
| 4,5 | 14.6 | 15 | 11.6 | 13 |
| 5,6 | 10.0 | | 10.7 | 12 |
| 6,7 | 9.1 | | | 10 |

Note: the deMon calculations were made using the first structures for each hydrated proton cluster shown in Figs. 4-6;
HF is from Newton [52] using 4-31G with zero-point correction;
$exp_1$ is the experimental enthalpy from Y.K. Lau et al. [62];
$exp_2$ is the experimental enthalpy from Kebarle et al. [63];
$(0,1)_1$ is the deMon result with basis set superposition error correction;
$(0,1)_2$ is the deMon result with basis set superposition error correction and zero-point correction using the TZVP basis set [64] which has triple-zeta quality with field-induced polarization functions.

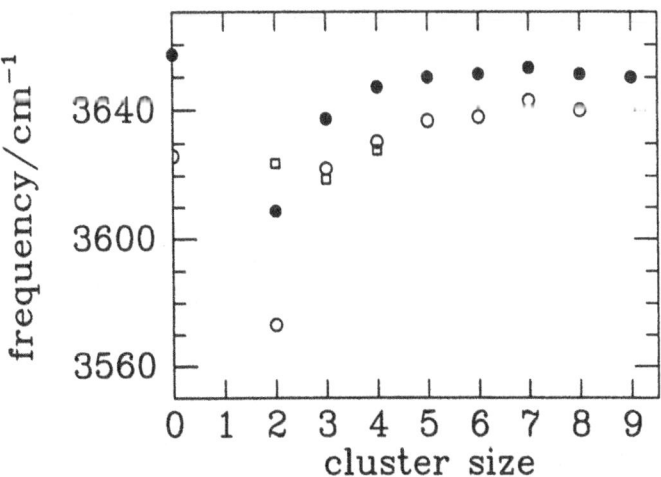

**Fig. 9.** $H_2O$ symmetric stretching frequencies as a function of the number of water molecules in the hydrated proton clusters (zero refers to the isolated water molecule). The open circles - present study, solid circles - experiment of Yeh et al. (J. Chem. Phys., **91**, 7319 (1989)) and open squares - Remington and Schaefer (unpublished results).

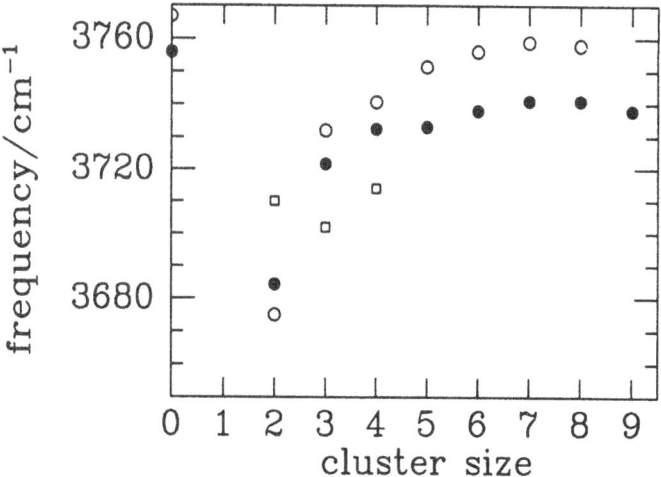

**Fig. 10.** $H_2O$ asymmetric stretching frequencies for hydrated proton clusters. The symbols are as in Fig. 9.

This is because the frequencies are due to symmetric or asymmetric stretches of $H_2O$ in the second solvation shell. As the cluster size grows, the $H_2O$ stretching frequencies approach the values of isolated $H_2O$. Both experimental and deMon results certainly show this trend.

## 5 Discussion

We have shown that PFI-ZEKE spectroscopy can be combined with density functional theory to reveal vibrational information on small transition-metal clusters. By comparing the PFI-ZEKE spectra with theoretical DFT and Franck-Condon calculations, we have been able to determine the geometric conformation of the cluster. The geometrical structure for $Nb_3O$ and $Nb_3O^+$ is planar, $C_{2v}$, with edge-bound oxygen atoms. For $Nb_3C_2$ the structure is a trigonal bipyramid geometry with each of the carbon atoms triply bridging to each face of the trigonal frame. For this system, the cationic cluster adopts $D_{3h}$ symmetry whereas the neutral cluster has lower symmetry ($C_s$) resulting from the Jahn-Teller distortion. The quality of the agreement between the experimental spectrum and the theoretical simulations is sufficient to assign the geometry of the (electronically) complex transition-metal clusters, triniobium oxide and triniobium dicarbide. With these studies, it has been possible, for the first time, to establish unequivocally the geometry and vibrational structure of a gas phase transition-metal cluster with more than three atoms.

DFT calculations were carried out to study the structure, spectroscopic, thermodynamic and dynamic properties of hydrated proton clusters. We find that if a positively charged proton is introduced to a water cluster, the proton tends to react with one or two water molecules to form a very stable hydronium ion, $H_3O^+$ or $H_5O_2^+$. This significantly alters the ring structure bridged by hydrogen bonds found in water molecules in the water clusters. For small clusters, the $H_3O^+$-centered structure has lower energy. When the hydrated proton clusters become larger the energy difference becomes very small. For $H_{15}O_7^+$ and $H_{17}O_8^+$, the $H_5O_2^+$-centered structures have lower energy than those centered on $H_3O^+$. This can be of importance in determining the proton transfer mechanism in clusters and in solution.

DFT calculations yield reliable vibrational frequencies for hydrated proton clusters compared with recent experimental results for all the cluster structures we optimized. The DFT hydration energy is found to be in good agreement with experimental data. Molecular dynamics simulation is very useful in generating anharmonic vibrational frequencies for small clusters. The frequencies obtained are quite close to the frequencies calculated by traditional quantum chemical methods.

In this chapter, we have illustrated how DFT calculations can be used to study structural, spectroscopic and other properties for two very interesting classes of clusters, namely, one, involving transition-metal atoms where the electron correlation and exchange are traditionally very hard to treat, and another with weak interactions of only a few kcal/mol. With increasing computing power and improving DFT computational technology we are confident that DFT methods will continue to reveal the properties of these, and other, kinds of clusters. Along with, or sometimes combined with, the other electronic structure and dynamics methods described in this book, DFT will help to keep TAMC - the theory of atomic and molecular clusters - the vibrant and challenging field that has deservedly attracted such wide attention.

## Acknowledgments

Support of some of this work by the Natural Sciences and Engineering Research Council of Canada and by the Fondation FCAR of the government of Quebec is gratefully acknowledged. Computational facilities were provided by CERCA, the Services Informatiques de l'Université de Montréal and the Laboratorio de Visaulization Y Computo en Paralelo/UAM (Mexico). We are very grateful to Dr. Dong-Sheng Yang for his generous help in generating some of the figures.

## References

1. M.D. Morse, Chem. Rev., **86**, 1049 (1986); in *Advances in Metal and Semiconductor Clusters, Vol. 1. Spectroscopy and Dynamics* ( JAI, Greenwich, 1993 ).
2. Kenneth J. Klabunde, *Free Atoms, Clusters, and Nanoscale Particles*, Academic Press, Inc., San Diego , 1994.
3. L. Goodwin and D.R. Salahub. Phys. Rev. A. **47**, 774 (1993).
4. G.A. Jeffrey and W. Saengeri, *Hydrogen Bonding in Biological Structures*, Springer-Verlag, New York, 1991.

5. G.W. Robinson, P.J. Thistlethwaite and J. Lee, J. Phys. Chem., **90**, 4224 (1986).
6. E.N. Baker and R.E. Hubbard, Prog. Biophys. Molec. Biol., **44**, 97 (1984).
7. Y. Cha, C.J. Murray and J.P. Klinman, Science, **243**, 1325 (1989); J. Rucker, Y. Cha, T. Jonsson, K.L. Grant and J.P. Klinman, Biochemistry, **31** 11489 (1992).
8. A. Warshel, in *Computer Modeling of Chemical Reactions in Enzymes and Solutions* ( Wiley, New York, 1991 ).
9. D. Borgis, S. Lee and J.T. Hynes, Chem. Phys. Lett. **19**,162 (1989); D. Borgis, G. Tarjus and H. Azzouz, J. Phys. Chem. **96**, 3188 (1992); D. Li and G. Voth, J. Phys. Chem. **95**, 10425, (1991).
10. D. Borgis and J.T. Hynes, J. Chem. Phys., **94**, 3619 (1991).
11. D. Laria, G. Ciccotti, M. Ferrario and R. Kapral, J. Chem. Phys., **97**, 378, 1992; J. Lobaugh and G.A. Voth, Chem. Phys. Lett., **198**, 311 (1992).
12. S. Scheiner, Acc. Chem. Res., **18**, 174 (1985).
13. K. Luth and S. Scheiner, J. Chem. Phys., **97**, 7507 (1992); Z. Latajka and S. Scheiner, J. Mol. Struc., **234**, 373 (1991).
14. D.Q. Wei and D.R. Salahub, J. Chem. Phys., **101**, 7633 (1994).
15. D.Q. Wei and D.R. Salahub, J. Chem. Phys., **106**, 6086(1997).
16. L.I. Yeh, M. Okumura, J.D. Myers, J.M. Price and Y.T. Lee, J. Chem. Phys., **91**, 7319 (1989).
17. J.M. Price, Ph.D. Thesis, Lawrence Berkeley Laboratory, University of California, 1990.
18. F. Sim, A. St-Amant, I. Papai and D.R. Salahub, J. Am. Chem. Soc., **114**, 4391 (1992).
19. D.Q. Wei, H. Guo, M. Leboeuf and D.R. Salahub, in preparation.
20. D.R. Salahub, R. Fournier, P. Mlynarski, I. Papai, A. St-Amant and J. Ushio, in *Density Functional Methods in Chemistry*, edited by J. Labanowski and J. Andzelm (Spinger, Berlin, 1991); A. St-Amant and D.R. Salahub, Chem. Phys. Lett., **169**, 387 (1990).
21. "DEMON. User's guide, version 1.0 beta", Biosym Technologies, San Diego (1992).
22. a) P. Hohenberg and W. Kohn. Phys. Rev., **B136**, 864 (1964); b) W. Kohn and L.J. Sham. Phys. Rev., **A140**, 1133 (1965).
23. P. Dahl and J. Avery, Eds., *Local Density Approximation in Quantum Chemistry and Solid State Physics* ( Plenum, New York, 1984 ); J.K. Labanowski and J.W. Andzelm, Eds. *Density Functional Methods in Chemistry* ( Springer-Verlag, New York, 1991 ).
24. S. Huzinaga, Ed. *Gaussian Basis Sets for Molecular Calculations*, (Elsevier,Amsterdam,1984).
25. N. Godbout, D.R. Salahub, J. Andzelm and E. Wimmer, Can. J. Chem., **70**, 560 (1992).
26. H. Sambe and R.H. Felton, J. Chem. Phys., **62**, 1122 (1975).
27. B.I. Dunlap, J.W.D. Connolly and J.R. Sabin. J. Chem. Phys. **71**, 3396, 4993 (1979).
28. S.H. Vosko, L. Wilk and M. Nusair. Can. J. Phys. **58**,1200 (1980).
29. J.P. Perdew, Phys. Rev., **B33**, 8800 (1986), **B34**, 7406 (1986).
30. J.P. Perdew and Y. Wang. Phys. Rev. **B33**, 8800 (1986).
31. A.D. Becke. Phys. Rev., **B38**, 3098 (1988).
32. J. Andzelm, E. Radzio and D.R. Salahub. J. Chem. Phys. **83**, 4573 (1985); R. Fournier, J. Andzelm and D.R. Salahub. J. Chem. Phys. **90**, 6371 (1989).
33. R. Car and M. Parrinello, Phys. Rev. Letters, **50**, 55 (1985).

34. K. Laasonen, R. Car, C. Lee and D. Vanderbilt, Phys. Rev., **B43**, 6796 (1991); M. Tuckerman, K. Laasonen, M. Sprik, and M. Parrinello, J. Phys. Condens. Matt., **6**, A93 (1994), J. Chem. Phys., **103**, 150 (1995), J. Phys. Chem., **99**, 5749 (1995), and references therein.

35. R.N. Barnett and U. Landman, Phys. Rev., **B48**, 2081 (1993).

36. X. Jing, N. Troullier, D. Dean, N. Binggeli, J.R. Chelikowsky, K. Wu and Y. Saad, Phys. Rev., **B50**, 12234 (1994).

37. D.A. Gibson, I.V. Ionova and E.A. Carter, Chem. Phys. Lett., **240**, 261(1995).

38. H.L Friedman, *A Course in Statistical Mechanics*, Prentice-Hall, Englewood Cliffs, 1985.

39. S.J. Riley. in *Metal Ligand Interactions:From Atoms, to Clusters,to Surfaces*, edited by D.R. Salahub and N. Russo. NATO ASI Series C (Kluwer, Dordrecht, 1992), Vol. 378, p.17.

40. P.A. Hackett, S.A. Mitchell, D.R. Rayner and B. Simard. in *Metal Ligand Interactions: Structure and Reactivity*, edited by D.R. Salahub and N. Russo. NATO ASI Series C,**474**, 289 (Kluwer,Dordrecht, 1996).

41. D.S. Yang, A.M. James, D.M. Rayner and P.A. Hackett. J. Chem. Phys. **102**, 3129 (1995).

42. D.S. Yang, A.M. James, D.M. Rayner and P.A. Hackett. Chem. Phys. Lett. **231**, 177 (1994).

43. D.S. Yang, M.Z. Zgierski, D.M. Rayner, P.A. Hackett, A. Martinez, D.R.Salahub, P.N. Roy, T. Carrington, Jr. J. Chem. Phys. **103**, 5335 (1995).

44. D.S. Yang, M.Z. Zgierski, A. Berces, P.A. Hackett, P.N. Roy, A. Martinez, T. Carrington Jr, D.R. Salahub, R. Fournier, T. Pang and Ch. Chen, J. Chem. Phys., **105**, 10663 (1996).

45. A.M James, P. Kowalczyk, R. Fournier and B. Simard. J. Chem. Phys. **99**, 8504 (1993).

46. GAUSSIAN 92/DFT, Revision G.4, M.J. Frisch, G.W. Trucks, H.B. Schlegel, P.M.W. Gill, B.G. Johnson, M.W. Wong, J.B. Foresman, M.A. Robb, M.Head-Gordon, E.S. Replogle, R. Gomperts, J.L. Andres, K. Raghavachari, J.S. Binkley, J.J.P. Stewart, and J.A. Pople (Gaussian, Inc. Pittsburgh, PA, 1993).

47. C. Mijoule, Z. Latajka and D. Borgis, Chem. Phys. Lett., **208**, 364 (1993).

48. C.J. Tsai and K.D. Jordan, Chem. Phys. Lett., **213**, 181 (1993).

49. S.S. Xantheas, J. Chem. Phys., **100**, 7523, 1994, ibid, **102**, 4505, 1995.

50. C. Lee, H. Chen and G. Fitzgerald, J. Chem. Phys., **101**, 4472 (1994), ibid, **102**, 1266 (1995).

51. R. Remington and H.F. Schaefer(unpublished results), quoted in ref 15.

52. M.D. Newton, J. Chem. Phys., **67**, 5535 (1978).

53. L. Ojamäe, I. Shavitt and S.J. Singer, Int. J. Quantum Chem., **29**, 657(1995).

54. Y. Xie, R.B. Remington, and H F. Schaefer III, J. Chem. Phys., **101**, 4878 (1994).

55. G. Alagona, R. Cimiraglia and U. Lamanna, Theoret. Chim Acta (berl.), **29**, 93 (1973).

56. B.O. Roos W.P. Kraemer and G.H.F. Diercksen, Theoret. Chim Acta (berl.), **42**, 77 (1976).

57. M. J. Frisch, J.E. Del Berne, J.S. Binkley and H. F. Schaefer, J. Chem. Phys., **84**, 2279(1986).

58. L. Verlet, Phys. Rev., **159**, 98 (1967).

59. H.Z. Daneel, Z. Elektrochem., **11**, 249 (1905).

60. K.S. Pitzer, *Quantum Chemistry*, Pretice-Hall, Englewood Cliffs, NJ, 1961.

61. S. Yamabe, T. Minato and K. Hirao, J. Chem. Phys., **80**, 1576 (1984).

62. Y.K. Lau, S. Ikuta, and P. kebarle, J. Am. Chem. Soc., **104**, 1462 (1982).
63. P. kebarle, S.K. Searles and A. Zolla, J. Am. Chem. Soc., **89**, 6393 (1967).
64. J. Guan, P. Duffy, J.T. Carter, D.P. Chong, K.C. Casida, M.E. Casida and M. Wrinn, J. Chem. Phys., **98**, 4753 (1993).

# Ultrafast Structural Response and Nonlinear Fragmentation Dynamics of Small Clusters Induced by Optical Excitation

M. E. Garcia, H. O. Jeschke and K. H. Bennemann

Institut für Theoretische Physik, Freie Universität Berlin, Arnimallee 14,
14195 Berlin, Germany

**Abstract.** The ultrafast relaxation of small clusters immediately after ionization and photodetachment is theoretically studied. Microscopic models are proposed to describe the non-equilibrium dynamics of the clusters. As an example of strong structural response we determine for $Ag_3$ the time scale for the change from a linear to triangular structure after the photodetachment process $Ag_3^- \rightarrow Ag_3$. We show that the time-dependent change of the ionization potential reflects in detail the internal degrees of freedom, in particular coherent and incoherent motion. We demonstrate that the time scales for bond breaking and formation are temperature dependent. We compare with experiment and point out the general significance of our results. We also study the ultrafast fragmentation dynamics of small $Hg_n$ clusters. We determine the fragmentation-time distributions induced by ionization. A dramatic change in the non-equilibrium fragmentation behavior occurs when the temperature before ionization reaches the melting point of the neutral clusters. This new effect could one allow to determine "melting points" of small clusters by pump&probe experiments. The ultrafast dynamics depends nonlinearly on the initial atomic positions and velocities, reflecting the intrinsic chaotic behavior of small clusters. The application of the theoretical models presented here, to the study of other relevant processes like control of chemical reactions, is discussed.

## 1  Introduction

A fundamental problem in the physics of small clusters and molecules is the description of relaxation mechanisms in the sub-picosecond time-domain. The ultrafast dynamics (UD) of a cluster is usually induced by an ultrashort laser pulse which excites or ionizes the cluster, abruptly bringing it to a nonequilibrium-state. Although the short-time response of clusters after excitation with an ultrafast laser pulse has become a major subject of study in the physics of clusters and molecules, it is still unclear how the excess energy released by the excitation process is transferred to the various modes of the system. Due to recent developments in femtosecond spectroscopy and pump&probe techniques it is now possible to monitor the dynamics of optically excited states of small clusters and molecules. Many different pump&probe experiments have been performed to study the UD of excited and ionized clusters[1–4]. The experimental results are usually interpreted in terms of master-equations, assuming constant decay probabilities. This allows one to obtain some information about decay-times for fragmentation processes. However, many questions still remain open. For instance, not much is known about the time-scales for energy transfer from electronic to atomic degrees of freedom. Moreover, one of the most important problems which has not been addressed so far, is whether the induced fragmentation of a cluster is sensitive to the cluster's initial conditions like the temperature or the thermodynamical state (solid-like or liquid-like), and if it reflects an intrinsic nonlinear, chaotic behavior of the cluster[5, 6]. One of the purposes of this article is to address these point from a theoretical point of view.

A nice example for time-dependent bond breaking and bond formation is given by the short-time dynamics of $Ag_3$ clusters formed by ultrashort photodetachment of $Ag_3^-$. Recently, this dynamics has been monitored in a pump&probe experiment performed on mass selected $Ag_3^-$ clusters[3]. The initially negatively charged clusters were neutralized through photodetachment by the pump pulse and after a delay ionized by the probe pulse in order to be detected. Due to the large differences in the equilibrium geometries of linear $Ag_3^-$[7], and obtuse isosceles triangular $Ag_3$[8], the ultrashort photodetachment process puts the neutralized trimer in an extreme non-equilibrium situation. As a consequence, a structural relaxation process occurs. The experimental signal of the $Ag_3^+$ yield was measured as a function of the delay time $\Delta t$ and the frequency of the laser pulses. For a frequency slightly above the ionization potential of $Ag_3$ a sharp rise of the signal is observed at approximately 750fs. After a maximum is reached, there is a saturation of the signal, which then remains constant for at least 100ps, which is the longest time delay used in the experiment. New features appear for higher frequencies. Again the signal increases sharply, reaches a maximum and then decreases to a constant value. A preliminary interpretation of these results uses the Franck-Condon-principle[3]. The first laser pulse creates a neutral linear silver trimer which begins to bend, passes through the obtuse isosceles triangle equilibrium geometry of the neutral $Ag_3$ and comes to a turning point near the equilateral equilibrium geometry of the positive ion[8]. After rebounding, the neutral trimers start pseudorotating through their three equivalent obtuse isosceles equilibrium geometries. This would explain the saturation behavior of the signal. However, this would mean that the pseudorotations have an extremely long mean lifetime. This seems improbable to us. Furthermore, it is not clear why the signal changes as a function of the frequency of the laser pulse.

In this article we perform a theoretical analysis of the physics underlying the ultrafast dynamics of $Ag_3$ clusters produced by photodetachment. In particular, we analyze the time evolution of the ionization potential and the dependence of the dynamics on the initial temperature of the clusters. We show that the experimental results can be explained using a physical picture which can be generally applied to other ultrashort-time processes.

Another important example of short-time response to perturbations is the fragmentation dynamics of weakly bonded clusters, like van der Waals clusters.

The ionization of a van der Waals cluster induces dramatic changes in its electronic structure. As a result, a relaxation process takes place, in which a certain amount of energy $\delta E$ is transferred from the electronic to the atomic degrees of freedom. Due to the different energy scales between the electronic and the atomic systems, $\delta E$ is usually very large in terms of thermal kinetic energy of the atoms. Therefore, melting and even fragmentation of the cluster is expected for rare-gas clusters and also small divalent-metal clusters, like $Hg_n$ clusters, which have been shown to be van der Waals bonded[9, 10]. In fact, fragmentation of rare-gas clusters upon ionization was one of the first experimental observations in cluster physics[11]. However, no detailed analysis of the fragmentation mechanisms and determination of relaxation times was possible until recently. Rapid fragmentation of $Hg_n$ clusters upon excitation has been observed in recent work[2].

The dynamics of van der Waals clusters can, in general, be described by performing Molecular Dynamics (MD) simulations. Usually, MD simulations for neutral van der Waals clusters are performed by assuming parameterized pair potentials[12]. However, this description is no longer valid when these clusters are ionized, since many-body effects like polarization come into play. In this article we present a microscopical theoretical description of the ionization induced fragmentation of small clusters and show, for the first time, that the UD opens the experimental possibility of observing phase-transitions in clusters. We consider small mercury clusters as prototype for strong response to ionization.

The article is organized as follows. In Sect. 2 we develop microscopic models to calculate the electronic properties of the clusters and to describe the short-time dynamics. In Sect. 3 we present the results for the ultra-short-time dynamics of both $Ag_3$ upon photodetachment and $Hg_n$ clusters after ionization. At the end of each subsection we present a brief summary of the main results obtained and discuss improvements and future investigations.

## 2 Theory

In what follows, we describe the electronic degrees of freedom quantum mechanically, whereas the nuclear motion is treated classically. This is justified by the fact that the de Broglie wavelength of heavy atoms like Ag and Hg is much smaller than the relevant length scale for changes in the potential. The coupling between the electrons and the atomic degrees of freedom is given by the Born-Oppenheimer approximation. This means that we consider the energy of electronic state as a function of the atomic coordinates as defining an adiabatic potential energy surface (PES). The PES represents the interaction potential for the atoms. The derivative of the PES with respect to the coordinates determines the forces acting on the atoms. The short time dynamics of the clusters is thus obtained by integrating the classical equations of motion.

## 2.1 Calculation of the PES and forces

### Ground state of Ag₃

In order to determine the potential energy surface (PES) needed for the MD simulations, we start from a Hamiltonian of the form[13, 14]

$$H = H_{TB} + \frac{1}{2} \sum_{i \neq j} \phi(\mathbf{r}_{ij}), \tag{1}$$

where the tight-binding part $H_{TB}$ is given by

$$H_{TB} = \sum_{i,\alpha,\sigma} \varepsilon_{i\alpha} c^{+}_{i\alpha\sigma} c_{i\alpha\sigma} + \sum_{\substack{i \neq j, \sigma \\ \alpha, \beta}} V_{i\alpha j\beta} c^{+}_{i\alpha\sigma} c_{i\beta\sigma}. \tag{2}$$

Here, the operator $c^{+}_{i\alpha\sigma}$ ($c_{i\alpha\sigma}$) creates (annihilates) an electron with spin $\sigma$ at the site $i$ and orbital $\alpha$ ($\alpha = 5s, 5p_x, 5p_y, 5p_z$). $\varepsilon_{i\alpha}$ stands for the on-site energy, and $V_{i\alpha j\beta}$ for the hopping matrix elements. For simplicity, and since the $5s$ electrons are expected to be rather delocalized, we neglect the intraatomic Coulomb matrix elements. $\phi(\mathbf{r}_{ij})$ refers to the repulsive potential between the atomic cores $i$ and $j$. For the distance dependence of the hopping elements and the repulsive potential we write, following Pettifor et. al.[15],

$$V_{i\alpha j\beta}(r_{ij}) = V^{0}_{\alpha\beta} \left(\frac{r_0}{r_{ij}}\right)^{n} \exp\left[n\left(-\left(\frac{r_{ij}}{r_c}\right)^{n_c} + \left(\frac{r_0}{r_c}\right)^{n_c}\right)\right],$$

$$\phi(r_{ij}) = \phi^{0} \left(\frac{r_0}{r_{ij}}\right)^{m} \exp\left[m\left(-\left(\frac{r_{ij}}{r_c}\right)^{n_c} + \left(\frac{r_0}{r_c}\right)^{n_c}\right)\right] \tag{3}$$

where $r_{ij}$ is the distance between atoms $i$ and $j$, $r_0$ is the equilibrium nearest distance for the lattice, $V^{0}_{\alpha\beta}$ are the hopping coefficients, $n$ is the hopping exponent, $\phi^{0}$ is the pair potential coefficient, $m$ is the pair potential exponent and $n_c$ is the cutoff exponent. With this distance-dependence we are able to account properly for the bond breaking and bond formation processes.

By diagonalizing $H_{TB}$, (taking into account the angular dependence of the hopping elements[16]), and summing over the occupied states, we calculate as a function of the atomic coordinates the attractive parts of the electronic ground-state energies $E^{-}_{attr}$ and $E^{0}_{attr}$ of Ag₃⁻ and Ag₃⁰, respectively. Then, by adding the repulsive part of $H$ we obtain the PES, which we need to perform the MD simulations. In order to determine the forces acting on the atoms we make use of the Hellman-Feynman theorem. Thus, the $\alpha$-component of the force acting on atom $i$ $F_{i\alpha} = -\partial E/\partial r_{i\alpha}$ is given by

$$F_{i\alpha} = -\sum_{k}^{occ.} \left\langle k \left| \frac{\partial H_{TB}}{\partial r_{i\alpha}} \right| k \right\rangle - \frac{1}{2} \sum_{i \neq j} \frac{\partial \phi(\mathbf{r}_{ij})}{\partial r_{i\alpha}}. \tag{4}$$

Here the $|k\rangle$'s are the eigenstates of $H_{TB}$ and the first sum runs over occupied states.

## Ground state of $Hg_n^+$ Clusters

For the description of the electronic structure of small, van der Waals bonded $Hg_n$ clusters after ionization we use the ionic-core model[17], which assumes that when a van der Waals cluster is ionized, the hole created in the ionization process delocalizes only within a subcluster (ionic core) of $m$ atoms (usually $m = 2$), whereas the other $n - m$ atoms in the cluster remain neutral. This model, which just imposes a constraint on the charge distribution of the cluster ion, has also been used, in combination with different electronic theories, to calculate cohesive energies and ionization potentials of ionized rare-gas clusters[9, 18–20]. The physics underlying the assumption of partial hole delocalization is related to the interplay between kinetic energy of the hole and the polarization energy of the remaining neutral part of the cluster[18]. The formation of an ionic core upon ionization has its origin in the interplay between delocalization energy of the hole and polarization energy of the cluster. If the hole becomes completely delocalized, it cannot efficiently induce dipole moments. On the other hand, if it remains completely localized, the polarization energy becomes large, but the kinetic energy to be paid is also large. The dimer core is a compromise resulting from the interplay between both these competing processes. Although the dimer-core model is a crude assumption to describe the charge distribution and the dipole distribution within the ionized cluster, recent all-electron, quantum-chemical calculations of $He_n$ clusters show that the hole is really localized in a subcluster, which is a trimer for small sizes and a dimer for larger ones[21]. Since the atomic polarizability of Hg is much larger than that of He, this suggests that the dimer-core is the lowest energy configuration for $Hg_n^+$ clusters. Thus, we start from a Hamiltonian, describing the main interactions present in ionized van der Waals clusters, which has the form[18, 24, 25]

$$H = H_{vdW} + H_{core} + H_{Q-P} + H_{P-P}. \tag{5}$$

In Eq. (5), $H_{vdW}$ describes the van der Waals interactions in the whole cluster, $H_{core}$ the hole hopping within the ionic core, $H_{Q-P}$ the interatomic charge-dipole interactions between the ionic core and the remaining neutral part, and $H_{P-P}$ refers to the dipole-dipole interactions between the $n - m$ neutral atoms. In the following we assume $m = 2$ (dimer core). The many-body Hamiltonian $H$ can be approximated by an effective single-particle Hamiltonian if one writes the charge and dipole operators as

$$\mathbf{P}_k = \langle \mathbf{P}_k \rangle + \delta \mathbf{P}_k,$$
$$Q_l = \langle Q_l \rangle + \delta Q_l, \tag{6}$$

respectively, and neglects terms containing charge-dipole fluctuations $\delta Q_l \delta \mathbf{P}_k$[18]. The resulting effective Hamiltonian can be easily diagonalized, and for $m = 2$ the electronic ground-state energy of the ionized cluster for a given atomic structure is given by[22, 24, 25]

$$E = \frac{(\phi_1 + \phi_2)}{2} - \sqrt{t_{ss}^2(\mathbf{r}_{12}) + \frac{(\phi_1 - \phi_2)^2}{4}}$$

$$+ \epsilon \sum_{\substack{k,k' \\ k \neq k'}} \left[ \left(\frac{r_0}{r_{kk'}}\right)^{12} - 2\left(\frac{r_0}{r_{kk'}}\right)^6 \right] - \frac{1}{2} \alpha \sum_{k \in \{n-2\}} | \mathcal{E}_k^Q + \mathcal{E}_k^P |^2$$

$$+ \sum_{k \in \{n-2\}} \langle \mathbf{P}_k \rangle \cdot \mathcal{E}_k^Q + \frac{1}{2} \sum_{k \in \{n-2\}} \langle \mathbf{P}_k \rangle \cdot \mathcal{E}_k^P. \qquad (7)$$

Here, the first two terms refer to the energy of the ionic dimer, where $t_{ss}$ is the interatomic $ss$ hopping element and $\phi_l$ is the potential of the induced dipole distribution 'felt' by the hole on atom $l$ of the dimer core. The third term describes the van der Waals energy and the core-core repulsions, both represented by a Lennard-Jones potential. The fourth term refers to the energy of the neutral subcluster as a system of induced dipoles $\langle \mathbf{P}_k \rangle$. $\alpha$ is the atomic polarizability, whereas $\mathcal{E}_k^Q$ and $\mathcal{E}_k^P$ are the charge- and dipole fields felt by atom $k$. Finally, the last two terms stand for the double-counting energy arising from the decoupling procedure used to obtain the effective single-particle Hamiltonian[18]. In Eq. (7), the expectation values of the charge ($\langle Q_l \rangle$) and dipole ($\langle \mathbf{P}_k \rangle$) distributions are coupled through the equations

$$\langle \mathbf{P}_k \rangle = \alpha \, \mathcal{E}_k \left( \langle \mathbf{P}_{k'} \rangle_{k \neq k'}, \langle Q_l \rangle \right),$$

$$\qquad (8)$$

$$\langle Q_l \rangle = |e| \sum_\sigma \langle n_{ls\sigma}^h \rangle,$$

which have to be solved self-consistently[18]. In Eq. (8), $\mathcal{E}_k = \mathcal{E}_k^Q + \mathcal{E}_k^P$, and $\langle n_{ls\sigma}^h \rangle$ is the number of holes with spin $\sigma$, at atom $l$, orbital $s$ of the ionic dimer. Thus Eqs. (7) and (8) define the electronic energy $E$ of the cluster. The function $E(\mathbf{r}_1, .., \mathbf{r}_n)$ depends on the atomic coordinates, and defines (within the Born-Oppenheimer approximation) a potential energy surface (PES) for the motion of the atoms. Note, that a similar derivation to that presented above can be done for the case of doubly ionized clusters (under the assumption of dimer-core formation). The only difference is that $H_{core}$ must describe the dynamics of two holes, taking also into account the Coulomb interaction between them[23].

The $\mu$ component of the force acting on a given atom $i$ is given by

$$F_i^\mu = -\frac{\partial E}{\partial \mathbf{r}_i^\mu}. \qquad (9)$$

The quantities $\phi_1$, $\phi_2$, $t_{ss}$, $\mathcal{E}_k^Q$, $\mathcal{E}_k^P$, which define the energy changes upon ionization [Eq. (7)] depend explicitly on the atomic coordinates. $\langle \mathbf{P}_k \rangle$ and $\langle Q_l \rangle$ have an implicit coordinate dependence and their derivatives with respect to $\mathbf{r}_i^\mu$ contribute to $F_i^\mu$. The quantities $\partial/\langle \mathbf{P}_k \rangle \partial \mathbf{r}_i^\mu$ and $\partial \langle Q_l \rangle \partial \mathbf{r}_i^\mu$ are obtained as the solution of a linear system of equations[24, 25]. Note that, although the equations satisfied by $\langle \mathbf{P}_k \rangle$ and $\langle Q_l \rangle$ are nonlinear, their derivatives are related by linear equations, which simplifies the numerical calculation. The effect of ionization on the electronic structure of the $Hg_n$ clusters can be visualized with the help of Eq. (7). Assuming that ionization is a very fast process which occurs at $t = 0$, and that the width of the ionizing laser pulse is negligible compared to the time scale of the nuclear motion, one has for $t = 0^-$ a cluster with a given temperature and structure. Since the total charge of the cluster is zero, all terms of Eq. (7) except the third one are zero, and the PES is defined by the Lennard-Jones potential. At $t = 0^+$ ionization has already taken place. The cluster has a net charge and $E$ has changed abruptly. Although the atomic structure remains the same as for $t = 0^-$, the forces are now determined by a different PES, whose gradients are much larger than for $t = 0^-$. Thus, the kinetic energy of the cluster increases rapidly, indicating energy transfer from electronic to atomic degrees of freedom. The most important contribution to the PES comes from the second term of Eq. (7). Thus, the dimer core obtains a large amount of kinetic energy $\delta E_{deloc}$, released by the electronic system mainly through hole delocalization, which is later distributed among the other atomic degrees of freedom. The contribution coming from the polarization energy $\delta E_{pol}$ is also relatively large (in comparison to that of the attractive part of the Lennard-Jones potential), but smaller than $\delta E_{deloc}$. The dynamics of the clusters after ionization consists in a redistribution of the total relaxation energy $\delta E \simeq \delta E_{deloc} + \delta E_{pol}$ among the different degrees of freedom of the system.

It is important to comment again on the dimer-core assumption used here. The dimer-core model might not describe correctly the time scales for the electronic processes (polarization and hole hopping). However, these time scales are definitively much smaller than the time scale for atomic motion. Consequently, the electronic structure rearranges itself long before the atoms change their positions. Therefore, and for the purpose of our study, it is not essential to know how the dimer is formed. An extensive analysis of the electronic time scales for processes taking place in these systems is given in Ref. [26]

## 2.2 Analysis of the chaotic behavior

The equations of motion for the coordinates $\mathbf{r}_i(t_n)$ and the velocities $\dot{\mathbf{r}}_i(t_n)$ are solved iteratively from their values at $t_{n-1}$. Since clusters are many-body systems, one could in principle expect the trajectories to depend strongly on the initial conditions, reflecting chaotic dynamics. This nonlinear effect can be quantified with the help of the Lyapunov exponents. The largest exponent $\lambda$ provides a measure for the average divergence of two initially nearby trajectories in phase space

$$| \delta \mathbf{r}(t_N) | \sim e^{\lambda N \tau} | \delta \mathbf{r}(0) | . \tag{10}$$

Formally, $\lambda$ is given by

$$\lambda = \lim_{N \to \infty} \frac{1}{N} \log \left| \prod_{n=0}^{N-1} J(\mathbf{r}_n, \mathbf{p}_n) \right|, \tag{11}$$

where $J(\mathbf{r}_n, \mathbf{p}_n)$ is the Jacobian matrix for the map from the configuration at time $t$ to the configuration at time $t + \Delta t$, and which for the Verlet algorithm is written as [5].

$$J(\mathbf{r}_n, \mathbf{p}_n) = \begin{pmatrix} \mathbf{I} - \frac{\tau^2}{2m}\mathcal{H}(n-1) & \frac{\tau}{m}\mathbf{I} \\ -\frac{\tau}{2}\left\{\mathcal{H}(n-1) + \mathcal{H}(n)\left[\mathbf{I} - \frac{\tau^2}{2m}\mathcal{H}(n-1)\right]\right\} & \mathbf{I} - \frac{\tau^2}{2m}\mathcal{H}(n) \end{pmatrix},$$
(12)

where $[\mathcal{H}(n)]_{ij} = \partial^2 E_\nu / \partial r_i \partial r_j$ is the Hessian matrix, and $\mathbf{I}_{ij} = \delta_{ij}$.

Of course, an analysis of the influence of chaos on the short-time dynamics makes no sense for very low initial temperatures (before excitation), i.e., when the system is near the zero point energy and quantum fluctuations play the most important role. For higher, but still low, initial temperatures an interesting question arises, namely: what is the interplay between chaos and quantum fluctuations? It is not our purpose to address this question here. We shall present a study of the chaotic behavior for those initial temperatures for which a classical description of the atomic motion is valid.

## 3 Results

### 3.1 Short-time dynamics of Ag$_3$

We present here results for the dynamics of silver trimers upon photodetachment, using the model described in Sect. 2.1 The parameters of $H$ are determined in the following way[13, 14]. The on-site energies were obtained from atomic data[27]. We fit the parameters $V_{i\alpha j\beta}(\mathbf{r}_{ij})$ and the potential $\phi(\mathbf{r}_{ij})$ in order to reproduce the equilibrium bond lengths of the silver dimers $Ag_2^-$, $Ag_2$ and $Ag_2^+$ obtained by effective core potential-configuration interaction calculations[7, 8]. We assumed the hopping elements $V_{\alpha\beta}$ to fulfill Harrison's relations[28]. The best fit was obtained by the following parameters: $V_{sp}^0 = 0.954\text{eV}$, $r_c = 4.33\text{Å}$, $n_c = 2$, $m = 5.965$ and $A = 0.605\text{eV}$, where $r_c$ and $n_c$ refer to the cutoff radius and exponent, whereas $m$ and $A$ stand for the exponent and strength of the repulsive potential $\phi(\mathbf{r}_{ij})$[15]. Using these parameters we calculated the vibrational frequencies of the dimers, which compare reasonably well with experimental values[29] and quantum-chemical calculations[7, 8]. Then, we determined the equilibrium geometries of the ground states of the silver trimers $Ag_3^-$, $Ag_3$ and $Ag_3^+$ which again yielded excellent agreement with the all-valence electron calculations of Refs. [7] and [8]. Note, our hopping parameters are comparable to those of bulk silver[30]. This gives another justification for the neglect of Coulomb interactions.

The MD simulations are performed applying the Verlet algorithm in its velocity form. We used a time step of $\Delta t = 0.05 fs$. This ensures energy conservation up to $10^{-6}\text{eV}$ after $10^5$ time steps[13, 14]. The equilibrium structures were obtained by performing simulated annealing. Starting with the equilibrium geometry of $Ag_3^-$, we generate an ensemble of approximately 1000 clusters characterized by the ensemble temperature T, defined as the time average of the kinetic energy for a long trajectory ($\sim 10^6$ time steps)[12].

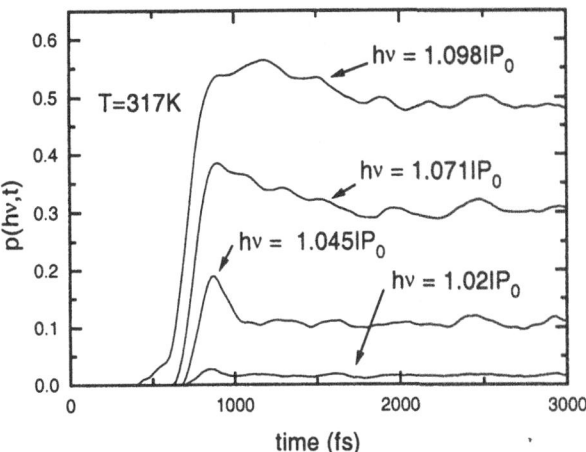

**Fig. 1.** Ionization probability $p(h\nu, t)$ for different values of $h\nu$, scaled by the ionization threshold $IP_0$. The initial temperature was $T = 317K$. The overall time- and frequency dependence compares well with the experimental results of Ref. 1. The most striking experimental facts, namely a sharp increase of the signal and a saturation for longer times are reproduced by our calculations.

In Fig. 1 we show $p(h\nu, t)$ for different values of $h\nu$, scaled with $IP_0$, which is the ionization threshold (ionization potential of $Ag_3$ in the equilateral equilibrium geometry of $Ag_3^+$). Note, slightly above $IP_0$, clusters can be ionized only after a delay of $t_0(h\nu) \simeq 750 fs$, where $p(h\nu, t)$ increases sharply. For longer times $p(h\nu, t)$ remains constant. For increasing energy the delay time $t_0(h\nu)$ decreases, and the signal displays the features observed in the experiment, namely, first a sharp increase, a maximum and then a decrease to a smaller value, which remains constant. For higher laser frequencies the maximum is broader and smaller relative to the long-time value of $p(h\nu, t)$.

These results can be understood physically as follows: upon photodetachment of a binding electron vibrational excitations occur, particularly for the central atom along the chain direction. The bond-lengths of $Ag_3^-$ are too small for the weaker bonded $Ag_3$ with only three electrons. The shape of the potential energy surface is responsible for the particular importance of the motion of the central atom. This motion dominates the ultrashort time response. Then, the slower thermally activated bending motion comes into play and yields triangular bonded $Ag_3$. The resultant bond formation is exothermic. The excess energy can in turn cause bond breaking or in the case of a uniform energy distribution a regular vibrational motion like pseudorotations.

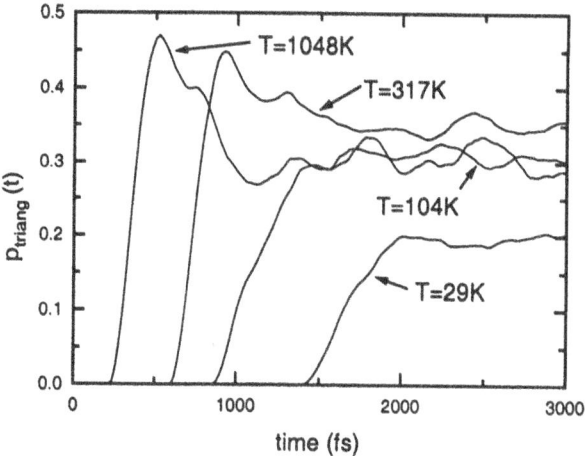

**Fig. 2.** Time dependence of the fraction of clusters $p(\phi_i(t))$ having a triangular structure with angles $\phi_i$ between 45 and 90 deg for different initial temperatures.

In order to demonstrate the temperature dependence of the bond formation
and bond breaking processes, we present in Fig. 2 results for the time dependent
probability $p_{triang}(t)$ of finding triangular clusters for different initial temperatures. $p_{triang}(t)$ is defined as the fraction of clusters having all angles between
45 and 90 deg. Comparison of the of $p_{triang}(t)$ for T = 317K with the curves
of Fig. 1 clearly shows that only these particular cluster geometries contribute
to the signal observed in the experiment[3]. Note that the onset of the abrupt
increase of $p_{triang}(t)$ shows a strong temperature dependence. With increasing
initial temperature the clusters bend faster. Also note that at low temperature
the maximum in $p_{triang}(t)$ disappears[13].

In Fig. 3 we show the time development of the (initially) large angle of the
silver trimer for the complete cluster ensemble. The distribution of the clusters
over different angles is indicated by grey scales. Black areas correspond to the
highest density of clusters, while in white areas no clusters are present. In the
initial cluster ensemble with a temperature of 317 Kelvin, the large angle is
in the range of 150 deg to 180 deg. After a time delay of only 800fs clusters
with geometries close to the equilibrium structure of $Ag_3$ are present in the
ensemble. Two effects contribute to this delay. The first one is that, owing to
the nonzero ensemble temperature, the clusters start their bending motion with
certain initial velocities, which are only slowly augmented by the energy gained
when descending the potential energy surface, because its slope is small as long
as the large angle is in the range of 180 to 120 deg. The second one is the already
mentioned oscillation of the central atom along the chain direction (and of course
smaller compensatory oscillations of the other two atoms). Because the slope of
the PES in this dimension for the more or less linear molecule is far greater than
the slope for the bending motion, the latter only becomes activated by the initial
temperature of the clusters. A prominent feature in Fig. 3 is the large black area

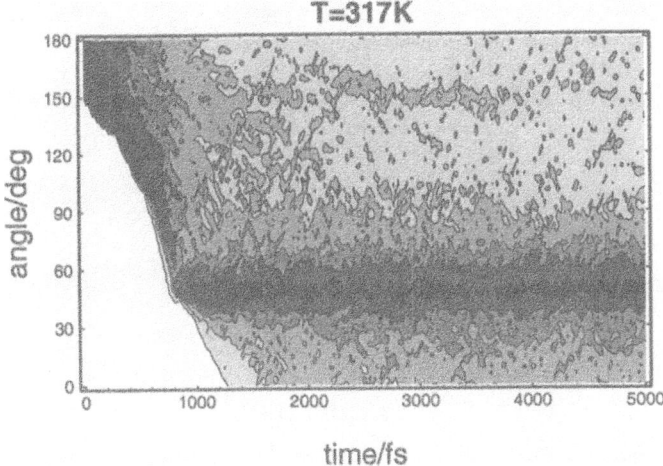

**Fig. 3.** Time-dependent distribution of the (initially) large angle of the silver trimer for the complete cluster ensemble. Black areas correspond to the highest density of clusters, while in white areas no clusters are present.

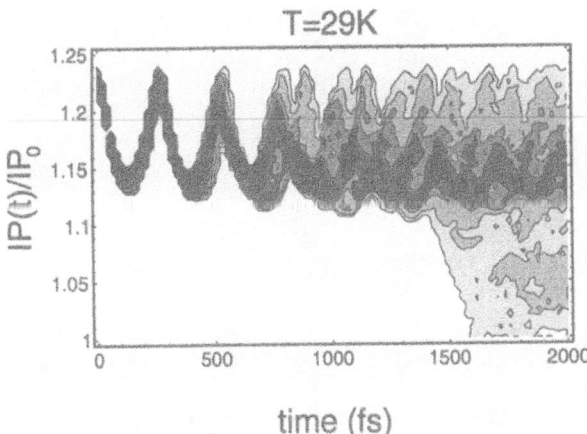

**Fig. 4.** Distribution of photodetached clusters as a function of time and IP for an initial temperature $T = 29K$. Black regions indicate large number of clusters, whereas no clusters are present in the white parts.

roughly in the time interval [800fs, 1800fs]. This corresponds to the maxima of $p(h\nu, t)$ in Fig. 1 and to the maxima seen in the experiment of Ref[3]. Physically this effect can be understood as resulting from the coherent movement of clusters in the ensemble. Starting with the photodetachment at $t = 0$, a significant part of the clusters oscillates and bends collectively, producing a number of clusters close to the equilibrium geometry of $Ag_3$ that cannot be reached again at greater times, because the temperature of the ensemble quickly destroys all coherence[14].

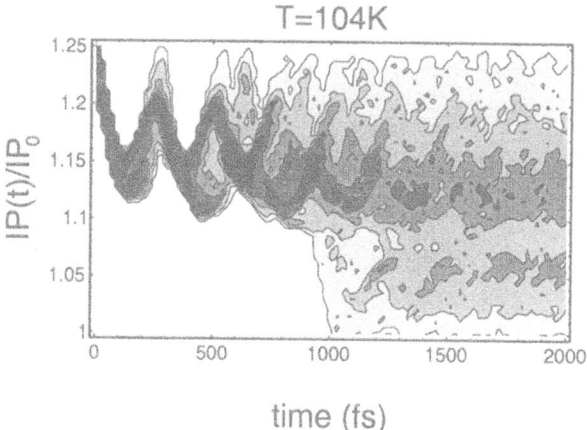

**Fig. 5.** Distribution of photodetached clusters as a function of time and IP for an initial temperature $T = 104$K.

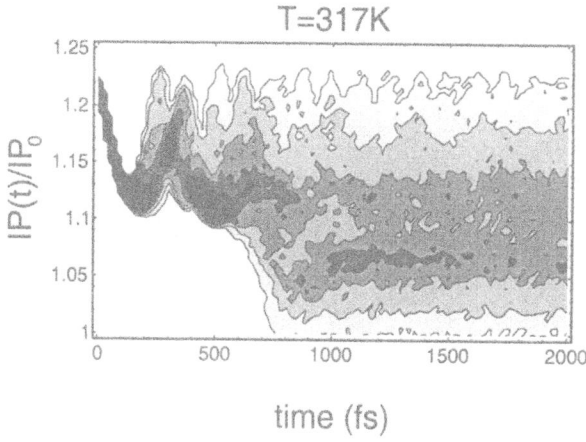

**Fig. 6.** Distribution of photodetached clusters as a function of time and IP for an initial temperature $T = 104$K.

Figs. 4,5 and 6 show contour plots for the time development of the distribution of the IPs of neutral silver trimers after photodetachment at $t = 0$. The most prominent feature of these pictures is the coherent oscillation of the IP of all clusters, which continues for about 700fs in the case of clusters starting at $T = 29$K and which is destroyed much more quickly at higher temperatures. These coherent oscillations of the IP are due to the internal vibrations of the linear chain before and during bending. High values of the IP correspond to a rather symmetric geometry, while the lower values stand for the turning points of the oscillation. The second feature to be noted are the areas in the plots with IPs below approximately $1.07IP_0$, that become populated only after time delays

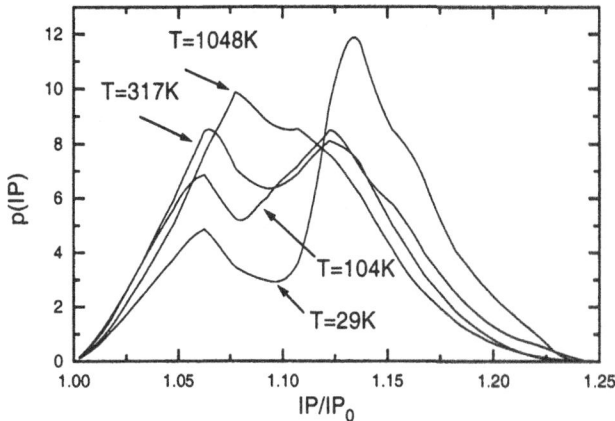

**Fig. 7.** Average distribution of neutral silver clusters over IPs for $2ps \leq \Delta t \leq 20ps$ after photodetachment.

of 1500, 700 and 300 fs respectively and which represent clusters that have bent to form geometries close to the equilibrium geometry of $Ag_3$.

With the help of Figs. 4-6 it is possible to explain the maximum of the IP as a result of the coherent vibrational motion of the clusters, i.e., while a large fraction of the clusters bend collectively, they still continue their vibrational motion, which results in a pronounced minimum of the IP in the same way as for the unbent trimers. This effect accounts for the dark areas in Fig. 6 for $IP(t)/IP_0$ approximately below 1.07 and is quantified by integration over energies in the calculation of $p(h\nu, t)$ [Fig. 1]. Since the clusters responsible for this effect have gained an average energy of 900K, while descending the PES towards the equilibrium structure of $Ag_3$, their coherent motion is quickly destroyed and this explains why no further maxima occur. This is supported by the fact that the cluster ensembles, starting out with temperatures of 29K and 104K (Figs. 4 and 5), do not show this maximum. Here, the coherent motion has already disappeared before the clusters bend toward the equilibrium geometry of $Ag_3$.

Figs. 4-6 reveal that for longer times ($t > 2000fs$) the clusters approach a time-independent distribution over the IPs or equivalently over the $Ag_3$ geometries. These asymptotic distributions are plotted in Fig. 7. Fig. 7 suggests that temperature can be used to design special distributions for the IP or for the cluster geometries present in the initial ensemble.

In Fig. 8 we have plotted the quantity $p(h\nu, t)$ as in Fig. 1(a), but for an initial temperature of $T = 29K$ and for a higher energy $h\nu = 1.17IP_0$. A new feature appears which consists in sharp oscillations of the probability between 0 and 1 within the time range $[0 \leq t \leq 700fs]$. These oscillations result from the dominant vibrational mode of the trimers for the first 700fs. This interesting effect was not observed in the experiment[3], because the laser energies used were not high enough.

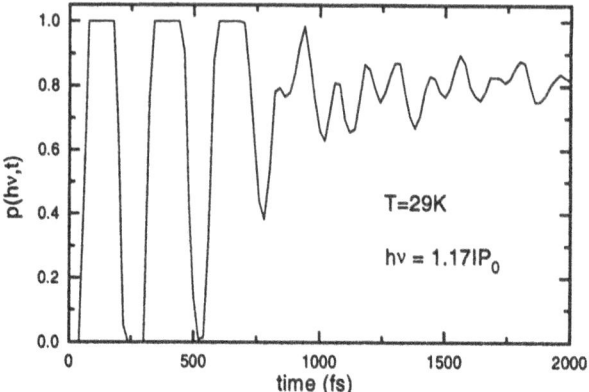

**Fig. 8.** Calculated $p(h\nu, t)$ for a large value of $h\nu$. Note that the internal vibration of the chain before and during bending can be detected.

The last important point to be investigated is the role of pseudorotations. In order to study these vibrational excitations we determined the autocorrelation function $G(\tau)$[14]

$$G(\tau) = \frac{\sum_k q_x(t_k)q_x(t_k + \tau)}{\sum_k q_x(t_k)q_x(t_k)}, \tag{13}$$

where $q_x$ is one of the normal coordinates of the triangular $Ag_3$ molecule[31]. If the pseudorotations had a long lifetime, $G(\tau)$ would show an oscillatory behavior. In the limit of infinite lifetime it must hold that $q_x \propto \sin(\omega_{PR}t)$, with $\omega_{PR}$ the pseudorotational frequency. However, as shown in Fig. 9(a), the correlations are strongly damped. This becomes clear by analyzing the Fourier transform of $G(\tau)$. The lifetime of the pseudorotations, determined from the width of the peak at half maximum, is only slightly shorter than the duration of one cycle. This shows that pseudorotations do not play an important role in the dynamics of the trimers. The high kinetic energy of the atoms seems to prevent the occurrence of a regular mode like pseudorotation. Instead, the saturation of the signal is a statistical effect induced by the temperature.

## Summary and Conclusions

By employing an electronic theory and MD simulations we have analyzed the femtosecond dynamics of $Ag_3$ upon photodetachment. Our main results are:
1) The photodetachment process induces a bending of the neutral trimer (originally in a linear configuration).
2) The bending motion is coherent until a characteristic time which depends on the initial temperature. This coherent dynamics explains the time evolution of the experimental signal during the first picosecond.
3) The time scale for bending is sensitive to the initial temperature. This is due to the fact that the motion of each trimer after detachment strongly depends on the initial positions and velocities of the silver atoms in the $Ag_3^-$ cluster.

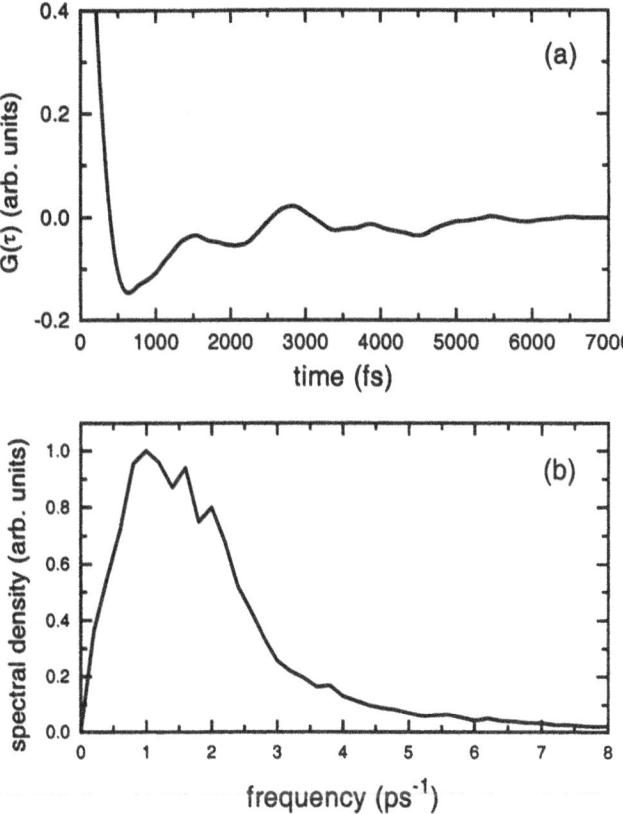

**Fig. 9.** (a) Behavior of the correlation function $G(\tau)$ (see text) showing the damping of the pseudorotations. (b) Spectral analysis of the pseudorotations. Note that the life-time is smaller than one pseudorotation period.

In our parametrization of the PES we found that both the intramolecular vibration (asymmetrical stretch mode) and the bending of the neutral trimer are excited by the photodetachment process. The excitation of the stretch mode can be explained by the removal of an electron from an antibonding orbital. The vibrational and bending motions in the neutral trimer take place on different time scales. The redistribution of excess energies among these motions depends on initial conditions, but also on the exact shape of the PES. Thus, further analysis regarding this point is desirable. So far, experiments cannot shed light on the time scales for activation of the different atomic degrees of freedom.

Our results show that a physically transparent theory, which can easily be extended to larger systems is able to account for all experimental observations[13, 14]. Most importantly, we find that the time scales for bond formation and breaking is affected by temperature[14]. Recent simulations based on a PES determined by quantum chemical techniques confirm this sensitivity of the response to the initial conditions[32, 33].

Our results might also be of general interest regarding laser control of chemical reactions. Fig. 7 suggests that temperature can be used to design special distributions for the IP or for the cluster geometries present in the initial ensemble. Fig. 7 shows, that by performing photodetachment and waiting until equilibrium is reached, one obtains a distribution of cluster structures which is strongly dependent on the initial temperature. Thus, by varying the initial temperature, one would be able to produce a desired distribution composed predominantly of chemically reactive structures[14].

## 3.2 Hg$_n$ clusters

### Fragmentation dynamics

We have performed molecular dynamics simulations on small ionized Hg$_n$ clusters ($3 \leq n \leq 13$). To solve the equations of motion we used the Verlet algorithm in velocity form, and determined the forces for each time step through the self-consistent procedure described in Sect. 2. For the distance dependence of the hopping element $ss$ ($t_{ss}(r_0) = 0.5$eV) we use the functional form proposed in Ref. [15]. $\varepsilon = 0.042$eV and $r_0 = 3.63$Å are the cohesive energy and the bond-length of Hg$_2$, respectively[34]. For the atomic polarizability we use $\alpha = 5.7$Å$^3$[35].

In Figs. 10 (a) and (b) we show results for the time evolution after ionization of a Hg$_3$ cluster at zero temperature[36]. In this cluster the large kinetic energy of the dimer core can be transferred to only one neutral atom, which is thus expected to be rapidly emitted. The short-time dependence of the kinetic energy per atom $E_{kin}(t)$ is shown in Fig. 10 (a). As expected, after ionization the cluster begins to oscillate, and the same occurs with $E_{kin}(t)$. The kinetic energy, which has a maximum of $\sim 0.5eV$ per atom, is concentrated in the dimer for $t \leq 1ps$. Therefore, the function $E_{kin}(t)$ goes to zero periodically, indicating the times at which the dimer motion reaches its turning points. At $t \sim 1ps$ the third atom begins to absorb kinetic energy from the dimer core and move, and this motion becomes superimposed on the dimer oscillations. For $t \geq 3.8ps$ a change in the time behavior occurs. $E_{kin}(t)$ continues to oscillate, but its minimum value is no longer zero but a constant $E_{kin}^0 \sim 0.05eV$. This indicates that one of the atoms, in this case the neutral one, has a time-independent kinetic energy and has left the cluster. The fragmentation process becomes clearer in Fig. 10(b), where the time-dependent distance of the neutral atom to the center of mass of the dimer core is shown. After a few oscillations the atom is emitted. By comparing Fig. 10 (a) with Fig. 10 (b) one observes that the frequency of the dimer oscillations is much larger than that of the motion of the neutral atom, reflecting the fact that the contribution from hole delocalization is larger than that one from charge-dipole interactions[25]. Fragmentation in this case seems to result from a decoupling of the dimer oscillation mode and the oscillation of the neutral atom. This explains the rapid ('cold') fragmentation. In Fig. 10 (c) we analyze fragmentation in the case of a Hg$_{13}$ cluster with initial zero temperature and icosahedral structure, for which we assume that after ionization the dimer

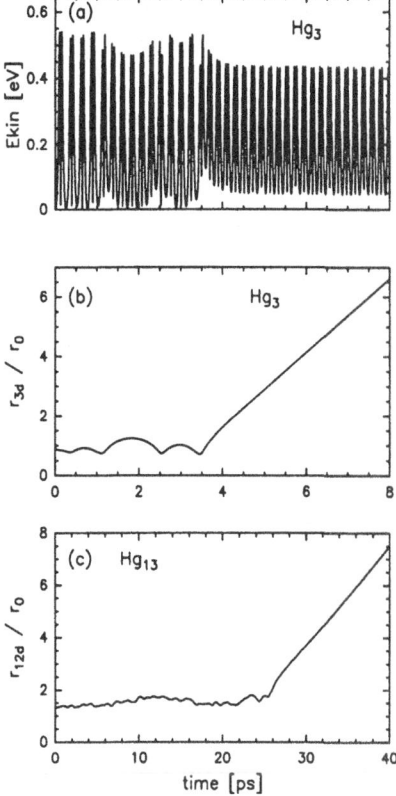

**Fig. 10.** Time evolution and fragmentation of $Hg_n$ clusters upon ionization. (a) Time dependence of the kinetic energy per atom for $Hg_3$ with an initial equilateral triangular structure. (b) Distance $R_{3d}$ from the emitted atom (3) to the dimer core (atoms 1 and 2)(see text). (c) Time dependence of the distance $R_{12d}$ from an emitted atom (12) to the dimer core for $Hg_{13}$. The dimer core is formed by atoms 2 and 3. Distances are scaled by $r_0$

core is situated on the surface. The distance of one of the neutral atoms to the center of mass of the dimer core shows a more complex dynamics than in the case of $Hg_3$. The kinetic energy transferred from the dimer core to the rest of the cluster is distributed among many degrees of freedom, and the fragmentation process takes much longer. In this case the first neutral atom is emitted $25ps$ after ionization. The oscillations of the emitted atom against the dimer core before fragmentation are irregular, indicating the redistribution of kinetic energy (thermalization). This results in a slow ('hot') fragmentation.

In order to demonstrate the importance of a self-consistent calculation of the PES, we show in Fig. 11(b) the time behavior of the positive charge on the ionized core for a $Hg_5^+$, which emits two atoms within the first 30ps of its relaxation process. Immediately after ionization, the hole starts hopping back and forth between the dimer-core atoms. This reflects the structural relaxation

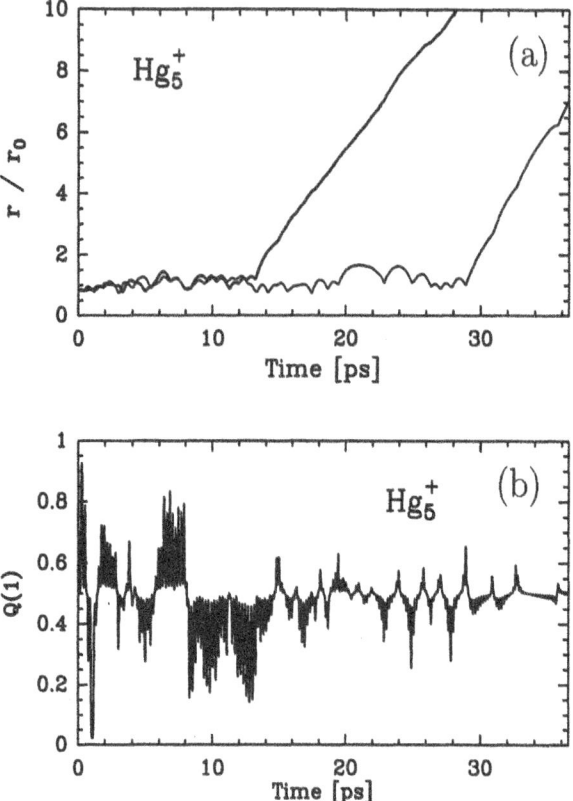

**Fig. 11.** Fragmentation dynamics of a $Hg_5^+$ cluster after ionization. Within the first 30ps after ionization two atoms have already left the cluster. The corresponding fragmentation (emission) times are 13ps and 29ps, respectively. Subfigure (a) shows the time dependence of the distances of the emitted atoms to the dimer core. In (b) the time dependence of the positive charge on one of the atoms of the dimer core is shown.

of the cluster, which influences the dipole and charge distributions. After the first atom leaves the cluster, the temperature of the cluster decreases, and the charge fluctuations become weaker. Emission of the second atom causes a further temperature decrease and the hole occupation becomes equal for both atoms of the dimer core.

In Fig. 12 we show the time dependence of the total polarization $\Pi = \sum_{k\in\{n-2\}} |\langle \mathbf{P}_k \rangle|^2$ of a $Hg_4^+$ cluster after ionization. $\Pi(t)$ reflects the time-dependent rearrangement of the dipole distribution during the relaxation process. After both neutral atoms have left the cluster, the induced dipole moments decrease due to the increase of the distance to the charge dimer, and consequently $\Pi(t)$ goes to zero[22, 25].

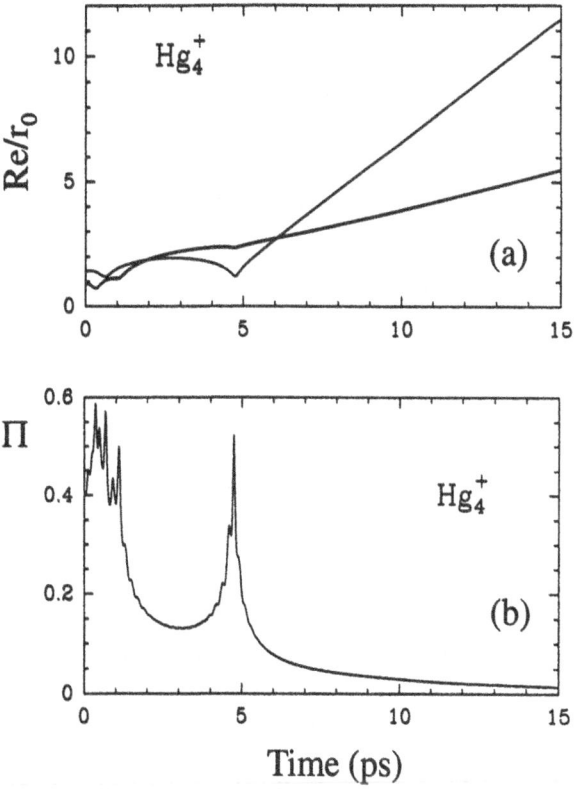

**Fig. 12.** Fragmentation dynamics of a $Hg_4^+$ cluster after ionization. Subfigure (a) shows the time dependence of the distances of the emitted atoms to the dimer core. In (b) the time dependence of the total dipole moment of the cluster is shown.

### Fragmentation Distributions: Nonlinear behavior

In order to describe real experimental conditions, we performed simulations at nonzero initial temperature. Thus, we obtained, for each cluster size and each initial temperature, a distribution of fragmentation-times given by the number of clusters whose fragmentation occurs at $\tau_F$. In Fig. 13 we show the normalized fragmentation-time distribution, $W(\tau_F)$, calculated for $Hg_6^+$ clusters with an initial temperature of $T = 40K$. Note that one can interpret the function $W(t)$ as the probability of fragmentation per unit time. Similar results to those of Fig. 13 have been obtained for $n = 3, 4, 5, 13$.

In Fig. 14(a) we show the normalized fragmentation-time distribution (FTD), $W(\tau_F)$, calculated for $Hg_3$ clusters with an initial temperature of $T = 40K$. Note that one can interpret the function $W(t)$ as the probability of fragmentation as a function of time. It becomes clear, that for $T \neq 0$ one cannot distinguish anymore between cold and thermal fragmentation as a function of the cluster size. For all sizes studied, both forms of fragmentation are present. The FTD of Fig. 14 (a) exhibits two general features common to all calculated distributions

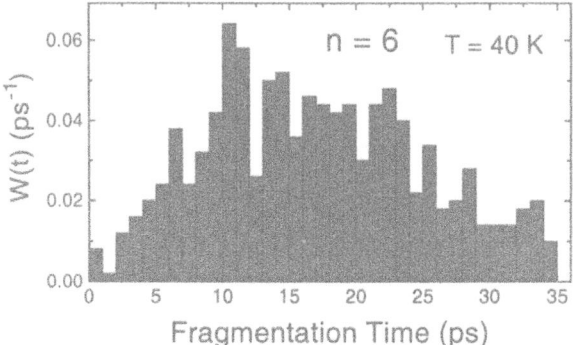

**Fig. 13.** Calculated fragmentation-time distribution $W(t)$ for $Hg_6^+$ clusters at $T = 40K$ before ionization. Note that $W(t)$ also represents the probability of fragmentation per unit time.

(s. also Fig. 13). It has a certain width (usually larger than 10ps), and its weight at short times is small. It is important to remark that the calculated decay-probability $W(t)$ is not constant, as is usually assumed. In contrast to what one would expect, the width of $W(t)$ does not decrease for decreasing temperature. Since lower temperature implies smaller displacements of the atoms around their equilibrium positions, this is a clear indication of a nonlinear dependence of $\tau_F$ on the initial conditions. For the sake of illustrating this effect, we determine the FTD for $Hg_3$ clusters at an extremely low temperature, for which the atomic displacements are of the order of $10^{-6}$Å. Note that this temperature corresponds to a kinetic energy smaller than that of the zero point motion of $Hg_3$ clusters and cannot be used to compare with experiment. In Fig 14(b) the resulting FTD is shown. The salient feature is a still a large width, the presence of a gap and several sharp peaks. This strange form of $W(t)$ is another indication of the intrinsic nonlinear behavior of the fragmentation dynamics. Small clusters are chaotic systems[5, 6]. This means that the trajectories in phase space are very sensitive to the initial conditions. Since the fragmentation times strongly depend on the trajectories of the ionized clusters on the PES, it is clear that they should exhibit nonlinear behavior. The various peaks of $W(t)$ in Fig 14 (b) correspond to fragmentation times which occur starting from many different initial configurations. We obtain this chaotic behavior after fragmentation for all cluster sizes studied ($n > 2$). A very small change of the initial conditions leads to completely different trajectories which involve different processes (e.g., cold- or thermal- or even no fragmentation). Note that in our simulations energy is conserved up to $10^{-4}$eV for $10^5$ time steps after ionization. Results of Fig 14 (b) serve just as an example, which is aimed to introduce chaotic effects in the short-time dynamics. Of course, at very low temperatures quantum effects must dominate and will tend to smear out the nonlinear features. But at low temperatures above the zero point motion, where the mean displacements of the

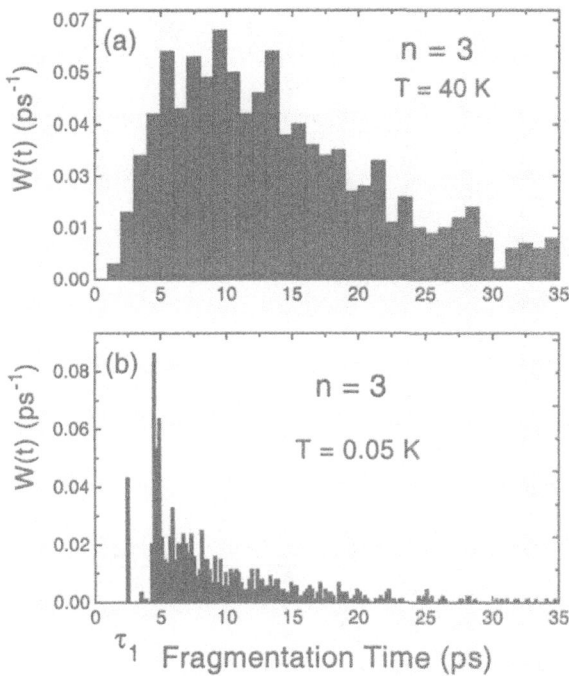

**Fig. 14.** Calculated ionization induced fragmentation-time distributions $W(t)$ for a) Hg$_3$ clusters at initially $T = 40$K and b) Hg$_3$ clusters at initially $T = 0.05$K.

atoms are still very small ($\sim 0.05$Å), the chaotic behavior might play a role in the fragmentation dynamics.

We found that the trajectories are strongly sensitive to the initial conditions. Even though total energy and angular momentum are conserved (up to at least 0.01% in the simulations we performed), an infinitesimal change in the initial coordinates leads to a completely different trajectory (see Fig. 15). The same occurs if the initial velocities are slightly changed. This is an indication of chaos. The most important consequence of this sensitivity to the initial conditions is that not only the trajectories, but also the *fragmentation times* strongly depend on the initial coordinates and velocities. Thus, fragmentation times cannot be estimated by using linear response considerations.

In Fig. 16 we show the temperature dependence of the (averaged) largest Lyapunov exponent for Hg$_4$ clusters before ($\lambda^0$) and after ($\lambda^+$) ionization. At a given initial temperature, the calculated $\lambda^+$ is more than one order of magnitude larger than $\lambda^0$. This clearly indicates that degree of nonlinearity is enhanced by the ionization process[24]. Due to the large energy $\delta E$ pumped into the atomic degrees of freedom, the cluster can explore a larger part of the PES, which leads to an increase of nonlinear behavior. Since $\delta E$ is of the order of $10^4 K$ it is also clear that $\lambda^+$ is almost independent of the initial temperature. This explains why the width $\Gamma$ of $W(t)$ does not decrease for decreasing temperature, and

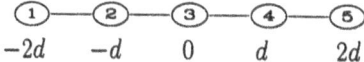

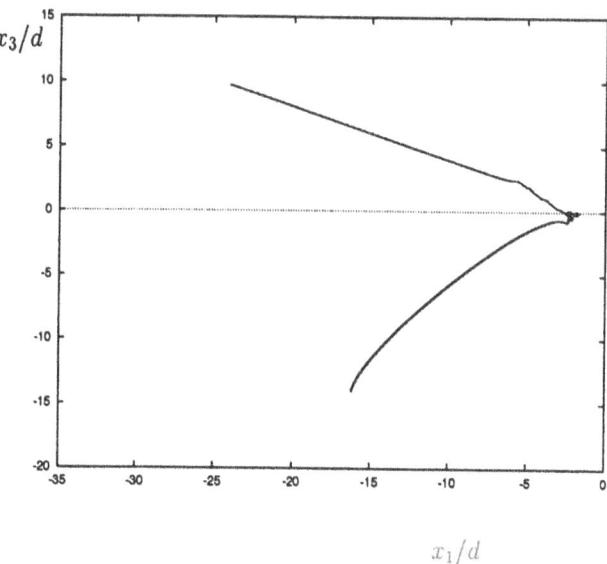

$x_1/d$

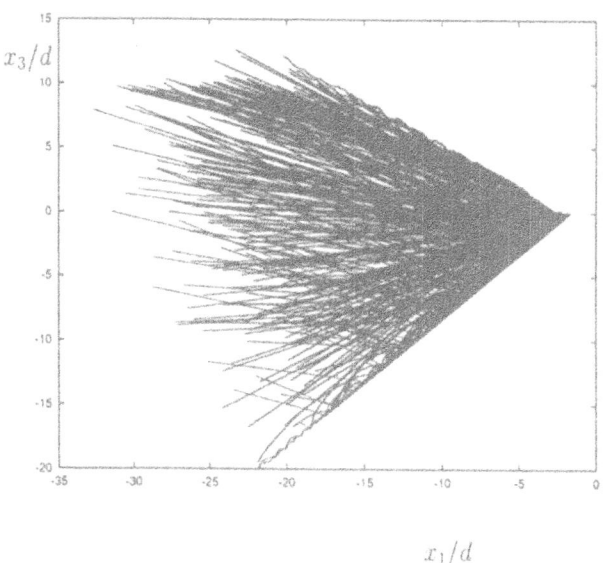

$x_1/d$

**Fig. 15.** Sensitivity of trajectories to the initial conditions. Upper subfigure: projection of two trajectories for linear $Hg_5$ clusters onto a plane $X_1(t)$, $X_3(t)$, defined by the coordinates of atoms 1 and 3. The initial structures differ by an infinitesimal change $\Delta X$ in the coordinates of atoms 4 and 5 ($\Delta X \sim 10^{-6}$ Å!). Distances are scaled by $d = r_0$. Lower subfigure: projection of 1000 different trajectories belonging to the same initial condition (up to changes of $\sim 10^{-6}$ Å).

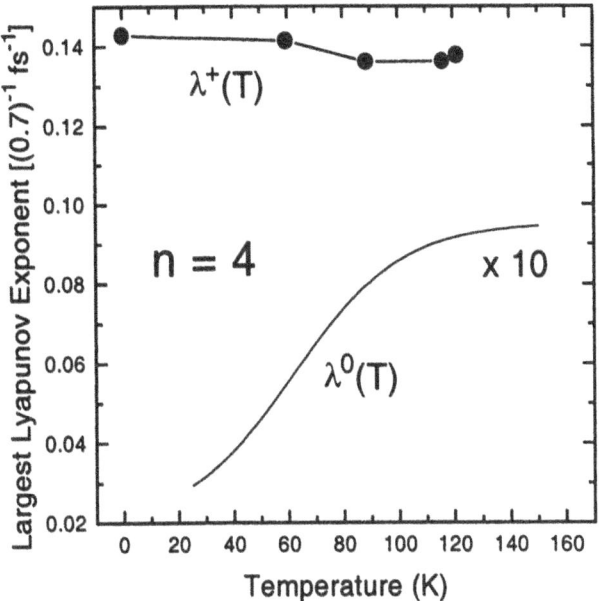

**Fig. 16.** Calculated temperature dependence of the largest Lyapunov exponent $\lambda(T)$ (in units of $\Delta t^{-1}$) for Hg$_4$ clusters before ($\lambda^0(T)$) and after ($\lambda^+(T)$) ionization. Note that the ionization process enhances the chaotic behavior of the clusters, by more than one order of magnitude.

shows that $\Gamma$ is mainly due to the nonlinear behavior. In contrast, $\lambda^0$ increases appreciably within the temperature range where the solid-like to liquid-like transition occurs, reflecting the increase of the nonlinear behavior. This has been also suggested for Ar$_n$ clusters[6].

## Temperature Effects: Sensitivity to Phase Transitions

For low temperatures, the calculated $W(t)$ and the corresponding values of $\langle \tau_F \rangle$ show only a weak temperature dependence. However, as shown in Fig. 17, we obtain a remarkable change in $W(t)$ in a temperature range which can be related to the solid-like to liquid-like transition of neutral clusters. This indicates a correlation between the energy transfer mechanisms after ionization and the thermodynamical state of the cluster before ionization.

For a more quantitative analysis of the temperature dependence of the average fragmentation times we have calculated the root-mean-square (rms) bond-length fluctuation

$$\delta = \frac{2}{n(n-1)} \sum_{i<j} \frac{\sqrt{(\langle r_{ij}^2 \rangle - \langle r_{ij} \rangle^2)}}{\langle r_{ij} \rangle}, \tag{14}$$

where $\langle ... \rangle$ means time-average over a long trajectory ($10^6$ time steps). For bulk material and for increasing temperature, $\delta$ shows typically a sharp increase at the

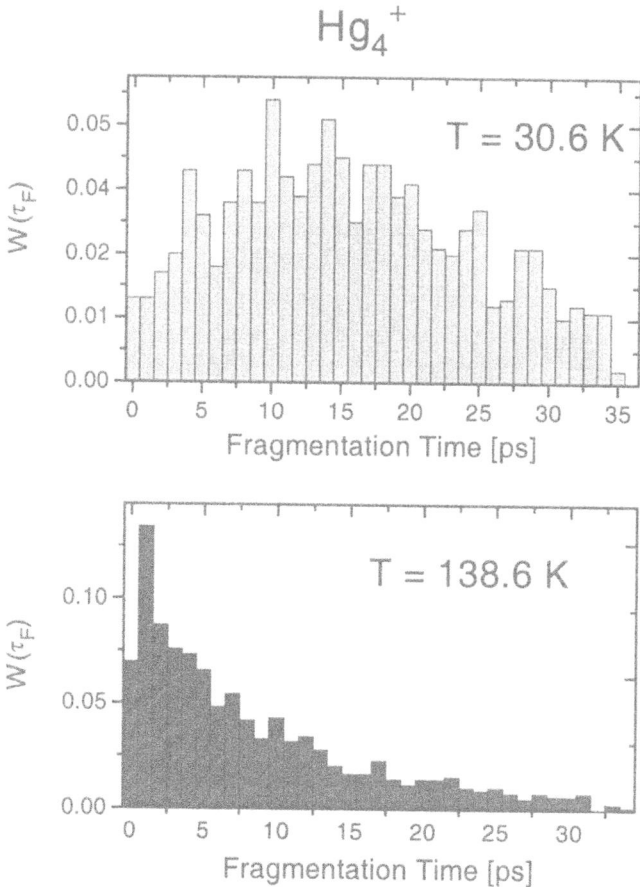

**Fig. 17.** Illustration of the changes induced on the fragmentation-time distributions when the initial temperature crosses a given 'critical' value.

solid-liquid transition, consistent with the Lindemann-criterion. In small clusters $\delta$ also characterizes the thermodynamical state[12]. In Figs. 18 (a), and (b) the temperature dependence of $\delta$ is shown for $Hg_n$ clusters ($n = 3, 13$). For $n = 3$ and 13, $\delta$ clearly shows a jump[25].

In Fig. 18 we also show the remarkable sensitivity of the average fragmentation times to the melting dynamics. Note that $\delta(T) \propto \langle \tau_F \rangle^{-1}(T)$. Such a correlated temperature dependence can be understood as follows. Fragmentation occurs on an average after a certain (incomplete) thermalization process. Thus, the distribution of the excess energy $\delta E$ among the different degrees of freedom leads to a homogeneous weakening of the bonds. This means that atoms (fragments) are emitted from a liquid-like cluster. If the cluster was in a solid-like state before ionization, part of the relaxation energy $\delta E$ has to be used first as latent heat. Only after melting can fragmentation occur. As a consequence, the

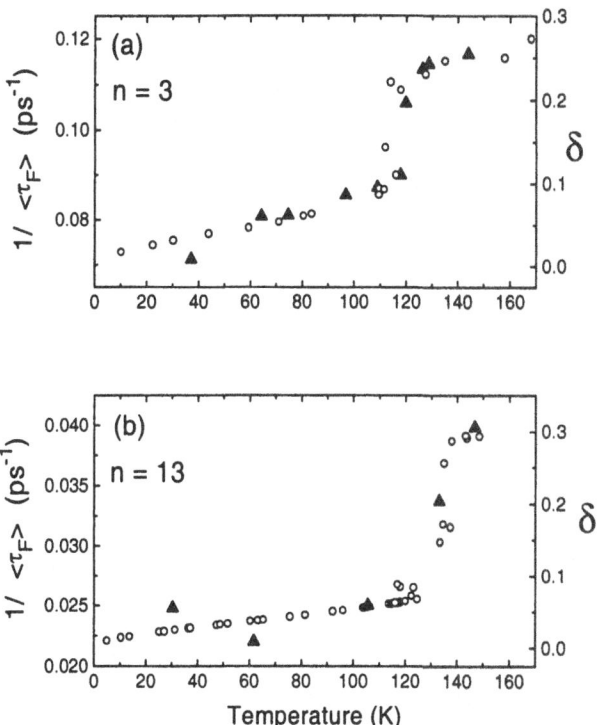

**Fig. 18.** Temperature dependence of the inverse average fragmentation times $\langle \tau_F \rangle^{-1}$ for a) $Hg_3^+$ and b) $Hg_{13}^+$ clusters (up triangles, left axis) and rms bond-length fluctuations (open circles, right axis) $\delta$ before ionization for a) $Hg_3$, and b) $Hg_{13}$ clusters. Note that the increase in $\langle \tau_F \rangle^{-1}$ characterizes the melting temperature.

mean fragmentation times are smaller for initially liquid clusters. Obviously, if $\delta E$ is smaller than the latent heat of the cluster, fragmentation may only occur above the melting point. Note that the above argument is independent of the bond character of the clusters considered. However, it is essential that a vertical ionization puts the cluster in a non-equilibrium situation. We believe that this effect can also be observed in metallic and covalent clusters.

## Summary and Conclusions

We have also studied the ionization induced ultrafast dynamics of small $Hg_n^+$ clusters. Nonlinear effects are found to be important. We have shown that the chaotic behavior of small clusters may play an important role for the ultrafast dynamics upon ionization[24].

We obtain a remarkable temperature dependence of $\langle \tau_F \rangle^{-1}(T)$,[25] which could allow one to determine the solid-like to liquid-like transition of small clusters by pump&probe experiments. We obtain a very small weight of the fragmentation-time distributions $W(t)$ at short times, which reflects a small probability for fragmentation at times shorter than the characteristic time for excess energy transfer. Thus, if one writes a master equation for the decay of the original clusters of size $n$, one obtains for the time-dependent number of clusters

$$N_n(t) = N_n(0)e^{-\int_0^t dt' W(t')}. \tag{15}$$

Moreover, since $dW(t)/dt \,|_{t=0} > 0$, it follows that $d^2 N_n/dt^2 \,|_{t=0} < 0$ in agreement with what is observed in pump probe experiments[1, 2], and in contrast with what one would obtain by assuming a constant $W$. Thus, from the experimental signal $N_n(t)$ one can obtain $W(t)$ and then $\langle \tau_F \rangle(T)$. We have shown that, even for the considered extreme case of strong hole localization, fragmentation occurs only after a certain thermalization process. Thus, we expect our results also to be valid for metallic and covalent clusters, where after ionization a weaker hole localization, and consequently a more homogeneous initial distribution of the excess energy $\delta E$, is present. It is interesting to compare the fragmentation behavior obtained in this paper with widely used kinetic theories as the RRKM theory[37, 38]. Unimolecular dissociation is usually described by a time-independent rate constant $k$, which depends on the available phase space for the original cluster. In the case considered in this paper, the cluster is put by the ionization process in such an extreme non-equilibrium situation, that the rate constant $W(t)$ depends on time. The time during which this first cooling process takes place can be estimated from the width of the function $W(t)$. For larger times after ionization $W(t)$ should converge to a small and constant value $W(t \to \infty) = k$, compatible with the RRKM predictions. The so called life-time distribution $P(\tau)$, used in the theory of unimolecular reactions[37, 39], would be given, in our case, by $P(\tau) = W(\tau) \exp\left[-\int_0^\tau dt' W(t')\right]$. This is reduced to the widely used expression $P(\tau) = k \exp(-k\tau)$ only for a constant fragmentation probability per unit time.

The theory presented here could serve as a basis for the study of other relevant problems. For instance, recent short-time experiments on graphite[40] show that it is in principle possible to obtain liquid carbon by exciting a large number of electron-hole pairs with an ultrashort laser pulse. In order to describe such processes, an extension of our theory is needed in order to take into account explicitly the laser pulse and the interplay between one-photon and two-photon absorption.

# 4 Acknowledgments

We thank Dr. D. Reichardt for useful discussions and for his contributions to our theory. Part of this work has been done in collaboration with him.

# References

1. J. Manz and L. Wöste (eds.), *Femtosecond Chemistry*, Verlag Chemie, Heidelberg, 1995.
2. T. Baumert, R. Thalweiser, V. Weiss, E. Wiedemann and G. Gerber, Proc. of the Royal Netherlands Academy Colloquium on *Femtosecond Reaction Dynamics*, edited by Douwe A. Wiersma, **42**, 29 (1994).
3. S. Wolf, G. Sommerer, S. Rutz, E. Schreiber, T. Leisner, L. Wöste and R. S. Berry, Phys. Rev. Lett. **74**,4177 (1995).
4. D. W. Boo, Y. Ozaki, L. H. Andersen and W. C. Lineberger, J. Phys. Chem. A **101**, 6688 (1997).
5. R.J. Hinde, R.S. Berry and D.J. Wales, J. Chem. Phys. **96**, 1376 (1992); C. Amitrano and R.S.Berry, Phys. Rev. Lett. **68**, 729 (1992).
6. S.K. Nayak, R. Ramaswamy and Ch. Chakravarty, Phys. Rev. **E51**, 3376 (1995).
7. V. Bonacic-Koutecky, L. Cespiva, P. Fantucci, and J. Koutecky, J. Chem. Phys. **98**, 7981 (1993).
8. V. Bonacic-Koutecky, L. Cespiva, P. Fantucci, J. Pittner, and J. Koutecky, J. Chem. Phys. **100**, 490 (1994).
9. M.E. Garcia, G.M. Pastor and K.H. Bennemann, Phys. Rev. Lett. **67**, 1142 (1991).
10. M.E. Garcia and K.H. Bennemann, Computational Materials Science **2**, 481 (1994).
11. U. Buck and H. Meyer, Phys. Rev. Lett. **52**, 109 (1984); D. Kreisle, O. Echt, M. Knapp, and E. Recknagel, Phys. Rev. A **33**, 768 (1986).
12. See for example T.L. Beck, J. Jellinek and R. S. Berry, J. Chem. Phys. **87**, 545 (1987).
13. H. O. Jeschke, M. E. Garcia and K. H. Bennemann, J. Phys. **B 29**, L545 (1996).
14. H. O. Jeschke, M. E. Garcia and K. H. Bennemann, Phys. Rev. **A 54**, R4601 (1996)
15. L. Goodwin, A. J. Skinner and D. G. Pettifor, Europhys. Lett. **9**, 701 (1989).
16. J. C. Slater and G. F. Koster, Phys. Rev. **94**, 1498 (1954).
17. H. Haberland, Surf. Sci. **156**, 305 (1985).
18. M.E. Garcia, G.M. Pastor and K.H. Bennemann, Phys. Rev. B **48**, 8388 (1993); M.E. Garcia, G.M. Pastor and K.H. Bennemann, Z. Phys D **26**, 293-295 (1993).
19. P. J. Kuntz and J. Valldorf, Z. Phys. D **8**, 195 (1988).
20. M. Amarouche, G. Durand, and J. P. Malrieu, J. Chem. Phys. **88**, 1010 (1988).
21. F.Yu. Naumkin, P.J. Knowles and J.N. Murrell, Chemical Physics **193**, 27, (1995).
22. M. E. Garcia, D. Reichardt and K. H. Bennemann, in *Structures and Dynamics of Clusters*, Frontiers Science Series No. 16, edited by T. Kondow, K. Kaya and A. Terasaki, Universal Academic Press, Inc., Tokyo 1996, p. 55.;
23. W. Klaus, M.E. Garcia and K.H. Bennemann, Z. Phys. D **35**, 43 (1995).
24. M. E. Garcia, D. Reichardt and K. H. Bennemann, J. Chem. Phys. **107**, 9857 (1997).
25. M. E. Garcia, D. Reichardt and K. H. Bennemann, J. Chem. Phys. 109, 1101 (1998).
26. H. J. Kim and J. T. Hynes, J. Chem. Phys. **96**, 5088 (1992).
27. C. Moore, Atomic Energy Levels, Nat. Bur. Stand. Circ. No. 467 (1958).
28. W. A. Harrison, Phys. Rev. **B 24**, 5835 (1981).
29. J. Ho, K.M. Erwin and W.C. Lineberger, J. Chem. Phys. **93**, 6987 (1990).
30. D. A. Papaconstantopoulos, *Handbook of the Band Structure of Elemental Solids*, Plenum Press, New York and London, 1986.
31. J.L. Martins, R. Carr and J. Buttet, J. Chem. Phys. **78**, 5647 (1983).

32. M. Hartmann, J. Pittner, V. Bonacic-Koutecky, A. Heidenreich and J. Jortner, J. Chem. Phys. **108**, 3096 (1998).
33. M. Hartmann, A. Heidenreich, J. Pittner, V. Bonacic-Koutecky and J. Jortner, J. Phys. Chem. A **102**, 4069 (1998).
34. R. D. van Zee, S. C. Blankespoor and T. S. Zwier, J. Chem. Phys. **88**, 4650 (1988).
35. Handbook of Chemistry and Physics, $66^{th}$ Edition (CRC, Boca Raton, Florida,1985).
36. D. Reichardt, M. E. Garcia and K. H. Bennemann, Surface Review and Letters **3**, 567 (1996).
37. D. L. Bunker and W. L. Hase, J. Chem. Phys. **59**, 4621 (1973).
38. G. H. Peslherbe and W. L. Hase, J. Chem. Phys. **101**, 8535 (1994); J. Chem. Phys. **105**, 7432 (1996).
39. W. Forst, in *Theory of unimolecular reactions*, edited by E. M. Loebl, Academic Press, New York, 1973.
40. D.H. Reitze and H. Ahn and M.C. Downer, Phys. Rev. **B 45**, 2677, (1992).

# Electron Dynamics in Metal Clusters

C. Guet and L. Plagne

Département de Recherche Fondamentale sur la Matière Condensée
CEA-Grenoble,
17 rue des Martyrs, F–38054 Grenoble Cedex 9, France

## 1 Introduction

Metallic clusters of between ten and a few thousand atoms offer the possibility to study the combined dynamics of electrons and ions in conditions that differ dramatically from those prevailing in bulk metal. During the last decade, clusters of simple metal atoms, such as sodium, have been the subject of both experimental and theoretical focus; quite robust features have emerged; for reviews see for instance [1, 2]. Their electronic and optical properties are understood at first order in terms of the quantal arrangement of delocalized electrons (one per atom for alkali-metal clusters) moving in a mean field resulting from their mutual interaction and a smooth positive background ensuring neutrality. One sees a metal cluster as a mesoscopic Fermi liquid, the relevant parameter being the number of interacting fermions. This discrete feature leads to a shell structure just as in atoms and nuclei. The similarities between metal clusters and atomic nuclei are indeed striking, the most spectacular one being the occurrence of ground state deformations, the spherical symmetry being broken as a result of an unfilled upper electronic shell. It is the electrons, not the ions, that regulate the variations in shape and stability.

The long mean free path makes that electron–electron scattering plays a relatively minor role in the physics of conduction in a bulk metal. The situation might be quite different for a metal cluster. Here, the typical size of the cluster, $R = r_s n^{1/3}$, ($r_s$ is the Wigner–Seitz radius of the bulk and $n$ the number of delocalized electrons) of the order of 10 Å is smaller than the bulk electron mean free path by at least one order of magnitude. Consequently one expects contributions to relaxation phenomena from collisions of electrons with the walls of the confining mean field potential. The relative importance of this relaxation mechanism ought to be compared to the contributions brought by scattering of electrons by ions and their thermal vibrations. This is the long term purpose of the present theoretical efforts that are described in this paper. The as yet unknown electron–phonon coupling, which controls the heat transfer between electrons and ions, is the key quantity to predict most features of metal cluster physics ranging from the decay of plasmons to the physics of photo-ionization, fragmentation and collisions.

Small amplitude electron dynamics which lies at the heart of the optical response of alkali-metals has been widely investigated. For these clusters there is clear experimental evidence that the delocalization of the valence electrons occurs already at the smallest size. Consequently, one expects that the response to a weak external electromagnetic field will lie in two domains that are well separated in frequency: one that is associated with the localized bound charges on the ions and another associated with the free charges represented by the valence electrons. Thus the leading features of "metallic" optical response can be confidently worked out in the jellium approximation. Indeed, since the pioneering calculation by Eckardt [3], the collective optical excitations of various alkali-metal clusters have been successfully understood within this simple approximation. Both the local density random phase approximation [3–5] and the random phase approximation with exact exchange [6] have proved to be powerful theoretical methods to account for the particle–hole interactions that lead to the strong collective response. Nevertheless a detailed agreement between theory and experimental data requires explicit treatment of the ion cores by nonlocal electron–ion pseudo-potentials and discrete ionic sites [7–9]. We shall however in the following stick to the jellium approximation.

Delocalized electrons in alkali-metal clusters experience large amplitude dynamics in physical processes such as either irradiation by an intense ultra-short laser pulse (today, pulses of less than 100 fs are available) or the passage through (or near) the cluster of a charged particle with sufficient velocity. As to ongoing experiments, we shall consider mostly this collision process.

Let us for the time being see the cluster as a small piece of condensed matter. When an ion penetrates matter the medium responds to the passing charge, as mobile electrons tend to screen the disturbance. If the charge velocity, $v_p$, is of the order of the Fermi velocity, core electrons play only a very minor role in slowing down the passing ion and it is reasonable to just consider the collective response of delocalized valence electrons. Dynamic screening by these electrons leads to plasmon oscillations and particle–hole excitations and eventually strong electronic excitations. For a finite cluster the excitation process lasts essentially during the time of passage of the ion, $\tau_{\text{pass}} \approx \frac{2R}{v_p}$, which is of the order of a few fs. One expects a strong electronic excitation to develop during a short time, higher than the typical electronic motion timescale of about 1 fs, but much smaller than the ionic vibrational timescale of order of 1ps. The ions effectively remain frozen as the ion passes by.

How a strong electronic excitation occurring on a timescale of a few femtoseconds relaxes into vibrational heating, melting and possibly fragmentation of a metal cluster remains a fully open problem. The relaxation process is governed by both electron–electron collisions and electron–phonon coupling. Both basic mechanisms ought to be different from their counterparts in bulk metal because of the finite size of the cluster.

The present paper is mainly devoted to discussing theoretical descriptions of the onset of electronic excitations induced by the sudden action of the strong transient electric field. We shall confine ourselves to the electron dynamics disregarding the coupling of electrons to phonons, thus to a period of time of a few tens of fs. Having in mind clusters of about or more than 100 atoms, purely quantum mechanical treatments are almost numerically prohibited even in such approximations as the local density approximation. Semiclassical approximations are indeed appealing in the context of large clusters and large electronic excitations. We shall describe one such method based on the Vlasov equation and illustrate its possibilities and limitations on a few examples. In harmony with recent experimental developments we shall focus attention on strong electronic excitations that occur within less than 5 femtoseconds in the course of a collision between an (highly charged) ion and a sodium cluster at velocities of the order of the Fermi velocity

Finally we shall discuss prospects that are eagerly awaited. Having a reasonable control of the pure electron dynamics, it is also necessary to account for the electron–phonon coupling in order to have a full comprehension of relaxation phenomena in strong and ultra-fast electronic excitations of metal clusters.

## 2  Electron Dynamics in Alkali-Metal Clusters

The response of a jellium cluster to a strong external disturbance acting for a short time in a non-perturbative regime is ideally given by solving the time-dependent many-body Schrödinger equation:

$$i\hbar \frac{\partial \Psi}{\partial t} = H\Psi \tag{1}$$

$$H = \sum_i^n \left[ -\frac{\hbar^2}{2m} \nabla_i^2 + V_b(\mathbf{r}_i) + V_{ext}(\mathbf{r}_i, t) \right] + \sum_{i<j}^n \frac{e^2}{r_{ij}} \tag{2}$$

where H is the full Hamiltonian for $n$ electrons in mutual Coulomb interaction in the presence of the neutralizing one-body potential, $V_b(\mathbf{r})$, and a one-body external time-dependent potential, $V_{ext}(\mathbf{r}, t)$.

It is impossible and actually unnecessary to try to solve the above equation. A standard but powerful simplification is obtained within a mean field approximation such as the time-honored Hartree–Fock approximation. Here one assumes, in order to comply with the Pauli exclusion principle, that the time-dependent many-body wavefunction, $\Psi$, is an antisymmetrized product (Slater determinant) of single particle orbitals, $\psi_i(\mathbf{r}, t)$ such as:

$$\Psi(t) = e^{-\frac{i}{\hbar} E_0 t} \| \psi_i(\mathbf{r}, t) \|, \tag{3}$$

where $E_0$ is the ground state Hartree–Fock (HF) energy in the absence of the external field, that is:

$$E_0 = \sum_{i \leq f} \left\langle i \left| -\frac{\hbar^2}{2m} \nabla^2 + V_b(\mathbf{r}) \right| i \right\rangle + \frac{1}{2} \sum_{i,j \leq f} \left\langle ij \left| \frac{e^2}{|\mathbf{r} - \mathbf{r}'|} \right| ij - ji \right\rangle, \tag{4}$$

with index $i$ (or $j$) standing for HF occupied orbitals $\varphi_i(\mathbf{r})$, and $f$ being the last occupied one. The one-electron wavefunctions propagate in time according to the set of integro-differential equations:

$$i\hbar\frac{\partial\psi_i(\mathbf{r},t)}{\partial t} = [h_0(\mathbf{r},t) + v[\rho(\mathbf{r},t)]]\,\psi_i(\mathbf{r},t),\tag{5}$$

where the one-body potential,

$$h_0(\mathbf{r},t) = -\frac{\hbar^2}{2m}\nabla^2 + V_b(\mathbf{r}) + V_{\text{ext}}(\mathbf{r},t),\tag{6}$$

contains the explicitly time dependent external potential, and $v[\rho(\mathbf{r},t)]$ is the self-consistent electron–electron Hartree–Fock mean field potential. The latter is itself the sum of the direct Coulomb potential, $v_d$, which satisfies Poisson's equation:

$$\nabla^2 v_d = 4\pi e \sum_i |\psi_i(\mathbf{r},t)|^2\tag{7}$$

and the non-local Coulomb exchange term.

One can expand the wavefunction, $\psi_i(\mathbf{r},t)$, in terms of the eigenfunctions, $\varphi_k(\mathbf{r})$, of the static HF potential:

$$\psi_i(\mathbf{r},t) = A\left[\varphi_i(\mathbf{r}) + \sum_{m>f} C_{mi}(t)\varphi_m(\mathbf{r})\right],\tag{8}$$

where $A$ is a normalization constant, and $C_{mi}(t)$ is the particle–hole amplitude of all unoccupied HF orbitals, $m$, to the initial occupied one, $i$.

The linearization of equations (5) leads to the Random Phase Approximation with exact Exchange (RPAE). The RPAE equations were derived in the context of atomic physics a long time ago [10] and used to describe collective effects in photo-ionization of atoms. The same formalism was later applied to work out the dipole response of metal clusters [6, 11]. Practically, the RPAE equations are solved in a particle–hole configurational space by diagonalization a huge matrix. The RPAE, which allows one to account unambiguously for a class of well defined correlation diagrams (ring diagrams), in both the ground state and the excited states, is a powerful method to describe collective harmonic excitations built on a Hartree–Fock ground state.

Alternatively a very similar formalism can be derived within the Local Density Approximation of the Density Functional Theory (DFT) [5]. Here, one gets the ground state orbitals from the Kohn–Sham equations (practically this means replacing the non-local exchange potential by an effective density dependent exchange-correlation potential, $v_{\text{xc}}[\rho]$, derived from DFT). The RPA equations themselves contain the density dependent residual interaction. This formalism, quite similar to what had been derived in nuclear physics [12], has also proved successful in the context of metal jellium clusters [5]. Exactly the same physics can also be worked out by directly solving the associated frequency-dependent Green's function equations [3, 4]. Unfortunately the numerical solution of the RPA equations in three dimensional space is highly involved for either method.

Interestingly enough the time-dependent Kohn–Sham equations, thus similar to equations (5), with the LDA exchange-correlation instead of the Fock potential, have recently been solved in real time and three dimensional space in a perturbative regime [13]. The aim was to study the dipole response of large lithium clusters (147 atoms) beyond the jellium approximation by keeping track of a supposedly known ionic geometry. Beyond the fact that these calculations confirmed earlier studies of important nonlocal pseudo-potential effects [7–9], the 'real time' method applied to small amplitude dynamics appeared to be more efficient than matrix or Green's function methods.

Following the trails open in nuclear physics a few decades ago in the study of large amplitude dynamics [14], Calvayrac and collaborators [15] recently addressed the study of the electron response of small jellium clusters (up to 10 electrons) to a sudden time-dependent electric field in a non-linear regime. They performed a numerical integration of the time-dependent Kohn–Sham equations (TDLDA) in axial symmetry. Note that a full three-dimensional TDLDA description of ion–cluster collisions was carried out just recently [37]. This is the approach that, insofar as it can be practically extended to large clusters without restricted symmetries, the semi-classical method of next section aims to approximate.

For further convenience, note that the TDLDA equations can be expressed as the time evolution of the one-body density operator, $\hat{\rho}$:

$$i\hbar \frac{\partial \hat{\rho}}{\partial t} = \left[ \hat{h} \left[ \hat{\rho} \right], \hat{\rho} \right] \tag{9}$$

where $h\left[\rho\right] = h_0 + v\left[\rho\right]$, is the self-consistent LDA Hamiltonian. Applying the Wigner transformation to this equation, one obtains [16]:

$$\frac{\partial f(\mathbf{r}, \mathbf{p}, t)}{\partial t} + \frac{\mathbf{p}}{m} \boldsymbol{\nabla}_r f(\mathbf{r}, \mathbf{p}, t) - \frac{2}{\hbar}(V_b + V_{\text{ext}} + v) \sin\left( \frac{\hbar}{2} \overleftarrow{\boldsymbol{\nabla}}_r \overrightarrow{\boldsymbol{\nabla}}_p \right) f(\mathbf{r}, \mathbf{p}, t) = 0. \tag{10}$$

The Wigner transform, $f(\mathbf{r}, \mathbf{p}, t)$, of the one-body density is expressed in terms of the single electron orbitals as:

$$f(\mathbf{r}, \mathbf{p}, t) = \int d^3 s \exp(-i s.\mathbf{p}) \sum_{i=1}^{N} \psi_i(\mathbf{r} - \mathbf{s}/2, t) \, \psi_i^\dagger(\mathbf{r} + \mathbf{s}/2, t). \tag{11}$$

It is a sort of phase space density distribution. However this distribution is purely quantum mechanical since it is just an other representation of the time-dependent one-body density; as such it can be negatively defined in some region of the phase space. In the same manner, the equation (10) is merely another representation of the TDLDA equation, but it is specially suited for semi-classical approximations. This representation, as well as other representations and semi-classical expansions, have been under thorough investigation; for extensive discussions see Ref. [18].

# 3   Semi-Classical Dynamics of Electrons in Metal Clusters

In the present approach, one solves the Vlasov equation, that is simply the semi-classical limit ( $\hbar \to 0$ ) of the TDLDA equation (10). The one-body phase space density, $f(\mathbf{r}, \mathbf{p}, t)$ evolves in time as:

$$\frac{\partial f(\mathbf{r}, \mathbf{p}, t)}{\partial t} + \frac{\mathbf{p}}{m} \boldsymbol{\nabla}_r \, f(\mathbf{r}, \mathbf{p}, t) - \boldsymbol{\nabla}_r (V_{\mathrm{b}} + V_{\mathrm{ext}} + v) . \boldsymbol{\nabla}_p \, f(\mathbf{r}, \mathbf{p}, t) = 0; . \quad (12)$$

The Vlasov equation can also be written as:

$$\frac{\partial f(\mathbf{r}, \mathbf{p}, t)}{\partial t} = \{h, f\}, \quad (13)$$

which follows directly the rule of correspondence from quantum to classical equations of motion which in the Heisenberg picture substitutes the commutator divided by $i\hbar$ for the Poisson bracket. The one-body classical Hamiltonian, $h$, is simply given by:

$$h(\mathbf{r}, \mathbf{p}, t) = \frac{\mathbf{p}^2}{2m} + V_{\mathrm{b}}(\mathbf{r}) + V_{\mathrm{ext}}(\mathbf{r}, t) + v \left[ \rho (\mathbf{r}) \right], \quad (14)$$

where the self-consistent potential, $v[\rho]$, acting on an electron at position $\mathbf{r}$ is made of two contributions: the Hartree potential $V_{\mathrm{H}}(\mathbf{r}) = \int \rho(\mathbf{r}')|\mathbf{r} - \mathbf{r}'|^{-1} d^3 \mathbf{r}'$, accounting for the mutual direct Coulomb interaction of the valence electrons, and the exchange-correlation density-dependent potential in LDA, $V_{\mathrm{xc}} \left[ \rho (\mathbf{r}) \right]$.

As a mean field equation with no explicit collision term the original Vlasov equation (12) does not create or destroy correlations. Thus it preserves the Pauli correlations built in the initial phase space distribution. For the sake of consistency, the latter must be a stationary solution of the Vlasov equation, which is achieved by the Thomas–Fermi distribution:

$$f(\mathbf{r}, \mathbf{p}, t = 0) = \frac{2}{(2\pi \hbar)^3} \Theta \left( E_{\mathrm{F}} - (V_{\mathrm{bkg}} + v) - \frac{p^2}{2m} \right), \quad (15)$$

where $E_{\mathrm{F}}$ is the Fermi energy. Practically one constructs the initial distribution $\rho(r)$ by iteratively solving the Thomas–Fermi equation until convergence is reached. In order to ensure best consistency with the Gaussian pseudo-particle integration method described just below, the Gaussian folding procedure also applies to the initial mean field potential. Eventually, one obtains the initial phase space density as:

$$f(r, p, 0) = \begin{cases} 2 \text{ if } |p| \le p_{\mathrm{F}}(|r|) \\ 0 \text{ if } |p| > p_{\mathrm{F}}(|r|) \end{cases} \quad (16)$$

where the local Fermi momentum:

$$p_{\mathrm{F}}(r) = (E_{\mathrm{F}} - (V_{\mathrm{bgk}} + v))^{\frac{1}{2}} = (3\pi^2 \rho(r))^{\frac{1}{3}} \quad (17)$$

is determined by the normalization condition:

$$\int f(r,p,0)d^3rd^3p = n. \tag{18}$$

Although no fundamental principle prevents using the semi-classical Vlasov equation as long as the initial Thomas–Fermi equation is properly built (in accordance with the Liouville theorem), the numerical integration of (12) is the practical obstacle. Within the scope of present computer possibilities one cannot conceive systematic direct six-dimensional lattice integrations. Instead we use the test particle method.

The one-body distribution $f(r,p,t)$ is approximated by a sum of a large number of $r$ and $p$ -folding Gaussian functions, $g_r$ and $g_p$, of standard deviation $\sigma_r$ and $\sigma_p$ respectively, centered at $r_k$ and $p_k$:

$$f(r,p,t) = \frac{n}{M} \sum_{k=1}^{M} g_r(r - r_k(t)) \, g_p(p - p_k(t)), \tag{19}$$

with the number of test particles, $M$, chosen according to required numerical accuracy. The time evolution described by the Vlasov equation can then be approximated by some classical dynamics of these test particles. Insert (19) into (12) and perform either an $r$ or $p$ integration in order to respectively obtain for any test-particle, $k$, the following pair of equations:

$$\frac{dr_k}{dt} = \frac{p_k}{m}$$
$$\frac{dp_k}{dt} = -\int \nabla_r(V_{bkg} + V_{ext} + v) \, g_r(r - r_k(t)d^3r. \tag{20}$$

The Hamilton dynamics of these $M$ particles within a mean field, $(V_{bkg}+V_{ext}+v)$, is thus a possible representation of the Vlasov dynamics. Notice that only the $r$-like folding appears. The folding width is optimally adjusted to the mesh size of the three-dimensional grid on which the numerical integration is carried out. Note that the width in $p$-space is irrelevant here since the potential is independent of velocity.

Note that the equivalence would be exact for an infinite number of test particles. It is interesting to underline the strong analogy with the problem of $n$ classical particles in an external field and mutual interaction which can be described at the one-body mean-field approximation by the Vlasov equation. Here, inversely the Vlasov equation is expressed as the Hamilton dynamics of $M$ $(M \gg n)$ particles interacting with an effective interaction folded with the test-particle distribution.

Although the original Vlasov equation (12) contains no collision term, the discretization (19) replaces it with a classical many-body problem. This brings unwanted fluctuations that may evolve into the Pauli forbidden phase space. This is possible since there is no explicit constraint to remain in the Pauli allowed domain, $0 \leq f(r,p) \leq 2$. The initial Fermi–Dirac distribution will decay irreparably into a classical Boltzmann distribution. This spurious increase of temperature leads to spontaneous ionization of the cluster via unwanted losses of test particles. The practical challenge is thus to delay such a decay well beyond the simulation time. A large amount of work has actually been devoted to the reliability of the Vlasov equation and the test-particle method to describe the dynamics of fermions, either nucleons or electrons. For more details and a critical review of the test-particle methods, see [18, 35].

The numerical procedure used in the present work, allows us to safely rely on simulations extending up to 100 fs (for details see Ref. [36]). This is in sharp contrast with previous works which claim a very fast decay of the order of a few fs. To reach such a high level of stability requires using a large number of test particles, $M$. This number depends on the maximum space that one needs to properly account for physical processes under investigation. Briefly, most of the simulations reported below (for clusters with a few hundred electrons) have used $\simeq 10^6$ Gaussian test particles with a folding width $\sigma_r = 1a_0$.

In order to cope with the classical dynamics (20) of such a large number of charged pseudo-particles under the requirement of recalculating the self-consistent potential at each time step, we developed some specific computational techniques that have been described elsewhere [25]. Briefly, the $10^6$ Gaussian test-particles are initially distributed inside a $128^3$ cubic grid to simulate the initial phase space density. Their time evolution is computed using a standard Verlet method with a time step of $\simeq 0.01$ fs adjusted on both the projectile and the electron Fermi velocity. It has proved convenient to work out the mean field potential on a B-spline basis set with the Hartree part being calculated as the solution of Poisson's equation. Calculations are carried out on the CRAY T3E at Grenoble taking maximum advantage of parallelization techniques.

## 4 Femtosecond Electron Dynamics in Ion–Metal Cluster Collisions

We shall illustrate the possibilities that the above theoretical scheme offers in relation to the interaction of (multiply) charged ions with metal clusters at intermediate velocities (near the Fermi energy). We choose a $Na_n$ ($n = 196$) cluster, which in the jellium approximation has a closed shell electronic structure. The distribution of ions supposedly at rest is approximated by a spherical homogeneous distribution of charge of radius, $R = 23.23\ a_0$, ($r_s = 4a_0$), thus providing a background electrostatic potential:

$$v_{\text{bkg}}(r) = \begin{cases} -\frac{n}{2R}\left[3 - \left(\frac{r}{R}\right)^2\right] & r \leq R, \\ -\frac{n}{r} & r > R, \end{cases} \tag{21}$$

The jellium ground state one-body space density $\rho(r)$ is obtained in the local density approximation from the solution of the Thomas–Fermi equation in spherical symmetry. The potential entering the self-consistent Thomas–Fermi equation is again the sum of the background potential, the Hartree potential, and the exchange-correlation potential. The latter is described in terms of the density functional given by Gunnarsson and Lundqvist [20]. Alternatively on may solve the Kohn–Sham equations and smear most of the fluctuations of quantal origin. The supposedly point-like passing ion of charge $Q$ provides a time-dependent external potential :

$$V_{\text{ext}}(\mathbf{r}) = Q|\mathbf{r} - \mathbf{r}_{\text{p}}(t)|^{-1}, \tag{22}$$

where $\mathbf{r}_{\text{p}}(t)$ is the projectile classical position at time $t$.

First attempts to use the Vlasov equation method to calculate the electronic response of a jellium cluster during and just after the collision have recently been reported both for large clusters ($n \sim 200$) [21, 22] and for small clusters ($n \leq 10$) [23]. Calculations are motivated by new experimental possibilities to measure energy losses, ionization rates, and vibrational excitations in ion–cluster collisions [24].

At variance with the present calculations, the calculations of Refs. [21, 22] used Dirac $\delta$ distributions instead of Gaussians for the test particle representation. It appeared that within these conditions, unwanted loss of particles became significant after about 10 fs for a number of test particles of the order of 10 per electron [21]. The only way to get a longer decay time is to considerably increase the number of test particles $M$. A much higher number of Dirac $\delta$ test particles (more than 6000 particles per electron) were actually used in a second work [22]; it allowed one to get calculations stable enough for a reliable analysis of electron emission over a timescale up to about 50 fs. A further improvement is brought by smearing each test particle over some phase space volume by a Gaussian convolution as discussed above. Such an approach combined with a number of test particles of 500 per electron was used to study the dynamical response of $\text{Na}_9^+$ jellium clusters over a time up to 50 fs [23]

The calculations for $\text{Na}_{196}$ that will be presented further on are quite numerically demanding. One can achieve an excellent stability with essentially no spurious loss of test particles insofar as one uses Gaussian test particles with a folding width $\sigma_r = 1a_0$, and a number of test particles of the order of $10^6$. Note that the width in $p$-space is irrelevant here since the potential is independent of velocity. In order to cope with the classical dynamics (20) of such a large number of charged pseudo-particles under the requirement of recalculating the self-consistent potential at each time step we developed some specific computational techniques that will be described elsewhere [25]. Briefly, $10^6$ Gaussian test-particles are initially distributed inside a $64^3$ cubic grid to simulate the initial phase space density. Their time evolution is computed using a standard leap frog method with a time step of 0.02 fs. It has proved convenient to work out the mean field potential on a B-spline basis set with the Hartree part being calculated as the solution of the Poisson's equation.

Typically one can simulate the Vlasov electron dynamics during about 50 fs. As an illustration we show in Fig. 1 a series of snapshots of the electronic density distribution during the collision of a 500 keV kinetic energy $\text{Xe}^{25+}$ ion with the $\text{Na}_{196}$ jellium cluster. This corresponds to a relative velocity of 0.39 a.u., which is close to the Fermi velocity. The impact parameter is $45\,a_0$ i.e.

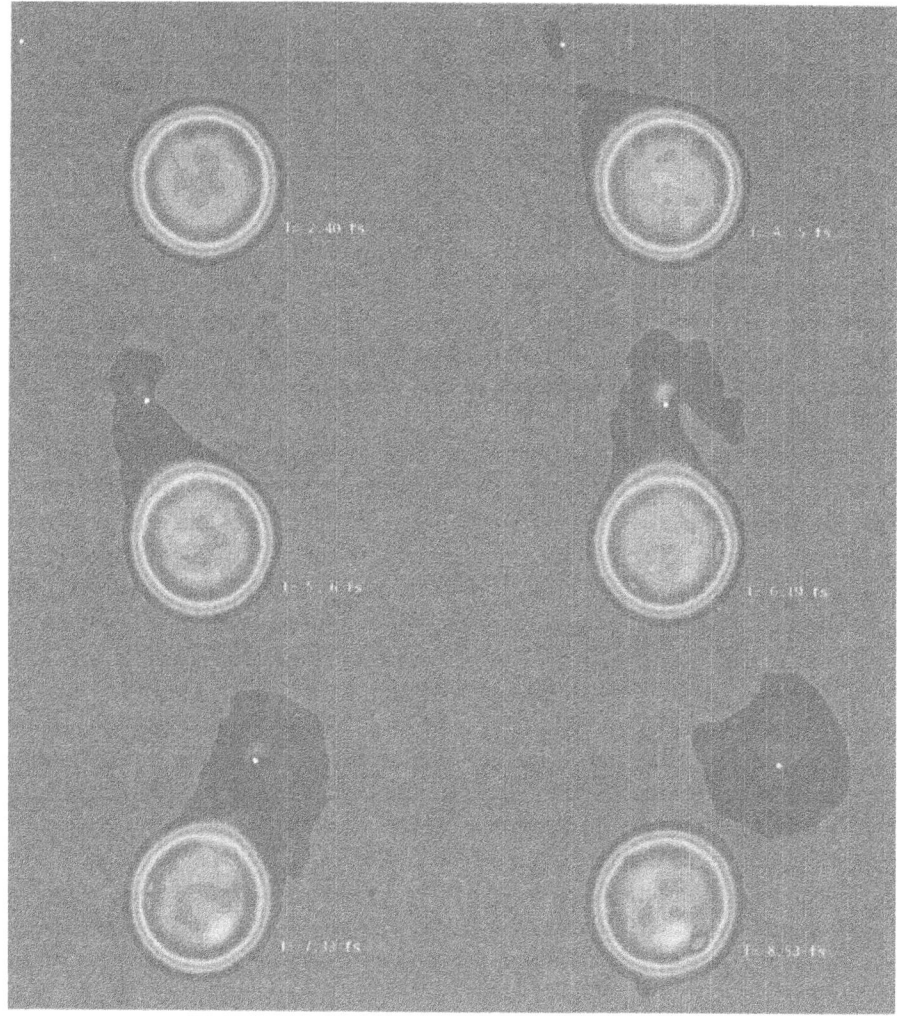

**Fig. 1.** Snapshots of the electron density during the collision between a 500 keV $Xe^{25+}$ ion and a $Na_{196}$ jellium cluster. Impact parameter: 45 $a_0$. The intensity scales down from light to dark.

twice the cluster radius. As the projectile approaches the cluster, it distorts the electron distribution considerably until test particles jump from the cluster to the ion. As the collision proceeds further, a flow of electronic fluid develops between target and projectile and strong collective oscillations set in. After about 4 fs the partners are separated far enough that no electronic transfer occurs any more. Both are highly excited as demonstrated by the large collective motions that they experience.

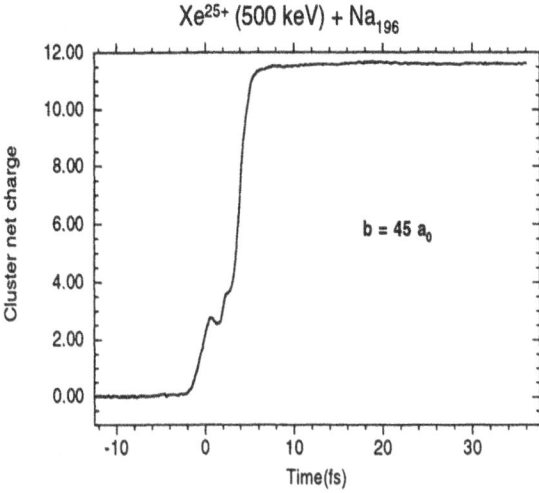

**Fig. 2.** Time evolution of the net charge of a $Na_{196}$ jellium cluster as it experiences the passage of a 500 keV $Xe^{25+}$ ion . Impact parameter: $45\,a_0$.

Particle emission is a stringent test for numerical stability. Figure 2 shows the time evolution of the residual charge of the $Na_{196}$ jellium cluster as the 500 keV kinetic energy $Xe^{25+}$ passes by. No test particle leak is observed in the unperturbed ground state at times exceeding any time scale here relevant (say at most 100 fs). We define the ionization rate as the number of test particles which get over a radial distance of ten times the cluster radius; there is thus an extra practical delay between the onset of electron emission and the onset of electronic density fluctuations triggered by the approaching ion. One sees that the electron emission is fast and develops only during the passage of the projectile. Within about 5 fs, 12 electrons are removed from the cluster (most of them are actually trapped by the projectile). Once the projectile moves away from the interaction zone, the charge state (+12) of the cluster remains perfectly stable. Beyond the numerical stability, one understands this feature as a consequence of the subsequent high ionization potential that the bound electrons cannot overcome, although they have been slightly excited during the process. In such a peripheral collision the ionization arises mainly from a direct transfer of electrons to the projectile ion. This transfer occurs because the potential seen by an electron outside the cluster is lowered enough to completely eliminate the barrier that otherwise prevents the electrons near the Fermi surface in the cluster from escaping. The situation is quite different for a central collision. As an example Fig. 3 shows the time evolution of the residual charge of the same $Na_{196}$ jellium cluster as a 20 keV kinetic energy $H^+$ goes through at zero impact parameter. Here, 5 electrons are emitted during the passage. One observes again a perfect charge stability before and after collision.

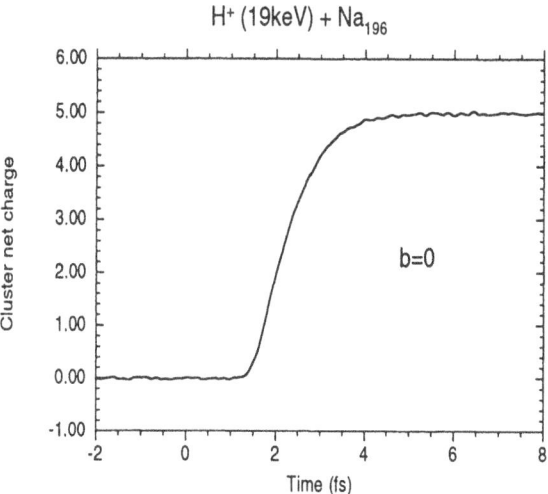

$H^+$ (19keV) + $Na_{196}$

b=0

**Fig. 3.** Time evolution of the net charge of a $Na_{196}$ jellium cluster as it suffers the passage of a 19 keV $H^+$ ion . Central collision.

Another observable of interest is the kinetic energy loss. This quantity is plotted as a function of time in Fig. 4 and Fig. 5. In a peripheral collision with a highly charge ion ($Xe^{25+}$) the projectile is first accelerated by its image charge when approaching, then slows down and loses up to 25 eV in a very short space of time (2 fs), and again accelerates as it leaves the interaction zone, see Fig. 4. Eventually there is a net gain of energy due to the long range repulsion between both charged partners, which asymptotically goes to about 45 eV in the present case. Note that the strong oscillations of kinetic energy seen as the ion moves away are not numerical but are due to the collective oscillations of the electron clouds in both the ion projectile and cluster target and to their mutual coupling. This situation is quite different for a $H^+$ ion hitting a cluster in a central collision as can be seen from Fig. 5. Here the loss of energy, which varies linearly with time, occurs just during the crossing of the cluster. The physics at stake is merely the screening of a passing charge in an electron gas as worked out a long time ago; for an extensive review of the subject see Ref. [26]. According to the elementary theory by Lindhard [27], based on the homogeneous electron gas, the proton stopping power, $dE/dx$, is indeed proportional to the velocity, $v$:

$$\frac{dE}{dx} = k\,v, \tag{23}$$

where $k$ depends only on the Fermi velocity (all quantities in atomic units):

$$k = \frac{2}{3\pi}\left[\ln\left(\pi v_F + \frac{2}{3}\right) - \frac{3\pi v_F - 1}{3\pi v_F + 2}\right]\left(1 - \frac{1}{3\pi v_F}\right)^{-2}. \tag{24}$$

For sodium, we get $v_F = 0.48$ a.u. and thus $k = 0.0828$.

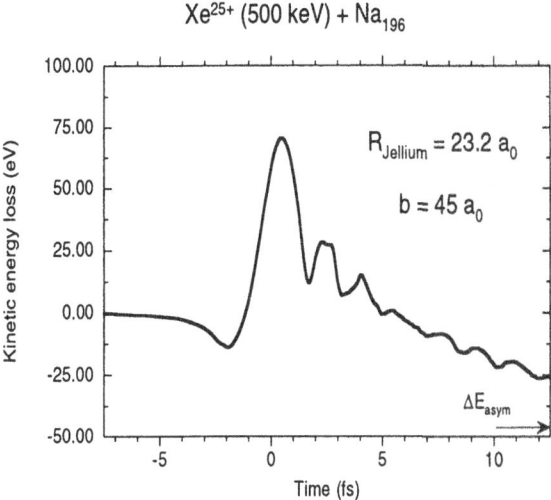

**Fig. 4.** Time evolution of the kinetic energy loss of a 500 keV $Xe^{25+}$ ion colliding with a $Na_{196}$ jellium cluster. Impact parameter: 45 $a_0$. The time origin coincides with the projectile being at the closest distance from the cluster.

In the case of a 19 keV proton crossing a slab of sodium of thickness twice the cluster radius, this formula yields an energy loss of about 91 eV, which is quite close to the present estimate. Notwithstanding the observed degree of accuracy, which surely is fortuitous, one expects a fair agreement insofar as the electronic density is indeed even inside the jellium cluster and as the Vlasov dynamics properly describes the collective behavior of the electron gas.

After the ion passage and the induced loss of particles, a highly charged excited cluster remains. The excitation energy is shared between plasmon-type collective motions and single particle excitations. In Fig. 6 we show the time evolution of the average displacement of electrons along the projectile axis ($x$):

$$\langle x(t) \rangle := M^{-1} \sum_{i=1}^{M} x_i(t), \tag{25}$$

and similarly along the impact parameter axis ($y$), during and after the peripheral collision with a $Xe^{25+}$ projectile. After a strong displacement of charge that develops in the direction of the approaching ion in about 1 fs, the restoring force from the positive background brings the electrons back to some average position, about which they establish harmonic oscillations. After about 10 fs, the amplitude of the dipole oscillation is much higher in the $y$-direction than in the $x$-direction and a quasi permanent dipole remains in the $x$-direction as a result of the polarization forces exerted by the escaping projectile. For a central collision with a $H^+$ ion, see Fig. 7, there is of course essentially no induced dipole oscilla-

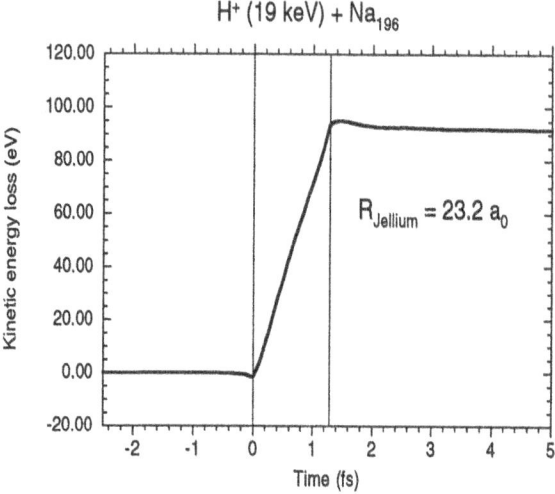

**Fig. 5.** Time evolution of the kinetic energy loss of a 19 keV $H^+$ion colliding with a $Na_{196}$ jellium cluster (central collision). The time origin coincides with the projectile passing the front jellium edge. Second vertical line is the time when the ion passes through the rear jellium edge.

tion along the $y$-direction. The time evolution of the $x$-displacement illustrates nicely the collective polarization that builds up to screen the passing charge. As soon as the proton has left ($\sim$3 fs) the $Na_{196}^{5+}$ keeps oscillating harmonically in the $x$-direction. The oscillation frequency corresponds to the expected dipole plasmon oscillation of a jellium clusters. Figure 8 shows the power spectra of $Na_{196}^{5+}$ and $Na_{196}^{12+}$ obtained as the Fourier transforms of the time-dependent dipole oscillations in the $H^+$ and $Xe^{25+}$ collisions, respectively. A single resonance emerges in both cases at a frequency of 3.2 eV ($Na_{196}^{5+}$) and 3.3 eV ($Na_{196}^{12+}$). These values are slightly smaller than the Mie dipole frequency of 3.4 eV. The shift is due to the spill-out of electrons outside the jellium edge, thus larger for the smaller charge. These values agree quite nicely with RPA calculations [11], giving us a further evidence of the validity of semi-classical dynamics à la Vlasov.

## 5 Prospects

Many physical problems pertaining to electron dynamics in metal clusters are still left open. The most important one is undoubtedly the coupling of the electrons to the ions. In order to illustrate the discussion, let us first consider the interaction of heavy charged particles with metals.

A swift ion (or a strong short laser pulse) is able to induce dramatic changes of material properties. As most of the energy is transferred to the excitation of the electron subsystem during a short time scale, the defects that are eventually

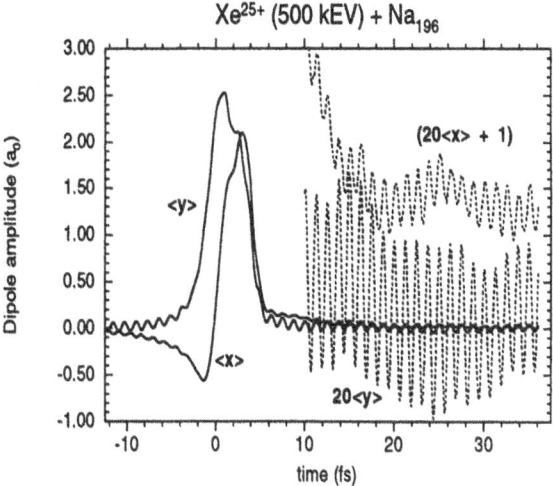

**Fig. 6.** Time evolution of the averaged displacement of electrons of a $Na_{196}$ jellium cluster along the projectile direction ($\langle x \rangle$) and perpendicular to it ($\langle y \rangle$) after collision with a 500 keV $Xe^{25+}$. Impact parameter: $45\, a_0$. The time origin coincides with the projectile being at the closest distance from the cluster. Magnification of oscillations: dashed lines.

observed, result from the heating of the ionic lattice by the highly excited electron gas. Schematically, the physical situation in the vicinity of the impact is that of a more or less equilibrized electron plasma set in contact with an as yet cold ionic system. Within a macroscopic one-dimensional continuum model for both electronic and ionic subsystems the time evolution of the electronic and lattice temperatures is given by two coupled non-linear differential equations with proper initial and boundary conditions [28, 29]:

$$C_e(T_e)\frac{\partial T_e}{\partial t} = \kappa_e \frac{\partial^2 T_e}{\partial z^2} - G(T_e - T_i) + f(z,t)$$
$$C_i(T_i)\frac{\partial T_i}{\partial t} = \kappa_i \frac{\partial^2 T_i}{\partial z^2} + G(T_e - T_i), \qquad (26)$$

where $T_e(z,t)$ and $T_i(z,t)$ are the electron and phonon temperatures respectively, and $z$ is the direction of propagation. The possibility of having one system at very high temperature whereas the other remains cold may occur because of the fact that the heat capacity $C_e$ for the electronic subsystem is smaller by orders of magnitude than the lattice heat capacity, $C_l$. The electron–phonon coupling constant, $G$, is the leading parameter that controls the rate of transfer from one subsystem to the other, and $\kappa_e$ ($\kappa_i$) is the electronic (ionic) thermal conductivity. The function $f(z,t)$ is the energy released in the metal.

Relaxation of non-equilibrium electrons in metal films had been investigated experimentally by means of femtosecond pump-probe spectroscopy [30–32]. The

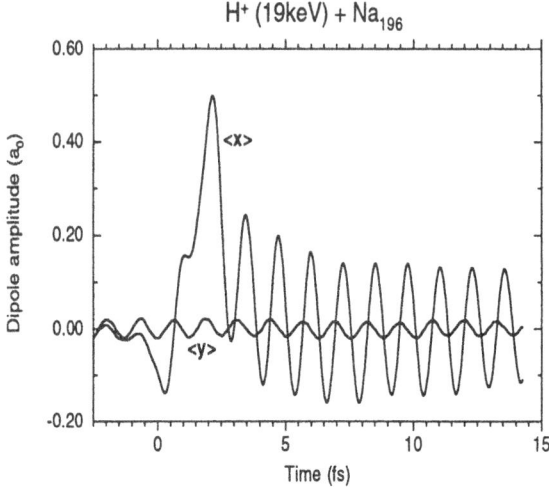

**Fig. 7.** Time evolution of the averaged displacement of electrons of a $Na_{196}$ jellium cluster along the projectile direction ($\langle x \rangle$) and perpendicular to it ($\langle y \rangle$) after collision with a 19 keV $H^+$. Central collision. The time origin coincides with the projectile passing the front jellium edge.

first *direct* measurement of electron temperature was carried out for gold films heated by a 400 fs visible laser pulse [33]. This experiment confirmed that the coupling between electrons and ions is weak on a subpicosecond timescale thus allowing one to highly excite the electron gas while keeping the lattice cold. Most important, the data showed that the initial electronic distribution is characterized by a hot tail and that this non-Fermi–Dirac distribution persists for a time at least as long as $\sim 1$ ps [33]. This result raises severe doubts about the validity of equation (26) for subpicosecond timescale processes, since it implicitly assumes that electrons have reached thermal equilibrium before any noticeable energy transfer has occurred. Conversely, it indicates that a proper theoretical treatment of the electron dynamics ought to consider electron–electron interactions in addition to electron–phonon collisions. We are not aware of such theoretical attempts to cope with highly excited metals at the subpicosecond level. Let us note that for a larger timescale, well above the ps level, equation (26) is reasonably valid and has been solved together with molecular dynamics simulations to describe laser-pulse irradiation of copper surfaces Cu(111), revealing the possibility for structural annealing of these surfaces through superheating [29].

As to metal clusters, one expects the above-mentioned competition between electron–electron and electron–phonon scattering to come forward. New features are likely to appear because of the finite size of the system, the electron mean free path actually being larger than the cluster size. In a recent beautiful experiment [34], the authors were able to investigate the dynamics of electrons in

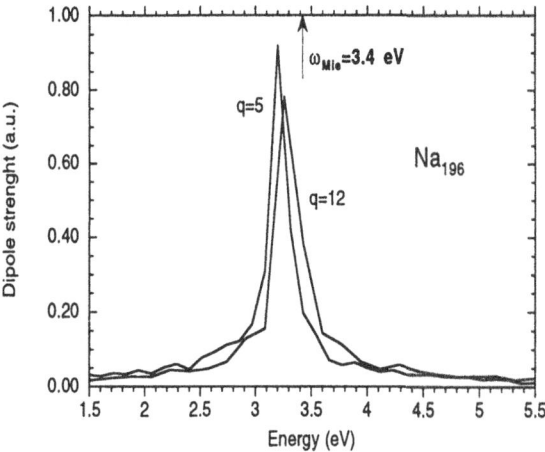

**Fig. 8.** Dipole response of a $Na_{196}$ jellium cluster after passage of an ion. Central collision with a 19 keV $H^+$ (residual charge is 5) , peripheral collision with a 500 keV $Xe^{25+}$ (residual charge is 12).

copper nanoparticles ( $\sim 10$ nm diameter) on a subpicosecond timescale by using femtosecond pump–probe spectroscopy and measuring time-dependent transmission. Typical cluster effects are observed. The strong Mie plasmon resonance acts to enhance a purely electron–electron relaxation: there is an exchange of energy between the collective surface plasmon and quasiparticle excitations (Landau damping) that lasts for $\sim 1$ ps and thus at the expense of slowing the thermalization into the lattice system.

The models which we have discussed in the previous section may serve as a good starting point for a theoretical investigation of relaxation processes in metal clusters on time scales remaining shorter than a few ps. In a first step, it would be quite informative to simply replace the jellium positive background by discrete ions at rest but interacting with the delocalized electrons through realistic pseudopotentials [7]. This refinement would allow us to better describe the electron–electron dynamics by naturally incorporating the scattering of electrons from the ions. This static approach, from the standpoint of the ions, is meaningful only for a time duration not exceeding a few hundred femtoseconds during which ions do not appreciably move. In order to also account for the electron–phonon coupling and eventually address the full problem of energy transfer and possible thermalization, the natural avenue is to relax the ions and let them move on classical trajectories and couple their dynamics with that of the electrons described à la Vlasov. Most theoretical and numerical tools are available today and just need to be adjusted one to another. It is reasonable to expect that major advances in these directions will soon be achieved.

# 6 Acknowledgments

We wish to thank Malte Gross for his major contribution to a first version of the Vlasov code. This work has been supported in part by a HCM grant of the European Community (project CHRX-CT94-0643) and in part by a NATO grant (CRG-950678).

# References

1. W. de Heer, Rev. Mod. Phys. **65**, 611 (1993).
2. M. Brack, Rev. Mod. Phys. **65**, 611 (1993).
3. W. Ekardt, Phys. Rev. Lett. **52,** 1925 (1984) ; W. Ekardt, Phys. Rev. B **31,** 6360 (195).
4. D. E. Beck, Phys. Rev. B **35**, 732 (1987)
5. C. Yannouleas, R. A. Broglia, M. Brack, and P. F. Bortignon, Phys. Rev. Lett. **63**, 255 (1989), C. Yannouleas, E. Vigezzi, and R. A. Broglia, Phys. Rev. A **44**, 5793 (1991).
6. C. Guet and W. R. Johnson, Phys. Rev. B **45**, 11 283 (1992).
7. S. A. Blundell and C. Guet,Z Phys.D **33**, 253 (1995), and Surface Review and Letters **3**, 519 (1996)
8. Ll. Serra, G. B. Bachelet, N. Van Giai, and E. Lipparini, Phys. Rev. **B48**, 14708 (1993)
9. K. Yabana and G. F. Bertsch, Z. Phys. D**32**, 329 (1995)
10. M.Ya. Amusia, N.A. Cherepkov, Case Stud. At. Phys. **5**, 47 ( 1975).
11. M. Madjet, C. Guet, and W. R. Johnson, Phys. Rev. A **51**, 1327 (1995).
12. D. J. Rowe, Rev. Mod. Phys.**40**, 153 (1968)
13. K. Yabana and G. F. Bertsch, Phys. Rev. B **54,** 4484 (1996).
14. H. Flocard, S. Koonin, and M. Weiss, Phys. Rev. C **17**, 1682 (1978).
15. F. Calvayrac, P. G. Reinhard, and E. Suraud, Phys. Rev. B **52**, R17056 (1995).
16. P. Ring and P. Schuck, *The nuclear many-body problem*, Berlin, Heidelberg, New York: Springer (1980).
17. L'Eplattenier, E. Suraud, and P. G. Reinhard, Ann. Phys. (N.Y.) **244**, 426 (1995).
18. P. G. Reinhard and E. Suraud, Ann. Phys. (N.Y.) **239**, 193(1995), Ann. Phys. (N.Y.) **239**, 216(1995),
19. C. Jarzynski, and G. F.Bertsch, preprint nucl-th/9507040, University of Washington.
20. O. Gunnarsson and B. I. Lundqvist, Phys. Rev. B **13**, 4274 (1976)
21. M. Gross and C. Guet, Z. Phys. D **33**, 289 (1995).
22. M. Gross and C. Guet, Phys. Rev.A **54**, R2547 (1996).
23. L. Feret, E. Suraud, F. Calvayrac, and P. G. Reinhard, J. Phys. B: At. Mol. Opt. Phys. **29**, 4477 (1996).
24. F. Chandezon, C. Guet, B. A. Huber, D. Jalabert, M. Maurel, E. Monnand, C. Ristori, and J. C. Rocco, Phys. Rev. Lett. **74**, 3784 (1995). F. Chandezon, C. Guet, B. A. Huber, D. Jalabert, M. Maurel, E. Monnand, C. Ristori, and J. C. Rocco, Surface Review and Letters **3**, 529 (1996).
25. L. Plagne, *Thesis*, Université de Grenoble, France, in preparation.
26. P. M. Echenique, F. Flores, and R. H. Ritchie, Solid State Physics **43**, 229 (1990).
27. J. Lindhard, K. Dan. Vindensk. Selsk. Mat-Phys. Medd. **8** 28 (1954).

28. S. I. Anisimov, B. L. Kapeliovich, and T. L. Perel'man, Sov. Phys. JETP **39**, 375 (1974).

29. H. Häkkinen and U. Landman, Phys. Rev. Lett. **71**, 1023 (1993).

30. G. L. Eesley, Phys. Rev. Lett. **51**, 2140 (1983).

31. H. E. Elsayed-Ali, T. B. Norris, M. A. Pessot, and G. A. Mourou, Phys. Rev. Lett. **58**, 1212 (1987).

32. R. W. Schoenlein, W. Z. Lin, J. G. Fujimoto, and G. L. Eesley, Phys. Rev. Lett. **58**, 1680 (1987).

33. W. S. Fann, R. Storz, H. W. K. Tom, and J. Bokor, Phys. Rev. Lett. **68**, 2834 (1992).

34. J. Y. Bigot, J. C. Merle, O. Cregut, and A. Daunois, Phys. Rev. Lett. **75**, 4702 (1995).

35. A. Domps, A.S. Krepper, V. Savalli, P.G. Reinhard, E. Suraud, Ann. Phys. (Leipz) **6**, 468 (1997).

36. L. Plagne and C. Guet, in preparation.

37. K. Yabana, T. Tazawa, Y. Abe, and P. Bozek, Phys. Rev. A **57**, R3165 (1998), K. Yabana, *Proc. int. symp. on Similarities and Differences between Atomic-Nuclei and -Clusters.* , Tsukuba, 1997, edited by Y. Abe et al, AIP Conf. Proc. No 416 (AIP, Woodbury, N.Y., (1997), p148).

# Accurate Phase Space Theory and Molecular Dynamics Calculations of Aluminum Cluster Dissociation

Gilles H. Peslherbe

Department of Chemistry and Biochemistry, Concordia University,
1455 De Maisonneuve Blvd. Ouest, Montréal, Québec, Canada H3G 1M8

and

William L. Hase

Department of Chemistry, Wayne State University, Detroit, MI 48202

**Abstract:** Calculations of phase space theory and molecular dynamics rate constants and product energy and angular momentum distributions for $Al_6$ and $Al_{13}$ dissociation are reviewed. The molecular dynamics simulations indicate the unimolecular dynamics of these clusters are ergodic and agree with the random lifetime assumption of Rice-Ramsperger-Kassel-Marcus (RRKM) theory. With anharmonicity effects properly included, orbiting transition state theory (OTS/PST) gives rate constants and product energy and angular momentum distributions in very good agreement with those obtained from the molecular dynamics simulations.

## 1. Introduction

Fluxional molecules are characterized by multiple minima (i.e. conformations) with similar potential energies, which are separated by transition states with low barriers [1-4]. Thus, a moderately excited fluxional molecule may undergo rapid transitions between its many different conformations, which may have a profound effect on its chemical dynamics. If transitions between the conformations are rapid, they will all contribute to the molecule's unimolecular dissociation rate constant. Many different conformations are expected for the products, but near the dissociation threshold only one set of product conformations is accessible. As the energy is increased, thresholds for other product conformations are reached.

Clusters are often highly fluxional [1], and their unimolecular fragmentation has been the focus of many experimental and theoretical studies [5-28]. Rice-Ramsperger-Kassel-Marcus (RRKM) theory [29] and other statistical models, such as phase space theory, have been used to analyze cluster unimolecular decay rates [12-16,19], binding energies [11-13], product size distributions [13] and product energy partitioning [14,15,17,18]. In experiments, the cluster internal energy distribution is not known and the internal energy has been estimated with statistical models [12]. In recent work [27], classical trajectory simulations were reported of the dissociation of randomly excited $Al_3$ clusters. Unimolecular rate constants and product energy distributions determined from the trajectories were found to be in good agreement with the predictions of RRKM theory and/or phase space theories [30-36].

In this chapter, we review our recent calculations of classical trajectory and phase space theory unimolecular rate constants and product energy distributions for $Al_6 \rightarrow Al_5 + Al$ and $Al_{13} \rightarrow Al_{12} + Al$ decomposition [37,38]. The potential energy function, used for the calculations, is a model derived from *ab initio* calculations [39], and is written as a sum of two-body Lennard-Jones (L-J) potentials

$$V_{ij} = \varepsilon \left[ \left(\frac{r_0}{r_{ij}}\right)^{12} - 2 \left(\frac{r_0}{r_{ij}}\right)^6 \right] \tag{1}$$

and three-body Axilrod-Teller (A-T) potential functions [40]

$$V_{ijk} = Z \varepsilon r_0^9 \frac{1 + 3 \cos \alpha_i \cos \alpha_j \cos \alpha_k}{\left( r_{ij} r_{jk} r_{ki} \right)^3} \tag{2}$$

where $\varepsilon = 26.52$ kcal/mol, $r_0 = 2.635$ Å, and $Z = 0.5$. Whereas the L-J/A-T potential introduces too much planarity in all structures, other analytic potential energy surfaces [41,42] were derived with the goal of reproducing bulk properties and do not describe the small clusters as well as does the L-J/A-T potential.

The $Al_n$ analytic potential energy functions, including the L-J/A-T potential, do not distinguish the different symmetries and spin multiplicities for the various electronic states. The potentials were derived with the goal of approximating the energies and structures of the various minima regardless of their electronic state. Since there should be extensive state mixing for fluxional species like the $Al_n$ clusters, it seems reasonable to include all - or many - cluster electronic states in an average way with a single model potential energy surface, and not explicitly consider the electronic transitions [9].

The combined L-J/A-T potential gives multiple minima and transition states for $Al_n$ clusters and some for $Al_6$ and $Al_{13}$ are shown in Figures 1 and 2. Internuclear separations, binding energies, unimolecular dissociation energies, vibrational frequencies and moments of inertia for most potential minima of $Al_n$ clusters (n=2,3,5,6,12,13) have been listed elsewhere [27,37]. The general chemical dynamics

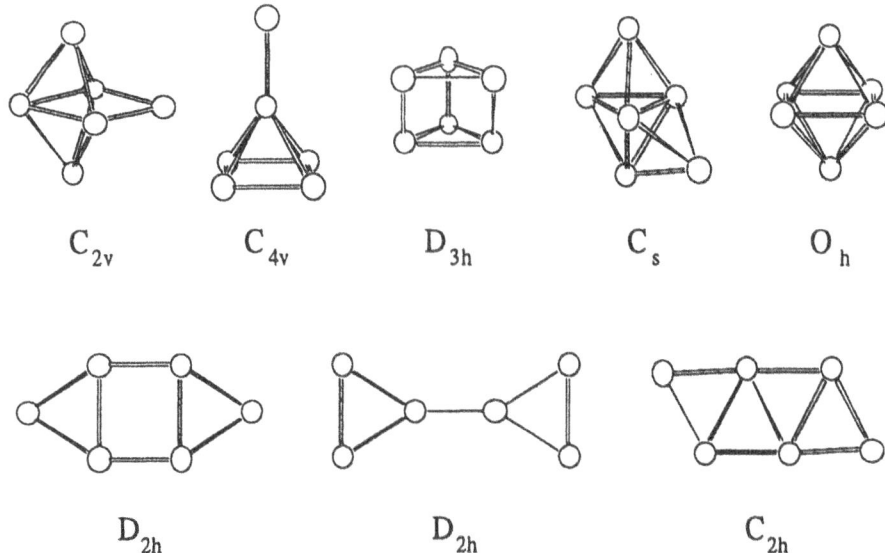

**Figure 1.** Some Al$_6$ cluster stationary points on the potential energy surface

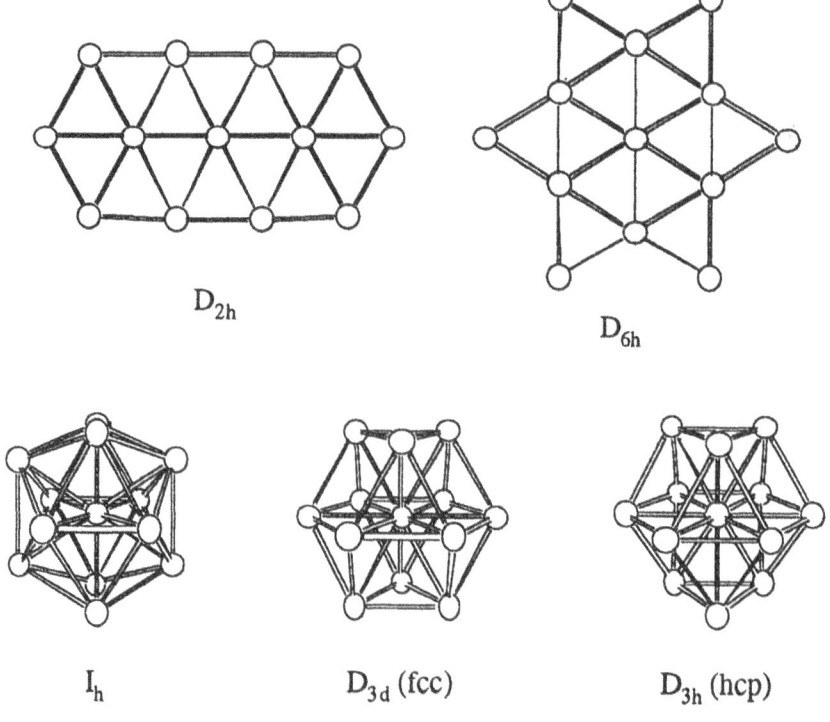

**Figure 2.** Some Al$_{13}$ cluster stationary points on the potential energy surface

computer program VENUS96 [43] was used to determine properties of the stationary points, which were found with the globally convergent Broyden quasi-Newton optimization technique [44]. No attempt was made to establish an exhaustive list of the stationary points for these clusters since the number of potential energy minima is thought to increase exponentially with cluster size [1,2,45-49]. One of the $Al_6$ $D_{2h}$ structures in Figure 1 is a transition state structure. The remaining structures in Figures 1 and 2 are for potential energy minima.

The parameters for the L-J/A-T potential were chosen to fit *ab initio* data for small $Al_n$ clusters, such as $Al_2$ and $Al_3$. Thus, the $Al_5$ and $Al_6$ potential energy minima on the L-J/A-T analytic potential energy surface are in much better agreement with the *ab initio* data than are those for the $Al_{12}$ and $Al_{13}$ clusters [39]. The *ab initio* calculations predict minimum energy octahedral and icosahedral structures for $Al_6$ and $Al_{13}$, respectively, whereas the analytic L-J/A-T potential gives planar minimum energy structures. The L-J/A-T potential gives an octahedral $Al_6$ structure only slightly higher in energy than the planar $C_{2h}$ minimum energy structure, while for $Al_{13}$ this potential gives an icosahedral structure significantly higher in energy than the planar $D_{2h}$ minimum energy geometry. The A-T three-body term induces too much planarity in the $Al_n$ minimum energy structures and, thus, the L-J/A-T analytic potential energy function should only be considered as a model for the larger clusters. However, the important and correct attribute [29] of the L-J/A-T potential is that it leads to a large number of low-lying potential energy minima for the $Al_n$ clusters.

It is necessary to establish whether statistical theories are valid for interpreting the unimolecular dissociation of clusters. One way to do this is to compare the results of molecular dynamics simulations with the predictions of statistical theories - on the same potential energy surface - as is done here for $Al_6$ and $Al_{13}$ and elsewhere [26] for the model metal-atom clusters $M_n$ (n=12-14).

## 2. Phase Space Theories

Phase space theories are RRKM microcanonical transition state theories which assume the decomposition of a molecule or a collision complex is governed by the phase space available to each product under strict conservation of angular momentum and energy [30-36]. The general RRKM unimolecular rate constant, as a function of total energy E and angular momentum quantum number J, is given by

$$k(E,J) = \frac{L^{\ddagger}}{h} \frac{N^{\ddagger}(E,J)}{\rho(E,J)} \tag{3}$$

where h is Planck's constant, $N^{\ddagger}(E,J)$ is the sum of states for the active degrees of freedom in the transition state, $L^{\ddagger}$ is the reaction channel statistical factor or reaction path degeneracy, and $\rho(E,J)$ is the density of states for the active degrees of freedom in the reactant. To calculate $N^{\ddagger}(E,J)$, phase space theories assume a separable product long-range Hamiltonian of the form

$$ H = H_{v1} + H_{v2} + T_{r1} + T_{r2} + \frac{\ell^2}{2\mu r^2} + E_t + V_{inter} \tag{4} $$

where the $H_v$'s and $T_r$'s are the vibrational Hamiltonians and rotational kinetic energies of the reaction fragments, $\ell$ is the orbital angular momentum quantum number, $\mu$ is the reduced mass of the fragments, r their center-of-mass separation, $E_t$ the relative translational energy projected on the fragments center-of-mass separation, and $V_{inter}$ is the fragments intermolecular potential. The unimolecular system total angular momentum $\mathbf{J}$ is the vector sum of the rotational angular momenta for the reaction fragments and the orbital angular momentum $\boldsymbol{\ell}$. In the following, it is assumed for simplicity that the product pairs consist of one product with rotational angular momentum $\mathbf{j}$ and an atom. In the formulation of phase space theories for product pairs which do not include an atom, both product rotational angular momenta have to be considered. The rotational quantum number j is related to the classical rotational angular momentum $\mathbf{j}$ by semiclassical quantization, i.e. $|\mathbf{j}|=\sqrt{j(j+1)}\,\hbar$. Similar relationships hold for the total angular momentum and orbital momentum quantum numbers J and $\ell$.

In one version of phase space theory, developed by Klots [31,32] and which is identified here as PST, a loose transition state is assumed with properties identical to those of the reaction products. This loose transition state limit assumes that the reaction potential energy surface is of no importance in determining the unimolecular rate constant. When recasting the PST expression for the unimolecular rate constant with total energy E and angular momentum quantum number J in the general microcanonical transition state theory formalism of Eq. (3), one obtains for the "transition state" sum of states

$$ N^{\ddagger}(E,J) = \int_0^{E-E_0} \rho_V^{\ddagger}(E-E_0-E_{tr})\, \Gamma_{ro}^{\ddagger}(E_{tr},J)\, dE_{tr} \tag{5} $$

where $\rho_V^{\ddagger}(E)$ is the vibrational density of states of the products at energy E, $E_0$ is the energy difference between reactants and products, $\Gamma_{ro}^{\ddagger}(E,J)$ is the product sum of rotational-orbital states with rotational-orbital energy less than or equal to E, and $E_{tr}$ is the sum of product translational and rotational energies. Note that in PST the product total energy is written as

$$E_\infty = E - E_0 = E_t + E_r + E_v = E_{tr} + E_v = E_t + E_{int} \tag{6}$$

where $E_\infty$ is the energy in excess of the classical threshold, $E_v$ is the product vibrational energy, $E_{int}$ is the product internal vibrational/rotational energy, $E_t$ is the product relative translational energy, and $E_r$ is the product rotational energy.

The differential total phase space volume accessible to the system [35], as a function of product translational and rotational energy, product rotational and orbital quantum numbers j and $\ell$, is given by

$$\rho_v^\ddagger(E_\infty - E_t - E_r)\, \rho_r^\ddagger(E_r, j)\, dj\, d\ell\, dE_r\, dE_t \tag{7}$$

with the product rotational density of states $\rho_r^\ddagger(E_r, j)$ given by

$$\rho_r^\ddagger(E_r, j) = \frac{\partial}{\partial E_r}\, \Gamma_r^\ddagger(E_r, j) \tag{8}$$

where $\Gamma_r^\ddagger(E_r, j)$ is the sum of product rotational states with rotational energy equal or less than $E_r$ and rotational angular momentum j. By combining Eqs. (5)-(8), one can write the rotational-orbital sum of states as [36]

$$\Gamma_{ro}^\ddagger(E_{tr}, J) = \iint \Gamma_r^\ddagger(E_r, j)\, d\ell\, dj \tag{9}$$

This double integral has to be evaluated under strict conservation of total energy and angular momentum.

According to phase space theories, the probability of forming products with given properties is proportional to the total phase space volume accessible to the system [35], whose differential form is given by Eq. (7). The probability of forming products with translational and rotational energy $E_{tr}$ is thus the integral of the aforementioned differential volume with fixed $E_{tr}$, and this leads to the following expression for the normalized kinetic energy release probability density

$$P_{EJ}(E_{tr}) = \frac{\rho_v^\ddagger(E - E_0 - E_{tr})\, \Gamma_{ro}^\ddagger(E_{tr}, J)}{N^\ddagger(E, J)} \tag{10}$$

Similarly, if one introduces the product rotational-orbital density of states as

$$\rho_{ro}^{\ddagger}(E_t,E_r,J) = \iint \rho_r^{\ddagger}(E_r,j) \, d\ell \, dj \qquad (11)$$

the probability densities for translational and rotational energy are given respectively by

$$P_{E,J}(E_t) \; \alpha \quad \int \rho_v^{\ddagger}(E-E_0-E_{tr}) \, \rho_{ro}^{\ddagger}(E_t,E_r,J) \, dE_r \qquad (12)$$

$$P_{E,J}(E_r) \; \alpha \quad \int \rho_v^{\ddagger}(E-E_0-E_{tr}) \, \rho_{ro}^{\ddagger}(E_t,E_r,J) \, dE_t \qquad (13)$$

Similar expression can be derived for rotational quantum number probability densities.

Chesnavich and Bowers [35,36] modified the phase space theory model by assuming: (1) an orbiting transition state located at the centrifugal barrier;[33-36,50] and (2) that orbital rotational energy at this transition state is converted into relative translational energy of the products. These assumptions are strictly valid if the long-range potential $V_{inter}$ in Eq. (4) is isotropic and, thus, only depends on r. In PST the rotational-orbital sums and densities of state are evaluated at the product asymptotic limit; as a result, the centrifugal barrier term and the intermolecular potential in the Hamiltonian of Eq. (4) vanish and the sum of $\ell$-states is found solely from the angular momentum conservation condition. In contrast, for the orbiting transition state model of phase space theory (OTS/PST), the rotational-orbital sums and densities of state are evaluated at the orbiting transition state, which yields additional constraints on the values of the rotational and orbital quantum numbers in the above integral equations. This also has the effect of adding the height of the centrifugal barrier to $E_t$ to obtain the product translational energy. In OTS/PST, it is customary to represent the product isotropic long-range interaction potential as a function of the center-of-mass separation with a $-C/r^n$ potential. Analytic solutions can be derived for the rotational-orbital sums and densities of state when n=4 (ion-molecule interaction) and 6 (molecule-molecule interaction) [35,36]. PST expressions for the rotational-orbital sums of state have also been reported for various reaction product pairs (i.e. sphere-sphere, sphere-atom, etc.), but mainly in the $J \approx 0$ limit [31].

In the next section, both PST and OTS/PST are used to predict the rate constants and product energy distributions of cluster monoatomic evaporation. The relevant PST and OTS/PST expressions for rotational-orbital sums and densities of state, i.e. for product pairs consisting of an atom and a linear, symmetric or spherical top, are given in the Appendix.

OTS/PST can be viewed as a variational unimolecular rate theory which minimizes the reaction flux versus E, J and $\ell$. For an isotropic long-range potential, the resulting transition state will be at the centrifugal barrier. The transition state sum of states for a particular E, J and $\ell$ is then

$$N^{\ddagger}(E,J,\ell) = \int\limits^{E_{\infty}-\varepsilon^{\ddagger}} \rho^{\ddagger}_t(E_{\infty}-E_{tr}) \int \Gamma^{\ddagger}(E_r,j) \, dj \, dE_{tr} \qquad (14)$$

where $\varepsilon^{\ddagger}$ is the height of the centrifugal barrier with respect to the asymptotic limit and $\Gamma^{\ddagger}(E_r,j)$ is the rotational sum of states at the orbiting transition state. Thus, $N^{\ddagger}(E,J)$ in OTS/PST is simply the integral of $N^{\ddagger}(E,J,\ell)$ in Eq. (14) over $\ell$; i.e.

$$N^{\ddagger}(E,J) = \int N^{\ddagger}(E,J,\ell) \, d\ell \qquad (15)$$

It is of interest to point out that, in contrast to OTS/PST which locates a transition state and calculates the unimolecular rate constants as a function of E, J and $\ell$, the flexible model of variational RRKM theory [51-54] calculates a variational unimolecular rate constant only as a function of E and J. The transitional modes' sum of states for the flexible model includes contributions for all possible values of orbital angular momentum $\ell$, so that the variational transition state of the flexible model is the "best" compromise of all variational transition states as a function of E, J and $\ell$. As a result, for an isotropic intermolecular potential, OTS/PST gives a variational k(E,J) less than or equal to that of flexible variational RRKM theory [27]. This relationship is similar to that between k(E,J) calculated for the harmonic vibrator variational transition state model with the K quantum number treated as either active or adiabatic [55].

Furthermore, the flexible variational RRKM model approaches PST as the energy is lowered and the separation between the product fragments becomes very large at the variational transition state. This is because the $\ell^2/2\mu r^2$ and $V_{inter}$ terms in Eq. (4) become negligible, so that the flexible Hamiltonian becomes

$$H = H_{v1} + H_{v2} + T_{r1} + T_{r2} + E_t \qquad (16)$$

which is the actual Hamiltonian of PST [31]. PST and the flexible variational RRKM model use different sets of conjugate coordinates and momenta to determine phase space volumes for the degrees of freedom that are treated classically. However, both models use the same conservation laws, i.e. total energy and that of angular momentum, and when the flexible Hamiltonian approaches that of PST, the flexible variational RRKM rate constant approaches that of PST.

In the flexible transition state model of variational RRKM theory the long-range potential is not assumed to be isotropic as in OTS/PST and the actual anisotropic long-range potential is used for $V_{inter}$ in Eq. (4). Using the anisotropic potential has the general effect of tightening the transition state and reducing the sum of states in comparison to what is found for the isotropic potential [56]. However, for $Al_3 \rightarrow Al_2 + Al$ dissociation, OTS/PST and the flexible transition state derived from the anisotropic potential gave nearly identical unimolecular rate constants [27]. As a result, OTS/PST is expected to give accurate rate constants for $Al_6$ and $Al_{13}$ dissociation.

# 3. Importance of Anharmonicity in the $Al_n$ Density of States

Because of the fluxional character of $Al_n$, it is important to consider anharmonicity when calculating its density of states [27,37]. Nosé constant temperature dynamics [57], combined with the multiple histogram method [15,58], was used to calculate the relative anharmonic density of states for $Al_5$, $Al_6$, $Al_{12}$ and $Al_{13}$ versus energy [37]. The multiple potential minima for the clusters may contribute to their anharmonic density of states. These minima arise from the different conformations and the n! *identical* or *degenerate* [59] configurations for each conformation. The Nosé dynamics does not distinguish between different conformations and identical configurations. Since there are potential energy barriers for transitions between identical configurations, the Nosé dynamics is expected to sample more identical configurations as the temperature (i.e. energy) is increased. Thus, all n! identical configurations for a particular conformation may contribute to the density of states at high energy, while only one configuration for the conformation with the deepest potential minimum may contribute to the density of states at low energy.

To transform the above relative density of state curves to absolute anharmonic densities of state, it is necessary to evaluate, for each cluster, the absolute anharmonic density of states for at least one energy. For $Al_5$ and $Al_6$, this was done [37] by using

Monte Carlo methods [60] to solve the multidimensional phase integral in Cartesian space for the clusters' anharmonic Hamiltonian [61,62]

$$N(E) = \frac{1}{h^s} \underset{H \leq E}{\iint \ldots \int} dq \, dp \tag{17}$$

The calculations were performed at relatively high energies, 60-140 kcal/mol above the deepest potential energy minimum and contributions from identical configurations were included in the resulting N(E); i.e. the N(E) were not divided by n!. This approach seems consistent with the relative density of state obtained from the Nosé dynamics. The density of states is obtained by differentiating the sum of states N(E) with respect to energy.

Following previous work [15,27,63] the adiabatic switching method [64] was used to calculate the absolute density of states for $Al_{12}$ and $Al_{13}$ [37]. This method is based on the Hertz invariance principle [65] which says that, if an ergodic Hamiltonian is changing slowly in time, its energy shell is an adiabatic invariant. Thus, adiabatically switching a Hamiltonian, into a reference Hamiltonian, maps a microcanonical ensemble on an initial hypersurface into a continuous family of surfaces, each of which is also at constant energy and, via Liouville's theorem [66], has the identical phase volume. The idea is to use a reference Hamiltonian for which the classical density of states is known precisely. If the final energies of the trajectories, which are switched into the reference Hamiltonian, do not exhibit a large dispersion, the anharmonic density of states is just that of the reference system with the average final energy. For a harmonic oscillator (ho) reference Hamiltonian this is expressed as

$$\rho(E) = \rho_{ho}(\overline{E}_{ho}) \tag{18}$$

This approach was used [37] to calculate anharmonic densities of state for $Al_{12}$ and $Al_{13}$ at an energy approximately 150 kcal/mol above each cluster's deepest potential energy minimum. Identical configurations are expected to contribute to these densities of state in a manner similar to that for the relative densities of state from the Nosé dynamics.

The importance of anharmonicity for the aluminum clusters is illustrated in Table I, where the ratio of the anharmonic and harmonic densities of state, versus energy, are listed for $Al_6$ and $Al_{13}$. The anharmonic densities of states were calculated as described above, and include contributions from different conformations as well as identical configurations. The harmonic densities were calculated for the lowest energy conformation, which has a $C_{2h}$ structure for $Al_6$ and a $D_{2h}$ structure for $Al_{13}$ [37]. The depth of these potential minima, with respect to the product asymptotic limit, are -43.8 and -56.2 kcal/mol for the $C_{2h}$ and $D_{2h}$ structures, respectively. At an energy equal to the cluster dissociation threshold, i.e. $E_\infty=0$, the anharmonic density is 56 and 4,600 times larger than the harmonic density for $Al_6$ and $Al_{13}$, respectively.

# 4. Unimolecular Rate Constants and Product Energy Distributions

### 4.1. Phase Space Theory Calculations

The theoretical procedure described in Section 2 and the anharmonic densities of state given in Section 3 were used to calculate OTS/PST rate constants and product energy distributions for $Al_6 \rightarrow Al_5 + Al$ and $Al_{13} \rightarrow Al_{12} + Al$ dissociation [37,38]. The harmonic rate constants are significantly larger than the anharmonic values, with the difference decreasing with increase in energy (see Table II). This is because anharmonicity in the transition state sum of states becomes more important, as the energy is increased, and begins to cancel the effect of anharmonicity in the reactant cluster's density of states. The effect of angular momentum on the OTS/PST rate constant was investigated for $Al_6$ dissociation. Increasing the total angular momentum quantum number J from 0 to 160 (which corresponds to a rotational temperature of ~1000 K) while keeping the $Al_6$ vibrational energy constant, has only a very small effect on the OTS/PST rate constant as well as the molecular dynamics rate constant (see below). The rate constants are rather insensitive to the choice of J and depend mainly on the total rotational/vibrational energy of the clusters [37].

The reaction path degeneracy factor, $L^\ddagger$ in Eq. (3), was assumed to equal unity in calculating the OTS/PST rate constants. As discussed above, the number of configurations for each conformation, accessed by the transition state and excited clusters, are included in the anharmonic densities of state. The $L^\ddagger$ factor then depends on the symmetry numbers for the cluster and transition state conformations contributing to the unimolecular dissociation. In a more complete analysis than the one presented here, the $L^\ddagger$ factor would be energy dependent.

OTS/PST was also used to calculate energy and angular momentum distributions for the products of $Al_6 \rightarrow Al_5 + Al$ and $Al_{13} \rightarrow Al_{12} + Al$ dissociation [38]. The internal energy $E_{int}$ and angular momentum quantum number j distributions for the $Al_5$ product are shown in Figure 3 for an $Al_6$ energy $E_\infty$=40 kcal/mol.

**Table I.** Classical Densities of State for $Al_n$ (n=6,13) [a]

| | $Al_6$ | | | $Al_{13}$ | | |
|---|---|---|---|---|---|---|
| $E_\infty$[b] | $\rho_h$[c] | $\rho_{anh}$[d] | $\rho_{anh}/\rho_h$ | $\rho_h$[c] | $\rho_{anh}$[d] | $\rho_{anh}/\rho_h$ |
| - 40.0 | 0.1 | 0.8 | 8 | $1.8 \times 10^9$ | | |
| - 20.0 | $7.5 \times 10^7$ | $1.2 \times 10^9$ | 16 | $1.1 \times 10^{20}$ | $1.2 \times 10^{23}$ | 1,100 |
| 0.0 | $6.2 \times 10^{10}$ | $3.5 \times 10^{12}$ | 56 | $1.4 \times 10^{26}$ | $6.5 \times 10^{29}$ | 4,600 |
| 20.0 | $3.9 \times 10^{12}$ | $5.2 \times 10^{14}$ | 133 | $2.4 \times 10^{30}$ | $3.5 \times 10^{34}$ | 14,600 |
| 40.0 | $5.2 \times 10^{13}$ | $1.9 \times 10^{16}$ | 365 | $4.2 \times 10^{33}$ | $6.6 \times 10^{38}$ | $1.6 \times 10^5$ |
| 60.0 | $8.3 \times 10^{14}$ | $3.2 \times 10^{17}$ | 385 | $1.8 \times 10^{36}$ | $3.5 \times 10^{42}$ | $1.9 \times 10^6$ |
| 80.0 | | | | $2.8 \times 10^{38}$ | $3.9 \times 10^{45}$ | $1.4 \times 10^7$ |
| 100.0 | | | | $2.3 \times 10^{40}$ | $1.4 \times 10^{48}$ | $6.1 \times 10^7$ |
| 120.0 | | | | $1.1 \times 10^{42}$ | $2.4 \times 10^{50}$ | $2.2 \times 10^8$ |
| 140.0 | | | | $3.3 \times 10^{43}$ | $2.2 \times 10^{52}$ | $6.7 \times 10^8$ |
| 160.0 | | | | $7.5 \times 10^{44}$ | $1.2 \times 10^{54}$ | $1.6 \times 10^9$ |

a. Energies in kcal/mol and densities of state in states per $cm^{-1}$.
b. Energy in excess of the product classical asymptotic limit.
c. Harmonic density of states for the lowest energy minimum (see text).
d. Anharmonic density of states.

**Table II.** Comparison of Anharmonic OTS/PST and Molecular Dynamics Rate Constants for $Al_6$ and $Al_{13}$ Dissociation

| $E_\infty$[a] | Rate Constants ($ns^{-1}$) | |
|:---:|:---:|:---:|
| | OTS/PST [b] | Molecular Dynamics [c] |
| $Al_6 \rightarrow Al_5 + Al$ | | |
| 30 | 5.20 (143) | $4.8 \pm 0.1$ |
| 40 | 47.6 (621) | $22.8 \pm 0.3$ |
| 50 | 155 (1637) | $67.6 \pm 0.6$ |
| 60 | 282 (3266) | $148.7 \pm 1.3$ |
| 70 | 484 (5533) | $308.2 \pm 5.1$ |
| 80 | 807 (8274) | $432.9 \pm 8.6$ |
| $Al_{13} \rightarrow Al_{12} + Al$ | | |
| 85 | 0.002 (4.20) | 0.01 |
| 110 | 0.1 (67.9) | 0.35 |
| 135 | 3.80 (252) | 3.9 |
| 160 | 20.4 (905) | 15.8 |
| 185 | 25.3 (2558) | 46.5 |

a. Energy in excess of the product classical asymptotic limit, in kcal/mol.

b. The anharmonic OTS/PST rate constants were calculated for the spherical top-atom model. The symmetric top-atom model gives nearly the same results [37]. The rate constants given in parentheses are harmonic OTS/PST values for the spherical top-atom model, estimated from the harmonic densities for the lowest energy minima for both the products and reactants.

c. Rate constants of cluster dissociation, obtained by microcanonical normal mode sampling around various $Al_n$ minima. Ensembles contain 200 trajectories and rate constants are obtained by a linear least squares fit of the $Al_n$ population $vs.$ time.

## 4.2. Molecular Dynamics Calculations

Microcanonical normal mode sampling [67] was used to prepare initial microcanonical ensembles about the $C_{2h}$, $C_{2v}$, and $O_h$ minima of the $Al_6$ cluster and about the $D_{2h}$ and $D_{3d}$ (fcc) minima of the $Al_{13}$ cluster. Classical trajectories were used to study the dissociation of each of these ensembles. The lifetime of a cluster was defined by the last inner turning point in the dissociation fragment center-of-mass relative velocity [37]. After a cluster had dissociated, the product energies and rotational angular momentum quantum numbers were calculated for the reaction fragments.

The lifetime distribution, for dissociation of one of the ensembles of excited $Al_6$ and $Al_{13}$ clusters, is given by

$$P(t) = \frac{1}{N(0)} \frac{d\,N(t)}{dt} \qquad (19)$$

where $N(0)$ is the total number of clusters in the ensemble and $N(t)$ is the number of clusters which remain undissociated at time t. RRKM theory assumes the random lifetime distribution [68]

$$P(t) = k\exp(-kt) \qquad (20)$$

where k is the microcanonical unimolecular rate constant. The lifetime distributions for each ensemble sampled for $Al_6$ and $Al_{13}$ are of the form defined by Eq. (20). A representative distribution is given in Figure 4 for $Al_6$ dissociation. For each ensemble, the t=0 intercept of P(t) is equal to the rate constant in the exponential. In addition, the rate constants for $Al_6$ and $Al_{13}$ are independent of the specific potential energy minima sampled in the initial conditions and only depend on the total energy. Apparently, the intramolecular dynamics of the $Al_6$ and $Al_{13}$ clusters is sufficiently chaotic that a microcanonical ensemble is quickly prepared over the complete phase space of the clusters. This phase space will include all accessible conformations and identical configurations. The central assumption of RRKM theory, which states that all accessible molecular states are occupied with equal probability, seems to be valid for $Al_6$ and $Al_{13}$ dissociation, which implies that the classical motion of these clusters is chaotic and their Hamiltonian systems presumably ergodic. The same conclusion was reached earlier for $Al_3$ dissociation [27].

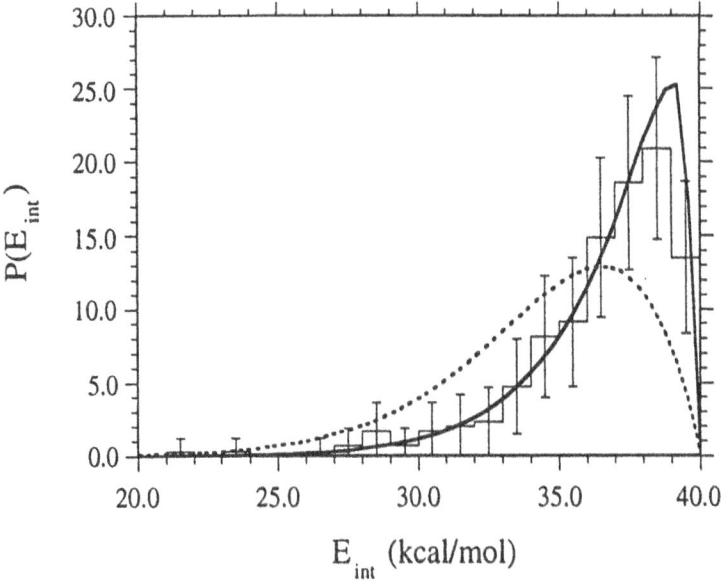

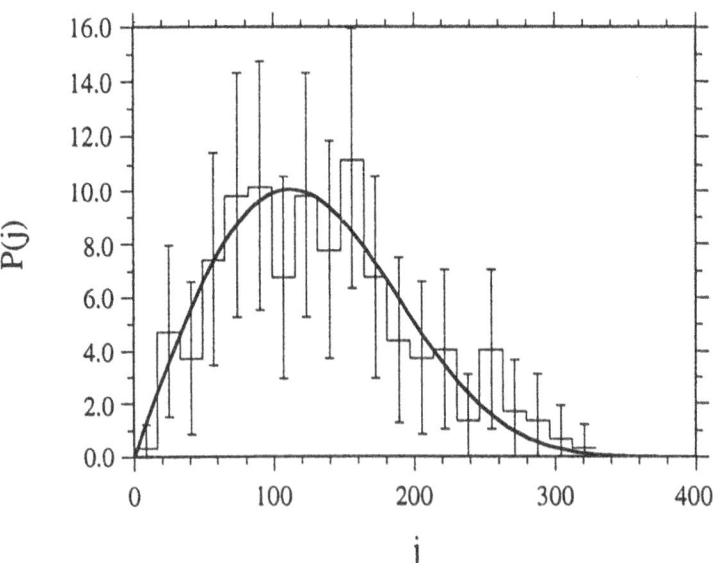

**Figure 3:** Product internal energy ($E_{int}$) and angular momentum (j) distributions for $Al_6 \rightarrow Al_5 + Al$ unimolecular fragmentation, $E_\infty=40$ kcal/mol. The histograms are distributions obtained from trajectory calculations. The solid thick lines represent the OTS/PST predictions and the dashed lines those of the Engelking model [13].

Since sampling about different minima of $Al_6$ and $Al_{13}$ at fixed energy E gives the same rate constants within statistical uncertainties, the P(t) calculated for the initial ensembles about the different minima were combined to determine a microcanonical rate constant for the cluster. The resulting k(E) values are listed in Table II. For the most part they are seen to agree with the anharmonic OTS/PST rate constants to within a factor of two. This is an impressive result, since the harmonic OTS/PST rate constants are one to three orders of magnitude different from the molecular dynamics rate constants. The largest difference between the anharmonic OTS/PST and molecular dynamics rate constants is a factor of 4-5 for $Al_{13}$ dissociation at $E_\infty$ of 110 and 85 kcal/mol. This difference may arise from our assumption of $L^\ddagger=1$ in calculating the OTS/PST rate constant and/or the small dissociation probability for $Al_{13}$ at low energies, which makes it more difficult to determine an accurate molecular dynamics rate constant.

At low energy monomer evaporation is the major dissociation pathway for the aluminum clusters, but as the energy is increased, dimer and trimer evaporation becomes non-negligible. For example, for $Al_6$ dissociation dimer evaporation contributes 1% at $E_\infty$=30 kcal/mol, but 11% at $E_\infty$=80 kcal/mol [37].

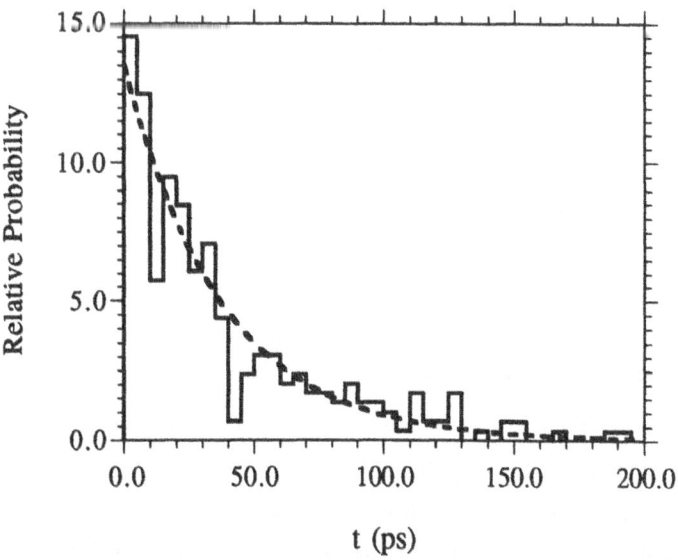

**Figure 4:** Trajectory $Al_6 \rightarrow Al_5 + Al$ lifetime distributions. The histogram plot represents the number of $Al_6$ dissociations per time interval. The dashed line represents the random lifetime prescription.

All the product energy and angular momentum distributions determined from the molecular dynamics simulations are in very good agreement with the predictions of OTS/PST. Representative distributions are given in Figure 3 for the $Al_5$ internal energy $E_{int}$ (i.e. rotational and vibrational) and angular momentum j following $Al_6 \rightarrow Al_5 + Al$ dissociation at $E_\infty = 40$ kcal/mol. The approximate statistical model of Engelking [13] does not fit the molecular dynamics $P(E_{int})$ distribution as well as OTS/PST. More detailed comparisons between the molecular dynamics and statistical theory product energy and angular momentum are given elsewhere [38].

## 5. Summary

From the calculations reviewed here and presented elsewhere [37,38], the following conclusions may be made regarding the unimolecular dissociation of $Al_3$, $Al_6$, and $Al_{13}$:

1. Anharmonic effects are important in calculating accurate sums and densities of state for $Al_n$ clusters. In a purely classical picture anharmonicity arises in part from the multitude of potential minima for the cluster's different conformations and their identical configurations. More work needs to be done to unravel the contributions from the minima for the different conformations and identical configurations. This will undoubtedly require identifying all the potential minima [59] and determining the potential energy barriers for transitions between the minima.

2. The classical unimolecular dynamics of the $Al_n$ clusters is ergodic. Preparing initial microcanonical ensembles, at the same total energy, about different potential minima for a cluster yields lifetime distributions which are the same within statistical uncertainties and agree with the prediction of RRKM theory. At low energies the clusters dissociate by monomer evaporation. Dimer and trimer evaporations become more important at higher energies.

3. OTS/PST is a good model for describing dissociation of the $Al_n$ clusters. OTS/PST with anharmonic effects included gives rate constants and product energy and angular momentum distributions in very good agreement with the results from the molecular dynamics calculations. OTS/PST and flexible transition state variational RRKM rate constants were compared for $Al_3 \rightarrow Al_2 + Al$ dissociation and found to be nearly identical.

**Acknowledgements:** This work was supported by the National Science Foundation.

# Appendix: PST and OTS/PST Rotational-Orbital Sums and Densities of State for Atom-Rigid Rotor Product Pairs

### General PST Expressions for Atom-Rigid Rotor Product Pairs

The rotational-orbital sum and density of states for product pairs which include one monatomic fragment are given by

$$\Gamma_{ro}^{\ddagger}(E_{tr},J) = \iint \Gamma_r^{\ddagger}(E_r,j)\, d\ell\, dj \tag{A1}$$

and

$$\rho_{ro}^{\ddagger}(E_t,E_r,J) = \iint \rho_r^{\ddagger}(E_r,j)\, d\ell\, dj \tag{A2}$$

respectively, where all terms are defined in Table A1. When the k rotational quantum number of the product is introduced, the rotational-orbital sum of states becomes

$$\Gamma_{ro}^{\ddagger}(E_{tr},J) = \iiint_{E_{tr}} dk\, d\ell\, dj \tag{A3}$$

while the rotational-orbital density of states can be expressed as

$$\rho_{ro}^{\ddagger}(E_t,E_r,J) = \frac{\partial\, \gamma_{ro}^{\ddagger}(E_t,E_r,J)}{\partial E_r} \tag{A4}$$

where $\gamma_{ro}^{\ddagger}(E_t,E_r,J)$ has been introduced as

$$\gamma_{ro}^{\ddagger}(E_t,E_r,J) = \iiint_{E_t,E_r} dk\, d\ell\, dj \tag{A5}$$

For a linear-atom product pair, there is of course no k quantum number, and it is just omitted in the above expressions. The triple integral in Eq. (A3) is evaluated under the condition that the translational-rotational energy is equal or less than $E_{tr}$, while the one in Eq. (A5) is evaluated under the condition that both the translational energy is equal or less than $E_t$ and the rotational energy is equal or less than $E_r$. Both integrals are evaluated under strict conservation of energy and conservation of

**Table A1.** Glossary of Important Terms in PST and OTS/PST

| | |
|---|---|
| J | Total angular momentum quantum number |
| j | Product rotational quantum number |
| $\ell$ | Orbital angular momentum quantum number |
| E | Total energy |
| $E_r$ | Product rotational energy |
| $E_t$ | Product translational energy |
| $E_{tr}$ | Product translational-rotational energy |
| $B_r$ | Product rotational constant |
| $B_k$ | Symmetric top product rotational constant associated with the k quantum number |
| $\Gamma_{ro}^{\ddagger}(E_{tr},J)$ | Rotational-orbital sum of states at translational-rotational energy $E_{tr}$ and total angular momentum quantum number J |
| $\gamma_{ro}^{\ddagger}(E_t,E_r,J)$ | Rotational-orbital sum of states at translational energy $E_t$, rotational energy $E_r$ and total angular momentum quantum number J |
| $\rho_{ro}^{\ddagger}(E_t,E_r,J)$ | Rotational-orbital density of states at translational energy $E_t$, rotational energy $E_r$ and total angular momentum quantum number J |
| $\Gamma_r^{\ddagger}(E_r,j)$ | Product rotational sum of states at rotational energy $E_r$ and rotational quantum number j |
| $\rho_r^{\ddagger}(E_r,j)$ | Product rotational density of states at rotational energy $E_r$ and rotational quantum number j |

angular momentum, whose vector expression is

$$\mathbf{J} = \mathbf{j} + \boldsymbol{\ell} \tag{A6}$$

and which translates in terms of quantum numbers by the triangular inequality

$$|J - j| \le \ell \le J + j \tag{A7}$$

The product rotational energy is expressed as

$$E_r = B_r\, j^2 \tag{A8}$$

for a linear or spherical top, where $B_r$ is the rotational constant. The sum of product rotational states $\Gamma_r^\ddagger(E_r,j)$ for linear and spherical tops is just their rotational degeneracy (spatial degeneracy excluded), which can be expressed as $(2j+1)^{2s-1}$, where $s=1/2$ for a linear and 1 for a sphere. On the other hand, the symmetric top rotational energy is given by

$$E_r(j,k) = B_r j^2 + B_k k^2 \qquad (A9)$$

and $\Gamma_r^\ddagger(E_r,j)=\min\{(2j+1),(2\sqrt{(E_r-B_r j^2)/B_k}+1)\}$ for a symmetric top depends on the relative magnitude of $E_r$ and j.

**Orbiting Transition State Restrictions**

In OTS/PST the rotational-orbital sums and densities of state are evaluated at the centrifugal barrier instead of the asymptotic product limit of PST. For a long-range isotropic potential of the form

$$V(r) = \frac{-C}{r^n} \qquad (A10)$$

where r is the product center-of-mass separation, the height of the centrifugal barrier, with respect to the asymptotic potential limit, is given by [33]

$$\varepsilon^\ddagger = \frac{\ell^\alpha}{\Lambda} \qquad (A11)$$

where $\quad \alpha = \dfrac{2n}{n-2} \quad$ and $\quad \Lambda = \dfrac{2\,(n\mu)^{n/(n-2)}\,C^{2/(n-2)}}{n-2} \qquad (A12)$

with $\mu$ the products reduced mass. The centrifugal barrier represents the minimum product relative translational energy for a particular $\ell$, i.e. $E_t \geq \varepsilon^\ddagger$, and this leads to additional constraints specific to OTS/PST

$$\ell_{max} = (\Lambda E_t)^{1/\alpha} \quad \text{and} \quad j_{max} = \sqrt{(E-E_t)/B_r} \qquad (A13)$$

The translational-rotational energy condition in OTS/PST is given by

$$E_r(j,k) + \frac{\ell^\alpha}{\Lambda} \leq E_{tr} \qquad (A14)$$

The minimum translational-rotational energy is non zero in OTS/PST for $J\neq0$, and it is found as the minimum $E_{tr}$ capable of generating an $\ell$-j pair capable of producing J, i.e. $\ell_{min}^{\alpha}/\Lambda + B_r(J-\ell_{min})^2$ where $\ell_{min}$ is determined by [35]

$$\alpha \frac{\ell_{min}^{\alpha-1}}{\Lambda} - 2 B_r (J-\ell_{min}) = 0 \tag{A15}$$

## Analytic Expressions for Zero Total Angular Momentum (J=0)

Since j=$\ell$ for the J=0 case, double integrals over j and $\ell$ can be replaced by a single integral over either j or $\ell$.

### Linear-Atom and Sphere-Atom

By solving Eq. (A3) for the rotational-orbital sum of states, one obtains

$$\Gamma_{ro}^{\ddagger}(E_{tr},0) = j_{max}^{2s}(E_{tr}) \tag{A16}$$

where $j_{max}(E_{tr})$ is just $\sqrt{E_{tr}/B_r}$ in PST, but is found from

$$B_r j_{max}^2 + \frac{j_{max}^2}{\Lambda} = E_{tr} \tag{A17}$$

for OTS/PST. The PST rotational-orbital density of states is not trivial to obtain, but Eq. (A4) can be solved directly by noting that the rotational density of states $\rho_r^{\ddagger}(E_r,j)$ for the linear and spherical tops is $(2j+1)^{2s-1}\delta(E_r-B_rj^2)$, where $\delta$ represents the Dirac delta-function. Making use of the jacobian of the transformation in Eq. (A8), one obtains

$$\rho_{ro}^{\ddagger}(E_t,E_r,j) = (2j+1)^{2s-2} \tag{A18}$$

where j is found from Eq. (A8). The latter expression is equivalent to $E_r^{s-1}$ [21,32]. In OTS/PST the rotational-orbital density of states is the same as shown in Eq. (A18) as long as j is smaller than both $j_{max}$ and $\ell_{max}$ in Eq. (A13); it is zero otherwise.

### Symmetric Top-Atom

The integral in Eq. (A3) for the symmetric top with J=0 just represents the surface of an ellipse with semi-axes of length $\sqrt{E/B_r}$ and $\sqrt{E/B_k}$, so that the PST expressions for the rotational-orbital sum and density of states are

$$\Gamma_{ro}^{\ddagger}(E_{tr},0) = \frac{\pi\, E_{tr}}{\sqrt{B_r B_k}} \qquad (A19)$$

and

$$\rho_{ro}^{\ddagger}(E_t,E_r,0) = \frac{\pi}{\sqrt{B_r B_k}} \qquad (A20)$$

For OTS/PST, the surface of the ellipse is truncated since not all values of j are allowed. Eq. (A3) can be solved by using Green's theorem to replace the integration over a surface by a contour integral, and this yields

$$\Gamma_{ro}^{\ddagger}(E_{tr},0) = \int_{0}^{j_{max}(E_{tr})} \left( 2 \sqrt{\frac{E_{tr}-B_r j^2 - j^{\alpha}/\Lambda}{B_k}} + 1 \right) dj \qquad (A21)$$

where $j_{max}(E_{tr})$ is found from Eq. (A17). The latter integral usually has to be solved numerically. Similarly, one can obtain

$$\gamma_{ro}^{\ddagger}(E_t,E_r,0) = \frac{1}{\sqrt{B_r}} \left[ j_m \sqrt{E_r - B_r j_m^2} + \sqrt{B_r E_r}\, \sin^{-1} \sqrt{\frac{B_r j_m^2}{\sqrt{E_r}}} \right] \qquad (A22)$$

where $j_m$ is the minimum of $j_{max}$ and $\ell_{max}$ in Eq. (A13). Note that if $j_m$ reaches the $\sqrt{E_r/B_r}$ limit ($\Lambda\to\infty$), the latter equation is equivalent to its PST counterpart. $\rho_{ro}^{\ddagger}(E_t,E_r,0)$ is just the derivative of $\gamma_{ro}^{\ddagger}(E_t,E_r,0)$ with respect to $E_r$.

## General Expressions for Non-Zero Total Angular Momentum (J≠0)

### Linear-Atom and Sphere-Atom

Solving for successive integrations over k, $\ell$ and j in Eq. (A3) yields the following PST expressions

$$\Gamma_{ro}^{\ddagger}(E_{tr},J) = \begin{cases} \dfrac{4s}{2s+1} \left(\dfrac{E_{tr}}{B_r}\right)^{s+1/2} & \text{if } E_{tr} \le B_r J^2 \\[2ex] \dfrac{4s}{2s+1}\, J^{s+1/2} + (2J+1)\left[\left(\dfrac{E_{tr}}{B_r}\right)^{2s} - J^{2s}\right] & \text{if } E_{tr} \ge B_r J^2 \end{cases} \qquad (A23)$$

and

$$\rho_{ro}^{\ddagger}(E_t,E_r,0) = \begin{cases} (2j+1)^{2s-1} & \text{if } E_r \le B_r J^2 \\ (2J+1)\,(2j+1)^{2s-2} & \text{if } E_r \ge B_r J^2 \end{cases} \qquad (A24)$$

Note that Eq. (A18) is just a particular case (J=0) of these more general expressions. For OTS/PST, integration over k and j can be performed trivially and applying Green' theorem, as mentioned before, one obtains

$$\Gamma_{ro}^{\ddagger}(E_{tr},J) = \int_0^{\ell^-} (J+1)^{2s}\,d\ell \;+\; \int_{\ell^-}^{\ell^+}\left(\frac{E_{tr}}{B_r} - \frac{\ell^\alpha}{\Lambda B_r}\right)^s d\ell \;+\; \int_{\ell^+}^{\ell^0} |J-\ell|^{2s}\,d\ell \qquad (A25)$$

where $\ell^-$ and $\ell^+$ are the positive roots of $\ell^\alpha/\Lambda + B_r(J-\ell)^2 = E_{tr}$ if $E_{tr} \le B_r J^2$, and $\ell^+$ is the positive root of the previous equation and $\ell^-$ is the positive root of $\ell^\alpha/\Lambda + B_r(J+\ell)^2 = E_{tr}$ if $E_{tr} \ge B_r J^2$. Analytic solutions can be easily derived for each of the integrals in Eq. (A25), except for the middle integral for the linear-atom case with usual values of $\alpha=3$ or 4 (n=6 or 4 for the potential parameter, respectively). However, such analytic solutions will not be listed here, since there is a large variety of them, depending on the relative magnitudes of J, $\ell^-$ and $\ell^+$.

The integration over j in Eq. (A2) is straightforward, recalling that the rotational density of states $\rho_r^{\ddagger}(E_r,j)$ for the linear and spherical tops is $(2j+1)^{2s-1}\delta(E_r-B_r j^2)$. Further integration over the allowed values of $\ell$ yields

$$\rho_{ro}^{\ddagger}(E_t,E_r,J) = (2j+1)^{2s-2}\left[\min\left\{\ell_{max}, J+j\right\} - |J-j|\right] \qquad (A26)$$

for allowed values of j, i.e. values of j smaller than $\min\left\{j_{max}, J+\ell_{max}\right\}$, and larger than $j_{min}$, which is 0 if $\ell_{max} > J$ or $J-\ell_{max}$ if $\ell_{max} \le J$. The rotational-orbital density of states is zero otherwise. In the above, $j_{max}$ and $\ell_{max}$ are found from Eq. (A13) and j is found from $E_r$ with Eq. (A8). Note that, when $\Lambda\to\infty$, $\ell_{max}$ also goes to infinity, and Eq. (A26) is similar to the PST results of Eq. (A24).

### Symmetric Top-Atom

Solutions for the rotational-orbital sums and densities of state for the symmetric top can be derived but are particularly tedious, and direct numerical integration is trivial for a three-dimensional problem. Numerical integration of the rotational-orbital sum of states $\Gamma_{ro}^{\ddagger}(E_{tr},J)$ in Eq. (A3) is performed by triple quadrature, successively over k with boundaries $|k| \le j$ and $E_r(j,k) \le E_{tr} - \varepsilon^{\ddagger}(\ell)$, over $\ell$ with boundaries $|J-j| \le \ell \le J+j$ and $E_{tr} \ge \varepsilon^{\ddagger}(\ell)$, and over j with boundaries 0 and $\sqrt{E_{tr}/B_r}$.

Note that for PST, $\varepsilon^\ddagger$ is not included in the boundary for integration over k, and integration over $\ell$ can be performed analytically in a trivial manner. Similarly, numerical integration of the rotational-orbital sum of states $\gamma_{ro}^\ddagger(E_t,E_r,J)$ in Eq. (A5) is also performed by triple quadrature, successively over k with boundaries $|k| \leq j$ and $E_r(j,k) \leq E_r$, over $\ell$ with boundaries $|J-j| \leq \ell \leq J+j$ and $\ell \leq \ell_{max}$, and over j with boundaries 0 and $\sqrt{E_r/B_r}$. The rotational-orbital density of states $\rho_{ro}^\ddagger(E_t,E_r,J)$ is just the derivative of $\gamma_{ro}^\ddagger(E_t,E_r,J)$ with respect to $E_r$.

The above shows that numerical solutions can be obtained easily for this particular case. However, it is of interest to point out that, for most of the previous cases, analytic closed forms prevail, since some of the integrals could not be solved numerically, e.g. integrals containing Dirac delta-functions.

# References

[1] D.J. Wales, *Science* **271**, 925 (1996).

[2] K.D. Ball, R.S. Berry, R.E. Kunz, F.-Y. Li, A. Proykova and D.J. Wales, *Science* **271**, 963 (1996).

[3] J.P.K. Doye and D.J. Wales, *Science* **271**, 484 (1996).

[4] J. Uppenbrink and D.J. Wales, *J. Chem. Phys.* **96**, 8520 (1992); *ibid.* **98**, 5720 (1993).

[5] J.M. Soler, J.J. Saenz, N. Garcia and O. Echt, *Chem. Phys. Lett.* **109**, 71 (1984).

[6] J.J. Saenz, J.M. Soler and N. Garcia, *Surf. Sci.* **156**, 121 (1985).

[7] T.D. Märk, *Int. J. Mass Spectrom. Ion Processes* **79**, 1 (1987).

[8] J.E. Campana, *Mass Spectrom. Rev.* **6**, 395 (1987).

[9] J.R. Heath, Y. Liu, S.C. O'Brien, Q.-L. Zhang, R.F. Curl, F.K. Tittel and R.E. Smalley, *J. Chem. Phys.* **83**, 5520 (1985); P.J. Brucat, L.-S. Zheng, C.L. Pettiette, S. Yang and R.E. Smalley, *J. Chem. Phys.* **84**, 3078 (1986).

[10] U. Ray, M.F. Jarrold, J.E. Bower and J.S. Kraus, *J. Chem. Phys.* **91**, 2912 (1989).

[11] S. Wei, W.B. Tzeng and A.W. Castleman Jr., *J. Chem. Phys.* **92**, 332 (1990); *ibid.* **93**, 2506 (1990); S. Wei, K. Kilgore, W.B. Tzeng and A.W. Castleman Jr., *J. Phys. Chem.* **95**, 8306 (1991); Z. Shi, V. Ford, S. Wei and A.W. Castleman Jr., *J. Chem. Phys.* **99**, 8009 (1993).

[12] C. Bréchignac, Ph. Cahuzac, J. Leygnier and J. Weiner, *J. Chem. Phys.* **90**, 1492 (1989); C. Bréchignac, Ph. Cahuzac, J. Leygnier, R. Pflaum, J.Ph. Roux and J. Weiner, *Z. Phys. D* **12**, 199 (1989); C. Bréchignac, H. Busch, Ph. Cahuzac and J. Leygnier, *J. Chem. Phys.* **101**, 6992 (1994).

[13] P.C. Engelking, *J. Chem. Phys.* **85**, 3103 (1986); *ibid.* **87**, 936 (1987).

[14] F.G. Amar and S. Weerasinghe, in *Mode Selective Chemistry*, edited by J. Jortner et al. (Kluwer Academic Publishers, 1991); S. Weerasinghe and F.G. Amar, *Z. Phys. D* **20**, 167 (1991).

[15] S. Weerasinghe and F.G. Amar, *J. Chem. Phys.* **98**, 4967 (1993).

[16] S.D. Bosanac and J.N. Murrell, *Chem. Phys. Lett.* **291**, 64 (1998).

[17] C.E. Klots, *J. Chem. Phys.* **83**, 5854 (1985); *Nature* **327**, 222 (1987); *J. Phys. Chem.* **92**, 5864 (1988); *Int. J. Mass Spectrom. Ion Processes* **100**, 457 (1990); *Z. Phys. D* **21**, 335 (1991); *J. Phys. Chem.* **99**, 1748 (1995).

[18] A.J. Stace and A.K. Shukla, *Chem. Phys. Lett.* **85**, 157 (1982); A.J. Stace, *J. Chem. Phys.* **85**, 5774 (1986); A.B. Jones, C.A. Woodward, J.F. Winkel and A.J. Stace, *Int. J. Mass Spectrom. Ion Processes* **133**, 83 (1994).

[19] G.F. Bertsch, N. Oberhofer and S. Stringari, *Z. Phys. D* **20**, 123 (1991); C.E. Roman and I.L. Garzon, *Z. Phys. D* **20**, 163 (1991).

[20] D.H.E. Gross, *Rep. Prog. Phys.* **53**, 605 (1990); D.H.E. Gross and P.A. Hervieux, *Z. Phys. D* **35**, 27 (1995); D.H.E. Gross, *Czech. J. Phys.* **48**, 736 (1998).

[21] C.E. Klots, *Z. Phys. D* **20**, 105 (1991); *J. Chem. Phys.* **98**, 1110 (1993).

[22] R.W. Smith, *Z. Phys. D* **21**, 57 (1991).

[23] R.N. Barnett, U. Landman and G. Rajagopal, *Phys. Rev. Lett.* **67**, 3058 (1991); R.N. Barnett, U. Landman, A. Nitzan and G. Rajagopal, *J. Chem. Phys.* **94**, 608 (1991).

[24] U. Rothlisberger and W. Andreoni, *J. Chem. Phys.* **94**, 8129 (1991); U. Rothlisberger, W. Andreoni and P. Giannozzi, *J. Chem. Phys.* **96**, 1248 (1992).

[25] C. Rey, L.J. Gallego, M.P. Iniguez and J.A. Alonso, *Physica B* **179**, 273 (1992).

[26] M.J. Lopez and J. Jellinek, *Phys. Rev. A* **50**, 1445 (1994).

[27] G.H. Peslherbe and W.L. Hase, *J. Chem. Phys.* **101**, 8535 (1994).

[28] A.A. Shvartsburg, K.M. Ervin and J.H. Frederick, *J. Chem. Phys.* **104**, 8458 (1996); A.A. Shvartsburg, J.H. Frederick and K.M. Ervin, *J. Chem. Phys.* **104**, 8470 (1996).

[29] T. Baer and W.L. Hase, *Unimolecular Reaction Dynamics. Experiments and Theory* (Oxford University Press, New York, 1996).

[30] J.C. Light, *J. Chem. Phys.* **40**, 3221 (1964); J.C. Light and J. Lin, *J. Chem. Phys.* **43**, 3209 (1965); P. Pechukas and J.C. Light, *J. Chem. Phys.* **42**, 3281 (1965); J. Lin and J. Light, *J. Chem. Phys.* **45**, 2545 (1966); P. Pechukas, J.C. Light and C. Rankin, *J. Chem. Phys.* **44**, 794 (1966); J.C. Light, *Discuss. Faraday Soc.* **44**, 14 (1967).

[31] C.E. Klots, *J. Phys. Chem.* **75**, 1526 (1971); *Z. Naturforsch., Teil A* **27**, 553 (1972).

[32] C.E. Klots, *J. Chem. Phys.* **64**, 4269 (1976); C.E. Klots, D. Mintz and T. Baer, *J. Chem. Phys.* **66**, 5100 (1977).

[33] E. Nikitin, *Theor. Exp. Chem.* **1**, 83 (1965).

[34] B.C. Eu and J. Ross, *J. Chem. Phys.* **44**, 2467 (1966).

[35] W.J. Chesnavich and M.T. Bowers, *J. Am. Chem. Soc.* **98**, 8301 (1976); *ibid.* **99**, 1705 (1977); *J. Chem. Phys.* **66**, 2306 (1977).

[36] W.J. Chesnavich and M.T. Bowers, in *Gas Phase Ion Chemistry*, edited by M.T. Bowers (Academic Press, New York, 1979).

[37] G.H. Peslherbe and W.L. Hase, *J. Chem. Phys.* **105**, 7432 (1996).

[38] G.H. Peslherbe and W.L. Hase, manuscript in preparation.

[39] L.G.M. Pettersson, C.W. Bauschlicher and T. Halicioglu, *J. Chem. Phys.* **87**, 2205 (1987).

[40] B.M. Axilrod and E. Teller, *J. Chem. Phys.* **11**, 299 (1943).

[41] S. Erkoc, *Phys. Stat. Sol. (b)* **152**, 447 (1989); *ibid.* **161**, 211 (1990); *Z. Phys. D* **19**, 423 (1991); Z. El-Bayyari and S. Erkoc, *Phys. Stat. Sol. (b)* **170**, 103 (1992).

[42] R.L. Johnston and J.Y. Fang, *J. Chem. Phys.* **97**, 7809 (1992).

[43] W.L. Hase, R.J. Duchovic, X. Hu, A. Komornicki, K. Lim, D.-h Lu, G.H. Peslherbe, K.N. Swamy, S.R. Vande Linde, A.J.C. Varandas, H. Wang and R.J. Wolf, *VENUS96*, QCPE **16**, 671 (1996).

[44] W.H. Press, S.A. Teukolsky, W.T. Vetterling and B.P. Flannery, *Numerical Recipes, the Art of Scientific Computing, Second Edition* (Cambridge University Press, Cambridge, England, 1992).

[45] M.R. Hoare and J. McInnes, *J. Chem. Soc. Faraday Trans.* **61**, 12 (1976).

[46] P.A. Braier, R.S. Berry and D.J. Wales, *J. Chem. Phys.* **93**, 8745 (1990).

[47] C.J. Tsai and K.D. Jordan, *J. Phys. Chem.* **97**, 11227 (1993).

[48] J.P.K. Doye and D.J. Wales, *J. Chem. Phys.* **102**, 9659 (1995).

[49] T. Raz, U. Even and R.D. Levine, *J. Chem. Phys.* **103**, 5394 (1995).

[50] E. Gorin, *Acta Physicockim. U.R.S.S.* **9**, 691 (1938).

[51] D.M. Wardlaw and R.A. Marcus, *Chem. Phys. Lett.* **110**, 230 (1984); *J. Chem. Phys.* **83**, 3462 (1985); *J. Phys. Chem.* **90**, 5383 (1986); *Adv. Chem. Phys.* **70**, 231 (1988).

[52] S.J. Klippenstein and R.A. Marcus, *J. Chem. Phys.* **87**, 3410 (1987); *J. Phys. Chem.* **92**, 3105 (1988); *ibid.* **92**, 5412 (1988).

[53] S.J. Klippenstein, *Chem. Phys. Lett.* **170**, 71 (1990); *J. Chem. Phys.* **94**, 6469 (1991); *ibid.* **96**, 367 (1992); *Chem. Phys. Lett.* **214**, 418 (1993).

[54] S.C. Smith, *J. Chem. Phys.* **95**, 3404 (1991); *ibid.* **97**, 2406 (1992); *J. Phys. Chem.* **97**, 7034 (1993).

[55] L. Zhu and W.L. Hase, *Chem. Phys. Lett.* **175**, 117 (1990); E.E. Aubanel, D.M. Wardlaw, L. Zhu and W.L. Hase, *Int. Rev. Phys. Chem.* **10**, 249 (1991).

[56] X. Hu and W.L. Hase, *J. Phys. Chem.* **93**, 6092 (1989).

[57] S. Nosé, *Mol. Phys.* **52**, 255 (1984); *J. Chem. Phys.* **81**, 511 (1984).

[58] C. Bichara, J.-P. Gaspard and J.-C. Mathieu, *J. Chem. Phys.* **89**, 4339 (1988); P. Labastie and R.L. Whetten, *Phys. Rev. Lett.* **65**, 1567 (1990).

[59] J.P.K. Doye and D.J. Wales, *Science* **271**, 484 (1996); M.A. Miller and D.J. Wales, *Mol. Phys.* **88**, 533 (1996).

[60] Y.A. Shreider, in *The Monte Carlo Method: The Method of Statistical Trials*, edited by Y.A. Shreider (Pergamon Press, Oxford, 1966).

[61] W. Forst, *Chem. Rev.* **71**, 339 (1971).

[62] W.L. Hase, *J. Chem. Ed.* **60**, 379 (1983).

[63] M. Watanabe and W.P. Reinhardt, *Phys. Rev. Lett.* **65**, 3301 (1990).

[64] W.P. Reinhardt, M. Watanabe and R. Waterland, in *Proceedings of the International Conference on Classical Dynamics in Atomic and Molecular Physics, Brioni, Yugoslavia, 1988*, edited by T. Grozdanov, P. Grujic and P. Krstic (World Scientific, Singapore, 1988); W.P. Reinhardt, *J. Mol. Struct.* **223**, 157 (1990).

[65] P. Hertz, *Ann. Phys.* **33**, 537 (1910).

[66] D.A. McQuarrie, *Statistical Thermodynamics* (University Science Books, Mill Valley, Ca., 1973).

[67] W.L. Hase and D. Buckowski, *Chem. Phys. Lett.* **74**, 284 (1980).

[68] D.L. Bunker, *J. Chem. Phys.* **37**, 393 (1962); *ibid.* **40**, 1946 (1964); D.L. Bunker, *Theory of Elementary Gas Reaction Rates* (Pergamon Press, London, 1966).

# Electronic Structure of Bimetallic Clusters Based on Alkali Elements

Julio A. Alonso and María J. López

Departamento de Física Teórica, Universidad de Valladolid, 47011 Valladolid, Spain

**Abstract.** In this paper we present a review on the alloying effects on the energetic, structural, electronic, and optical properties of small bimetallic clusters. Density Functional Theory is used to obtain the ground state electronic structure and the optical response of the clusters. The stability of mixed alkali metal clusters can be understood in terms of the electronic shell model. Electronic shell closing effects also appear in alkali clusters doped with higher valence metals, although reordering in the electronic levels occurs in some cases giving rise to different magic numbers. Ab initio calculations support this view. The influence of the geometrical structure (mixed or segregated) on the collective electronic response of bimetallic clusters is analyzed. Finally we study clusters which may have relevance for understanding the properties of liquid ionic alloys.

## 1  Introduction

Experiments for clusters of simple $sp$-metals have revealed many interesting properties of these species. Perhaps the most striking one is the existence of magic numbers in the abundance spectrum. In the case of alkali metal clusters those magic numbers correspond to sizes $n = 2, 8, 18, 20, 34, 40, ...$, and have been explained by the formation of electronic shells in a smooth confining effective potential well that raises rather abruptly at the cluster surface [1]. Laser light can excite collective oscillations of the valence electron cloud against the positive ionic background, and those collective oscillations have some similarity to the "giant dipole resonance" in nuclei [2]. The clusters of noble metal atoms share some of the properties of those formed by $sp$-elements [3], although they also display interesting differences arising from the presence of $d$-cores [4]. Those differences become pronounced for the case of clusters of typical transition elements, having unfilled $d$-shells. Instead of electronic shells, these clusters form compact atomic arrangements with a predominance for icosahedral and cuboctahedral structures. Those structures have been revealed by reactivity experiments [5], electron diffraction [6] and photoemission studies [7], and in the case of magnetic metals such as Nickel, by deflection in a Stern-Gerlach apparatus [8]. Motivated by those experiments, extensive theoretical work on noble metal and transition metal clusters has also been done.

In contrast, work covering experimental or theoretical aspects of bimetallic clusters is scarce. Bimetallic clusters can be considered as "microalloys". The richness of phases and structures in the Temperature-Composition phase diagram of binary alloys suggests that a similar richness, or even enhanced, could be found in the study of those microalloys containing only a few, or a few hundred atoms. In this paper we review the work on clusters formed by mixed alkali species and alkali clusters containing higher valent impurities.

## 2 Structure of Mixed Alkali Metal Clusters

The observed variations of the abundance of pure alkali clusters ($Li_n$, $Na_n$, $K_n$, etc.) with cluster size $n$ have been explained by the electronic-shell model [1]. In this model, the valence electrons move in a confining spherically symmetric effective potential well, $V_{eff}$. Within the Density Functional Theory (DFT) [9] this potential well can be expressed as a sum

$$V_{eff} = V_{jellium} + V_{Hartree} + V_{XC} , \qquad (1)$$

where the last two terms, $V_{Hartree}$ and $V_{XC}$ give the classical electron-electron and exchange-correlation contributions respectively. The first term gives the contribution from the ionic background. It is convenient for alkali clusters to simulate this background by a homogeneous distribution of positive charge (jellium model). Selfconsistent calculations then show that electrons group together in highly degenerate shells. The shell filling order is $(1s)^2$ $(1p)^6$ $(1d)^{10}$ $(2s)^2$ $(1f)^{14}$ $(2p)^6$ $(1g)^{18}$..., where the maximum occupancy of each shell is given as a superscript, and clusters with filled shells become very stable when the gap to the next "empty" shell is large. Consequently drastic changes in the binding energy of the cluster occur after shell closing. The number of electrons for which shell-closing occurs, $N_e$ = 2, 8, 18, 20, 34, 40, 58, 92, ... agree with the observed steps in the abundance mass spectra or in the measured ionization potentials (IP) [1]. The structure in the size region between main steps is explained by an extension of the model allowing for spheroidal deformations of the clusters [10].

Although much work has been done for alkali metal clusters containing a single alkali element, relatively few experiments have been performed for alkali mixtures. Introducing a monovalent impurity (or a number of them) in an alkali cluster represents a weak perturbation, so one expects only small changes in the properties of the cluster. The most complete work is that of Kappes et al. [11], who performed supersonic beam expansions of a mixture of Lithium and Sodium vapours. The most salient results of the experiment are: a) mixed Li-Na clusters are produced, in spite of the fact that Li and Na do not form solid alloys and present a large miscibility gap in the liquid phase; b) the abundance magic numbers are the same as for pure $Li_n$ or $Na_n$; c) Lithium enrichment is observed in the clusters, compared with the composition of the mixed vapours.

The fact that the abundance magic numbers do no change is easily understood. The ionic pseudopotentials of the alkali metals are weak and the sizes of the ion cores are relatively small compared to the atomic volumes. Then, a jellium with the averaged background density describes well the mixed clusters. However, this averaged jellium model is unable to give any information about the distribution of atoms within the cluster. This can be improved by using the Spherically Averaged Pseudopotential (SAPS) model [12]. Alkali ions are well described by local pseudopotentials $V_l(r)$, and the total ionic pseudopotential of the cluster can be written

$$V_{\text{ions}}^{\text{Ps}}(\mathbf{r}) = \sum_j V_l(|\mathbf{r} - \mathbf{R}_j|) , \tag{2}$$

where $\{\mathbf{R}_j\}$ indicates the set of ionic positions. When the global cluster shape is not far from spherical, a reasonable approximation is to substitute $V_{\text{ions}}^{\text{Ps}}(\mathbf{r})$ by its spherical average about the cluster center

$$V_{\text{ions}}^{\text{Ps}}(\mathbf{r}) \longrightarrow V^{\text{SAPS}}(r) . \tag{3}$$

This simplification drastically reduces the computational effort since we now have a system of interacting electrons moving in a spherically symmetric external potential well. The distribution of atoms within the cluster can be obtained by minimizing the total energy with respect to variations in the atomic positions $\{\mathbf{R}_j\}$. The search for minima can be done using steepest descent or simulated annealing techniques. In this process the ion-ion repulsion energy

$$E_{\text{ion-ion}} = \sum_{i \neq j} U(|\mathbf{R}_i - \mathbf{R}_j|) \tag{4}$$

is calculated without recourse to any geometrical approximation (geometrical restrictions only apply to the electron-ion interaction).

One can set limits of validity for the SAPS model. The clusters should not be too small because open electronic shells cause the arrangement of ions to deform away from spherical symmetry. On the other hand, planar surface facets form in sufficiently large clusters. The range of intermediate sizes is then the best adapted to the SAPS model.

The model has been applied to Li-Na clusters [13]. Figure 1 shows the calculated radial atomic distribution in $Li_5Na_{11}$. The cluster is built as a single atomic shell with radius $R \simeq 7$ a.u., formed by fifteen atoms enclosing a single Na atom at the center. The picture is very similar to that in one-component alkali clusters [12]. This seems to be a general feature. The calculations show that strong similarities exist between the structures of $Li_mNa_n$ clusters with a fixed number $m + n$ of atoms and varying concentrations of Li and Na. Another discernible feature is that some of the Na atoms are at a slightly higher distance from the cluster center than the Li atoms. This effect occurs generally for other Li-Na clusters and it can be viewed as a manifestation of the tendency to phase separation observed in the bulk solid and liquid Li-Na alloys. This demixing tendency is much less drastic in small Li-Na clusters, where Li and Na atoms coexist in the same layer. The surface, where the atoms have a larger freedom to accommodate themselves compared to the bulk, is responsible for the reduction of the demixing tendency. This has been corroborated by calculating the heat of solution $\Delta H_{\text{sol}}$ of a Li impurity in Na clusters. $\Delta H_{\text{sol}}$ is negative, and this explains why Na-Li clusters form in the supersonic expansion experiments.

Finally, the observed enrichment in Li is explained by studying the heat of the reaction

$$Li_m Na_n + Li \longrightarrow Li_{m+1} Na_{n-1} + Na .$$ (5)

The calculated heat is negative, indicating that the substitution of Na by a Li atom in an exchange collision in the beam is favorable.

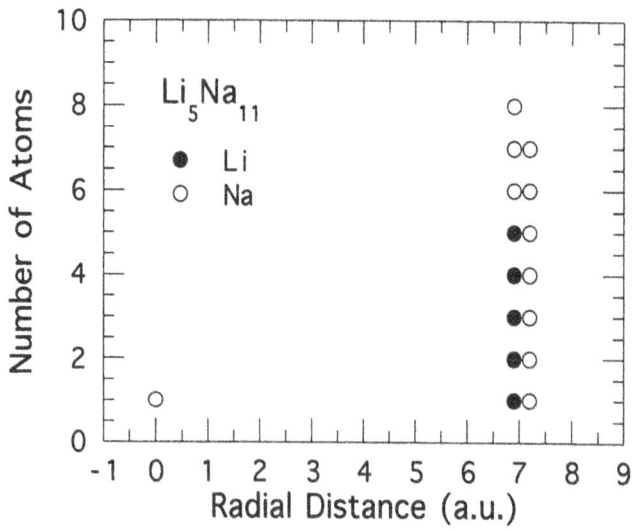

**Fig. 1.** Radial distribution of atoms in $Li_5 Na_{11}$ measured from the cluster center.

The difference between the Wigner-Seitz radii of Na and Li is $\Delta R_{WS}$(Na-Li) = 0.72 a.u., and this difference increases for Na-K and Na-Cs pairs: $\Delta R_{WS}$(K-Na) = 0.87 a.u. and $\Delta R_{WS}$(Cs-Na) = 1.81 a.u. Since the properties of alkali metals vary smoothly with atomic radius, one can expect that the weak segregation tendency found in Li-Na clusters will be enhanced in Na-K and Na-Cs. The radial atomic distribution in Na-K clusters differs drastically from the Li-Na case. For instance, in $Na_{10}K_{10}$ the Na and K atoms form well separated layers [14]: the Na layer has a radius $R \sim 6$ a.u. and the K layer, or surface layer, has $R \sim 10$ a.u. A tendency for K segregation at the surface of a $Na_{10}K_{10}$ cluster was also predicted by ab initio molecular dynamics (MD) simulations performed by Ballone et al. [15]. The tendency of the heavier element to occupy the cluster surface is common to Li-Na [13], Na-K [14] and Na-Cs [16] clusters, and is driven by the lower surface energy of the heavier element. This tendency is stronger in Na-K than in Li-Na. A difference between bulk Li-Na and Na-K alloys is that an ordered stoichiometric compound ($Na_2K$) forms in the second alloy. In fact, the calculated radial atomic distribution of large clusters like $Na_{34}K_{34}$ shows that not all the K atoms are segregated on the outer region of the cluster [14]; instead, there is an alternation of atomic layers which perhaps can be interpreted as a precursor of the ordering tendency in the bulk solid alloy. Na-Cs clusters show similar features [16]: alternation of Na and Cs layers, and the presence of Cs at the surface. Those features are again consistent with the existence of the solid compound $Na_2Cs$. In spite of these structural effects, the electronic configuration of Na-K and Na-Cs clusters remains rather simple, and the electronic magic numbers are the same as for the pure unmixed clusters. Our results are also consistent with Monte Carlo simulations of the liquid-vapor interface of Na-Cs alloys [17]. The simulations yield segregation of Cs atoms to the surface.

New structural features introduced by a third component have been studied for $Na_{10}K_{10}Cs_n$ clusters with varying $n$ [18]. As expected, Cs segregates to the cluster surface. The location of the K atoms also deserves discussion. Let us begin with $Na_{10}K_{10}$, which is a cluster formed by two layers: the Na atoms form the internal one and the surface layer only contains K atoms. When Cs atoms are added to this cluster, those atoms prefer surface sites. The geometrical information is included in Fig. 2, which shows the positions of the different atomic layers, each layer being identified by its mean radius. For small $n$ both the K and the Cs atoms can be viewed as forming the surface, although the mean radius of the Cs layer is a little bit larger than the radius of the K layer. This indicates a highly corrugated surface. Then, as additional Cs atoms are added, more and more surface sites become occupied by Cs atoms, and an increasing amount of K atoms lose direct access to the surface. In such a case, migration of the K atoms towards the inner part of the cluster is observed. For $n = 33$, all the K atoms have migrated to the inner region of the cluster, and this is now formed by a Cs surface layer and an inner layer containing the K and Na atoms. Even more, for $n > 33$ this inner layer is formed by two sublayers with very similar radii: the K atoms form the innermost sublayer and the Na atoms the other. A striking inversion of the location of the K and Na sublayers has occurred. The peculiar behavior of the K atoms is a consequence of the interplay between two effects: one is the surface effect and the other is an electronegativity effect. The surface effect is responsible for the presence of K atoms at the surface when the number $n$ of Cs atoms is small. However, when the K atoms lose direct access to the surface, the electronegativity effect comes into play. Electronic charge flows from regions of low electronegativity to regions of high electronegativity and this provides an "ionic-like" bonding contribution. The electronegativities of the alkalis increase in the order Cs→K→Na. A sequence of layers (from inside to outside) K-Na-Cs becomes more favorable to maximize electronegativity effects than a sequence Na-K-Cs since Na-Cs contacts have a larger electronegativity difference than K-Cs contacts.

# 3 Collective Electronic Response in Mixed Alkali Metal Clusters

Surface segregation of one atomic species also affects the collective electronic response to laser radiation. As an example we show in Fig. 3 the calculated

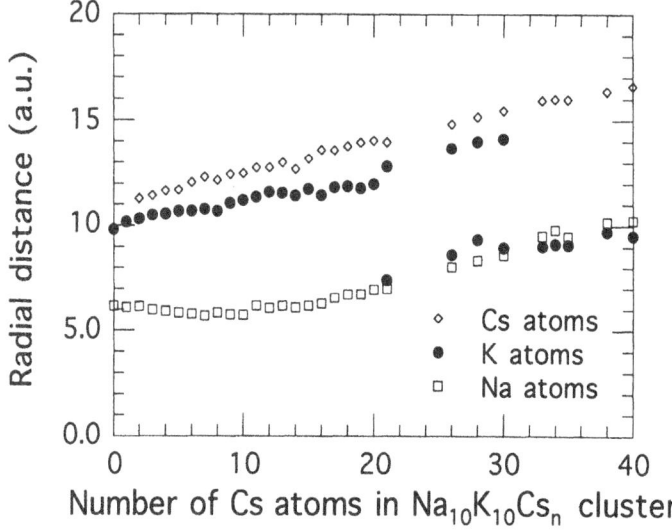

**Fig. 2.** Radial location of the different atomic species in $Na_{10}K_{10}Cs_n$.

photoabsorption cross section of $Na_{20}K_{20}$ in its ground state [14]. The structure of this cluster, obtained from the SAPS model, can be viewed as formed by three layers. The twenty K atoms form the surface layer, which has a radius $R \sim 12.5$ a.u., and the two inner layers, with radii 6 a.u. and 10.5 a.u., are formed by eleven and eight Na atoms respectively. The additional Na atom occupies the center position. The two outer layers have widths of about 1 a.u. The calculation of the photoabsorption cross section was done using the time dependent density functional theory (TDLDA). The photoabsorption spectrum shows a resonance peak at 2.1 eV, which represents a collective dipole excitation of the valence electron cloud oscillating against the ionic background. The tail of the resonance extends up to 3 eV and concentrates a sizable amount of oscillator strength. This broadening is due to the interaction of the collective resonance with single particle-hole transitions, some of the most important ones being transitions from the HOMO level to the continuum (the right arrow indicates the ionization threshold). Comparing the position of the collective resonance with those calculated for pure K and Na clusters [19, 20] using the same method, we observe that the resonance frequency in $Na_{20}K_{20}$ is closer to that in pure K clusters, which we interpret as a manifestation of the fact that the cluster surface, whose electron density contributes most to the collective excitation, is formed by K atoms.

The cluster geometry can be modified to see how this affects the surface plasmon. If the positions of some Na and K atoms are simply interchanged preserving the cluster geometry, the shape of the calculated photoabsorption spectrum changes little, although a shift of the resonance peak to higher energies occurs as the surface becomes depleted in K and enriched in Na. The same effect

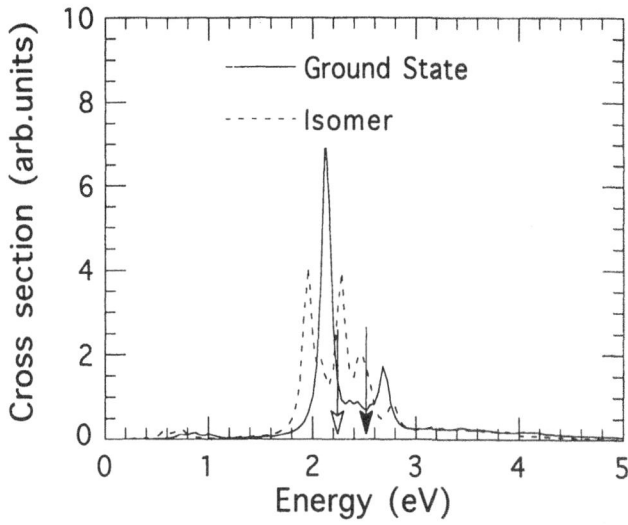

**Fig. 3.** Calculated photoabsorption cross sections for two isomers of $Na_{20}K_{20}$. The continuous curve corresponds to the ground state geometry. The left and right arrows indicate the ionization thresholds of the isomer and ground state, respectively.

has been predicted for other $Na_nK_n$ clusters [21]. A drastic geometrical change can be simulated by calculating the spectrum for one of the isomers found in the process of searching for the absolute energy minimum. The dashed curve in Fig. 3 corresponds to an isomer having fewer atoms in the innermost and surface layers and more in the middle layer compared to the ground state. The energy of this isomer is 0.12 eV per atom above the ground state. The plasmon resonance is now broadened and fragmented because the ionization threshold (the left arrow in the figure) interacts more strongly with the plasmon. This ionization threshold is lower than that for the ground state, reflecting the lower stability of the isomer. The optical response of $Li_mNa_{8-m}$ and $Li_mNa_{20-m}$ clusters has been studied by Balbás et al. [22, 23].

In summary the shape of the photoabsorption spectrum is sensitive to a) the cluster geometry and b) the degree of segregation. Comparison between measured and calculated spectra may be useful to elucidate the structure of mixed clusters produced in the gas phase experiments. This comparison has already been done for homoatomic clusters [24].

Very small clusters have been studied by ab initio methods. Furthermore, the atomic structure of small clusters formed by alkali atoms adapts itself to the shape of the valence electron cloud [25] and consequently deviates from spherically symmetric shapes. Motivated by measurements of the optical absorption of $LiNa_3$ and $Li_2Na_2$ [26], Bonačić-Koutecký et al. have performed Configuration Interaction (CI) calculations for those two clusters [27], and Dahlseid et al. [28]

have studied the whole family $Li_mNa_{4-m}$ $(m = 0 - 4)$ by the Hartree-Fock (HF) method. The optical response was modelled in both cases via CI calculations for excited electronic states. Both studies give planar rhombic forms as the most stable structures: slightly distorted for $LiNa_3$ and $Li_3Na$, and undistorted for $Li_4$, $Na_4$ and $Li_2Na_2$. This insensitivity to relative composition agrees with the general finding of the SAPS model for larger clusters.

Energetically low lying isomers were also discovered in each $Li_mNa_{4-m}$ case, corresponding to different ways of arranging the $m$ Li atoms and the $(4 - m)$ Na atoms in the four vertices of the rhombus. Although the ground state atomic structure is insensitive to $m$, the photoabsorption spectrum is very sensitive. The spectrum of $Na_4$ resembles that obtained from the Mie-Drude theory for an ellipsoidal droplet with three different axes; but as Na atoms are substituted by Li, the deviations from the Mie-Drude theory become increasingly large. The measured spectra of $LiNa_3$ and $Li_2Na_2$ are well explained by the ab initio calculations. In $Li_2Na_2$ the comparison between the experimental absorption spectrum and the spectra calculated for the three possible isomers allows to confirm that the best agreement is obtained for the predicted lowest energy isomer. However, the spectra predicted for the two singlet isomers of $LiNa_3$ are very similar and it was not possible to distinguish which isomer or whether a combination of both singlet isomers contributes to the measured spectrum.

# 4 Divalent Impurities in Alkali Metal Clusters

Supersonic expansion experiments performed by Kappes and coworkers with mixed metal vapors have shown that some changes in the magic numbers occur when small Na or K clusters are doped with divalent impurities: Ba, Sr, Eu, Ca, Yb, Mg, Zn dopants in Na, and Mg, Hg, Zn dopants in K [29, 30]. These authors concentrated on the case of clusters containing a single dopant atom. For some impurities a new magic number was found, corresponding to ten valence electrons in the cluster, for instance $Na_8Mg$ and $K_8Zn$. At the same time the magic number corresponding to eighteen valence electrons vanishes. Baladrón and Alonso [31] have noticed that the single coordinate $\Delta n_+ = n_+(\text{impurity}) - n_+(\text{host})$ allows to separate the impurities in two subsets. Values of $\Delta n_+$ roughly higher than 0.008 a.u. induce changes in the magic numbers. This is the case for Mg and Zn impurities in Na and for Mg, Hg and Zn impurities in K. In contrast, there are no changes for $\Delta n_+ < 0.008$ a.u., which is the case for Ba, Sr, Eu, Ca and Yb impurities in Na (monovalent impurities also fit in this group). Above, $n_+(\text{host})$ and $n_+(\text{impurity})$ are the values of the positive background densities of pure host and impurity solids in the jellium model.

The success of the coordinate $\Delta n_+$ immediately leads to a description of the cluster which is a simple extension of the spherical jellium model. The impurity, placed at the center of the host cluster, is described by a jellium background with density $n_+(\text{impurity})$ and radius $R_{\text{Imp}}$, the Wigner-Seitz radius of the pure impurity crystal. Then another jellium background with density $n_+(\text{host})$ surrounds the impurity, and the radius $R$ of the whole spherical cluster is determined by $R_{\text{Imp}}$, $R_{\text{Host}}$ and the number of host atoms. Selfconsistent DFT calculations [31] for this jellium-on-jellium model show that the binding energy of the s-type electrons ($L = 0$) increases because of the deep depression of the effective potential in the central region of the cluster caused by the impurity, whereas the binding energies of other electrons ($L \geq 1$) change very little. Consequently the magnitude of the gap between de $2s$ shell and the $1d$ shell decreases with increasing $\Delta n_+$. When $\Delta n_+$ becomes roughly 0.008 a.u., reordering of these electronic subshells occurs, and the filling order changes to $(1s)^2(1p)^6(2s)^2(1d)^{10}$. Clusters with ten valence electrons adopt the closed-shells configuration $(1s)^2(1p)^6(2s)^2$, which accounts for the experimental observations. The magic number 20 still remains present but there is no shell closing at 18 electrons. Yeretzian [30] has arrived at the same conclusions using a parametrized Wood-Saxon potential modified to account for the presence of the central impurity and Zhang et al. [32] noticed the $2s - 1d$ inversion in their pseudopotential ab initio calculations. For larger sizes and strongly attractive impurities, Yeretzian predicts an inversion between the $2p$ and $1f$ shells which would give rise to a new magic number of 26 valence electrons and would cause the magic number 40 to vanish. Experiments for the relevant host-impurity combinations in this size range have not yet been performed.

Some small anomalies with respect to the general trends discussed have been observed in the experiments [30], and have been ascribed to geometrical effects beyond the jellium-type models [30, 33]. Also, better agreement with experiment is achieved by relaxing one of the constraints of the jellium-on-jellium model. Yannouleas et al. [34] have varied the density $n_+$(impurity) of the jellium sphere representing the impurity, and correspondingly its radius, $R_{\text{Imp}}$, while preserving $n_+$(host) equal to the bulk value. The value of $n_+$(impurity) was fixed by requiring a better matching of the calculated magic numbers with experiment. This leads in some cases to values of the "electronic" Wigner-Seitz radius $r_s$ differing substantially from the bulk values. For instance, the electronic Wigner-Seitz radius of bulk Zn is $r_s(\text{Zn}) = 2.31$ a.u. while the optimized one for the Zn impurity in Na is 1.15 a.u.. This indicates a much larger attractive potential at the central impurity site. The electronic Wigner-Seitz radius is related to the jellium density $n_+$ of a material by $r_s = (4\pi n_+/3)^{-1/3}$.

The optical response of doped alkali clusters has also been studied for this cluster model using the Random Phase Approximation (RPA) [34]. For the case of $Na_8Zn$ the calculated optical spectrum is characterized by two closely spaced lines at 2.87 eV, carrying 26% of the total strength, and a stronger line at 2.57 eV which carries 42% of the strength. This occurs for $r_s(\text{Zn}) = 1.15$ a.u., and it is in good agreement with the experimental double peak, formed by a higher energy component at 2.97 eV which carries a smaller amount of strength than the lower energy component at 2.63 eV. In contrast, increasing the $r_s(\text{Zn})$ toward values more appropriate to bulk Zn produces a single line and then a worsening of the results. The fragmentation which occurs at $r_s(\text{Zn}) = 1.15$ a.u. is due to a degeneracy between the plasmon and the $2s \rightarrow 2p$ and $1p \rightarrow 3s$ particle-hole transitions [34]. A strong downwards shift of those transitions (with respect to the transitions for $r_s(Zn) \simeq 2.31$ a.u.) is required for this degeneracy to develop. The effect of the Zn impurity on the optical response is noticeable: the spectrum of $Na_8$ is characterized by a single line at 2.53 eV.

To test the reliability of the two-step jellium model Yannouleas et al. have carried out parallel molecular orbital calculations for $K_8Mg$, $Na_8Zn$ and $K_6Mg$, and have compared the electronic structures resulting from the two approaches. In the molecular calculations the geometry was assumed to be a body centered cube for $K_8Mg$ and $Na_8Zn$, and a centered octahedron for $K_6Mg$. For the jellium calculations the densities $n_+$(impurity) and $n_+$(host) were taken equal to the corresponding bulk values. For those two geometries the molecular electronic configurations are of the type $(1a_{1g})^2(1t_{1u})^6(2a_{1g})^6$, and $(1a_{1g})^2(1t_{1u})^6$, respectively. Comparison with the jellium calculations is based on the correspondence $1s \rightarrow 1a_{1g}$, $1p \rightarrow 1t_{1u}$, $2s \rightarrow 2a_{1g}$, and $1d \rightarrow (1d_t + 1d_e)$. To illustrate the validity of this correspondence the molecular wave functions were decomposed in spherical harmonics with respect to the cluster center. The coefficient $c_L$ of the spherical harmonic with the maximum contribution is given in Tab. 1 for $Na_8Zn$. The correspondence is very good. The Table also lists the binding energies. Both calculations reproduce the downward shift of the $2s$ level below the $1d$ level. The agreement between the electronic configurations extends also to $K_6Mg$ and $K_8Mg$. Ab initio calculations by Koutecký et al. [35, 36, 37, 38] on $Li_nBe$, $Li_nMg$ and $Na_nMg$ are consistent with this conclusion. The ground state electronic configurations are $(1a_{1g})^2(1t_{1u})^6$ for the $A_6M$ centered octahedron, $(1a_1')^2(1e_1')^4(1a_2'')^2(2a_1')^1$ for the $A_7M$ regular pentagonal bipyramid and $(1a_{1g})^2(1t_{1u})^6(2a_{1g})^2$ for the $A_8M$ centered cubic form, respectively.

Table 1. Binding energy and character of the molecular orbitals in $Na_8Zn$. The jellium orbitals are given for comparison [34].

| Molecular | | | Jellium | |
|---|---|---|---|---|
| | $\epsilon$(eV) | $c_L$ | | $\epsilon$(eV) |
| $1a_{1g}$ | -7.41 | 0.995 (L=0) | $1s$ | -6.05 |
| $1t_{1u}$ | -3.89 | 0.985 (L=1) | $1p$ | -3.68 |
| $2a_{1g}$ | -2.81 | 0.985 (L=0) | $2s$ | -2.71 |
| $1d_t$ | -1.88 | 0.980 (L=2) | $1d$ | -2.02 |

In spite of the explanation of the magic number $N_e = 10$, both calculations (two step jellium and molecular) have some difficulties to account for other features observed in the mass spectra. In particular the absence of a peak at $N_e = 8$ when the peak at $N_e = 10$ is present (the only case when both peaks are present in the experimental spectrum is $Na_nMg$). This is because both calculations exhibit comparable gaps between the $1p$ and $2s$ levels and between the $2s$ and $1d$ levels. The discrepancy indicates that the actual $2s$ level is probably situated even lower than the calculations suggest, probably close to the $1p$ level. In the framework of the two-step-jellium model it suffices to increase the value of $n_+$(impurity) to produce the necessary downward shift of the $2s$ level and this is the reason for the modification introduced by Yannouleas et al. [34]. In the framework of molecular orbital calculations a better agreement between theory and experiment can be reached by relaxing the geometrical constraints. Balbás et al. [22] have calculated the optical response of $Na_8Zn$ using the SAPS model (the Zn atom is at the cluster center). Two peaks are obtained at 2.68 eV and 2.9 eV respectively, and the second one was interpreted as a fragmentation of the surface plasmon induced by its proximity to the ionization threshold. This interpretation differs from that of Yannouleas et al. [34].

Röthlisberger and Andreoni [33] have performed ab initio molecular dynamics calculations for $Na_nMg$ ($n = 6 - 9, 18$) and have found that, in general, the Mg impurity is not placed at the cluster center, although it becomes increasingly surrounded by Na atoms as $n$ grows. This produces electronic configurations that do not adapt so well to the jellium picture as if Mg is at the center. The situation is more complex and $1d - 2s$ mixing is present, reflecting the lowering of the symmetry introduced by doping. The same authors also point out that when the impurity is sufficiently smaller than the host atoms, the impurity prefers to sit at the cluster center, thus validating the two-step-jellium energy level structure.

## 5 Higher Valent Impurities

Impurities of valence higher than two lead to a stronger perturbation of the effective potential, and one can expect larger changes in the cluster properties compared to the case of monovalent or divalent impurities. Cheng, Barnett and Landman [39] have studied Lithium clusters containing an Al impurity. This atom has a valence three. These authors performed local-spin-density (LSD) Kohn-Sham calculations, coupled with classical molecular dynamics for the ions. Non local norm-conserving pseudopotentials were used for the electron-ion interactions [40]. The equilibrium structures of the $Li_nAl$ clusters are three-dimensional for $n \geq 3$. This is in contrast with pure $Li_n$ clusters, for which two-dimensional structures prevail up to $n = 6$ [41]. For $n \leq 5$ the Aluminium atom does not occupy an internal position. However, for larger clusters the Al atom is internally located, surrounded by the Li atoms. The binding energy of the cluster (per Li atom), $E_b/n$, the evaporation energy of a Lithium atom, $\Delta E_{ev}(n) = E(Li_{n-1}Al) + E(Li) - E(Li_nAl)$, and the ionization potential, $IP$, display a monotonic increase for $1 \leq n \leq 5$ and then a drop at $n = 6$, which is particularly sharp in $\Delta E_{ev}$ and $IP$. This indicates a unique stability of the $Li_5Al$, which has 8 valence electrons. The structure of this cluster is roughly the same as that for $Na_5Pb$ shown in Fig. 4, with the Al atom a little more distant from the plane of four Li atoms. The analysis of the orbital eigenvalues shows a doubly occupied orbital at $\sim -7.5$ eV with $s$-orbital character (about the Al site) and a manifold of three closely spaced and doubly occupied states with predominantly $p$ character at $\sim -3.5$ eV (that is, separated 4 eV from the $s$-like level). The valence charge density and the individual orbitals indicate that the bonding is covalent, characterized by charge accumulation in the regions connecting the Al and Li ions. Cheng et al. then suggest that the electronic structure of $Li_n Al$ clusters with $n \leq 5$ reflects the localized "atomic-like" nature of the orbitals, and that the shell-closing properties of $Li_5 Al$ are associated to the closing of the Al $3p$ states, albeit somewhat perturbed by the Lithium environment via hybridization, bonding and crystal-field effects. Further addition of Li atoms to the $Li_5 Al$ "core" leads to a picture consistent with the delocalized shell model, with the ordering of the $1d$ and $2s$ levels reversed from that corresponding to the homonuclear jellium model, just as for the case of very attractive divalent impurities already discussed in Section 4 above.

The picture obtained for $Li_5 Al$ clusters suggests that $Li_{5n} Al_n$ clusters may exhibit properties characteristic of an assembly of $Li_5 Al$ units. Indeed, the calculations of Cheng et al. predict that $Li_{10} Al_2$ is formed by two slightly distorted $Li_5 Al$ subunits, bonded to each other along the square bases of the subunits. The analysis of the eigenvalue spectrum also resembles the bonding between two closed-shell molecules.

$Li_n Al$ clusters have also been studied by Nehete et al. [42] using density functional molecular dynamics based on an orbital-free (OF) functional. In this method the electronic kinetic energy is represented in terms of an approximate functional containing the local density Thomas-Fermi term and a density gradient correction, namely (using Hartree atomic units)

$$T[\rho] = F(N_e) \, T_{\text{TF}}[\rho] \; + \; T_{\text{grad}}[\rho] =$$

$$F(N_e)\frac{3}{10}(3\pi^2)^{2/3} \int \rho(\mathbf{r})^{5/3} \, d^3r \; + \; \frac{\lambda}{8} \int \frac{\nabla\rho(\mathbf{r}) \, \nabla\rho(\mathbf{r})}{\rho(\mathbf{r})} \, d^3r \; . \qquad (6)$$

The factor $F(N_e)$ in front of the Thomas-Fermi term is sensitive to the number of valence electrons $N_e$ in the cluster [43] (the calculations use pseudopotentials). For the gradient term, the usual values $\lambda = 1/9$ and $\lambda = 1/5$ were tested, and it was verified that the interatomic densities and binding energies are sensitive to $\lambda$. The geometries of $Li_n Al$ were found to agree with those determined by Cheng et al. In particular, the clusters are three dimensional for $n \geq 3$, and the Al atom is trapped inside the cluster for $n = 6$. These results are encouraging and should motivate more work with orbital-free functionals.

Another interesting case is that of alkali clusters containing group-14 atoms: C, Si, Ge, Sn, Pb. Supersonic expansion of lead-sodium vapor from a hot oven source led to the observation of an exceptionally abundant $Na_6 Pb$ cluster [44]. Several ab initio calculations [45, 46, 47, 48] have been performed to explain the nature of this species. Marsden [45] used the Hartree Fock method to optimize the geometries and electron correlation was introduced afterwards by approximate CI (Configuration Interaction). The other three calculations used DFT.

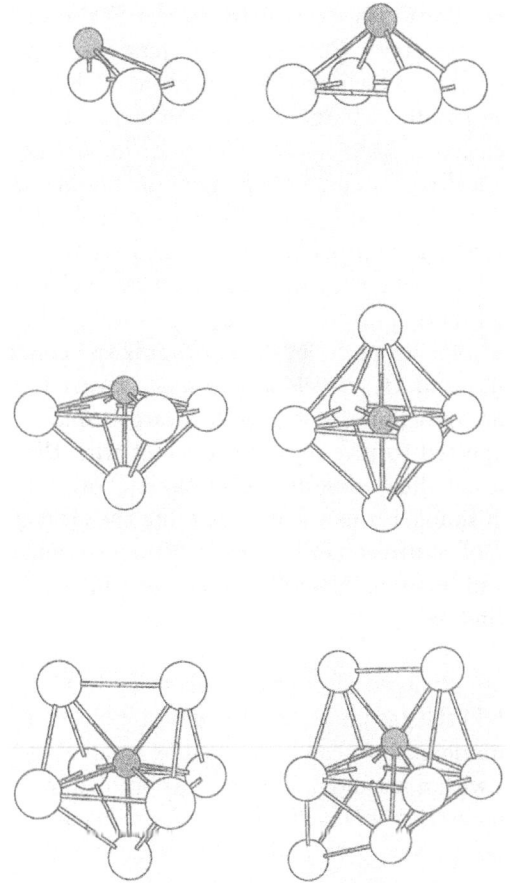

**Fig. 4.** Calculated lowest energy structures of $Na_n Pb$ clusters, $n = 3 - 8$ [46].

Chang et al. [46] and Balbás and Martins [48] employed Car-Parrinello ab initio molecular dynamics [49], with the LDA for exchange and correlation. On the other hand Schleyer [47] used the generalized gradient approximation (GGA) [50]. Pseudopotentials were employed in all the ab initio calculations. A related cluster, $Li_6 C$, had earlier been described computationally [51, 52] and later observed as a product formed from the vaporization of $C_2 Li_2$ [53].

The ab initio calculations predict that the Pb atom sits at the center of an octahedral cage formed by the Na atoms. The electronic configuration of the occupied valence orbitals is $(1a_{1g})^2 (1t_{1u})^6 (2a_{1g})^2$. The $1a_{1g}$ level, which is substantially more bound than the others, is localized on the central atom and dominated by the Pb $s$-orbital contribution. The large energy gap between the $1a_{1g}$ and the $1t_{1u}$ manifolds (near 7eV) is due to the very attractive $s$-part of the Pb pseudopotential. The gap between the manifold of degenerate $1t_{1u}$ levels and the $2a_{1g}$ level (HOMO) is 1eV, and the HOMO is filled with two electrons, so a gap, again of 1eV, develops between the HOMO and LUMO levels (the LUMO has $e_g$ symmetry). Contour plots show the almost spherical shape of the total pseudo-charge density [48], consistent with the closed-shell nature of the cluster. The bonding charge density, defined as the difference between the total charge density and the superposition of the atomic charge densities, also displays a spherically symmetric shape (that is, no directionality effects) and is maximal in the region between the Pb and the shell of Na atoms. The results adapt well to the picture provided by the shell model, and the stability of the cluster can be viewed as arising from its structure of closed electronic shells. The highest occupied molecular orbitals $(2a_{1g})$ are delocalized. But the $(1a_{1g})^2$ and the $(1t_{1u})^6$ electrons are more localized towards the cluster center and the gap between the $1a_{1g}$ and $1t_{1u}$ levels is also much larger than in a simple jellium model. A jellium-on-jellium model with a large $\Delta n_+ = n_+(\text{Pb}) - n_+(\text{Na}) = 0.016$ a.u. is expected to give results consistent with the ab initio calculations. A localized "atomic-like" description of the $t_{1u}$ orbitals is perhaps the most accurate one: in simple terms we are observing the closing of the valence $p$-shell of the Pb atom, of course somewhat perturbed by the surrounding Na atoms via hybridization and bonding. Overall, the bonding in $Na_6Pb$ is a mixture of ionic and metallic bonding.

The structure of closed electronic shells helps to understand the high abundance of $Na_6Pb$ in the gas phase experiments [44] but by itself does not provide a complete explanation. Much can be learned by comparing the binding energies of clusters $Na_nPb$ with $n$ around 6. The results obtained by Chang et al. [46] for the ground state geometries of $Na_nPb$, $n = 3-8$, are given in Fig. 4. Again we stress that the calculations were performed using ab initio total energy pseudopotential methods [54] and the energy was minimized with a Car-Parrinello-like scheme [49, 55] in a supercell geometry with a simple cubic cell of lattice constant 34 a.u., a plane wave basis set with 8 Rydbergs kinetic energy cutoff, and a single $\Gamma$ k-point. For each of the clusters several reasonable starting geometrical configurations were selected, and for each of these the atomic arrangement was relaxed until equilibrium was achieved. Up to $n = 7$ each added Na atom binds directly to the Pb atom and the coordination of the Pb atom increases. For $n = 3, 4, 5$ the Pb atom is on the outside of the cluster to allow the Na atoms to come in closer contact with one another and bind, albeit weakly. The lowest energy structure of $Na_6Pb$ is the octahedron, which for the first time has the Pb atom in the interior of the cluster. The growth continues as before, with a seventh atom binding to the Pb, but at this stage, the Pb atom seems to be fully coordinated by seven Na atoms because the additional Na atom in $Na_8Pb$ is not bound to the Pb atom but begins to form a second Na shell. The lowest energy structures found agree with those obtained by Balbás and Martins [48] (who did a more intensive search using simulated annealing) except for $Na_7Pb$, although these authors also find Pb in the interior for $n \geq 6$. Marsden [45] has found other structures for $n = 7, 8$ and a small distortion, which lowers the symmetry, for $n = 4$. The evaporation energies $E_v$, which are defined in terms of total cluster energies by the equation

$$E_v(\mathrm{Na}_n\mathrm{Pb}) = E(\mathrm{Na}_{n-1}\mathrm{Pb}) + E(\mathrm{Na}) - E(\mathrm{Na}_n\mathrm{Pb}) , \qquad (7)$$

are given in Tab. 2. The evaporation energies fluctuate as a function of $n$ and pronounced maxima occur for $\mathrm{Na}_4\mathrm{Pb}$ and $\mathrm{Na}_6\mathrm{Pb}$. The former is also a closed-shell cluster with a HOMO-LUMO gap larger than the gap in $\mathrm{Na}_6\mathrm{Pb}$ [48].

**Table 2.** Energies (in eV) required to evaporate a Na atom from the $\mathrm{Na}_n\mathrm{Pb}$ cluster.

| cluster | Ab initio [46] | SAPS [64] |
|---------|----------------|-----------|
| $\mathrm{Na}_3\mathrm{Pb}$ |      | 2.03 |
| $\mathrm{Na}_4\mathrm{Pb}$ | 2.04 | 2.58 |
| $\mathrm{Na}_5\mathrm{Pb}$ | 1.19 | 1.17 |
| $\mathrm{Na}_6\mathrm{Pb}$ | 1.58 | 1.85 |
| $\mathrm{Na}_7\mathrm{Pb}$ | 0.76 | 0.07 |
| $\mathrm{Na}_8\mathrm{Pb}$ | 1.33 | 0.58 |
| $\mathrm{Na}_9\mathrm{Pb}$ |      | 0.62 |

A plausible explanation of the enrichment of the mass spectrum at $\mathrm{Na}_6\mathrm{Pb}$ is as follows [46]. In the supersonic expansion experiments the clusters grow by aggregation of atoms following collisions. The conditions in the experiment of Yeretzian et al. seem to be such that the $\mathrm{Na}_n\mathrm{Pb}$ clusters remain rather small, not much larger than $\mathrm{Na}_6\mathrm{Pb}$. This is because the concentration of Pb in the source is not too small (larger than 10%) and Pb atoms compete with one another to bind Na atoms, which prevents formation of clusters with many Na atoms. However, during the growth stage the clusters become hot (growth is an exothermic process) and can cool down by evaporating Na atoms. Since the evaporation energy increases by a factor of 2 between $\mathrm{Na}_7\mathrm{Pb}$ and $\mathrm{Na}_6\mathrm{Pb}$, it is reasonable to expect that the evaporation cascade practically stops at $\mathrm{Na}_6\mathrm{Pb}$, and in this way the population of this cluster becomes highly enriched in the beam.

$\mathrm{Li}_6\mathrm{C}$ and $\mathrm{Na}_6\mathrm{Pb}$ are the only two observed clusters in a broad class $\mathrm{A}_6\mathrm{M}$, where M = C, Si, Ge, Sn, Pb and A = Li, Na, K, Rb, Cs. This family has been studied by Marsden [45] and Schleyer [47]. Their structure is a centered octahedron, with the single exception of $\mathrm{Li}_6\mathrm{Sn}$. Elementary notions of covalent radii do not yield usefully accurate M-A bond length estimates. The bonding in the entire family is similar, with considerable ionic character and cage AA interactions. Most of the clusters are quite floppy, and the most rigid one is $\mathrm{Li}_6\mathrm{C}$, which is exceptionally stable (1108 KJ/mole, relative to separated atoms [45]). The binding energy decreases as the alkali atom changes from Li to Cs.

This reflects the residual alkali-alkali bonding. On the other hand by changing the tetravalent atom, the binding energies decrease in the order $A_6C > A_6Si \approx A_6Ge > A_6Sn \approx A_6Pb$, and this stability order follows the variation of the Allred-Rochow electronegativity scale [56] down group 14. This means that the stability of $Na_6Pb$ is relatively low compared with many other $A_6M$ systems. Hence, a large number of clusters in this family await experimental discovery.

The ionization potentials (IP) of Cs clusters containing one or more Oxygen impurities have been measured by Bergman et al. [57], and calculations of the electronic properties of $Cs_nO$ clusters have been performed using the SAPS model [58, 59, 60]. In those calculations the Oxygen atom was assumed to be at the center of the cluster, and all core and valence electrons of the Oxygen were considered. The electronic configuration of the Oxygen atom is $(1s)^2(2s)^2(2p)^4$. The $2p$ shell becomes filled in the cluster by the valence electrons donated by the Cs atoms, and the electronic shells above the $(2p)^6$ shell become filled, as the number $n$ of Cs atoms increases, in the following order: $3s, 3p, 1d, 1f, 4s, 4p, 1g, 2d, ....$ The onset of the $4s$ shell is practically degenerate with $1f$ and the same happens for the $1g$ and $4p$ shells. Consequently, pronounced shell effects (associated to large HOMO-LUMO gaps) only occur for $n = 10, 20, 36$ and 60. The shell closings become reflected in the calculated ionization potential, that displays pronounced drops at those particular sizes [58]. The predicted drops show up in the experimental IP [57]. In summary, the Oxygen atom in $Cs_nO$ forms an $O^{2-}$ anion and the remaining $n - 2$ valence electrons of the cluster behave according to the shell model. The experiments indicate the same behavior for clusters with more Oxygen atoms: that is, clusters $Cs_nO_x$ have $n - 2x$ nearly free electrons. A second aspect of interest concerns the structural effects induced by the impurity. The population of the coordination shell around the Oxygen atom is remarkable independent of $n$. With only a few exceptions this shell is formed by six Cs atoms in octahedral $(O_h)$ arrangement around the impurity. The six fold coordination and the O-Cs bond length (about 5 a.u.) agree with the corresponding features of solid $Cs_2O$. Finally, let us recall that the octahedral structure is also adopted by $A_6M$, where M is an element of the C group and A is an alkali atom. The difference with respect to the $Cs_6O$ case is that the highest electronic shell is only partially filled in this cluster.

# 6   Clusters of Interest for Liquid Ionic Alloys

The solid alloys of Sn or Pb with alkali metals (except Lithium) [61] follow an interesting building principle developed by Zintl [62]. As a consequence of the large electronegativity difference the outer electron of the alkali is transferred to one Pb atom. The resulting $Pb^-$ anion is isoelectronic with the P or As atoms, which form tetrahedral molecules $P_4$ and $As_4$. By analogy $Pb^-$ anions group together in these solid alloys forming charged tetrahedra $(Pb_4)^{4-}$ surrounded by larger, oppositely directed, alkali tetrahedra. The role of the alkali cations is restricted to keeping the $(Pb_4)^{4-}$ anions apart. Neutron diffraction also provides evidence in favor of the existence of the Zintl ions in the liquid Pb-alkali alloys

[61]. The electrical resistivity shows a sharp maximum at 50% Pb in liquid K-Pb, Rb-Pb and Cs-Pb. For Li-Pb and Na-Pb the maximum occurs at 20% Pb, although a small shoulder at 50% appears also for Na-Pb. The stoichiometry of the maximum in Li-Pb is consistent with that of the octet compound $Li_4Pb$, and the change of stoichiometry with increasing atomic number of the alkali is consistent with a transition to a Zintl configuration. Other properties like density, specific heat and thermodynamic stability also reveal the change of stoichiometry. The picture emerging from those studies is the formation in the liquid of clusters with compositions $Li_4Pb$, $Na_4Pb$ on one hand, and $Na_4Pb_4$, $K_4Pb_4$, $Rb_4Pb_4$, $Cs_4Pb_4$ on the other, but the nature of those clusters may be more complex than in the gas phase. Since the clusters are interacting with the liquid medium, an individual cluster may not necessarily be a long-lived species. Ab initio MD simulations of the liquid alloys provide the best theoretical tool to give microscopic information about the existence and stability of those clustered species [63]. However, the study of the relevant "isolated" clusters can also provide useful insights.

SAPS calculations for $A_nPb$ (A = Li, Na, K, Rb, Cs) [64] give results in full agreement with the ab initio predictions for $Na_nPb$. A relative maximum of the evaporation energy is obtained for $n = 6$ (see Tab. 2 for the Na case). But, more important for the present purposes, another maximum is found at $n = 4$, with an evaporation energy larger than for $n = 6$. In the environment provided by the liquid alloy in equilibrium at a given temperature, all $A_nPb$ clusters can compete with each other and we expect $A_4Pb$ to be more abundant because of the energy to remove a Na atom from this cluster is largest. The high stability of $A_4Pb$ is also a shell-closing effect. With the Pb atom at the cluster center (this is the prediction of the SAPS model) the valence electrons of the cluster arrange in shells which become occupied in the order $1s, 1p, 2s, 1d, ....$ For $Li_4Pb$ and $Na_4Pb$ the electrons adopt the closed-shells configuration $(1s)^2(1p)^6$, and a substantial gap (1.26 eV in $Na_4Pb$) separates the $1p$ shell from the empty $2s$ shell. The ab initio calculations [46, 48] support the shell closing picture, in spite of a less symmetrical geometry (see Fig. 4).

Since $A_4Pb_4$ (A = Na, K, Rb, Cs) appear to form in the liquid alloys, $A_nPb_4$ clusters have been studied as a function of $n$, again using the SAPS model [64]. The Pb atoms arrange as a $Pb_4$ tetrahedron surrounded by an outer layer of alkali atoms (in the particular case $n = 4$, the alkali atoms form an opposite tetrahedron, precisely as in the solid alloys). The calculated evaporation energy of an alkali atom

$$E_v(A_nPb_4) = E(A_{n-1}Pb_4) + E(A) - E(A_nPb_4) \qquad (8)$$

predicts a sharp drop between $A_4Pb_4$ and $A_5Pb_4$, that is, at the Zintl stoichiometry. The cloud of 20 valence electrons of $A_4Pb_4$ adopts in this model the closed-shell configuration $(1s)^2(1p)^6(2s)^2(1d)^{10}$, and the $1d$ shell is separated from the empty $2p$ shell by a substantial gap giving a high stability to $A_4Pb_4$.

The electronic structures of $Na_4Pb_4$, $K_4Pb_4$ and $Rb_4Pb_4$ with the double tetrahedron structure have been examined [65] using the all-electron self-consistent field multiple scattering $X\alpha$ (SCF-MS-$X\alpha$) method [66]. The calculations predict an electronic configuration of closed shells. With symmetry labels belonging to the $T_d$ group this configuration is $(1a_1)^2(1t_2)^6(2a_1)^2(2t_2)^6(1e)^4$, and one can notice the splitting of the $1d$-jellium shell into $2t_2$ and $1e$ shells by the tetrahedral symmetry. The calculations have also analyzed the ionicity. The occupied electronic levels have little contribution from the alkali atoms, suggesting a substantial charge transfer to lead. Charge transfer $\Delta Q$ is a concept rather difficult to quantify and several measures of $\Delta Q$ were considered. One giving quite reasonable results is obtained by integrating $\Delta\rho(r) = \rho(r) - \rho_{ni}(r)$, that is

$$\Delta Q = 4\pi \int_{r_o}^{\infty} r^2 \Delta\rho(r)dr , \qquad (9)$$

where $\rho(r)$ is the spherically averaged cluster density and $\rho_{ni}(r)$ is a density formed from "non interacting" $Pb_4$ and $A_4$ fragments (with the same geometry as in the fully interacting cluster. $r_o$ is a radial distance separating regions of positive and negative values of $\Delta\rho(r)$. Values $\Delta Q = 1.9$ e, $2.7$ e and $2.6$ e are obtained for $Na_4Pb_4$, $K_4Pb_4$ and $Rb_4Pb_4$ respectively, which are, however, lower than the ideal $\Delta Q = 4$ e assumed in the chemical models. The analysis of the electronic levels of the cluster and fragments allows also to understand how the lowest unoccupied manifold of $Pb_4$ ($1e$-levels) becomes fully occupied by the electrons donated by the alkali atoms. The difference between the ideal and calculated ionicities accounts for the bonding between the alkali atoms and $Pb_4$. This bonding is very weak and the charge transferred contributes to strengthen the Pb-Pb bonds.

In summary, the SAPS calculations predict that clusters of composition $A_4Pb$ and $A_4Pb_4$ are very stable and support the possibility of those clusters forming in the liquid alloy. The stability can be viewed as arising from the filling of electronic shells in the effective potential well. A debatable point that deserves further attention is the lifetime of the clusters in the alloy.

Still, the SAPS results leave some open questions and we have recently performed ab initio calculations using Car-Parrinello methods to get further insight into this type of clusters [67, 68]. One of the relevant results is that one can now explain the presence of octet clusters $Li_4Pb$, $Na_4Pb$ in the liquid alloys of Pb with Li and Na, and the absence of octet clusters in alloys of the heavy alkalis (K, Rb, and Cs). Figure 5 gives the evaporation energy for the series of $Li_nPb$, $Na_nPb$, and $K_nPb$ clusters, defined as in eq. (7). The results confirm the picture for $Na_nPb$ obtained in Table 2, with an absolute maximum of $E_v$ for $Na_4Pb$. This maximum is even more pronounced for the Li-series. However, in the K-series, $K_4Pb$ is no longer highly stable with respect to other clusters, and we can confidently extrapolate these results and predict that $Rb_4Pb$ and $Cs_4Pb$ will be even less distinguished in the corresponding Rb- and Cs-series. These features account for the pronounced effects observed at 20% Pb in the liquid alloys with Li and Na, and the absence of effects at this concentration in the alloys of K, Rb, and Cs. Only in the Li and Na cases are the octet clusters, $Li_4Pb$ and $Na_4Pb$, highly stable with respect to competing complexes and only in these two cases can one expect a substantial amount of octet clustering in the liquid.

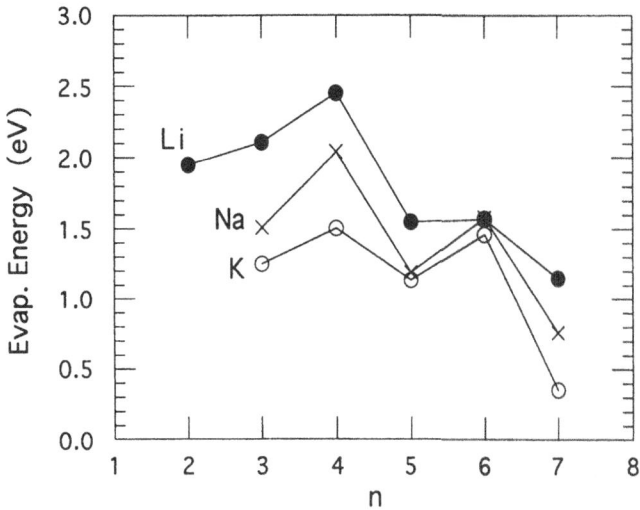

**Fig. 5.** Energy to evaporate an alkali atom from $Li_nPb$, $Na_nPb$, and $K_nPb$ clusters.

Another relevant result is a fundamental explanation of the stability of $A_4Pb_4$ clusters [67] that confirms their description as Zintl clusters. The ground state geometry of a free $Pb_4$ cluster is a rhombus. But a low lying isomer exists with the structure of a tetrahedron. The interesting characteristic of this tetrahedral isomer is that is has a doubly-degenerate LUMO not far from the HOMO level, and a "large" gap between the LUMO and the next higher level. This means that filling the LUMO with four electrons may produce a highly stable cluster. This is, in fact, what happens when alkali atoms are successively attached to $Pb_4$. The first alkali atom added does not alter the rhombus structure of $Pb_4$. Adding a second alkali atom deforms the rhombus into a butterfly-like structure. With three alkali atoms added, the structure of $Pb_4$ is already a deformed tetrahedron with the three alkali atoms capping three faces of the tetrahedron. Finally, four alkali atoms cap the four faces, respectively, of a perfect $Pb_4$ tetrahedron, and the expected large energy gap between occupied and unoccupied states fully develops.

These calculations nicely explain the high stability of the Zintl $A_4Pb_4$ clusters and justify their presence in both solid and liquid alloys. However, one has still to explain why Li alloys are an exception. The absence of $Li_4Pb_4$ clusters in the alloys is due to an atomic size effect. The surrounding alkali cations act as a buffer preventing different $Pb_4$ units from coming too closely near each other. This shielding effect is well developed in the alloys of the heavy alkalis (Na, K, Rb, and Cs) because the ionic radii of $Na^+$, $K^+$, $Rb^+$, and $Cs^+$ are large. But, in contrast, the ionic radius of $Li^+$ is small and the buffer ions do not prevent the $Pb_4$ units from interacting with one another. This interaction between Pb atoms from different $Pb_4$ tetrahedra allows for the distortion and breaking of the tetrahedra, leading to the instability of the clustered phase and to the development of a different, more stable, atomic arrangement with no Zintl clusters [68].

## Acknowledgements

This work has been supported by DGICYT (Grant No. PB95-0720-C02-01) and Junta de Castilla y León (Grant No. VA25/95).

# References

1. W. A. de Heer, W. D. Knight, M. Y. Chou, and M. L. Cohen, Solid State Phys. **40**, 93 (1987)
2. M. Brack, Rev. Mod. Phys. **65**, 677 (1993)
3. I. Katakuse, T. Ichihara, Y. Fujita, T. Matsuo, T. Sakurai, and H. Matsuda, Int. J. Mass Spectrom. Ion Proc. **67**, 229 (1985)
4. J. Tiggesbäumker, L. Köller, K. H. Meiwes-Broer, and A. Liebsch, Phys. Rev. A **48**, R1749 (1993)
5. E. K. Parks, L. Zhu, J. Ho, and S. J. Riley, J. Chem. Phys. **100**, 7206 (1994)
6. D. Reinhardt, B. D. Hall, D. Ugarte, and R. Monot, Z. Phys. D **26**, 576 (1993)
7. M. Pellarin, B. Baguenard, J. L. Vialle, J. Lermé, M. Broyer, J. Miller, and A. Pérez, Chem. Phys. Lett. **217**, 349 (1994)
8. S. E. Apsel, J. W. Emmert, J. Deng, and L. Blomfield, Phys. Rev. Lett. **76**, 1441 (1996)
9. *"Theory of the Inhomogeneous Electron Gas"*, Ed. N. H. March and S. Lundqvist, Plenum Press, New York, (1983)
10. K. Clemenger, Phys. Rev. B **32**, 1359 (1985); W. Ekardt and Z. Penzar, Phys. Rev. B **38**, 4273 (1988); C. Kohl, B. Montag, and P. G. Reinhard, Z. Phys. D **35**, 57 (1995)
11. M. M. Kappes, M. Schär, and E. J. Schumacher, J. Phys. Chem. **91**, 658 (1987)
12. M. P. Iñiguez, M. J. López, J. A. Alonso, and J. M. Soler, Z. Phys. D **11**, 163 (1989)
13. M. J. López, M. P. Iñiguez, and J. A. Alonso, Phys. Rev. B **41**, 5636 (1990)
14. A. Bol, G. Martin, J. M. López, and J. A. Alonso, Z. Phys. D **28**, 311 (1993)
15. P. Ballone, W. Andreoni, R. Car, and M. Parrinello, Europhys. Lett. **8**, 73 (1989)
16. A. Mañanes, M. P. Iñiguez, M. J. López, and J. A. Alonso, Phys. Rev. B **42**, 5000 (1990)
17. S. A. Rice, Z. Phys. Chem. **156**, S53 (1988)
18. A. Bol, J. A. Alonso, J. M. López, and A. Mañanes, Z. Phys. D **30**, 349 (1994)
19. A. Rubio, L. C. Balbás, and J. A. Alonso, Phys. Rev. B **45**, 13657 (1992)
20. A. Rubio and Ll. Serra, Z. Phys. D **26**, S118 (1993); Phys. Rev. B **48**, 18222 (1993)
21. A. Bol, J. A. Alonso, and J. M. López, Int. J. Quantum. Chem. **56**, 839 (1995)
22. L. C. Balbás, A. Rubio, and M. B. Torres, Z. Physik D **31**, 269 (1994)
23. L. C. Balbás, A. Rubio, and M. B. Torres, Computat. Mater. Sci. **2**, 509 (1994)
24. A. Rubio, J. A. Alonso, X. Blase, L. C. Balbás, and S. G. Louie, Phys. Rev. Lett. **77**, 247 (1996)
25. J. L. Martins, J. Buttet, and R. Car, Phys. Rev. B **31**, 1804 (1985)

26. S. Pollack, C. R. C. Wang, T. A. Dahlseid, and M. M. Kappes, J. Chem. Phys. **96**, 4918 (1992)

27. V. Bonačić-Koutecký, J. Gaus, M. Guest, and J. Koutecký, J. Chem. Phys. **96**, 4934 (1992)

28. T. A. Dahlseid, M. M. Kappes, J. A. Pople, and M. A. Ratner, J. Chem. Phys. **96**, 4924 (1992)

29. M. M. Kappes, M. Schär, C. Yeretzian, U. Heiz, A. Vayloyan, and E. Schumacher, *Physics and Chemistry of Small Clusters*, NATO ASI Series B, vol. 158 (Ed. P. Jena, B. K. Rao, and S. N. Khanna), Plenum Press, New York (1987) p. 263

30. C. Yeretzian, J. Phys. Chem. **99**, 123 (1995)

31. C. Baladrón and J. A. Alonso, Physica B **154**, 73 (1988)

32. S. B. Zhang, M. L. Cohen, and M. Y. Chou, Phys. Rev. B **36**, 3455 (1987)

33. U. Röthlisberger and W. Andreoni, Chem. Phys. Lett. **198**, 478 (1992)

34. C. Yannouleas, P. Jena, and S. N. Khanna, Phys. Rev. B **46**, 9751 (1992)

35. W. Pewestorf, V. Bonačić-Koutecký, and J. Koutecký, J. Chem. Phys. **89**, 5794 (1988)

36. P. Fantucci, V. Bonačić-Koutecký, W. Pewestorf, and J. Koutecký, J. Chem. Phys. **91**, 4229 (1989)

37. V. Bonačić-Koutecký, P. Fantucci, C. Fuchs, J. Koutecký, and J. Pittner, Z. Phys. D **26**, 17 (1993)

38. V. Bonačić-Koutecký, L. Češpiva, P. Fantucci, C. Fuchs, J. Koutecký, and J. Pittner, J. Chem. Phys. **186**, 275 (1994)

39. H.-P. Cheng, R. N. Barnett, and U. Landman, Phys. Rev. B **48**, 1820 (1993)

40. N. Troullier and J. L. Martins, Phys. Rev. B **43**, 1993 (1991)

41. V. Bonačić-Koutecký, P. Fantucci, and J. Koutecký, Chem. Rev. **91**, 1035 (1991)

42. D. Nehete, V. Shah, and D. G. Kanhere, Phys. Rev. B **53**, 2126 (1996)

43. P. K. Acharya, L. J. Bartolotti, S. B. Sears, and R. G. Parr, Proc. Natl. Acad. Sci. USA **77**, 6978 (1980)

44. C. Yeretzian, U. Röthlisberger, and E. Schumacher, Chem. Phys. Lett. **237**, 334 (1995)

45. C. J. Marsden, Chem.Phys. Lett. **245**, 475 (1995)

46. J. Chang, M. J. Stott, and J. A. Alonso, J. Chem. Phys. **104**, 8043 (1996)

47. P. von R. Schleyer and J. Kapp, Chem. Phys. Lett. **255**, 363 (1996)

48. L. C. Balbás and J. L. Martins, Phys. Rev. B **54**, 2937 (1996)

49. R. Car and Parrinello, Phys. Rev. Lett. **55**, 2571 (1985)

50. A. D. Becke, Phys. Rev. A **38**, 3098 (1988); J. Chem. Phys. **98**, 5648 (1993)

51. P. von R. Schleyer, in: New horizons of quantum chemistry, eds P.O. Löwdin and B. Pullman (Reidel, Dordrecht, 1983) p. 95

52. E. U. Würthwein, K. D. Sen, J. A. Pople, and P. von R. Schleyer, Inorg. Chem. **22**, 496 (1983)

53. H. Kudo, Nature **355**, 432 (1992)

54. M. C. Payne, M. P. Teter, D. C. Allan, T. A. Arias, and J. D. Joannopoulos, Rev. Mod. Phys. **64**, 1045 (1992)

55. R. Stumpf and M. Scheffler, Comp. Phys. Commun. **79**, 447 (1994)

56. A. L. Allred and E. J. Rochow, J. Inorg. Nucl. Chem. **5**, 264 (1958)

57. T. Bergman, H. Limberger, and T. P. Martin, Phys. Rev. Lett. **60**, 1767 (1988)

58. U. Lammers, A. Mañanes, G. Borstel, and J. A. Alonso, Solid State Commun. **71**, 591 (1989)

59. U. Lammers, G. Borstel, A. Mañanes, and J. A. Alonso, Z. Phys. D **17**, 203 (1990)

60. A. Mañanes, J. A. Alonso, U. Lammers, and G. Borstel, Phys. Rev. B **44**, 7273 (1991)
61. W. van der Lugt, Phys. Scripta T**39**, 372 (1991)
62. E. Zintl and G. Woltersdorf, Z. Elektrochem. **41**, 876 (1935)
63. G. A. deWijs, G. Pastore, A. Selloni, and W. van der Lugt, J. Chem. Phys. **103**, 5031 (1995)
64. L. M. Molina, M. J. López, J. A. Alonso, and M. J. Stott, Ann. Physik **6**, 35 (1997)
65. R. Pis Diez, M. P. Iñiguez, J. A. Alonso, and J. A. Aramburu, J. Molec. Struct. Theochem. **330**, 267 (1995)
66. K. H. Johnson, Adv. Quantum Chem. **7**, 143 (1973)
67. L. M. Molina, M. J. López, A. Rubio, J. A. Alonso, and M. J. Stott, Int. J. Quantum Chem. **69**, 341 (1998)
68. J. A. Alonso, L. M. Molina, M. J. López, A. Rubio, and M. J. Stott, Chem. Phys. Lett. **289**, 451 (1998)

# Alloy Clusters: Structural Classes, Mixing, and Phase Changes

Julius Jellinek and Evgueni B. Krissinel*

Chemistry Division, Argonne National Laboratory, Argonne, IL 60439, USA

**Abstract.** Results on structural and dynamical properties of model nickel-aluminum alloy clusters are presented and analyzed. A classification scheme of the structural forms is introduced. New general definitions of the mixing energy and mixing coefficient are formulated, and the role of mixing in defining the relative stability of the structural forms is pointed out. The dynamical features are analyzed over a broad energy range. The peculiarities of the solid-to-liquid-like transition in clusters with different stoichiometric composition are described and correlated with the composition-specific features in the energy spectra of the corresponding structural forms. An analytical model, which allows one to estimate a variety of structural and energy characteristics of two-component clusters directly from the parameters of the potential, is presented. An important merit of the model is that it makes transparent the role of the interactions between like atoms vs. those between unlike atoms in defining these characteristics.

## 1 Introduction

The explosion of activity, theoretical and experimental, in the field of atomic and molecular clusters produced an impressive body of knowledge on clusters of different sizes and materials. The scope of the studies continues to expand, both in terms of variety of systems and properties investigated, and degree of sophistication of the concepts and techniques applied. The progress and the achievements are not always uniform, which is, at least in part, attributable to the differences in the degree of complexity of the different systems and issues. The theoretical and experimental studies not always evolve "synchronously", but they provide a constant challenge and stimulus for each other.

Metal clusters are among the more complex and interesting ones from both cognitive and practical points of view. The term "metal clusters" is used to designate cohesive assemblies of atoms of those elements that are metals in bulk quantities. These assemblies, however, frequently behave, as judged by at least some of their properties, differently from the corresponding metals. For example, the optical properties of small mercury clusters indicate an evolution in the nature of the interatomic interactions from a van der Waals-type, through covalent, and eventually to metallic as the clusters grow in size [1]. Understanding the evolution of different properties of metal clusters towards, and eventually into, those characteristic of metals, is one of the most intriguing and challenging

---

* Permanent address: Institute for Water and Environmental Problems, Siberian Branch, Russian Academy of Sciences, Barnaul 656099, Russia

aspects of these systems. The challenge is only exacerbated by the fact that, in general, the properties vary with size differently not only in different classes of metals (alkali, transition, or noble), but also from one metal to another within the same class. [In contrast, different van der Waals clusters, as mimicked quite realistically by Lennard-Jones potentials, obey the principle of corresponding states [2], and, consequently, many of their properties, especially the qualitative ones, are similar.]

Theoretical (cf., e.g., [3–46] and citations therein) and experimental (cf., e.g., [3–6, 47–72] and citations therein) investigations of metal clusters cover their geometric, electronic (including optical and magnetic), thermal, and chemical properties. The theoretical studies are based on either first principles approaches (different quantum chemistry techniques, the density functional theory, and combinations of the two) [7–18], or models (e.g., jellium and tight-binding) [19–25] and fitted semiempirical potentials [26–42]. The majority of these studies are devoted to homogeneous (one-component) systems. Two-component (alloy) metal clusters have also been investigated (cf., e.g., [43–46, 66–72]), but to a considerably lesser degree. A possible reason for this is the added complexity of these heterogeneous systems. Their properties are defined not only by interactions between like atoms ("homointeractions"), but also those between unlike atoms ("heterointeractions"). The many-body nature of the metallic cohesion (a consequence of the characteristic delocalization of the electrons) makes an adequate description of even the homointeractions a challenging task [73]. Reproduction, or mimicking, of the heterointeractions is only a more complex problem. In general, these latter depend on and change with not only the types of the metals, but also with their relative concentrations. Experimental data on the concentration dependence are largely unavailable.

Another issue is the abundance of the structural forms. Whereas in homogeneous clusters a difference in structure means a difference in the geometric arrangement, or packing, of the atoms (we abstract ourselves from permutational isomers of the same overall geometry), in heterogeneous clusters different structures can be obtained by different distributions of the two types of atoms between the sites of the same geometric arrangement. In addition, one also has to consider different stoichiometric makeups of a two-component cluster. As a result, the numbers of possible structural forms of two-component clusters are considerably larger than those of their one-component counterparts. In what follows we use the term "isomer" to designate different geometric forms of a cluster, and utilize the term "homotops" ("the same topography") to label those structural forms of a heterogeneous cluster that have the same geometry, but differ by the placement of the two types of atoms at the sites of that geometry.

The choice of the component metals and their relative concentrations are those "knobs", additional to the size, that could be used to "tune" the various physicochemical properties of bimetallic clusters to the desired specifications. For such a tuning, however, a sufficiently detailed, atomic-level understanding of the properties of these heterogeneous systems is required. The fundamental questions include the following: 1) What are the (globally or locally) stable structural forms of a two-component cluster of a given size, materials and composition (stoichiometry)?; 2) How many homotops are associated with different isomers?; 3) What is the relationship between the number of homotops and the stoichiometric composition?; 4) What defines the energy ordering of the different structural forms, and can one introduce a hierarchical classification of these forms?; 5) What defines the propensity of the two component materials to mix or to segregate in a cluster?; 6) What is an adequate quantitative measure of the degree of mixing or segregation?; 7) What is the relative role of the homointeractions vs. those of the heterointeractions in defining the different structural and dynamical (thermal) properties of two-component, in our case bimetallic, clusters?; 8) What novel concepts and techniques may be needed to allow for more adequate and/or efficient analyses and description of two-component systems?

Below we present an overview of our recent studies of structural and dynamical properties of model nickel-aluminum clusters [45]. We use a literature potential for the description of the homo (nickel-nickel and aluminum-aluminum) and hetero (nickel-aluminum) interactions. Our emphasis is on the questions listed above rather than on the accuracy of the potential. Our primary goal is to formulate and to apply novel and general analyses tools and techniques and to establish generic trends and correlations in properties of bimetallic (more generally, two-component) clusters. These tools and techniques can then be applied and the trends and correlations can be searched for with any potential, including those that will be developed in the future. The next section outlines the theoretical and methodological background. In Sect. 3, results on structural properties are presented and analyzed, and a classification scheme of the structural forms is introduced. In Sect. 4, the dynamical features are described and correlated with the peculiarities of the configurational energy spectra of the relevant structural forms. An analytical model that predicts and explains many of the properties originally deduced from numerical simulations is presented in Sect. 5. A summary is given in Sect. 6.

## 2 Theoretical and Methodological Aspects

We use the many-body Gupta-like potential [74],

$$V = \sum_{i=1}^{N} \left\{ \sum_{j=1(j\neq i)}^{N} A_{ij} \exp\left(-p_{ij}\left(\frac{r_{ij}}{r_{ij}^{o}} - 1\right)\right) \right.$$
$$\left. - \left[\sum_{j=1(j\neq i)}^{N} \xi_{ij}^{2} \exp\left(-2q_{ij}\left(\frac{r_{ij}}{r_{ij}^{o}} - 1\right)\right)\right]^{1/2} \right\}, \quad (1)$$

to mimic the interatomic interactions in a mixed nickel-aluminum cluster. In (1), N is the total number of atoms in the cluster; $i$ and $j$ label the individual atoms; $A_{ij}$, $\xi_{ij}$, $p_{ij}$, $q_{ij}$, and $r_{ij}^{o}$ are adjustable parameters, the values of which depend on the type of the $i$th and the $j$th atoms; and $r_{ij}$ is the distance between the *i*th and the *j*th atoms. The values of the parameters for mixed nickel-aluminum systems are adopted from [75] and are listed in Table 1. The same potential, only with parameters appropriate for gold-copper systems, has been utilized in [44]. As mentioned, we use the Gupta-like potential as a model and consider, as a paradigm, $Ni_nAl_m$ clusters with all possible nonnegative integer values of $n$ and $m$ such that $n + m = 13$. When $n = 0$ or $m = 0$ the potential describes the pure $Al_{13}$ or $Ni_{13}$ cluster. The parameters of interactions between like, i.e., nickel-nickel and aluminum-aluminum, atoms are fitted to properties of pure nickel and aluminum, respectively. The icosahedron emerges as the lowest energy structure of both $Ni_{13}$ (cf. [26, 28, 29, 33, 35]) and $Al_{13}$. The distances from the center of the clusters to the surface, as defined by the parameters in Table 1, are 2.43 Å and 2.66 Å, respectively. These agree well with the results of other studies [76, 77]. The binding energies are 2.67 eV/atom for $Ni_{13}$ and 2.60 eV/atom for $Al_{13}$. The corresponding values obtained in different electronic structure calculations range from 2.91 to 4.26 eV/atom for $Ni_{13}$ [76], and from 2.77 to 3.21 eV/atom for $Al_{13}$ [77].

**Table 1.** Parameters of the Gupta-like potential for nickel-aluminum alloys

| Parameters | Ni-Ni | Al-Al | Ni-Al |
|---|---|---|---|
| A (eV) | 0.0376 | 0.1221 | 0.0563 |
| $\xi$ (eV) | 1.070 | 1.316 | 1.2349 |
| p | 16.999 | 8.612 | 14.997 |
| q | 1.189 | 2.516 | 1.2823 |
| r° (Å) | 2.4911 | 2.8638 | 2.5222 |

Different geometric forms (isomers) of clusters can be obtained through simulated thermal quenching from configurations generated by molecular dynamics (MD) trajectories. For this one should use trajectories of sufficiently high energy and of sufficient length to assure a representative sampling of the cluster configuration space. Since the emphasis of our study is on techniques, trends, and concepts, we did not perform a systematic search for isomers of clusters with different stoichiometric composition. Instead, we performed such a search for pure $Al_{13}$ and $Ni_{13}$ and used the geometries of the first six stable isomers of $Al_{13}$ as templates for generating structures of the alloy clusters. (Overall 129 different isomers of $Al_{13}$ were obtained from 1500 initial configurations accumulated along a long trajectory corresponding to a temperature of about 1340 K; cf. below. For calibration we note that the melting temperature of $Al_{13}$ calculated with the parameters listed in Table 1 is about 850 K (cf. Table 2 below). We used the following quenching procedure to obtain the isomers. First, the kinetic energy of the cluster is extracted every 30-50 propagation steps until the reduction in the total energy in fifty consecutive extractions is less than $10^{-4}$ eV. Then the energy minimization is continued using the Broyden-Fletcher-Goldfarb-Shanno modification of the variable metric quasi-Newton method [78]. About 25 iterations are sufficient to arrive at the equilibrium configurational energy converged with the machine accuracy. Normal mode analysis is used to filter out stationary configurations that correspond to saddles, rather than to minima, of the potential energy surface. Only structures with all real characteristic frequencies are qualified as isomers).

The homotops of the alloy clusters are obtained for each of the six template isomers and for all the stoichiometric compositions. This is accomplished in the following way. All possible replacements of $n$ aluminum atoms by nickel atoms in an isomer of $Al_{13}$ are considered. The total number of such replacements is the number of different possible combinations of $n$ (or, equivalently, $m$) elements in a manifold of $n + m = 13$ elements. For example, when $n = 6$ the number of such combinations is 1716. However, because of a possible symmetry of the templates, not necessarily all combinations result in a different homotop.

A structure that is an equilibrium geometry of $Al_{13}$ becomes, in general, a nonequilibrium geometry when some of the aluminum atoms are replaced by nickel atoms. The equilibrium structures of the alloy clusters are obtained by, first, minimizing their energy through uniform scaling of all the interatomic distances and then applying the quenching procedure described above. Only those stationary configurations that correspond to minima of the potential energy surface are qualified as homotops. In our case, the equilibrated structures of the alloy clusters differ from those of the "parent" templates only by small relaxations. These relaxations do not change the overall atomic packing, although they may cause slight deviations from the overall symmetry of the template. We view such relaxed structures of an alloy cluster and the geometry of $Al_{13}$ with which they correlate as representing the same isomer. In some cases the relaxation, no matter how carefully it is performed, leads to a change in atomic packing. These are the cases when an isomeric form of $Al_{13}$ cannot support a stable homotop (or homotops) for certain values of $n$ and $m$. On the other hand, alloy clusters may, in general, possess equilibrium geometries that are not exhibited by their one-component counterparts. Apart from visual inspection of the geometries before and after equilibration, we use the adjacency matrix $\{A_{ij}\}$ to detect possible changes in the atomic packing. The elements $A_{ij}$ of this matrix are equal to one, if the $i$th and the $j$th atoms are first neighbors, and to zero otherwise. Some care should be exercised in this analysis because the adjacency matrix may, in principle, change as a consequence of the exchange in the positions of two (or more) atoms. Such exchanges do not alter the overall geometry, or isomeric form, of a cluster. In cases when the atoms that exchange their positions are of different type the change in the adjacency matrix may indicate a transition from one homotop to another. Such transitions are, however, unlikely in the course of the relaxation procedure because of the energy barriers involved. We did not observe changes in the adjacency matrix other than those caused by true changes in atomic packing.

An important characteristic of two-component systems is the propensity of the constituent elements to mix or to segregate. To characterize the degree of mixing, or segregation, one needs a quantitative measure. It has been suggested [79] to use the total number of bonds between pairs of unlike atoms that are first neighbors of each other as a measure of mixing in a two-component cluster. The limitations of such a measure, when one wants to correlate it with the "mixing energy" and the equilibrium configurational energy of the different structural forms (this latter defines the relative stability of these forms), is obvious. This measure establishes such a correlation only if and when a) the total configurational energy can be represented as a sum of pairwise interactions and b) the ranges of these interactions do not extend beyond the first nearest neighbors. In this case the number of "heterobonds" between first neighbors is (neglecting the possible small differences in the lengths of these bonds) indeed proportional to the "heteroenergy" and the mixing energy parts of the total configurational energy.

A general definition of the mixing energy and/or the degree of mixing should be applicable to systems described by either pairwise-additive or many-body potentials, which may be short- or long-range. Such a general and physically meaningful definition for the mixing energy $V_{\mathrm{mix}}$ of a two-component cluster $A_n B_m$ in a given configuration, as specified by the coordinates of its atoms, where A and B are the types of the atoms, is given by

$$V_{\mathrm{mix}} = V_{A_n B_m} - \left[ V_{A_n}^{(A_n A_m)} + V_{B_n}^{(B_n B_m)} \right] , \tag{2}$$

where $V_{A_n B_m}$ is the energy of the $A_n B_m$ cluster, $V_{A_n}^{(A_n A_m)}$ is the energy of the $A_n$ subcluster in the $A_n A_m$ cluster, and $V_{B_m}^{(B_n B_m)}$ is the energy of the $B_m$ subcluster in the $B_n B_m$ cluster. In definition (2), the one-component $A_n A_m$ and $B_n B_m$ clusters have the same configuration as $A_n B_m$, and the configurations of the $n$-atom and $m$-atom subclusters are the same in $A_n B_m$, $A_n A_m$ and $B_n B_m$. Since

$$V_{A_n B_m} = V_{A_n}^{(A_n B_m)} + V_{B_m}^{(A_n B_m)} , \tag{3}$$

where $V_{A_n}^{(A_n B_m)}$ and $V_{B_m}^{(A_n B_m)}$ are the energies of the $A_n$ and $B_m$ subclusters in the $A_n B_m$ cluster, the mixing energy can be expressed as

$$V_{\mathrm{mix}} = \left( V_{A_n}^{(A_n B_m)} - V_{A_n}^{(A_n A_m)} \right) + \left( V_{B_m}^{(A_n B_m)} - V_{B_m}^{(B_n B_m)} \right) . \tag{4}$$

$V_{\mathrm{mix}}$ is, therefore, the summed-up change in energy experienced by the $A_n$ and $B_m$ subclusters (or, ultimately, their individual atoms) when they are removed from the one-component $A_n A_m$ and $B_n B_m$ clusters, respectively, and brought together to form the $A_n B_m$ cluster. For (3) and (4) to be meaningful, the energy of a part of a cluster (subcluster) has to be well-defined. Of course, this condition is satisfied if the energy of each atom in a cluster is well-defined, which is indeed the case not only for the pairwise-additive but also all the many-body potentials introduced in the literature. It is clear that for pairwise-additive potentials that extend only to the first neighbors our definition of $V_{\text{mix}}$ becomes equivalent (proportional) to the number of bonds between first-neighbor atoms of unlike type. The (global) mixing coefficient $M$ can be defined as the fraction of the mixing energy in the total configurational energy,

$$M = \frac{V_{\text{mix}}}{V} \cdot 100\% \ . \tag{5}$$

The dynamical features of the nickel-aluminum 13-mers are investigated over a broad energy range represented by a fine grid. These features are deduced from constant energy MD simulations. The trajectories are propagated using the velocity version of the Verlet algorithm [80] for $5 - 8 \cdot 10^6$ steps with a step size of 2 fs. This step size assures conservation of the total energy within 0.01% in the longest runs. The initial conditions are chosen so as to supply no linear and angular momenta to the clusters. The temperature $T$ associated with each individual total energy is computed as

$$T = \frac{2\langle E_{\text{k}} \rangle_t}{(3N - 6)\text{k}} \ , \tag{6}$$

where $E_{\text{k}}$ is the (internal) kinetic energy of the cluster, k is the Boltzmann constant, and $\langle \ \rangle_t$ stands for time averaging over the entire trajectory. The analyses of the dynamics are performed in terms of the following quantities:

1. Caloric curve, i.e., time-averaged kinetic energy (per atom) as a function of total energy (per atom);
2. Relative root-mean-square (rms) bond length fluctuation $\delta$,

$$\delta = \frac{2}{N(N - 1)} \sum_{i<j} \frac{\left( \langle r_{ij}^2 \rangle_t - \langle r_{ij} \rangle_t^2 \right)^{1/2}}{\langle r_{ij} \rangle_t} \ , \tag{7}$$

as a function of total energy (per atom);
3. Specific heat $C$ (per atom) [81],

$$C = \left[ N - N\left( 1 - \frac{2}{3N - 6} \right) \langle E_{\text{k}} \rangle_t \langle E_{\text{k}}^{-1} \rangle_t \right]^{-1} \ , \tag{8}$$

as a function of total energy (per atom).

Prominent in the analyses of the dynamical features of alloy clusters are the effects of the stoichiometric composition and of the homotopic form chosen to represent the zero-temperature structure of an isomer.

# 3  Analysis of Structural Forms

The six isomers of $Al_{13}$ used as template geometries for generating the alloy nickel-aluminum clusters are shown in Fig. 1. Interestingly, for $Ni_{13}$, as defined by the values of the parameters for nickel in Table 1, only structures I, III, IV and V are stable isomers, and their energy ordering is I, V, IV and III. Structures II and VI correspond to saddle-point configurations of $Ni_{13}$. Isomers III, IV and V can be obtained from the icosahedral (ico) structure of isomer I by placing one of the surface atoms over a face and allowing for relaxations. The more distorted isomers II and VI also are derivatives of the ico structure, but, instead of over a face, a surface atom is placed over an edge. In isomer II the involved edge is one of those in the five-fold ring adjacent to the top apex atom. In isomer VI it is one of the edges connecting the top apex atom with an atom in the adjacent five-fold ring. The distortions are consequences of relaxations.

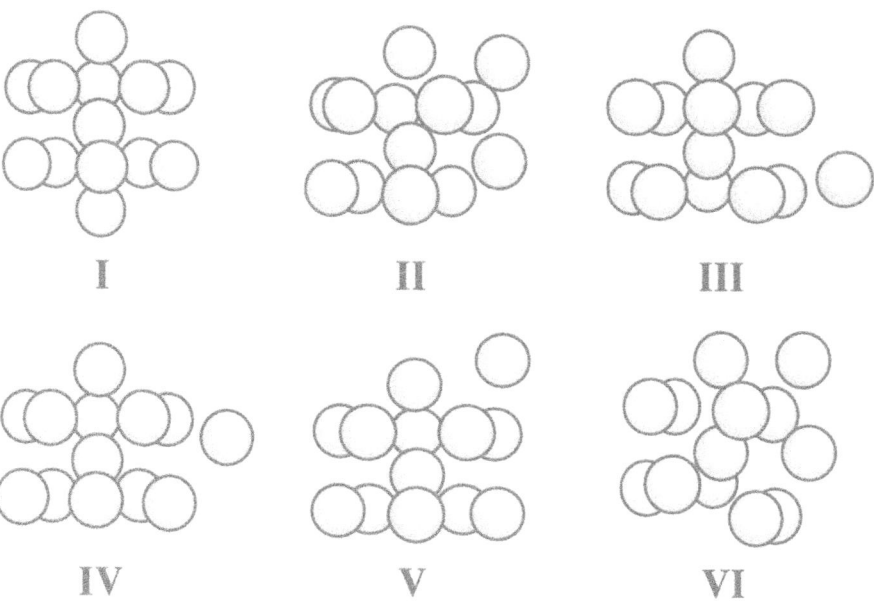

I    II    III

IV    V    VI

**Fig. 1.** The first six isomers of $Al_{13}$ as defined by the Gupta-like potential with parameters listed in Table 1. Their energies are: (I) -33.812 eV; (II) -33.085 eV; (III) -33.066 eV; (IV) -33.060 eV; (V) -33.037 eV; and (VI) -33.013 eV

As mentioned, not necessarily every possible replacement of $n$ aluminum atoms by nickel atoms in a given isomer of $Al_{13}$ produces a new homotop. Use of the adjacency matrix for identification of equivalent homotops would require introduction of a somewhat elaborate renumbering scheme. A visual inspection of the structures would be too cumbersome because of the possible large number of homotops (cf. below). For a fixed isomer, we use the proximity of the equilibrium energies of the homotops as an indicator of their equivalency. Barring accidental degeneracy, two equilibrated homotops are classified as equivalent (identical) if their energies differ by less then $10^{-8}$ eV (the smallest energy gap we detected between two neighboring homotops is of the order of $10^{-6}$ eV).

The spectra of equilibrium energies and the total numbers of homotops associated with the first four (of the six considered) isomers are shown in Figs. 2-5, which correspond to four different fixed compositions of the mixed $Ni_n Al_m$ 13-mer. The same information, but for the first two isomers of six different compositions, is shown in Fig. 6. For every isomer, the energy spectrum of the homotops is separated into two parts that correspond to classes with nickel and aluminum, respectively, as the central atom. The spectra are presented in the form of resolved energy levels for $Ni_{12}Al$ and $NiAl_{12}$, and as distributions of the energy levels for all the other compositions. Each isomer is represented by the corresponding homotop of the lowest energy, and these are displayed in the order of increasing energy. The Roman numerals establish a correspondence with the template geometries of Fig. 1. Geometries II and VI do not survive at all as equi-

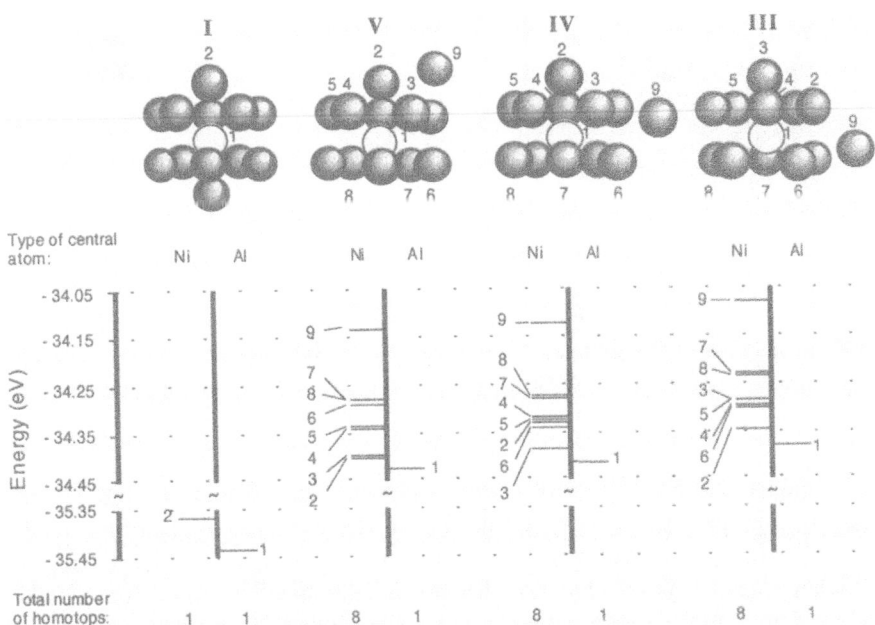

**Fig. 2.** Four isomers of $Ni_{12}Al$ and the energy levels of the homotops corresponding to them. Each isomer is represented by its lowest energy homotop (the darker spheres depict nickel, the lighter ones aluminum). The homotop energy levels are separated into classes defined by the type of the central atom. The numbers labeling the levels indicate the position of the aluminum atom in the corresponding homotop. The Roman numerals establish correlation with the template isomers of Fig. 1

librium structures of the $Ni_{12}Al$ cluster (Fig. 2). They are not isomers of $NiAl_{12}$ (Fig. 5) either, when a nickel atom occupies the central position. Incidentally, for no stoichiometric composition, other than $Ni_5Al_8$, does the template geometry VI emerge as one of the first four isomers; it is the third isomer of $Ni_5Al_8$, whereas the second isomer of this cluster has the geometry of the template II (cf. Fig. 6).

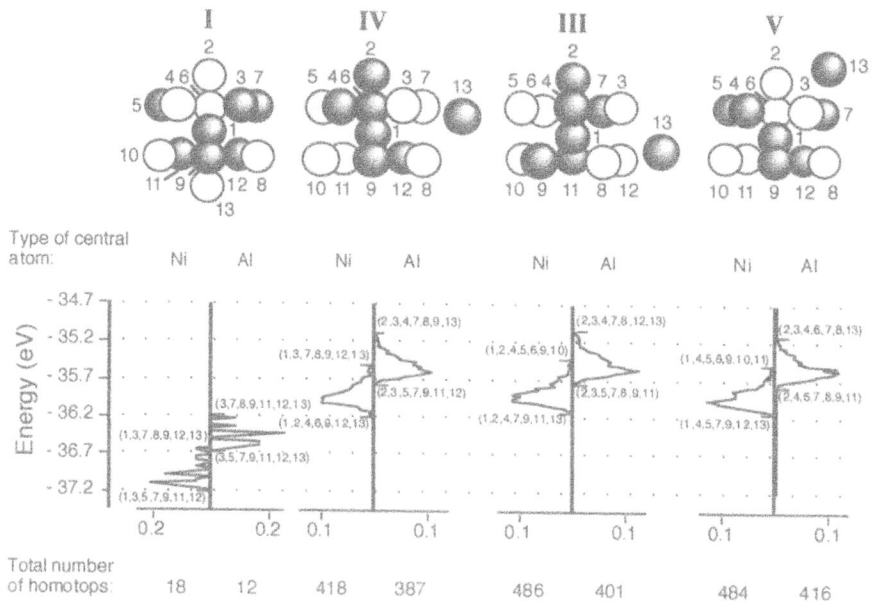

Fig. 3. The same as Fig. 2, but for $Ni_7Al_6$. Because of the large number of homotops in each class, (normalized) distributions of their energies are shown. The distributions (histograms) are obtained with a box size of 0.03 eV. The levels corresponding to the lowest and the highest energy homotop in each class are labeled. The numbers forming a label represent the positions of the nickel atoms in the corresponding homotop

Examination of Figs. 2-6 leads to the following general observations: 1) The icosahedron is the lowest energy isomer for all the possible compositions of the mixed $Ni_nAl_m$ 13-mer. However, as is clear from the figures, the ordering of the higher energy isomers changes with the stoichiometric composition; 2) The number of homotops corresponding to an isomer depends both on symmetry and composition of the isomer. As a rule, the lower the symmetry and the closer the composition to 50/50%, the larger the number of the corresponding homotops; 3) For all stoichiometric compositions, except $Ni_{12}Al$, nickel is the energetically favored type of the central atom; 4) For a fixed stoichiometric composition, the homotop energy spectra of the different isomers, at least those of lower energy, show some common features. The changes in these features with the composition are characterized by systematic trends. In $Ni_{12}Al$ the energies of the two ico homotops are separated by a large gap from those of the homotops corresponding to the higher energy isomers, but within each individual isomer the energy of the homotop with aluminum in the center is close to that (or the range of those) with nickel in the center. As the number of aluminum atoms in the cluster increases, the range of energies of the ico homotops moves closer to, and even overlaps with, the ranges of energies of the homotops that correspond to the higher energy isomers. This change is caused by that within each individual isomer the energy range corresponding to the class of homotops with aluminum in the center moves away ("upward") from that corresponding to the class of homotops with nickel in the center. An additional consequence of this change is that the distribution of the homotop energies of each isomer becomes bimodal. The two branches of the distribution represent the two classes of homotops as defined by the type of the central atom.

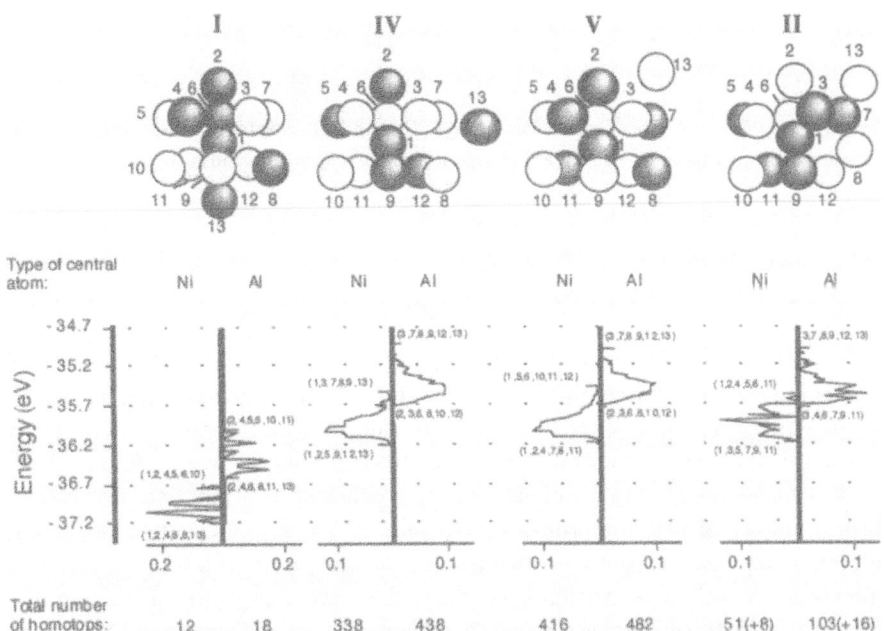

**Fig. 4.** The same as Fig. 3, but for $Ni_6Al_7$. The numbers with a "+" in parenthesis indicate the number of additional stationary homotopic structures, which correspond to saddles, rather than to minima, of the potential energy surface. These are not included in the distributions

An indication of the attribute responsible for the energy ordering of the homotops within each class is obtained by examining the labels of the homotop energy levels, either individual or those limiting the energy ranges. The exami-

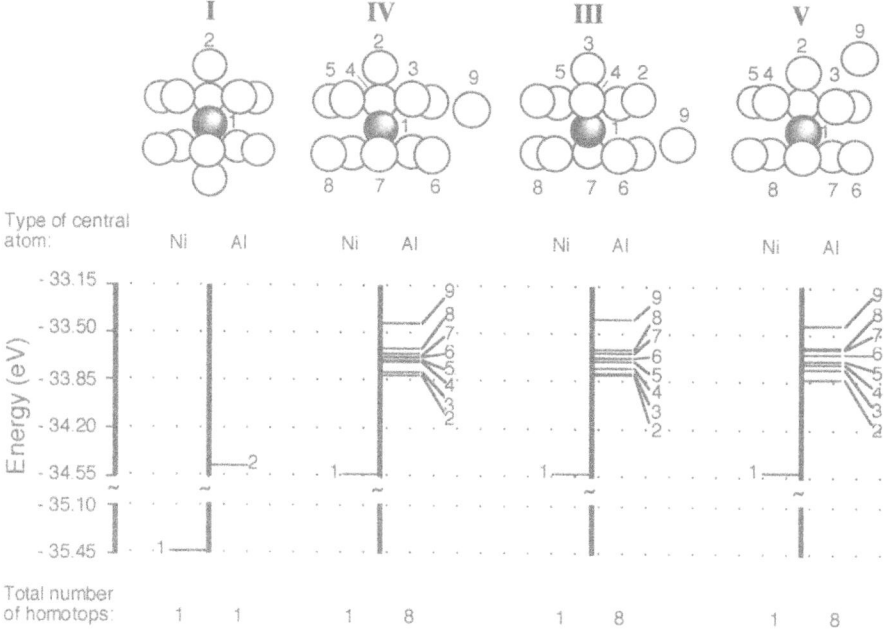

**Fig. 5.** The same as Fig. 2, but for NiAl$_{12}$. The number labeling each energy level represents the position of the nickel atom in the corresponding homotop

nation shows that the homotop of the lowest energy in a class is the one in which the nickel and aluminum atoms are mixed the most, as perceived for example visually, whereas the homotop of the highest energy is that in which the two types of atoms are segregated the most. That the degree of mixing (or segregation) is indeed the property which defines the energy ordering of the homotops within structural classes specified by a) the composition, b) the isomeric form, and c) the type of the central atom of a two-component cluster, follows from Fig. 7, which displays graphs of the mixing coefficient as a function of the equilibrium energy of the homotops. The graphs in the figure correspond to four different classes associated with two isomers of the Ni$_6$Al$_7$ cluster. They clearly indicate a general trend: within each class, the larger the degree of mixing of a homotop, as defined by the mixing coefficient, the lower its energy. The trend holds (almost) exactly for the classes of the ico isomer, in which most of the neighboring homotops are separated by energy gaps that are not too small. It holds only globally for the classes of the higher energy isomer, in which the energy separations between neighboring homotops are very small. Indeed, in this latter case 338 homotops (in the class with nickel as the central atom) and 438 homotops (in the class with aluminum as the central atom) are "squeezed" into ranges with widths of 0.557 eV and 0.722 eV, respectively. The violation of the rigorous monotonicity of the graphs on the local scale involving a couple or a

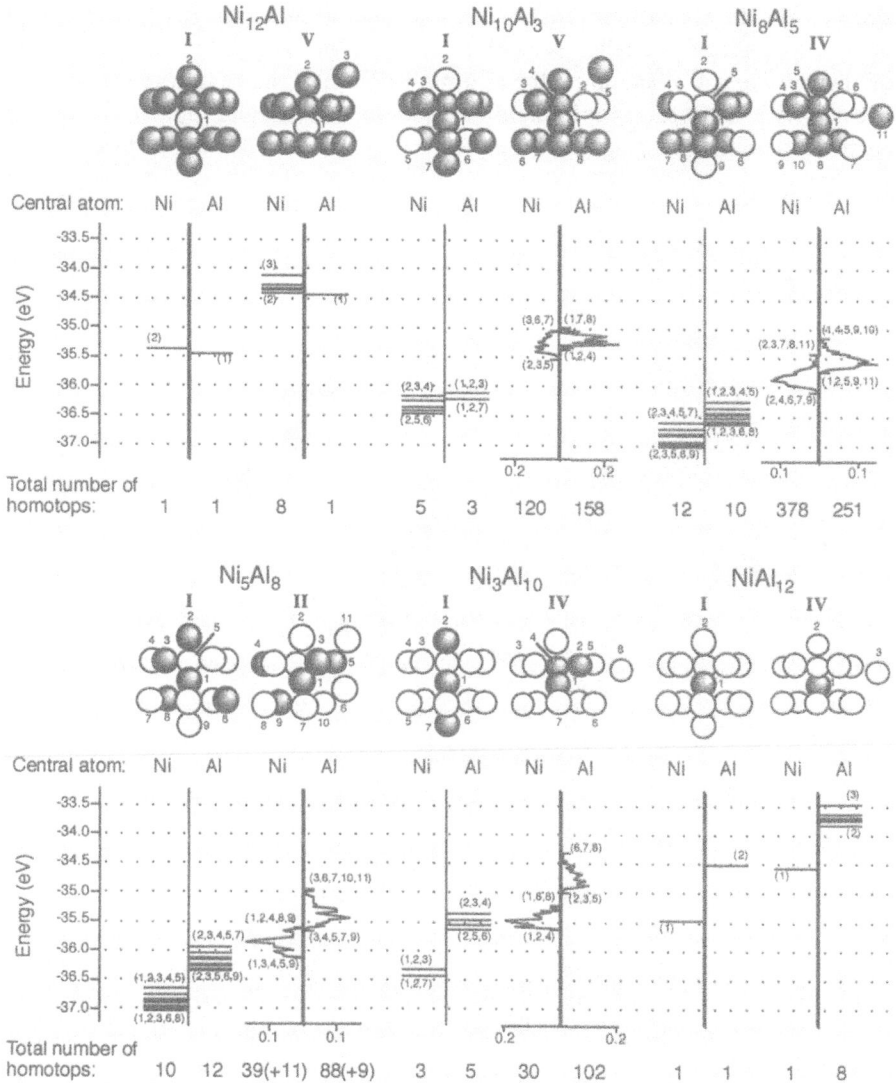

**Fig. 6.** The same as Figs. 2-5, but for the first two isomers of six different compositions of the mixed nickel-aluminum 13-mer. The numbers forming the label of the lowest and the highest energy homotop in each class indicate the positions of the aluminum atoms in the case of $Ni_{12}Al$, $Ni_{10}Al_3$, and $Ni_8Al_5$, and of the nickel atoms in the case of $Ni_5Al_8$, $Ni_3Al_{10}$, and $NiAl_{12}$. For each composition, the homotop energy spectrum of the second isomer is typical for other higher energy isomers as well

290

few neighboring homotops is a consequence of the mentioned equilibration relaxations in the structures and/or the possible additional sensitivity to the type
of the "secondary" (locally) central atom(s). Both change the energy and the
mixing coefficient of the homotops only a little. But the changes may be comparable with, or even exceed, the differences in the energies and/or the mixing
coefficients of neighboring homotops and cause a switch in their ordering.

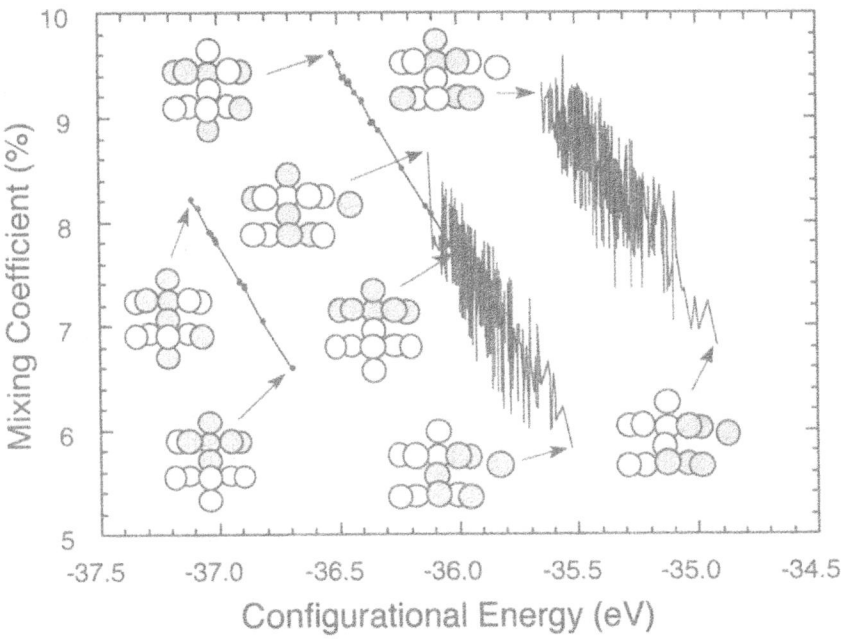

**Fig. 7.** Mixing coefficient as a function of the equilibrium configurational energy of the
homotops for four structural classes corresponding to the first two isomers of $Ni_6Al_7$.
The pictures show the lowest and the highest energy homotop in each class. The structures of these homotops correspond to, respectively, the highest and the lowest degree
of mixing

The propensity of two materials to mix (as in the present case) or to segregate in a cluster is a function of the parameters of the potential [cf. (2)-(4)]. But
the role of the degree of mixing (or segregation) as the property that defines the
overall energy ordering of the homotops within the structural classes specified
above is a general attribute. Our numerical results corroborate this conclusion
for all the isomers and compositions considered. Finally, a remark of a technical
nature: the notion of the central atom(s) used in the preceding discussion may
become problematic in cases of less symmetric isomers. One, however, can always identify an effective "central" atom(s) - that (or those) with the highest
coordination or the largest binding energy.

# 4 Analysis of Dynamical Features

The specificity of the structural forms (isomeric and homotopic) and the peculiarities of their energy spectra, as defined by the stoichiometric composition, translate into composition-specific dynamical features of the mixed clusters. Before discussing these features, we describe briefly the dynamical properties of the pure $Ni_{13}$ and $Al_{13}$. Figure 8 displays the graphs of the caloric curve, rms bond length fluctuation $\delta$, and specific heat $C$ for these clusters. The changes in the graphs with the energy - variations in the slope of the caloric curves, abrupt increases in the value of $\delta$, and existence of a maximum in the $C$ graphs - are characteristic of a solid-to-liquid-like transition in clusters. As discussed in numerous earlier studies (see, e.g., [9, 15, 28, 31, 34, 35, 82]), this transition takes place over a finite range of energy (or temperature) and involves one or more intermediate stages, such as isomerizations, (sequential) "coexistence" of liquidlike and solidlike clusters, partial melting, surface melting, etc. The abrupt increases in the $\delta$ graphs are analogs of the Lindemann criterion [83] for bulk melting.

A distinguishing feature of the $\delta$ graphs in Fig. 8 is that, in contrast not only to the bulk case but also to those of clusters of a variety of materials (including earlier descriptions of Ni13 with other potentials [28]), they show instead of one two abrupt changes. The processes giving rise to these changes can be identified by performing simulated thermal quenchings from many configurations generated by MD trajectories at different energies. Quenchings from configurations generated at energies below those corresponding to the first abrupt increase in the $\delta$ graphs all lead to the same, for $Ni_{13}$ and $Al_{13}$, respectively, ico structure; at these energies each cluster explores only the catchment area of its initial ico isomer. Quenchings from configurations generated at energies between those corresponding to the two abrupt increases in the $\delta$ values produce different isomers. All these, however, have the same, for $Ni_{13}$ and $Al_{13}$, respectively, central atom, as identified by a numerical label. At these energies the clusters undergo isomerization transitions that involve only their surface atoms, and the onset of these surface isomerizations is responsible for the first abrupt increase in the $\delta$ values. Configurations generated at energies above those corresponding to the second abrupt change in the $\delta$ graphs produce isomers in which the initial central atom may be replaced by a surface atom. The onset of these global isomerizations is the cause for the second abrupt increase in the $\delta$ values. The graphs of the specific heat display a maximum at an energy ("melting energy") that is somewhat higher than the energy of the onset of global isomerizations. This maximum indicates a transition to a liquidlike state, which is attained when the characteristic time-scale of structural changes becomes comparable to that of the intracluster vibrations [82]; the signature of the first-order melting phase transition in bulk materials is a discontinuity in the specific heat. The $C$ graphs in Fig. 8 do not display a maximum that could be correlated with the stage of surface isomerizations, which indicates that in $Ni_{13}$ and $Al_{13}$ these isomerizations do not evolve into the state of surface melting (cf. below).

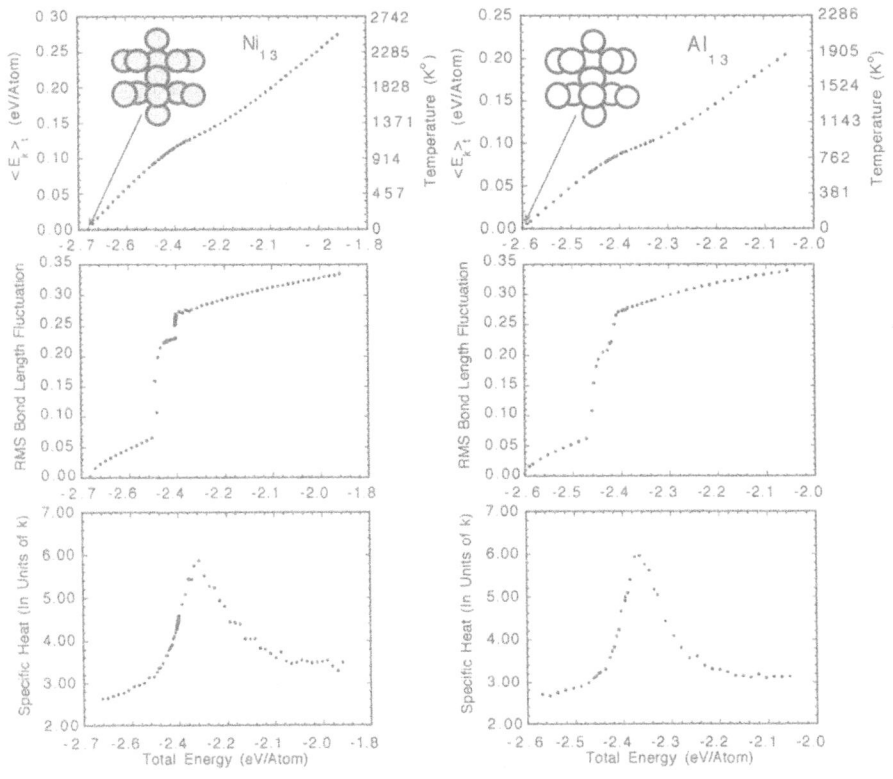

**Fig. 8.** Caloric curve, rms bond length fluctuation, and specific heat per atom (in units of the Boltzmann constant k) as a function of total energy per atom for Ni$_{13}$ and Al$_{13}$. The icosahedron was used as the zero-temperature form of the clusters

The dynamical features of the Ni$_n$Al$_m$ 13-mers with different compositions $(n,m)$ are shown in Figs. 9 and 10. The graphs of Fig. 9 are obtained using the lowest energy structure for each composition as the zero-temperature form of the mixed cluster. Examination of the graphs shows that the alloy clusters also undergo a meltinglike transition as their energy is increased, but the details (mechanisms) depend on the stoichiometric composition. Ni$_{12}$Al melts in a manner similar to that of Ni$_{13}$ and Al$_{13}$. The $\delta$ graph clearly displays the well-separated stages of surface and global structural changes, which may be isomeric and homotopic in character. The barrier for exchanging the central (in this case, aluminum) atom with a surface (in this case, nickel) atom in the ico geometry is very high. Energetically it is easier first to convert the ico geometry into an isomer (or isomers) with an incomplete shell by moving a surface atom into the second layer. Surface structural transitions are the only ones experienced by the cluster over a finite range of extra energy. In the course of these transitions the aluminum atom continues to occupy the central position. As the energy is increased further, the global structural changes get "switched on" as

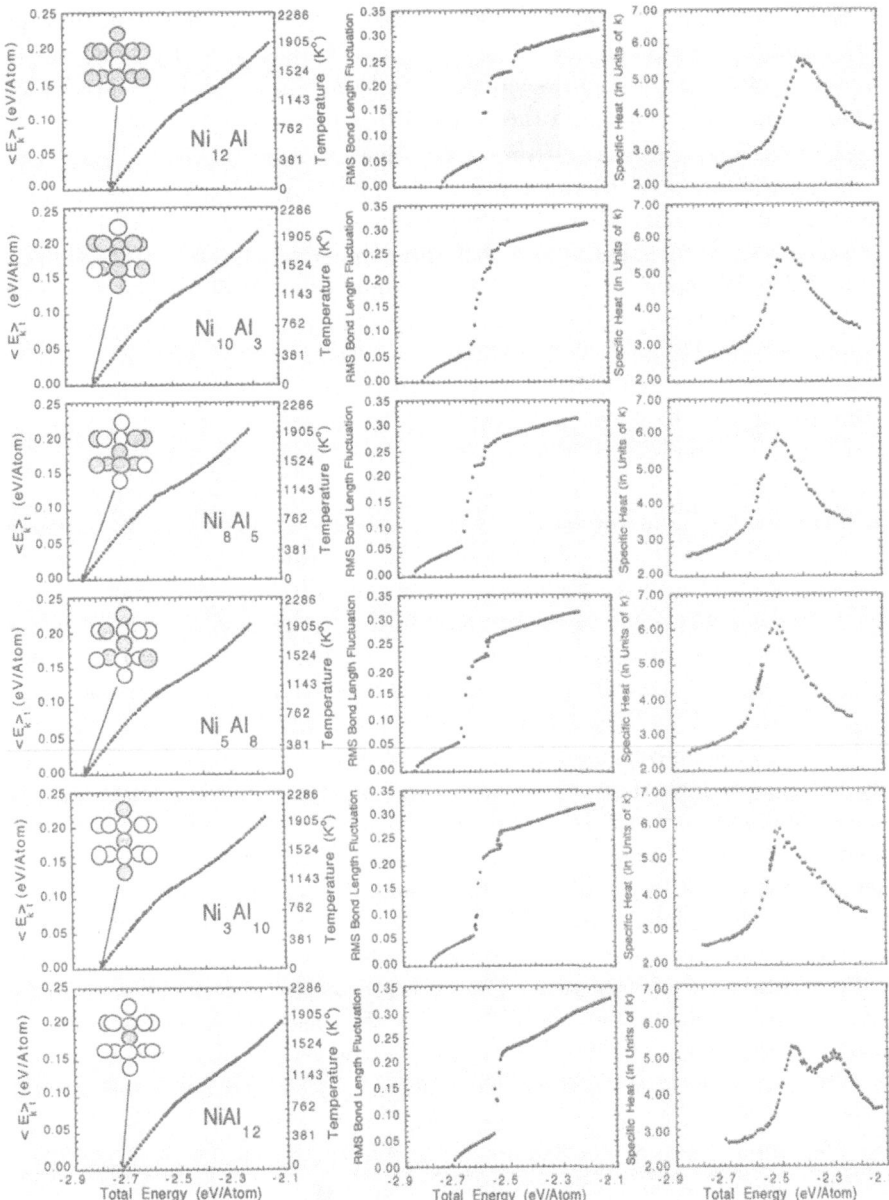

**Fig. 9.** The same as Fig. 8, but for six different compositions of the mixed nickel-aluminum 13-mer. The lowest energy homotop (shown) is used as the zero-temperature structure for each composition

well, and the aluminum atom can escape to the surface. The barrier for such an escape is lower in isomers with an incomplete shell. At still higher energies the cluster samples a larger number of its isomeric and homotopic forms, the rate of transitions between these forms increases, and the cluster gradually attains its liquidlike state (cf. the peak in the graph of the specific heat).

The described stagewise melting is also characteristic of other compositions of the nickel-aluminum 13-mer (cf. Fig. 9). The only differences are that for the other compositions the surface transitions do not involve a nickel atom, - the preferred central atom, - and that the number of the homotops available for sampling increases as the composition approaches 50/50%. The details of the dynamics, however, change in the limit of small number of nickel atoms. The graphs of the rms bond length fluctuation and of the specific heat for $NiAl_{12}$ (and $Ni_2Al_{11}$ - not shown) are different from those for the other compositions. There is only one abrupt change in the $\delta$ graph (a remnant of the second change can be identified as a variation in the slope of the curve at higher energies), whereas the $C$ graph displays two peaks. These features can be correlated with the patterns of the corresponding homotop energy spectra (cf. Fig. 5). As discussed in Sect. 3, within each isomer the single homotop of $NiAl_{12}$ with nickel in the center is separated from those with aluminum in the center by a large energy gap. On the other hand, the homotops with nickel in the center that correspond to the three higher energy isomers have very close energies. They become accessible almost simultaneously and remain the only ones sampled over quite a broad range of extra energy. The effect of the extra energy is to increase the rate of the surface isomerizations that interconvert these homotops. Eventually the rate becomes so high that the cluster is best described as a liquidlike shell of twelve aluminum atoms that encage the single nickel atom. This is the stage of surface melting. The signature of transition to this stage is the first (lower energy) peak in the graph of the specific heat. As the energy of the cluster is increased further, global structural changes become also accessible (the central nickel atom can now escape to the surface), and a new peak begins to develop in the graph of the specific heat. The expected second abrupt increase in the rms bond length fluctuation associated with the onset of global transitions is masked by the already large values of $\delta$ caused by surface melting. Eventually, the cluster attains the state of complete melting. The transition to this state is signified by the second (higher energy) peak in the graph of the specific heat. The same type of arguments also applies to the $Ni_2Al_{11}$ cluster, which exhibits a similar (albeit less distinctly expressed) behavior.

Yet another aspect of the dynamical features - the dependence on the homotop chosen to represent the zero-temperature form of a mixed cluster - is shown in Fig. 10. The figure displays graphs of the caloric curve, rms bond length fluctuation, and specific heat generated from two different homotops for each of the three stoichiometric compositions shown. The two homotops used as "generating structures" are the lowest energy representatives of the two ico classes with nickel and aluminum, respectively, as the central atom. Analysis of the graphs shows that for each composition there is an energy above which the dynamics generated from the two homotops become indistinguishable as judged by the three quantities. Above this energy the clusters "loose memory" of the generating structure. The energy of memory loss depends on the stoichiometric composition, and it defines, or at least is related to, the threshold at which dynamical changes in the type of the central atom become feasible. An interesting, and at first sight tantalizing, observation is that this energy, as defined by the graphs generated from the homotop with the less favored central atom (panels B), is lower than the energy of the onset of global structural changes, as defined by the graphs generated from the homotop with the more favored central atom (panels A); one of the consequences of this is the somewhat less distinct separation between the onsets of surface and global structural changes in the $\delta$ graphs in panels B. The reason for this "paradox" is that all the graphs are obtained from long but finite MD trajectories (alternatively, observations times), and it can be attributed either to the difference in the "volumes" of the basins of attraction of the structures with nickel and aluminum, respectively, as the central atom, or to peculiarities of the topography of the corresponding potential energy surfaces that make the transition over a barrier in a given direction more probable than in the reverse direction, or to both.

The signature of the loss of memory of the generating structure by the dynamics depends on the stoichiometric composition (cf. Fig. 10). This dependence also can be understood and rationalized in terms of the composition-specific energy spectra of the corresponding structural forms. In the case of an ico $NiAl_{12}$ with the preferred nickel atom in the central position, the surface structural transitions become observable at an energy of about -2.56 eV/atom. In the case when an aluminum atom occupies the central position, the same transitions get switched on at an energy which is approximately 0.04 eV/atom higher (cf. the $\delta$ graphs in panels A and B for $NiAl_{12}$ in Fig. 10). As indicated by the corresponding graphs of the caloric curve and the specific heat, an additional extra energy of less than 0.02 eV/atom is sufficient to allow for the global transition that replaces the central aluminum atom by the nickel atom. After this replacement the cluster behaves as if it were originally prepared with the nickel atom in the center: the corresponding graphs in panels A and B are identical for energies larger than -2.5 eV/atom. When MD trajectories at an energy of about -2.5 eV/atom are initialized from a generating structure with aluminum in the center, the dynamics move the nickel atom into the central position relatively early on, and then they never revisit, on the time-scale of the simulations, structures with aluminum in the center. The structures with nickel in the center dominate the dynamics at this energy, and, since their configurational energies are considerably lower than those of their counterparts with aluminum in the center (cf. Fig. 5), the time-averaged kinetic energy of the cluster drastically increases. The change is so abrupt that it appears as a finite discontinuity in the graphs of the caloric curve and the specific heat (panels B).

Similar arguments apply also to the $Ni_7Al_6$ cluster. The dynamics-driven replacement of a less preferred aluminum atom by a nickel atom in the center of the cluster at an energy of about -2.59 eV/atom causes a pronounced, albeit somewhat smaller and less abrupt, increase in the caloric curve (panel B). The signature of this replacement in the graph of the specific heat is a very narrow peak (represented by a single point), rather than a discontinuity, at an energy of about -2.59 eV/atom. The effects are somewhat attenuated, as compared to the case of the $NiAl_{12}$ cluster, because for $Ni_7Al_6$ the ranges of energies of the homotops in classes with nickel and aluminum, respectively, as the central atom are closer to each other (in fact, they overlap; cf. Fig. 3). The large number of homotops with a dense spectrum of configurational energies provides additional pathways for changing the type of the central atom from aluminum to the more preferred nickel, and this also contributes to the attenuation of the effect.

In the case of $Ni_{12}Al$, the identity of the generating homotop has only a minor influence at all energies. The reason for this is that for this cluster the energies of the homotops with nickel in the center are close to that of the corresponding homotop with aluminum in the center (cf. Fig. 2). The most noticeable effect of starting with a homotop with the less preferred nickel atom in the center is blurring of the distinction between the surface and global structural transitions in the $\delta$ graph (panel B). The causes for this blurring are the above mentioned "earlier" switching on of global transitions and the gradual inclusion in sampling of an increasing number of homotops with nickel in the center, which provide alternative pathways for changing the type of the central atom.

The preceding analysis establishes a correlation between the dynamical features of the mixed clusters and the composition-specific energy spectra of their equilibrium structural forms. In general, however, the dynamics depend not only on the minima but also the barriers of the potential energy surfaces. The fact that the discussed features of the dynamics can be rationalized in terms of the minima alone suggests that the distributions of the barriers of the corresponding potential energy surfaces follow the same trends as the distributions of their minima.

Table 2 presents the temperatures [cf. (6)] characterizing the onset of the different stages in the meltinglike transition of the pure and mixed nickel-aluminum 13-mers. The temperature $T^s$ of surface structural changes corresponds to the lowest-energy point in the first abrupt change in the $\delta$ graphs. The temperature $T^g$ of global structural transitions is defined for all clusters, except $NiAl_{12}$ and $Ni_2Al_{11}$, by the lowest-energy point in the second abrupt change in the $\delta$ graphs. For $NiAl_{12}$ and $Ni_2Al_{11}$, $T^g$ is specified by the energy of the minimum between the two maxima in the corresponding graphs of the specific heat. The energies of the first and the second maximum in these graphs define the temperatures $T_m^{(s)}$ of surface melting and $T_m^g$ of global (or complete) melting, respectively. For all the other compositions, $T_m^g$ is specified by the energy of the single maximum in the corresponding graph of the specific heat. Inspection of the table shows that the replacement of aluminum atoms by nickel atoms, even if only one, in $Al_{13}$ has a stabilizing effect: the temperatures of all the individual stages increase. It is particularly interesting that even the temperatures of surface melting in $NiAl_{12}$ and $Ni_2Al_{11}$ are higher than the melting temperature of the pure $Al_{13}$. An increase in the number of nickel atoms beyond one does not introduce new

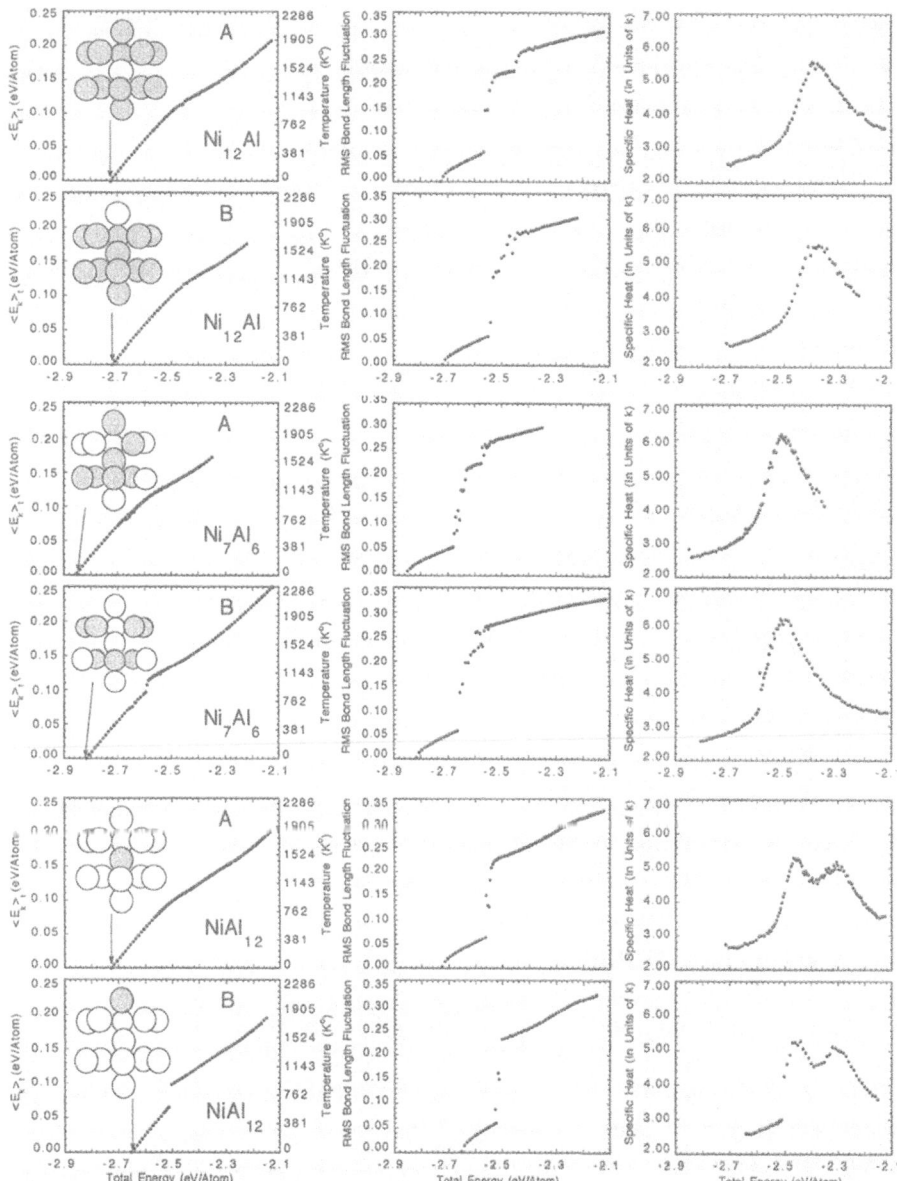

**Fig. 10.** The same as Fig. 9, but for three compositions and two different zero-temperature structures (shown) for each composition. The two structures are the lowest energy homotops of the icosahedral isomer with nickel and aluminum, respectively, as the central atom. Panels A and B correspond to dynamics generated from the lower and the higher energy homotop, respectively

**Table 2.** Temperatures (in K) associated with different stages in the meltinglike transition. $T^s$ and $T^g$ are the temperatures of the onset of surface and global structural changes, respectively. $T_m^{(s)/g}$ are the temperatures of the onset of surface (only in $NiAl_{12}$ and $Ni_2Al_{11}$) and global melting. See text for details

| Cluster | $T^s$ | $T^g$ | $T_m^{(s)/g}$ |
|---|---|---|---|
| $Al_{13}$ | 570 | 715 | 850 |
| $NiAl_{12}$ | 680 | 1180 | (985)/1365 |
| $Ni_2Al_{11}$ | 690 | 1220 | (1070)/1315 |
| $Ni_3Al_{10}$ | 680 | 1000 | 1085 |
| $Ni_4Al_9$ | 705 | 1010 | 1145 |
| $Ni_5Al_8$ | 710 | 1050 | 1180 |
| $Ni_6Al_7$ | 790 | 1135 | 1200 |
| $Ni_7Al_6$ | 735 | 1080 | 1200 |
| $Ni_8Al_5$ | 785 | 1030 | 1220 |
| $Ni_9Al_4$ | 780 | 1070 | 1220 |
| $Ni_{10}Al_3$ | 750 | 1015 | 1200 |
| $Ni_{11}Al_2$ | 775 | 1020 | 1215 |
| $Ni_{12}Al$ | 680 | 1040 | 1200 |
| $Ni_{13}$ | 810 | 1050 | 1190 |

trends, at least not systematic ones. The overall calibration of the values in Table 2 can be established by noting that the melting temperatures of the bulk aluminum and nickel are 933 K and 1726 K, respectively.

# 5 Analytical Model

All the structural and dynamical properties of a system modelled by a semiempirical potential are embedded in the functional form of the potential and the values of its parameters. The peculiar features of two-component systems are largely determined by the relative role of the heterointeractions vs. those of the homointeractions. We show here that for the Gupta-like potential one can formulate an analytical model which not only reproduces and explains a variety of properties of two-component clusters, but also displays in a transparent form the roles played by the different interactions in defining these properties.

The first element of the model is the assumption that one atom (of either type) can be identified as the central (or most coordinated) atom of the cluster. The remaining ("surface") atoms of a fixed type are assumed to have the same effective (average) energy; we refer to this latter in what follows simply as energy. The energy of the two types of surface atoms is, in general, different. We use

$U_\alpha(l|\gamma)$ to denote the energy of an atom of the type $\alpha$ (in our case, $\alpha =$ Ni or Al) at the position $l$ (= c or s, where "c" stands for "center" and "s" for "surface") in a cluster belonging to class $\gamma$ (= Ni or Al), where $\gamma$ specifies the type of the central atom. Then the energy $V(\gamma)$ of a homotop in the classes $\gamma =$ Ni and $\gamma =$ Al can be written as

$$V(\text{Ni}) = U_{\text{Ni}}(\text{c}|\text{Ni}) + (n-1)U_{\text{Ni}}(\text{s}|\text{Ni}) + mU_{\text{Al}}(\text{s}|\text{Ni}) \tag{9}$$

and

$$V(\text{Al}) = U_{\text{Al}}(\text{c}|\text{Al}) + (m-1)U_{\text{Al}}(\text{s}|\text{Al}) + nU_{\text{Ni}}(\text{s}|\text{Al}) \ . \tag{10}$$

The difference $\Delta V(\text{Ni}|\text{Al})$ between the energies of two homotops chosen from the two classes is given by

$$\begin{aligned}
\Delta V(\text{Ni}|\text{Al}) &= V(\text{Ni}) - V(\text{Al}) \\
&= \left\{ \left[ U_{\text{Ni}}(\text{c}|\text{Ni}) - U_{\text{Ni}}(\text{s}|\text{Ni}) \right] - \left[ U_{\text{Al}}(\text{c}|\text{Al}) - U_{\text{Al}}(\text{s}|\text{Al}) \right] \right\} \\
&\quad + \left[ U_{\text{Al}}(\text{s}|\text{Ni}) - U_{\text{Al}}(\text{s}|\text{Al}) \right]m + \left[ U_{\text{Ni}}(\text{s}|\text{Ni}) - U_{\text{Ni}}(\text{s}|\text{Al}) \right]n \ . \tag{11}
\end{aligned}$$

Equation (11) shows what and to which extent defines the energy mismatch $\Delta V(\text{Ni}|\text{Al})$. When applied to the lowest energy homotops in the two classes of a chosen isomer, (11) allows one to predict the preferred type of the central atom in that isomer for any composition. The preferred atom is Ni when $\Delta V(\text{Ni}|\text{Al}) < 0$, and it is Al when $\Delta V(\text{Ni}|\text{Al}) > 0$. The magnitude of $\Delta V(\text{Ni}|\text{Al})$ is in this case a measure of the relative stability of the two classes of homotops.

The next element of the model is the prescription for evaluation of the energies $U_\alpha(l|\gamma)$. In general, this prescription should be isomer-specific. Our goal here is to illustrate the approach. Therefore, we formulate it for the simplest case of the ico structure. The energies $U_\alpha(\text{c}|\alpha)$ and $U_\alpha(\text{s}|\gamma)$ of an atom in a mixed ico cluster can be expressed as

$$U_\alpha(\text{c}|\alpha) = R_{\alpha,\text{c}}^{\alpha,\text{s}}(\alpha)(N_\alpha - 1) + R_{\alpha,\text{c}}^{\beta,\text{s}}(\alpha)N_\beta - \sqrt{A_{\alpha,\text{c}}^{\alpha,\text{s}}(\alpha)(N_\alpha - 1) + A_{\alpha,\text{c}}^{\beta,\text{s}}(\alpha)N_\beta} \tag{12}$$

and

$$U_\alpha(\text{s}|\gamma) =$$

$$R_{\alpha,\text{s}}^{\gamma,\text{c}}(\gamma|0) + \sum_{f=1}^{3} \left\{ R_{\alpha,\text{s}}^{\alpha,\text{s}}(\gamma|f)L^{(f)} + \Delta R_\text{s}^{\alpha,\beta}(\gamma|f)\overline{M}_\alpha^{(f)}(\gamma|N_\alpha, N_\beta) \right\}$$

$$- \sqrt{A_{\alpha,\text{s}}^{\gamma,\text{c}}(\gamma|0) + \sum_{f=1}^{3} \left\{ A_{\alpha,\text{s}}^{\alpha,\text{s}}(\gamma|f)L^{(f)} + \Delta A_\text{s}^{\alpha,\beta}(\gamma|f)\overline{M}_\alpha^{(f)}(\gamma|N_\alpha, N_\beta) \right\}} \ , \tag{13}$$

where

$$\Delta R_\text{s}^{\alpha,\beta}(\gamma|f) = R_{\alpha,\text{s}}^{\beta,\text{s}}(\gamma|f) - R_{\alpha,\text{s}}^{\alpha,\text{s}}(\gamma|f) \ , \tag{14}$$

$$\Delta A_\text{s}^{\alpha,\beta}(\gamma|f) = A_{\alpha,\text{s}}^{\beta,\text{s}}(\gamma|f) - A_{\alpha,\text{s}}^{\alpha,\text{s}}(\gamma|f) \ ; \tag{15}$$

$N_\alpha$ and $N_\beta$ are the total number of atoms of the type $\alpha$ and $\beta$, respectively; $\gamma = \alpha, \beta$; $\beta \neq \alpha$; $f = 0$ stands for a pair of atoms one of which is in the central position; $f = 1, 2, 3$ designate pairs of surface atoms which are first, second, and third neighbors of each other; $L^{(f)}$, $f = 1, 2, 3$, is the total number of the $f$th surface neighbors of each surface atom (we took into account that all surface atoms of an ico isomer have the same number of first, second and third surface neighbors, respectively: $L^{(1)} = 5$, $L^{(2)} = 5$, and $L^{(3)} = 1$); and $\overline{M}_\alpha^{(f)}(\gamma|N_\alpha, N_\beta)$ is the average number of surface atoms of the type $\beta$ that are the $f$th neighbors of a surface atom of the type $\alpha$ in a cluster with a central atom of the type $\gamma$:

$$\overline{M}_\alpha^{(f)}(\gamma|N_\alpha, N_\beta) = \frac{M^{(f)}(\gamma|N_\alpha, N_\beta)}{N_\alpha - \delta_{\alpha\gamma}} \ , \tag{16}$$

where $M^{(f)}(\gamma|N_\alpha, N_\beta)$ is the total number of pairs of unlike surface atoms that are the $f$th neighbors of each other, and $\delta_{\alpha\gamma}$ is the Kronecker symbol. In (12)-(15), the $R_{\alpha,l}^{\sigma,k}(\gamma|f)$ and $A_{\alpha,l}^{\sigma,k}(\gamma|f)$ terms are, respectively, the repulsive and the (square of the) attractive exponents in the Gupta-like potential for pairs of atoms, one of the type $\alpha$ and the other of the type $\sigma$ ($= \alpha, \beta$), which occupy positions $l$ ($= $ c, s) and $k$ ($= $ c, s), respectively, and are separated by an average distance $\bar{r}_{\alpha,l}^{\sigma,k}(\gamma|f)$ in a cluster with the central atom of the type $\gamma$:

$$R_{\alpha,l}^{\sigma,k}(\gamma|f) = A_{\alpha\sigma} \exp\left(- p_{\alpha\sigma}\left(\frac{\bar{r}_{\alpha,l}^{\sigma,k}(\gamma|f)}{r_{\alpha\sigma}^0} - 1\right)\right) \ , \tag{17}$$

$$A_{\alpha,l}^{\sigma,k}(\gamma|f) = \xi_{\alpha\sigma}^2 \exp\left(- 2q_{\alpha\sigma}\left(\frac{\bar{r}_{\alpha,l}^{\sigma,k}(\gamma|f)}{r_{\alpha\sigma}^0} - 1\right)\right) \ , \tag{18}$$

where the average distance $\bar{r}$ depends, as indicated, on $\alpha, l, \sigma, k, \gamma$ and $f$, as well as the composition of the cluster. The numbers $M^{(f)}(\gamma|N_\alpha, N_\beta)$ are homotop-specific, i.e., they depend on the particular distribution of the two types of atoms between the sites of the chosen (in our case, ico) isomer.

The final element of the model is the prescription for assigning the characteristic distances $\bar{r}_{\alpha,l}^{\sigma,k}(\gamma|f)$ in ico nickel-aluminum 13-mers of different compositions. As discussed in Sect. 2, the equilibrium structures of the mixed ico 13-mers are, in general, slightly different from an ideal icosahedron. It turns out, however, that when one averages the first-, second-, and third-neighbor distances between all pairs of atoms on the surface of the cluster, as well as those between the central atom and all the surface atoms, one arrives at values which relate in a manner very close to that characteristic of an ideal icosahedron. The values of the average distances depend on the composition, the type of the central atom, and the homotop at hand. In the ico case, this latter dependence is very weak, and all homotops belonging to the same class are characterized by essentially the same average interatomic distances. Therefore, we represent the true equilibrium structures of the mixed 13-mers through ideal icosahedra with interatomic distances that depend on the composition and the type of the central atom. This representation further simplifies the model by eliminating the $\alpha$- and

$\beta$-dependence from the characteristic interatomic distances $\bar{r}_l^k(\gamma, f, m)$, where the number $m$ of the aluminum atoms is used to indicate the composition of the cluster. The characteristic distances for different compositions are obtained through linear interpolation between the average distances $\bar{r}_l^k$ computed for the lowest energy homotops of the $Ni_{12}Al$ and $NiAl_{12}$ clusters:

$$\bar{r}_l^k(\gamma|f|m) = \frac{(12 - m)\bar{r}_l^k(\gamma|f|m = 1) + (m - 1)\bar{r}_l^k(\gamma|f|m = 12)}{11} \quad , \quad m=1,...,12.$$

(19)

We also used even a more approximate approach, in which the $m$-dependence of the characteristic distances is obtained through linear interpolation between the distances $\bar{r}_l^k$ in the pure $Ni_{13}$ and $Al_{13}$ clusters:

$$\bar{r}_l^k(f|m) = \frac{(13 - m)r_l^k(f|m = 0) + m r_l^k(f|m = 13)}{13} \quad , \quad m=1,...,13.$$

(20)

This latter approach eliminates another dependence of the characteristic distances, specifically that on $\gamma$.

If needed, the model can be refined. For example, in obtaining the $m$-dependence of the characteristic distances one can use polynomial, instead of linear, interpolations. This, however, requires computing and incorporating in the interpolation procedure average distances for additional compositions. The separation of the atoms into a "central" and "surface" atoms can be generalized to structures more complex than the icosahedron: the atoms will have to be grouped into "classes of coordination similarity". The primary purpose of the model, however, is to aid in understanding the results of the exact computations, rather than in replacing them.

To illustrate the quality of the model, we compare some of its predictions with the results of exact computations. Figure 11 displays the graphs of $\Delta V(Ni|Al)$ as a function of the number of aluminum atoms in the ico 13-mer. In the exact computations, $\Delta V$ is obtained from the energies of the most stable homotops in the classes with nickel and aluminum, respectively, as the central atom. The values of $M^{(f)}$ used in the model calculations are those corresponding to these homotops. The graphs display the following features: 1) With the exception of the case when only first-neighbor interactions are taken into account in the model, the graphs of $\Delta V(Ni|Al)$ cross the zero value between $m = 1$ and $m = 2$. This is in accord with the fact mentioned earlier that, apart from the case of $Ni_{12}Al$, all 13-atom $Ni_nAl_m$ clusters prefer nickel as the central atom; 2) For $m \geq 2$, the magnitude of $\Delta V(Ni|Al)$ increases monotonically, which means that the relative stability of the class with nickel in the center, vs. that of the class with aluminum in the center, increases with the number of aluminum atoms in the cluster. This trend is reproduced by the model even when one considers only first-neighbor interactions; 3) Comparison of the graphs in panels A and B indicates that, as expected, the interpolation defined by (19) leads to a better reproduction of the exact results than that defined by (20).

It has been found [44] that in the case of 13-atom $Cu_nAu_m$ clusters mimicked by the Gupta-like potential gold is the preferred central atom for $m \leq 4$, and

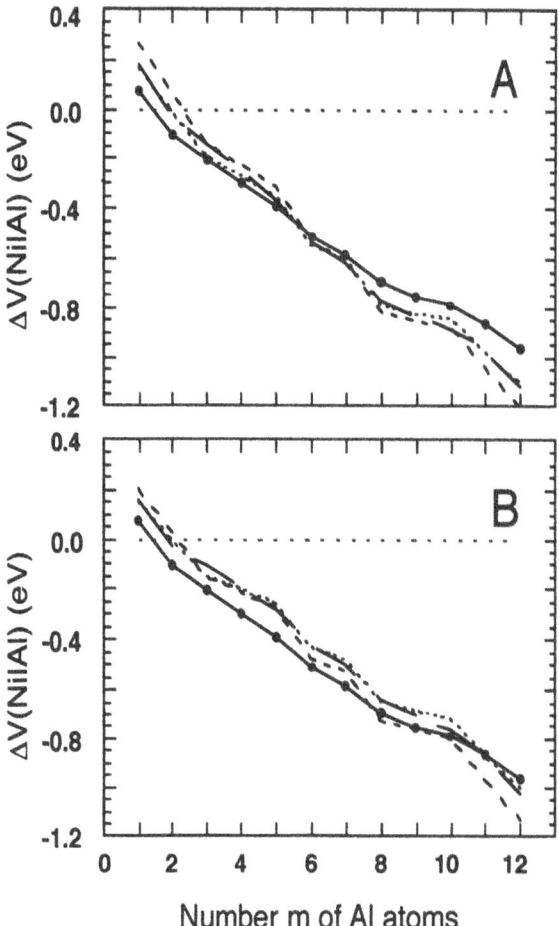

**Fig. 11.** The energy difference $\Delta V(\mathrm{Ni}|\mathrm{Al})$ between the icosahedral classes with nickel and aluminum, respectively, as the central atom plotted as a function of the composition. The solid line with the full circles represents exact computations. The other graphs correspond to the model with 1) only first-neighbor interactions (short-dashed line), 2) first- and second-neighbor interactions (long-dashed line), 3) first-, second-, and third-neighbor interactions (dotted line). Panels A and B correspond to interpolations defined by (19) and (20), respectively

copper for $m \geq 5$. We have checked that the described model, when used with the values of parameters appropriate for mixed copper-gold systems [75], reproduces this result as well. By eliminating the details of the local relaxations the model simplifies the computation of structural and energy properties of mixed clusters. Its most important merit, however, is that it makes transparent the comparative role of the interactions between like atoms vs. those between unlike atoms in defining these properties. Further applications of the model will be described elsewhere.

# 6 Summary

A review of our recent studies of two-component alloy clusters has been presented. As a paradigm we considered 13-atom nickel-aluminum systems modelled by the many-body Gupta-like potential. We have shown that a hierarchical systematics of the large number of possible structural forms of two-component clusters can be introduced by grouping them into classes specified by the stoichiometric composition, the isomeric (geometric) form, and the type of the central atom. We introduced new general definitions of the mixing energy and mixing coefficient, which are applicable to systems described by either pairwise-additive or many-body potentials, and have demonstrated that the overall energy ordering of the homotopic structures within each class is defined by the mixing coefficient.

The dynamical features were studied over a broad energy range using constant-energy molecular dynamics simulations. These revealed that the mixed clusters undergo a solid-to-liquid-like transition as their energy is increased. In many respects this transition is similar to that experienced by one-component clusters; it also takes place in stages. The nature of the stages is largely defined by the cluster composition. The stages exhibited by the nickel-aluminum 13-mers include surface structural transitions (isomeric and homotopic), global structural transitions, surface melting, and complete melting. The composition-dependent peculiarities of the dynamical behavior were correlated with and rationalized in terms of the composition-specific energy spectra of the corresponding structural forms. The analysis includes a discussion of the effect on the dynamics of the structure chosen to represent the zero-temperature form of the cluster.

Finally, we presented an analytical model that allows one to estimate a variety of structural and energy characteristics of two-component clusters directly from the parameters of the Gupta-like potential. Even more importantly, the model gives a clear picture of the roles played by the interactions between like atoms and those between unlike atoms in defining these characteristics.

### Acknowledgments

This work was performed under the auspices of the Office of Basic Energy Sciences, Division of Chemical Science, US-DOE, under Contract Number W-31-109-ENG-38. EBK also was supported by the NIS-IPP Program. We thank Drs. U. Salian and A. Goldberg for help with LaTeX.

## References

1. K. Rademann, B. Kaiser, U. Even, and F. Hensel, Phys. Rev. Lett. **59**, 2319 (1997);
   C. Brechignac, M. Broyer, P. Cahuzac, G. Delacretaz, P. Labastie, J. P. Wolf, and
   L. Wöste, ibid. **60**, 275 (1988); M. E. Garcia, G. M. Pastor, and K. Bennemann,
   ibid. **67**, 1142 (1991); K. Rademann, O. Dimopoulou-Rademann, M. Schlauf, U.
   Even, and F. Hensel, ibid. **69**, 3208 (1992); H. Haberland, B. von Issendorf, Y.

Yufeng, and T. Kolar, ibid. **69**, 3212 (1992); G. M. Pastor and K. Bennemann, in *Clusters of Atoms and Molecules*, Vol. 1, H. Haberland (Ed.), Springer-Verlag, Heidelberg, 1994; R. Busani, M. Folkers, and O. Cheshnovsky, Phys. Rev. Lett. **81**, 3836 (1998)

2. D. A. McQuarrie, *Statistical Mechanics*, Harper & Row, New York, 1976

3. *Physics and Chemistry of Small Clusters*, P. Jena, B. K. Rao, and S. N. Khanna (Eds.), Plenum Press, New York, 1987

4. *Elemental and Molecular Clusters*, G. Benedek, T. P. Martin, and G. Pacchioni (Eds.), Springer-Verlag, Berlin, 1988

5. *Physics and Chemistry of Finite Systems: From Clusters to Crystals*, Vols. 1 and 2, P. Jena, S. N. Khanna, and B. K. Rao (Eds.), Kluwer Academic Publishers, Dordrecht, 1992

6. *Clusters of Atoms and Molecules*, Vols. 1 and 2, H. Haberland (Ed.), Springer-Verlag, Heidelberg, 1994

7. U. Röthlisberger and W. Andreoni, J. Chem. Phys. **94**, 8129 (1991); U. Röthlisberger, W. Andreoni, and P. Giannozzi, ibid. **96**, 1248 (1992)

8. V. Bonačić-Koutecký, P. Fantucci, and J. Koutecký, Chem. Rev. **91**, 1035 (1991)

9. J. Jellinek, V. Bonačić-Koutecký, P. Fantucci, and M. Wiechert, J. Chem. Phys. **101**, 10092 (1994); V. Bonačić-Koutecký, J. Jellinek, M. Wiechert, and P. Fantucci, ibid. **107**, 6321 (1997); D. Reichardt, V. Bonačić-Koutecký, P. Fantucci, and J. Jellinek, Chem. Phys. Lett. **279**, 129 (1997)

10. R. Kawai, J. F. Tombrello, and J. H. Weare, Phys. Rev. A **49**, 4236 (1994); M.-W. Sung, R. Kawai, and J. H. Weare, Phys. Rev. Lett. **73**, 3552 (1994)

11. I. G. Kaplan, R. Santamaría, and O. Novaro, Int. J. Quantum Chem. **55**, 237 (1995); I. G. Kaplan, J. Hernandez-Cobos, O. Ortega-Blake, and O. Novaro, Phys. Rev. A **53**, 2493 (1996)

12. P. Calamanici, A. M. Köster, N. Russo, and D. R. Salahub, J. Chem. Phys. **105**, 9548 (1996); M. Castro, C. Jamorski, and D. R. Salahub, Chem. Phys. Lett. **271**, 133 (1997); A. Martinez, A. Vela, and D. R. Salahub, Int. J. Quantum. Chem. **63**, 301 (1997); C. Jamorski, A. Martinez, M. Castro, D. R. Salahub, Phys. Rev. B **55**, 10905 (1997)

13. D. A. Gibson and E. A. Carter, Chem. Phys. Lett. **271**, 266 (1997)

14. R. O. Jones, A. I. Lichtenstein, and J. Hutter, J. Chem. Phys. **106**, 4566 (1997)

15. J. Jellinek, S. Srinivas, and P. Fantucci, Chem. Phys. Lett. **288**, 705 (1998)

16. S. K. Nayak and P. Jena, Chem. Phys. Lett. **289**, 473 (1998); S. K. Nayak, S. N. Khanna, and P. Jena, Phys. Rev. B **57**, 3787 (1998)

17. A. Rytkonen, H. Hakkinen, and M. Manninen, Phys. Rev. Lett. **80**, 3940 (1998); J. Akola, H. Hakkinen, and M. Manninen, Phys. Rev. B **58**, 3601 (1998)

18. M. Hartmann, J. Pittner, V. Bonačić-Koutecký, A. Heidenreich, and J. Jortner, J. Chem. Phys. **108**, 3096 (1998); M. Hartmann, A. Heidenreich, J. Pittner, V. Bonačić-Koutecký, and J. Jortner, J. Phys. Chem. **102**, 4069 (1998)

19. M. Brack, Rev. Mod. Phys. **65**, 677 (1993)

20. A. Bulgac and C. Lewenkopf, Europhys. Lett. **31**, 519 (1995); J. M. Thompson and A. Bulgac, Z. Phys. D **40**, 462 (1997)

21. W. Ekardt, Z. Phys. B **103**, 305 (1997); J. M. Pacheco and W. Ekardt, ibid. **103**, 327 (1997)

22. C. Yannouleas and U. Landman, Phys. Rev. B **51**, 1902 (1995); J. Chem. Phys. **105**, 8734 (1996); J. Phys. Chem. A **102**, 2505 (1998)

23. S. M. Reimann, M. Koskinen, H. Hakkinen, P. E. Lindelof, and M. Manninen, Phys. Rev. B **56**, 12147 (1997); J. Kolehmainen, H. Hakkinen, and M. Manninen, Z. Phys. D **40**, 306 (1997)

24. R. Poteau, J.-L. Heully, and F. Spiegelman, Z. Phys. D **40**, 479 (1997)

25. A. N. Andriotis and M. Menon, Phys. Rev. B **57**, 10069 (1998); A. N. Antonis, M. Menon, G. E. Froudakis, Z. Fthenakis, and J. E. Lowther, Chem. Phys. Lett. **292**, 487 (1998)

26. K. Raghavan, M. S. Stave, and A. E. DePristo, J. Chem. Phys. **91**, 1904 (1989); M. S. Stave and A. E. DePristo, ibid. **97**, 3386 (1992); T. L. Wetzel and A. E. DePristo, ibid. **105**, 572 (1996)

27. H. Diep, S. Sawada, and S. Sugano, Phys. Rev. B **39**, 9252 (1989); S. Sawada and S. Sugano, Z. Phys. D **20**, 259 (1991); ibid. **24**, 37 (1992)

28. J. Jellinek and I. L. Garzón, Z. Phys. D **20**, 239 (1991); I. L. Garzón and J. Jellinek, ibid. **20**, 235 (1991); ibid. **26**, 235 (1993); in Ref. 5, Vol. 1, p. 405; Z. B. Güvenç, and J. Jellinek, Z. Phys. D **26**, 304 (1993); Z. B. Güvenç, J. Jellinek, and A. F. Voter, in Ref. 5, Vol. 1, p. 411

29. J. Jellinek and Z. B. Güvenç, Z. Phys. D **19**, 371 (1991); in *Mode Selective Chemistry*, J. Jortner, R. D. Levine, and B. Pullmann (Eds.), Kluwer Academic Publishers, Dordrecht, 1991, p. 153; in Ref. 5, Vol. 2, p.1047; in *Nuclear Physics Concepts in the Study of Atomic Cluster Physics*, R. Schmidt, H. O. Lutz, and R. Dreizler (Eds.), Springer-Verlag, Heidelberg, 1992, p. 169; Z. Phys. D **26**, 110 (1993); in *Topics in Atomic and Nuclear Collisions*, B. Remaud, A. Calboreanu, and V. Zoran (Eds.), Plenum Press, New York, 1994, p. 243; in *The Synergy Between Dynamics and Reactivity at Clusters and Surfaces*, L. J. Farrugia (Ed.), Kluwer Academic Publishers, Dordrecht, 1995, p. 217

30. F. Ercolessi, W. Andreoni, and E. Tossati, Phys. Rev. Lett. **66**, 911 (1991)

31. H.-P. Cheng and R. S. Berry, Phys. Rev. A **45**, 7969 (1992)

32. J. Uppenbrink and D. J. Wales, J. Chem. Phys. **98**, 5720 (1993), D. J. Wales and L. J. Munro, J. Phys. Chem. **100**, 2053 (1996)

33. M. J. López and J. Jellinek, Phys. Rev. A **50**, 1445 (1994)

34. J. García-Rodeja, C. Rey, L. J. Gallego, and J. A. Alonso, Phys. Rev. B **49**, 8495 (1994)

35. J. Jellinek, in *Metal-Ligand Interactions: Structure and Reactivity*, N. Russo and D. R. Salahub (Eds.), Kluwer Academic Publishers, Dordrecht, 1996, p. 325

36. A. Posada-Amarillas and I. L. Garzón, Phys. Rev. B **54**, 10362 (1996); I. L. Garzón and A. Posada-Amarillas, ibid. **54**, 11796 (1996); I. L. Garzón, I. G. Kaplan, R. Santamaria, and O. Novaro, J. Chem. Phys. **109**, 2176 (1998); I. L. Garzón, K. Michaelian, M. R. Beltran, A. Posada-Amarillas, P. Ordejón, E. Artacho, D. Sánchez-Postal, and J. M. Soler, Phys. Rev. Lett. **81**, 1600 (1998)

37. S. K. Nayak, B. Reddy, B. K. Rao, S. N. Khanna, and P. Jena, Chem. Phys. Lett. **253**, 390 (1996); S. K. Nayak, S. N. Khanna, B. K. Rao, and P. Jena, J. Phys. Chem. A **101**, 1072 (1997); S. K. Nayak, P. Jena, K. D. Ball, and R. S. Berry, J. Chem. Phys. **108**, 234 (1998)

38. B. Chen, M. A. Gomez, J. D. Doll, and D. L. Freeman, J. Chem. Phys. **105**, 9686 (1996); E. Curotto, A. Matro, D. L. Freeman, and J. D. Doll, ibid. **108**, 729 (1998); B. Chen, M. A. Gomez, J. D. Doll, and D. L. Freeman, ibid. **108**, 4031 (1998)

39. C. L. Cleveland, U. Landman, T. G. Schaaff, M. N. Shafigullin, P. W. Stephens, and R. L. Whetten, Phys. Rev. Lett. **79**, 1873 (1997); C. L. Cleveland, W. D. Luedtke,

and U. Landman, ibid. **81**, 2036 (1998); W. D. Luedtke and U. Landman, J. Phys. Chem. B **102**, 6566 (1998)

40. H. Grönbeck, D. Tománek, S. G. Kim, and A. Rosén, Chem Phys. Lett. **264**, 39 (1997)

41. D. H. E. Gross, M. E. Madjet, and O. Shapiro, Z. Phys. D **39**, 75 (1997); M. E. Madjet, P. A. Hervieux, D. H. E. Gross, and O. Shapiro, ibid. **39**, 309 (1997); O. Shapiro, P. J. Kuntz, K. Móhring, P. A. Hervieux, D. H. E. Gross, and M. E. Madjet, ibid. **41**, 219 (1997); D. H. E. Gross and M. E. Madjet, Z. Phys. B **104**, 541 (1997)

42. J. E. Hearn and R. L. Johnston, J. Chem. Phys. **107**, 4674 (1997); L. D. Lloyd and R. L. Johnston, Chem. Phys. **236**, 107 (1998)

43. S. B. Zhang, M. L. Cohen, and M. Y. Chou, Phys. Rev. B **36**, 3455 (1987); U. Röthlisberger and W. Andreoni, Chem. Phys. Lett. **198**, 478 (1992); C. Yannouleas, P. Jena, and S. N. Khanna, Phys. Rev. B **46**, 9751 (1992); V. Decoulon, F. A. Reuse, and S. N. Khanna, Phys. Rev. B **48**, 814 (1993); H.-P. Cheng, R. N. Barnett, and U. Landman, Phys. Rev. B **48**, 1820 (1993); V. Bonačić-Koutecký, P. Fantucci, C. Fuchs, J. Koutecký, and J. Pittner, Z. Phys. D **26**, 17 (1993); J. A. Alonso, Phys. Scripta **55**, 177 (1994); A. Bol, J. A. Alonso, and J. M. López, Int. J. Quantum Chem. **56**, 839 (1995)

44. M. J. López, P. A. Marcos, and J. A. Alonso, J. Chem. Phys. **104**, 1056 (1996)

45. J. Jellinek and E. B. Krissinel, Chem. Phys. Lett. **258**, 283 (1996); E. B. Krissinel and J. Jellinek, Int. J. Quantum Chem. **62**, 185 (1997); Chem. Phys. Lett. **272**, 301 (1997); J. Jellinek and E. B. Krissinel, in *Nanostructured Materials: Clusters, Composites, and Thin Films*, V. M. Shalaev and M. Moskovits (Eds.), American Chemical Society, Washington, 1997, p. 239; in *Novel Materials: Design and Properties*, B. K. Rao and S. N. Behera (Eds.), Nova Science Publishers, New York, 1998, p. 83

46. C. Rey, J. Garcia-Rodeja, and L. J. Gallego, Phys. Rev. B **54**, 2942 (1996)

47. C. L. Pettiette, S. H. Yang, M. J. Craycraft, J. Conceicao, R. T. Laaksonen, O. Cheshnovsky, and R. E. Smalley, J. Chem. Phys. **88**, 5377 (1988); K. J. Taylor, C. L. Pettiette-Hall, O. Cheshnovsky, and R. E. Smalley, ibid. **96**, 33129 (1992); J. Conceicao, R. T. Laaksonen, L.-S. Wang, T. Guo, P. Nordlander, and R. E. Smalley, Phys. Rev. B **51**, 4668 (1995)

48. M. F. Jarrold, U. Ray, J. E. Bower, and K. M. Creegan, J. Chem. Soc. Faraday Trans. **86**, 2537 (1990)

49. T. P. Martin, T. Bergmann, H. Göhlich, and T. Lange, J. Chem. Phys. **95**, 6421 (1991); S. Frank, N. Malinowski, F. Tast, M. Heinebrodt, I. M. L. Billas, and T. P. Martin, J. Chem. Phys. **106**, 6217 (1997); M. Heinebrodt, S. Frank, N. Malinowski, F. Tast, I. M. L. Billas, and T. P. Martin, Z. Phys. D **40**, 334 (1997)

50. G. Alameddin, J. Hunter, D. Cameron, and M. M. Kappes, Chem. Phys. Lett. **192**, 122 (1992); S. Haupt, J. Kaller, D. Schooß, D. Cameron, and M. M. Kappes, Z. Phys. D **40**, 331 (1997); O. Hampe, P. Gerhardt, S. Gilb, and M. M. Kappes, J. Chem. Phys. **109**, 3485 (1998)

51. D. C. Parent and S. L. Anderson, Chem. Rev. **92**, 1541 (1992)

52. L. Lian, C.-X. Su, and P. B. Armentrout, J. Chem. Phys. **96**, 7542 (1992); J. B. Griffin and P. B. Armentrout, ibid. **108**, 8062 (1998); J. Xu, M. T. Rodgers, J. B. Griffin, and P. Armentrout, ibid. **108**, 9339 (1998)

53. W. A. de Heer, Rev. Mod. Phys. **65**, 611 (1993); A. Hirt, D. Gerion, I. M. L. Billas, A. Châtelain, and W. A. de Heer, Z. Phys. D **40**, 160 (1997)

54. W. J. C. Menezes and M. B. Knickelbein, J. Chem. Phys. **98**, 1856 (1993); G. M. Koretsky and M. B. Knickelbein, ibid. **106**, 9810 (1997); M. B. Knickelbein and G. M. Koretsky, J. Phys. Chem. **102**, 580 (1998)

55. J. P. Bucher and L. A. Bloomfield, Int. J. Mod. Phys. **7**, 1079 (1993); S. E. Apsel, J. W. Emmert, J. Deng, and L. A. Bloomfield, Phys. Rev. Lett. **76**, 1441 (1996)

56. B. A. Collings, K. Athanassenas, D. Lacombe, D. M. Rayner, and P. A. Hackett, J. Chem. Phys. **101**, 3506 (1994); A. Bérces, P. A. Hackett, L. Lian, S. A. Mitchell, and D. M. Rayner, ibid. **108**, 5476 (1998); L. A. Brown and D. M. Rayner, ibid. **109**, 2474 (1998); D. B. Pedersen, J. M. Parnis, and D. M. Rayner, ibid. **109**, 551 (1998)

57. L.-S. Wang, H.-S. Cheng, and J. Fan, J. Chem. Phys. **102**, 9480 (1995); H. Wu, S. R. Desai, and L.-S. Wang, Phys. Rev. Lett. **77**, 2436 (1996); L.-S. Wang and H. Wu, Z. Phys. Chem. Bd. **203**, 45 (1998)

58. E. K. Parks, G. C. Nieman, and S. J. Riley, Surf. Sci. **355**, 127 (1996); E. K. Parks, G. C. Nieman, K. P. Kerns, and S. J. Riley, J. Chem. Phys. **107**, 1861 (1997); E. K. Parks, G. C. Nieman, K. P. Kerns, and S. J. Riley, ibid. **108**, 3731 (1998)

59. A. Ruff, S. Rutz, E. Schreiber, and L. Wöste, Z. Phys. D **37**, 175 (1996); S. Rutz, H. Ruppe, E. Schreiber, and L. Wöste, ibid. **40**, 25 (1997); L. Wöste, Z. Phys. Chem. Bd. **196**, 1 (1996); Š. Vajda, S. Wolf, T. Leisner, U. Busolt, L. H. Wöste, and D. J. Wales, J. Chem. Phys. **107**, 3492 (1997)

60. M. Andersson, J. Persson, and A. Rozén, J. Phys. Chem. **100**, 12222 (1996); L. Holmgren, M. Andersson, and A. Rozén, J. Chem. Phys. **109**, 3232 (1998)

61. J. Tiggesbäumker, L. Köller, and K.-H. Meiwes-Broer, Chem. Phys. Lett. **260**, 428 (1996); Ch. Lüder and K.-H. Meiwes-Broer, ibid. **294**, 391 (1998)

62. S. Fedrigo, T. L. Haslett, and M. Moskovits, Z. Phys. D **40**, 99 (1997); T. L. Haslett, K. A. Bosnick, and M. Moskovits, J. Chem. Phys. **108**, 3453 (1998)

63. B. Lang, A. Vierheilig, E. Weidenmann, H. Buchenau, and G. Gerber, Z. Phys. D **40**, 1 (1997)

64. M. M. Alvarez, J. T. Khoury, T. G. Schaaff, M. N. Shafigullin, I. Vezmar, and R. L. Whetten, Chem. Phys. Lett. **266**, 91 (1997); T. G. Schaaff, M. N. Shafigullin, J. T. Khoury, I. Vezmar, R. L. Whetten, W. G. Cullen, P. N. First, C. Gutiérrez-Wing, J. Ascensio, and M. J. Jose-Yacamán, J. Phys. Chem. B **101**, 7885 (1997)

65. C. Ellert, M. Schmidt, T. Reiners, and H. Haberland, Z. Phys. D **39**, 317 (1997); M. Schmidt, R. Kusche, W. Kronmüller, B. von Issendorff, and H. Haberland, Phys. Rev. Lett. **79**, 99 (1997); M. Schmidt, R. Kusche, B. von Issendorff, and H. Haberland, Nature **393**, 238 (1998)

66. R. W. Farley, P. Ziemann, and A. W. Castleman Jr., Z. Phys. D. **14**, 353 (1989); R. W. Farley and A. W. Castleman Jr., J. Am. Chem. Soc. **111**, 2734 (1989); R. W. Farley and A. W. Castleman Jr., J. Chem. Phys. **92**; 1790 (1990); Y. Yamada, H. T. Deng, E. M. Snyder, and A. W. Castleman Jr., Chem. Phys. Lett. **203**, 330 (1993); S. F. Cartier, B. D. May, and A. W. Castleman Jr., J. Chem. Phys. **104**, 3423 (1996); B. D. May, S. E. Kooi, B. J. Toleno, and A. W. Castleman Jr., ibid. **106**, 2231 (1997)

67. T. G. Taylor, K. F. Willey, M. B. Bishop, and M. A. Duncan, J. Phys. Chem. **94**, 8016 (1990); K. F. Willey, K. LaiHing, T. G. Taylor, and M. A. Duncan, ibid. **97**, 7435 (1993)

68. S. Nonose, Y. Sone, and K. Kaya, Z. Phys. D **19**, 357 (1991); K. Hoshino, T. Naganuma, Y. Yamada, K. Watanabe, A. Nakajima, and K. Kaya, J. Chem. Phys. **97**, 3803 (1992); A. Nakajima, K. Hoshino, T. Sugioka, T. Naganuma, T. Taguwa,

Y. Yamada, K. Watanabe, and K. Kaya, J. Phys. Chem. **97**, 86 (1993); K. Hoshino, T. Naganuma, K. Watanabe, A. Nakajima, and K. Kaya, Chem. Phys. Lett. **211**, 571 (1993); K. Hoshino, K. Watanabe, Y. Konishi, T. Taguwa, A. Nakajima, and K. Kaya, ibid. **231**, 499 (1994); K. Hoshino, T. Naganuma, K. Watanabe, Y. Konishi, A. Nakajima, and K. Kaya, ibid. **239**, 369 (1995)

69. W. J. C. Menezes and M. B. Knickelbein, Chem. Phys. Lett **183**, 357 (1991); Z. Phys. D **26**, 322 (1993)

70. S. Pollack, C. R. C. Wang, T. A. Dahlseid, and M. M. Kappes, J. Chem. Phys. **96**, 4918 (1992); T. A. Dahlseid, M. M. Kappes, J. A. Pople, and M. A. Ratner, ibid. **96**, 4924 (1992)

71. M. P. Andrews and S. C. O'Brien, J. Phys. Chem. **96**, 8233 (1992)

72. Š. Vajda, S. Rutz, J. Heufelder, P. Rosendo, H. Ruppe, P. Wetzel, and L. Wöste, J. Phys. Chem. A **102**, 4066 (1998)

73. M. J. López and J. Jellinek, J. Chem. Phys., to be published

74. R. P. Gupta, Phys. Rev. B **23**, 6265 (1981)

75. F. Cleri and V. Rosato, Phys. Rev. B **48**, 22 (1993)

76. G. Pacchioni, S.-C. Chung, S. Krüger, and N. Rösch, Chem. Phys. **184**, 125 (1994); F. A. Reuse and S. N. Khanna, Chem. Phys. Lett. **234**, 77 (1995); N. N. Lathiotakis, A. N. Andriotis, M. Menon, and J. Connoly, J. Chem. Phys. **104**, 992 (1996)

77. J.-Y. Yi, D. J. Oh, J. Bernholc, and R. Car, Chem. Phys. Lett. **174**, 461 (1990); H.-P. Cheng, R. S. Berry, and R. L. Whetten, Phys. Rev. B **43**, 10647 (1991); M. R. Pederson, in Ref. 5, Vol. 1, p. 861; X. G. Gong and V. Kumar, Phys. Rev. Lett. **70**, 2078 (1993); S. N. Khanna and P. Jena, Chem. Phys. Lett. **219**, 479 (1994)

78. W. H. Press, A. A. Teukolsky, W. T. Vetterling, and B. P. Flannery *Numerical Recipes*, Cambridge Univ. Press, London, 1992

79. G. E. López and D. L. Freeman, J. Chem. Phys. **98**, 1428 (1993)

80. W. C. Swope, H. C. Andersen, P. H. Berens, and K. R. Wilson, J. Chem. Phys. **76**, 638 (1982)

81. E. M. Pearson, T. Halicioglu, and W. A. Tiller, Phys. Rev. A **32**, 3030 (1985)

82. J. Jellinek, T. L. Beck, and R. S. Berry, J. Chem. Phys. **84**, 2783 (1986); R. S. Berry, T. L. Beck, H. L. Davis, and J. Jellinek, Adv. Chem. Phys. **70**, Part 2, 139 (1988)

83. I. Z. Fisher, *Statistical Theory of Liquids*, Univ. of Chicago Press, Chicago, 1966

# Quantum Mechanics of Hydrogen on Nickel and Palladium Clusters

M. A. Gomez[1], B. Chen[1], David L. Freeman[2], and J. D. Doll[1]

[1] Department of Chemistry, Brown University, Providence, RI 02912, USA
[2] Department of Chemistry, University of Rhode Island, Kingston, RI 02881, USA

**Abstract.** Within the broad class of metal-hydrogen systems, clusters are of particular importance. Their high surface to volume ratio makes them ideal candidates for catalytic applications. Surface and bulk studies have shown that transport and vibrational spectroscopy of hydrogen are very sensitive to substrate structure. The wide variety of geometries exhibited by clusters offers a noteworthy opportunity to examine the effect of substrate geometry on hydrogen. Further, hydrogen's small mass and uniquely large isotopic variation gives rise to a number of intrinsically quantum mechanical effects. For example, inverse isotope effects have been observed for hydrogen chemisorption on palladium clusters. Comparing classical and quantum mechanical Monte Carlo methods, the effects of quantum mechanics on cluster structure, population distribution, vibrational spectra, and rates for hydrogen motion are discussed for a single hydrogen on nickel and palladium clusters.

## 1 Introduction

The role of hydrogen on metal clusters is important both from a technological and a fundamental point of view. Technologically, their high surface to volume ratio makes them ideal candidates for catalytic applications. Since many applications involve hydrogen, an understanding of the properties of metal-hydrogen clusters is of special significance.

Fundamentally, hydrogen's small mass and uniquely large isotopic variation gives rise to intrinsically quantum mechanical effects. For example, Fayet, Kaldor, and Cox [1] notice an inverse isotope effect in the rate of chemisorption of $H_2$ and $D_2$ on palladium clusters. Such inverse isotope effects are also observed in the palladium bulk and the surface of Pd(111) [2, 3]. Further, since clusters provide a wide variety of geometries, they offer a noteworthy opportunity to examine the effect of substrate geometry on the behavior of hydrogen.

In a previous study [4] we examined the properties of small nickel and palladium clusters that contained a single hydrogen atom. This study revealed that quantum mechanical structures differ from classical ones for some palladium clusters with a hydrogen atom. The differences result from an inversion in energy ordering of isomers when zero-point effects are taken into account.

In this chapter, we discuss the effect of quantum mechanics on structure on larger nickel and palladium clusters. For the $Pd_7H$, we also discuss non-zero temperature population distributions, vibrational frequencies, and rates for hydrogen entering the cluster. This chapter is arranged as follows: Section 2 briefly describes and justifies the model potential used in this work. Section 3 discusses the classical structure of nickel and palladium clusters with a hydrogen atom. Section 4 explains how quantum mechanics changes the zero temperature structure. The next sections comprise a detailed case study of $Pd_7H$. In particular, non-zero temperature effects on isomer population distributions for this cluster are discussed in Sec. 5. Section 6 explains the quantum mechanical effects on the vibrational frequencies of $Pd_7H$. Next, Sec. 7 discusses quantum mechanical effects on rates for hydrogen entering the seven atom palladium cluster. Finally, in Sec. 8 we present some concluding remarks.

## 2    Potential

Throughout this study the embedded atom method (EAM) of Daw and Baskes [5] is used. While it is an incomplete description of the present cluster systems, using the same potential in the current work as in previous surface studies [3, 6, 7] enables comparison of hydrogen in cluster and surface environments. Although we have chosen not to use them, we note that more recent EAM parameterizations for nickel and hydrogen systems have been published by Rice et al. [8], and Wonchoba and Truhlar [9]. In addition, Jellinek and collaborators have extensively studied deuterium/nickel systems with the Voter-Chen potential [10] as well as other many-body potentials [11]. Fournier, Stave, and DePristo have also studied deuterium/nickel systems with Corrected Effective Medium (CEM) methods.[12] Curotto, Matro, Freeman, and Doll used an extended Huckel theory potential to study mono and di-hydrogenated nickel clusters.[13] Gronbeck, Tomanek, Kim, and Rosen studied the effect of hydrogen on the melting of palladium clusters using a many body alloy potential.[14] Further, Wolf et al. have developed more recent H/Pd EAM potentials [15]. We emphasize that none of the methods in the present study depend on the details of the potential. More complete estimates of the microscopic interactions can and will be used as they become available.

Even though EAM [5] is fit to bulk data, previous research shows that EAM is robust and yields at least qualitative data for the structure and binding energies of nickel clusters. Sequential nickel binding energies obtained by Stave and DePristo [12] using computationally more intensive empirical potentials are closer to the experimental results of Lian et al. [16] than those predicted with EAM [17]. However, the patterns in these energies are qualitatively similar [4]. Most of the structures predicted by EAM [4, 11, 17] and CEM [12] are consistent with those inferred by Parks et al. [18] from $N_2$ absorption data. In the EAM potential [5] used in the present study, the exceptions are $Ni_8$ and $Ni_{14}$.

# 3  Classical Structure

Classical minima are found using simulated annealing [19], and quantum annealing [20] on random configurations. About 200 random configurations are used for simulated annealing and quantum annealing. As a further check, the simplex method [19] is used on between 200 and 600 structures constructed with the minima of the previous size cluster plus an a new atom in a random position. For the purposes of the present discussion, the lowest potential energy structure found for a given system by these searches is referred to as the "classical global minimum." While appreciable care has been exercised in locating these minima, experience with such problems makes it clear that it is generally impossible to know with total certainty when the absolute potential minimum has been located.

The classical global minima for $Ni_nH$ with n=11-16 are shown in Fig. 1. Hydrogen prefers to bind on the outside of nearly completed coordination shells and inside coordination shells that are just starting. Enlarging a tetrahedral site to accommodate a hydrogen in nearly complete coordination shells compresses adjacent tetrahedral sites and breaks the symmetry of the cluster. In the incomplete shells, hydrogen binding usually occurs inside tetrahedral sites that do not have neighboring tetrahedral sites.

The global minima for the smaller nickel clusters are reproduced in Fig. 2 for comparison. A comparison of global minimum structures of nickel clusters with and without a hydrogen atom reveals rearrangements of the structure. Fig. 3 shows the structure of bare nickel clusters of seven to ten atoms. The global minimum potential energy structures of the seven and ten atom nickel clusters with the hydrogen rearrange to incorporate a larger binding site for the hydrogen, namely an octahedral site. In both of these cases, the second lowest minimum for the bare cluster has an octahedral site [4].

The global minima for $Pd_nH$ with n=11-15 have the same basic structures as the nickel clusters except $Pd_{11}H$. The global minimum of $Pd_{11}H$ is shown in Fig. 4. In this case, hydrogen is at the outside edge of the tetrahedral site. It should be noted that while hydrogen is outside the tetrahedral site for clusters of 12 and 13 palladium atoms just as in nickel, it is just barely outside. The fact that palladium can sometimes accommodate hydrogen inside while nickel can not is not surprising. Hydrogen readily percolates [21] through bulk palladium but must be "pounded" into nickel [9, 22]. This suggests that palladium can accommodate hydrogen in its smaller binding sites far better than nickel can. The lattice constant in palladium is 3.89 Å while in nickel it is 3.52 Å. The difference indicates that the lattice structure is more open in palladium than in nickel. A similar trend is seen in the clusters. Palladium clusters generally form larger sites than nickel.

This tendency for palladium to accommodate hydrogen better than nickel is also seen in the smaller palladium structures [4]. The global potential energy minimum structures for palladium clusters with a hydrogen atom are the same as those for nickel clusters with the exceptions of $Pd_nH$ with n = 7, 9, and 10 which are shown in Fig. 5.

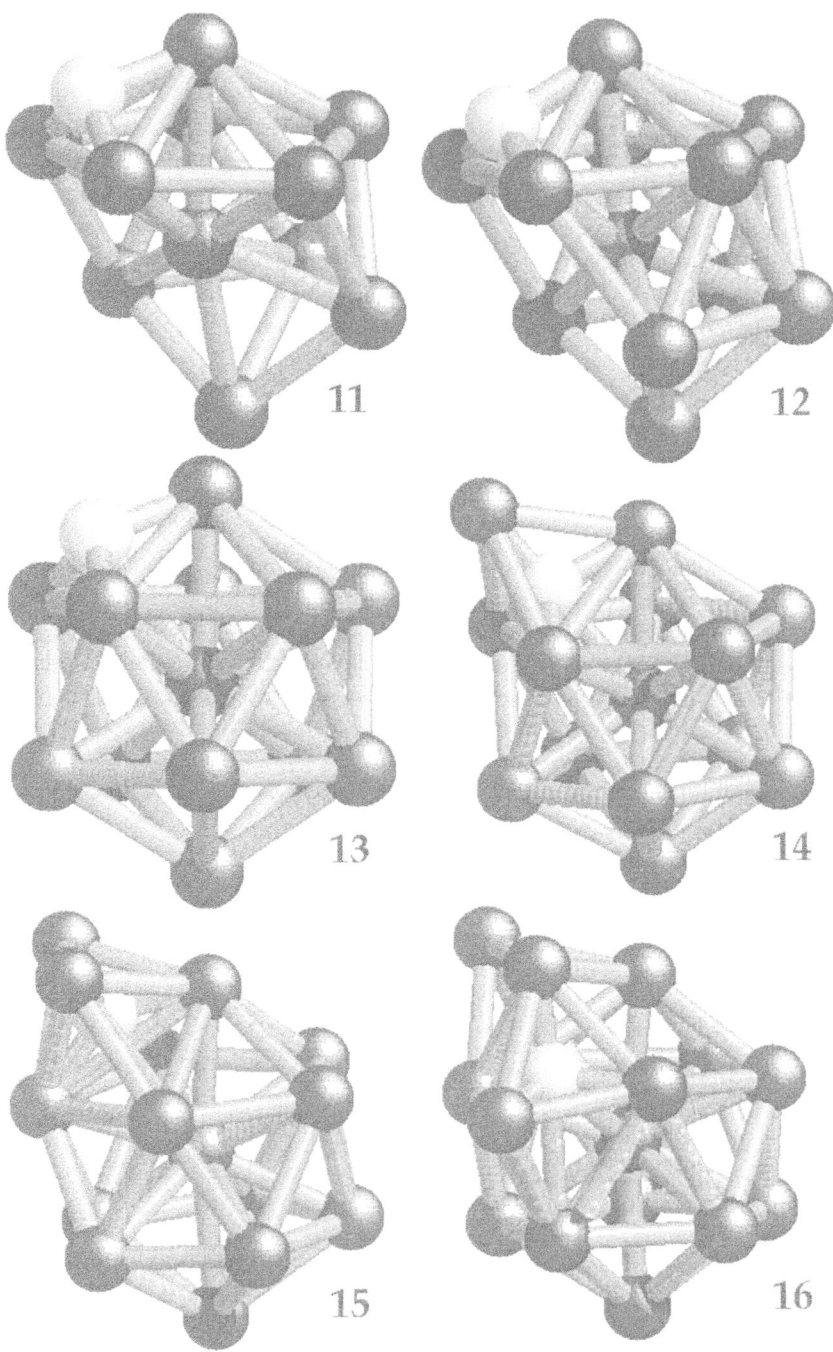

**Fig. 1.** $Ni_nH$ global minima for n=11-16

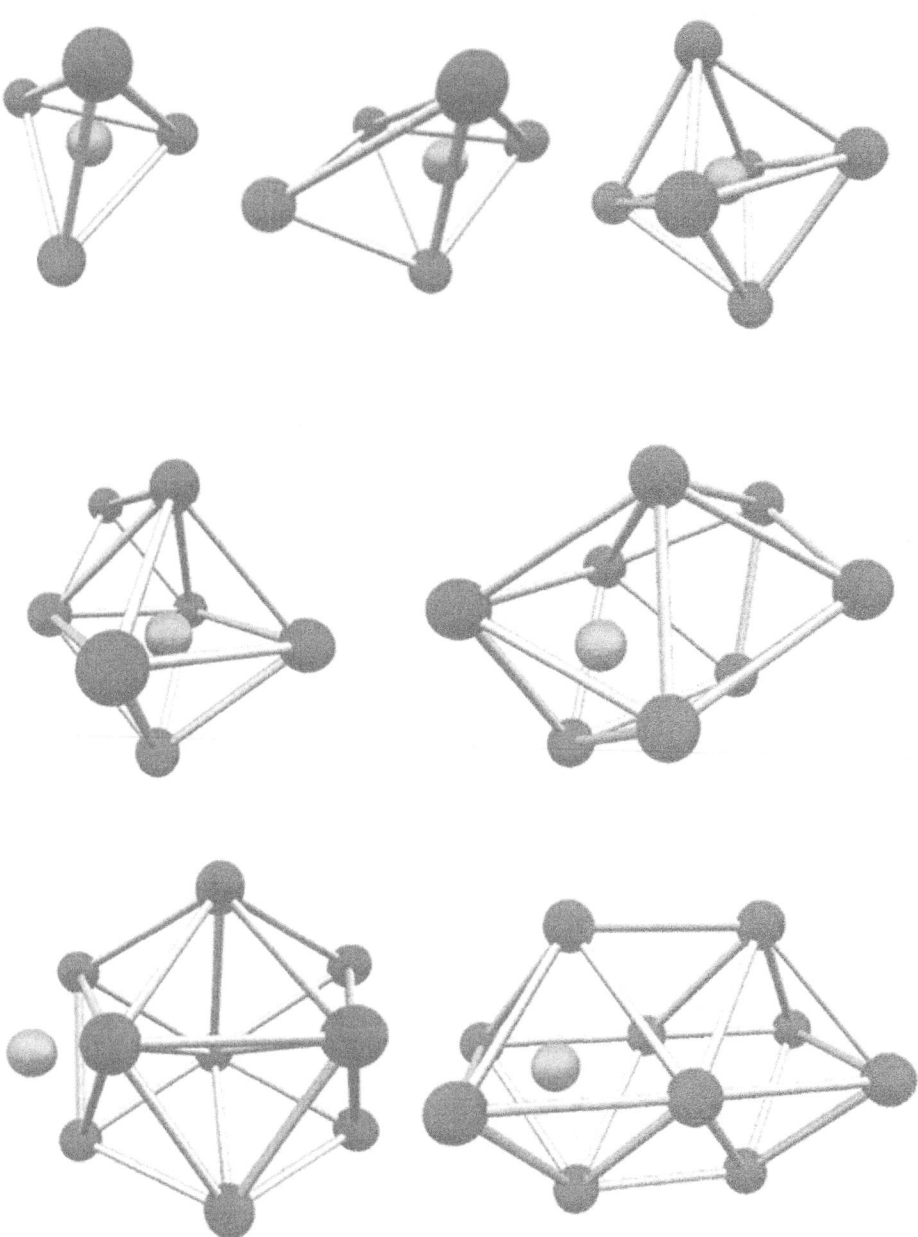

**Fig. 2.** Global minimum energy structures of $Ni_nH$ for n=4-10. For purposes of identification, the H atom is shown without bonds to the nickel atoms.

314

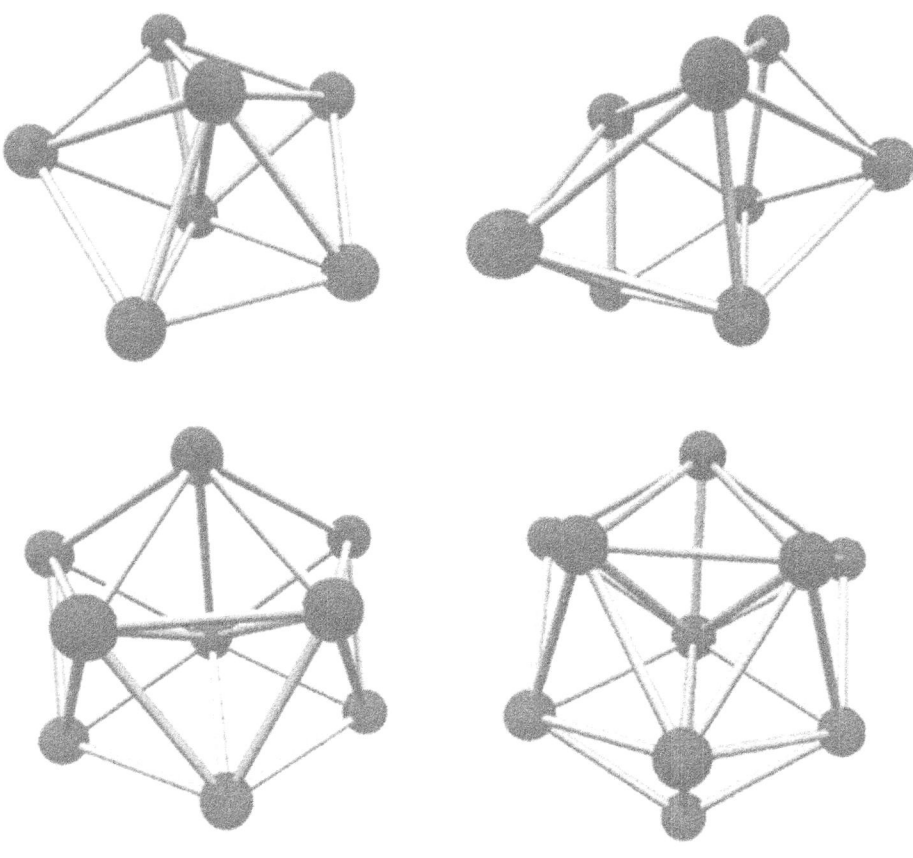

**Fig. 3.** Global Minimum energy structures for nickel clusters of sizes 7 to 10 atoms.

## 4 Quantum Mechanical Influence on Ground State Structure

Normal mode analysis, which reflects both potential energy minima and local curvature, provides a very useful tool for predicting the quantum mechanical energy ordering of different isomers. As seen in our previous paper [4], zero point energy effects can alter the energy ordering of isomers of some palladium clusters with a hydrogen atom. For example, normal mode analysis predicts that clusters $Pd_7H$ and $Pd_{10}H$ have different classical and quantum mechanical structures at zero temperature [4]. Diffusion Monte Carlo confirms this [4]. The basic quantum mechanical structure at zero temperature for $Pd_7H$ and $Pd_{10}H$ is that of the corresponding nickel clusters shown in Fig. 2.

In this chapter we apply this analysis to larger clusters. The addition of zero-point energy to these clusters and their close lying isomers reveals that ordering in energy for most clusters remains the same. However, for $Pd_{11}H$ and

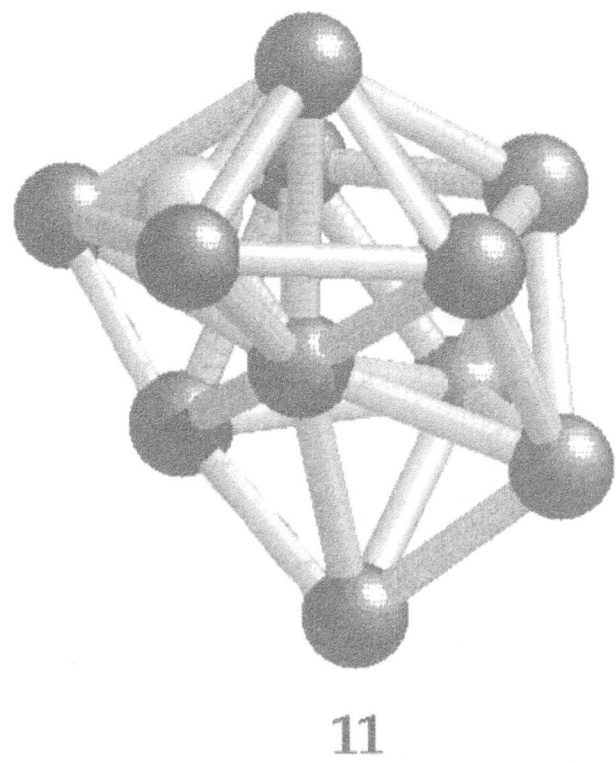

11

Fig. 4. Global minimum energy structure for $Pd_{11}H$

$Ni_{16}H$, the ordering is changed. While the global minimum structure for $Pd_{11}H$ has the hydrogen inside the tetrahedral site as shown in Fig. 4, there is a close lying isomer with the hydrogen outside the tetrahedral site just as in the nickel clusters shown in Fig. 1. When harmonic zero-point energy effects are taken into account, the energy of $Pd_{11}H$ with hydrogen on the inside at zero temperature goes from $-33.439$ eV to $-32.70$ eV. The energy of $Pd_{11}H$ with hydrogen outside the cluster goes from $-33.438$ eV to $-32.71$ eV. Normal mode analysis, which takes into account not only the potential energy of the cluster but also the curvature of the potential, indicates that when quantum mechanical effects are included hydrogen prefers binding outside $Pd_{11}$.

The first inversion in energy of nickel clusters with a hydrogen atom occurs for $Ni_{16}H$, the classical structure of which is shown in Fig. 1. The energy for the cluster with hydrogen bound inside the tetrahedral site at zero temperature goes from $-57.002$ eV to $-55.71$ eV. By contrast the energy for $Ni_{16}H$ with hydrogen bound outside the cluster at zero temperature goes from $-56.984$ eV

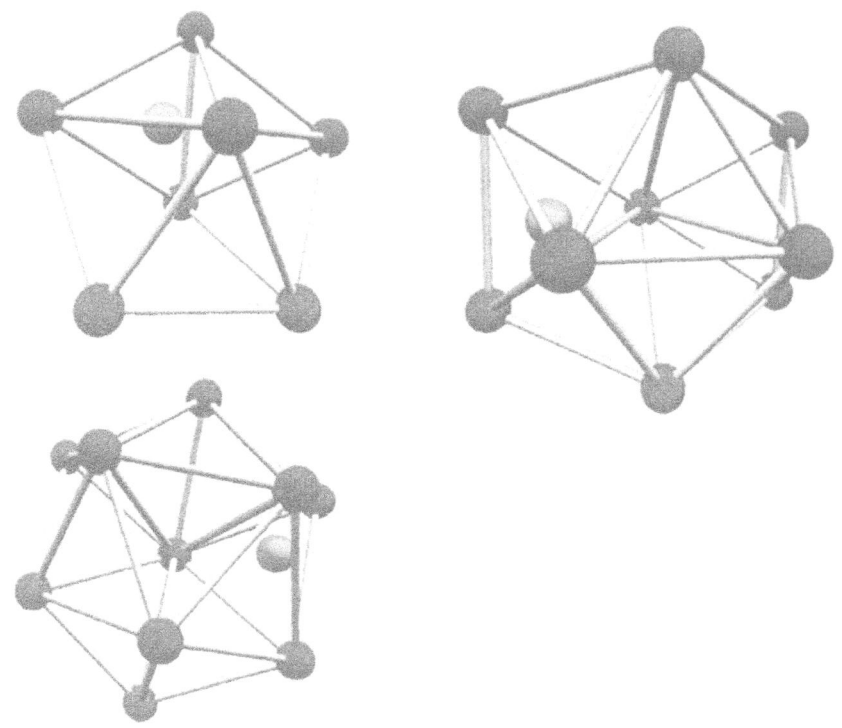

**Fig. 5.** Global Minimum energy structures for palladium clusters of sizes 7, 9, and 10 atoms with a hydrogen.

to $-55.73$ eV when zero point energy is taken into account. Hence, normal mode analysis predicts that the structure of $Ni_{16}H$ at zero temperature is the one shown in Fig. 6.

# 5 Non-zero Temperature Quantum Mechanical Structure Distributions

The study of cluster isomerization offers valuable insight into the nature of cluster dynamics [23, 24]. The relative populations of the stable structures are examined using the Path Integral Monte Carlo (PIMC) method coupled with quenching techniques [25]. The Monte Carlo configuration is quenched every 1200 moves via very fast simulated annealing [19].

Figure 7 presents the probability that the $Pd_7H$ explores non-ground state structures given that it is started in the ground state configuration at various temperatures. This probability is labeled probability of isomerization. The onset of isomerization is seen to occur at around 500 K. At lower temperatures, the

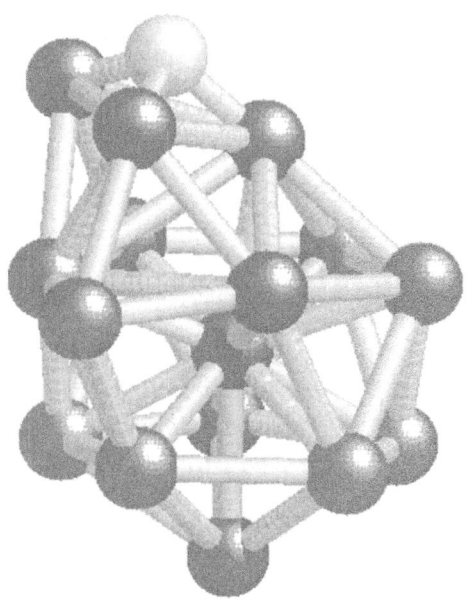

**Fig. 6.** Minimum harmonic ground state energy structure for $Ni_{16}H$

system remains localized in the vicinity of the initial isomeric forms. The probability of isomerization increases with temperature and approaches a constant value of 0.82 at temperatures higher than about 800 K. The onset temperature may be biased by the "quasi-ergodicity" problem. The low quantum tunneling rate prevents the observation of transition between isomers during the finite length of simulation. The strong temperature dependence of the quenching population indicates the activated nature of isomerization. Care must be exercised to assure proper low temperature sampling [26].

Figure 8 shows the temperature dependence of the isomer distribution for $Pd_7H$. As is shown there, the population of the ground state isomer approaches 100% for temperatures lower than 500 K. As the temperature increases, the atoms become more and more mobile. As a result, other isomer structures start to appear. Only the isomers with energies in the vicinity of the ground state energy play a significant role at lower temperatures where the quantum ground state structure dominates. It should be noted that the classical global minimum is not the ground state structure. This is revealed both by normal mode analysis and diffusion Monte Carlo [4]. Classically, $Pd_7H$ contains a pentagonal bipyramidal palladium skeleton with a hydrogen atom in one of its tetrahedral sites. When zero-point energy effects are incorporated, a palladium skeleton of a capped octahedron with hydrogen in the octahedral site is preferred. Both the classical

318

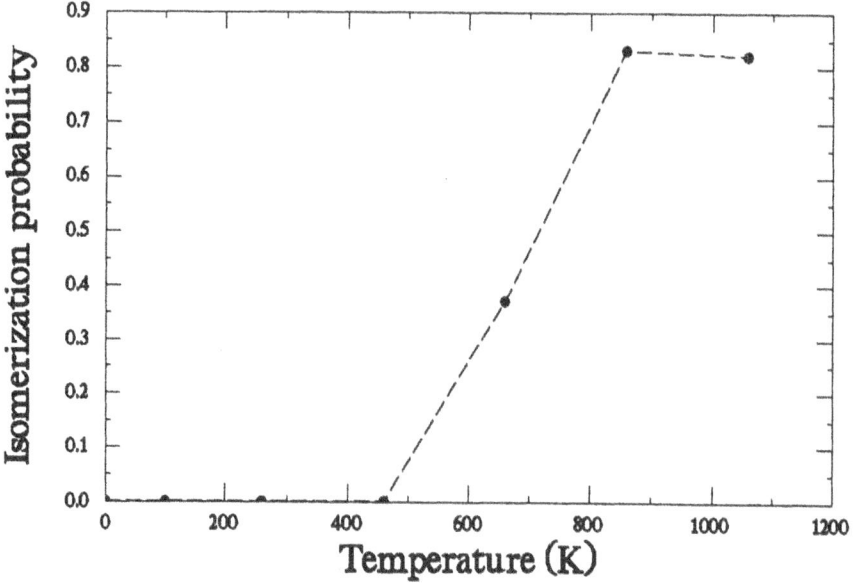

**Fig. 7.** Temperature dependence of isomerization probability of $Pd_7H$. The initial structure is the quantum ground state structure.

global minimum structure and the quantum mechanical ground state structure are shown in Fig. 9. At higher temperatures, quantum mechanics becomes less important. Ultimately, the classical Boltzmann distribution in which the classical global minimum dominates is reached.

# 6 Quantum Mechanical Effects on Vibrational Frequencies

The cluster, $Pd_7H$, is studied to show the effect of anharmonicities and quantum mechanics on frequencies. As seen in Fig. 9, the quantum and classical zero temperature structures for $Pd_7H$ differ. In order to focus on the effect of quantum mechanics on frequencies, both our classical and quantum spectra calculations use the same structure, namely the quantum mechanical ground state structure. First, we consider the zero temperature classical frequencies (i.e. the normal mode frequencies). Next, we look at non-zero temperature classical frequencies which are obtained from a molecular dynamics trajectory. Finally, we compare and contrast these with quantum mechanical frequencies.

The zero temperature classical frequencies are the normal mode frequencies of the system. The quantum mechanical ground state structure for this cluster (Figure 9) has hydrogen in a $C_{3v}$ environment. Consequently, the hydrogen

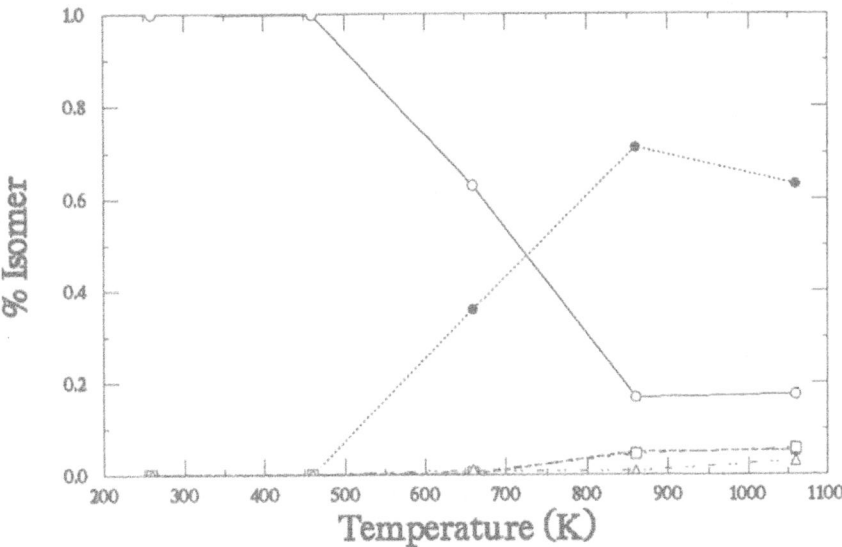

**Fig. 8.** Isomer distribution of Pd$_7$H as a function of temperature. The open circles represent the quantum ground state structure. The closed circles represent the classical global minimum structure. The lower two curves are the isomers with next closest energies.

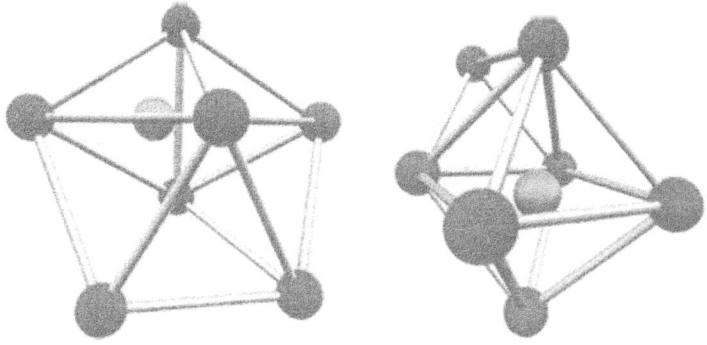

**Fig. 9.** The two lowest energy isomers of the seven atom palladium cluster with a hydrogen atom are shown. The classical global minimum is on the left. However, zero-point energy effects show that the structure on the right is representative of quantum mechanical ground state structure. This is confirmed via Diffusion Monte Carlo.

vibrational motions are the same for the cluster and the surface, a symmetric $A_1$ state and a doubly degenerate E state. Normal mode analysis reveals that hydrogen does contribute significantly to these three modes. The vibrational frequencies for hydrogen in the quantum mechanical ground state structure for $Pd_7H$ are 1591 cm$^{-1}$ for the doubly degenerate E modes, and 1228 cm$^{-1}$ for the $A_1$ mode. The Pd-Pd frequencies range from 80 to 402 cm$^{-1}$.

Comparing normal mode frequency estimates with classical spectra provides a qualitative measure of the anharmonicities in the potential. The classical spectra at 100 K shown in Fig. 10 is obtained by taking the Fourier transform of the hydrogen position autocorrelation function. The vibrational frequencies for hydrogen in the quantum mechanical ground state structure for $Pd_7H$ are 1572 cm$^{-1}$ for the doubly degenerate E modes and 1213 cm$^{-1}$ for the $A_1$ mode. Comparison with normal mode frequencies suggests that there are anharmonicities in the system. Since the frequency is lowered more in the E modes, these modes may be more anharmonic than the $A_1$ mode.

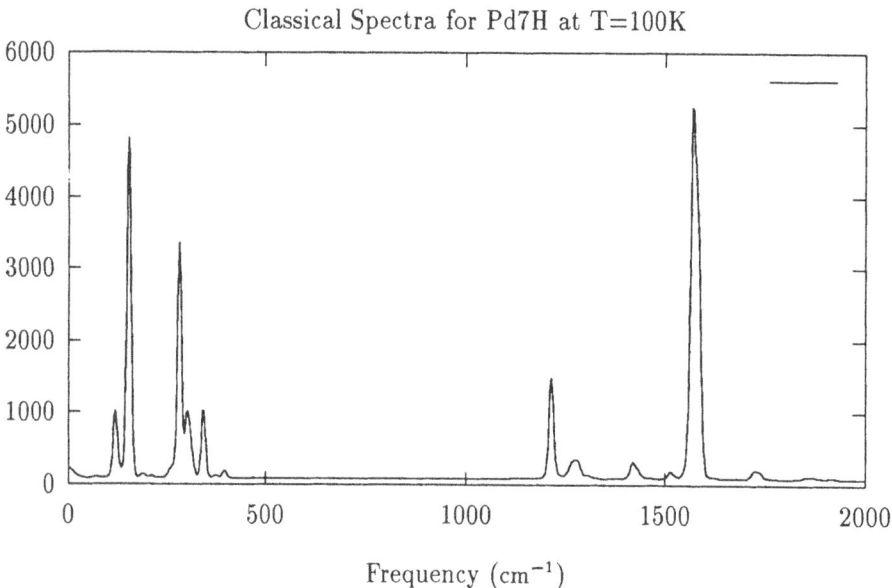

**Fig. 10.** Classical hydrogen vibrational spectra for $Pd_7H$ at 100 K. Results are obtained by Fourier transforming hydrogen position autocorrelation function data.

The quantum mechanical frequencies of the $Pd_7H$ cluster at 100 K are shown in Fig. 11. The spectra are obtained by inverting the imaginary time hydrogen position autocorrelation function via maximum entropy methods [27]. The position autocorrelation function is calculated using a Fourier Path integral approach [25]. The solid curve and dotted curve correspond to the doubly degenerate E state, and the dot-dashed line is the spectra for $A_1$ state. The calculated frequencies are lower than both the normal mode frequencies and the classical frequencies at 100 K. Quantum mechanically, the anharmonicities in the potential are probed more thoroughly since the zero-point energy causes the molecule to move in higher energy regions of the potential. The classical cluster would need a higher temperature to probe the same regions. The amount of shift is most notable for E state, about 15% from the normal mode frequency. The shift for A state is much smaller by comparison, only about 2% from the normal mode frequency. The smaller anharmonic effect for $A_1$ state is a consequence of its smaller zero-point energy contribution. The frequencies of the E and the $A_1$ modes are 1355 cm$^{-1}$ and 1200 cm$^{-1}$, respectively. The ordering is still the same but the difference in frequency is significantly smaller than the corresponding classical result.

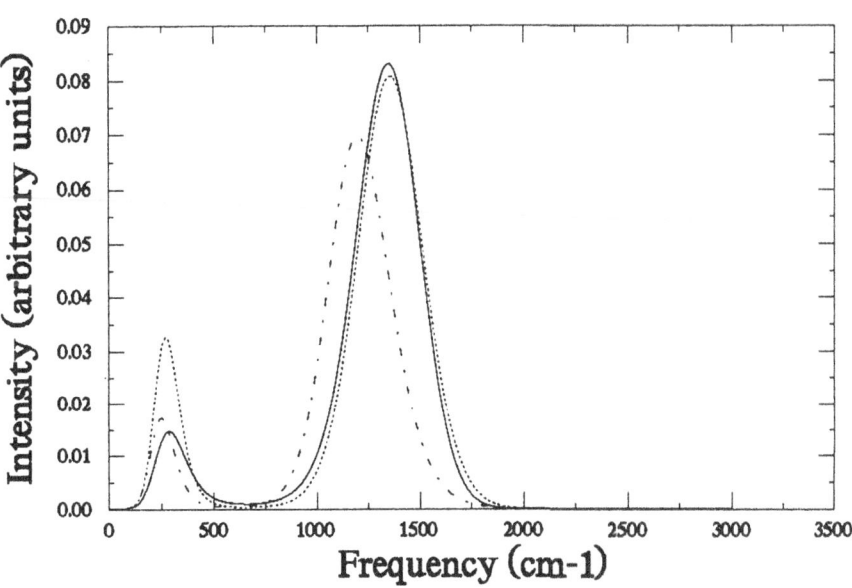

**Fig. 11.** The quantum mechanical spectra at 100K are shown for the doubly degenerate E state (solid and dotted curves) and the A state(dot-dashed curve).

## 7   Quantum Rate Effects

As noted in the introduction, inverse isotope effects are observed for hydrogen diffusion in bulk palladium and palladium clusters [1, 2, 3]. The hydrogen diffusion

322

studies of Rick, Lynch, and Doll suggest that the reason for the inverse isotope effect is highly constrained transition state sites [3]. The quantum mechanical zero temperature structure of the $Pd_7H$ cluster has very similar geometry to the Pd(111) surface sites. In particular, there are three-fold hollow, octahedral, and tetrahedral sites in sequence along an axis. As a result of the similarity in the sites of $Pd_7H$ and the Pd(111) surface, this cluster is a reasonable starting point for the study of unusual quantum rate effects in palladium clusters.

As was the case in discussing quantum effects on cluster structure, a zero-point analysis is a convenient starting point for the discussion of quantum effects on thermal rates. Figure 12 shows the potential profile of hydrogen entering the $Pd_7H$ isomer. From Fig. 12, we see that the classical barrier going from the three fold hollow surface site to the octahedral site is 0.110 eV while the barrier in the reverse direction is 0.436 eV. Zero point energies modify these results, yielding 0.100 and 0.560 eV, for the two barriers respectively. While too simplistic to be highly accurate, the present normal mode analysis serves to indicate that quantum mechanical effects can act to either increase or decrease a reaction barrier.

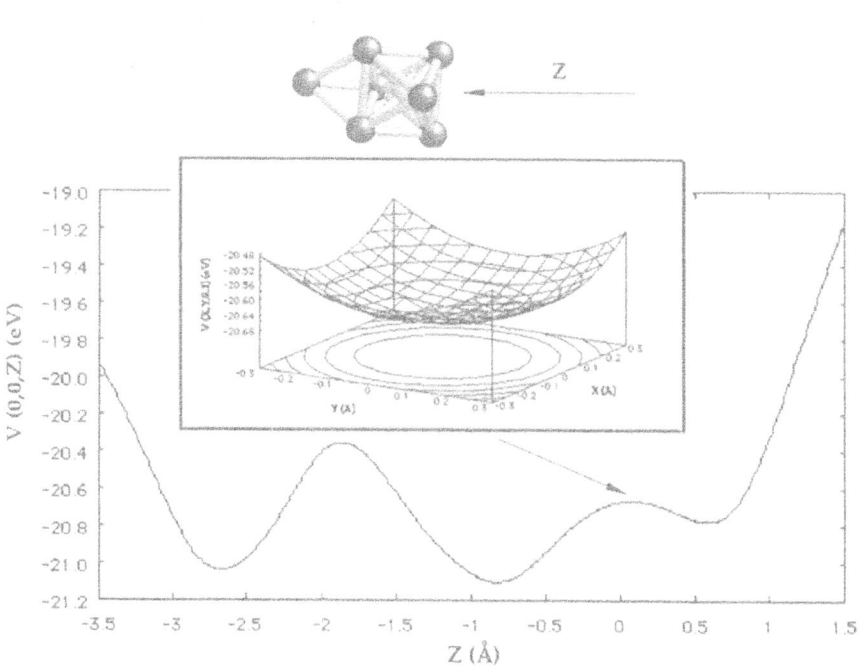

Fig. 12. Potential energy profile of hydrogen entering the cluster. At the barrier, the potential energy as a function of $x$ and $y$ with $z$ at the barrier is shown. The upward curvature indicates that zero point energy effects need to be taken into account at the barrier.

Transition state theory provides a useful framework with which to extend the previous analysis of quantum effects on rates. Classically, the transition state theory rate is given by,

$$k_{A\to}^{TST} = \frac{1}{2} \left( \frac{2kT}{\pi m} \right)^{1/2} < \delta(x-q) >_A \qquad (1)$$

where $x$ is the position and $q$ is the transition state. The average of the delta function is done over the Boltzmann distribution in the reactant system, A. Quantum mechanically, Miller, Chandler, and Voth [28] have shown that rate can be written approximately in the same form as Eq. 1 with the centroid variable, $x_c$, replacing the position, $x$. The average of the delta function is done over the quantum distribution in site A.

We performed the averages required by the classical and quantum rate calculations using Voter's displacement vector Monte Carlo method (DVMC) [3, 28, 29]. The DVMC method is a technique for computing ratios of partition functions. These calculations predict that the rate constant for hydrogen entering the seven atom palladium cluster at 300 K are $0.95 \pm 0.03$ ps$^{-1}$ classically and $0.4 \pm 0.2$ ps$^{-1}$ quantum mechanically. As with the previous surface studies [3], quantum mechanics is acting to reduce the rate of hydrogen transport into the "sub-surface" palladium regions.

# 8  Conclusion

This chapter has shown how quantum mechanics can affect structure, population distribution, vibrational spectra, and rates. Structures of similar classical potential energies can have significantly different ground state energies as a consequence of different local curvatures in the potential. Population distributions at low temperatures are significantly affected by these zero point energy effects. Quantum mechanically, vibrational frequencies corresponding to very anharmonic modes shift significantly to lower frequencies. Finally, quantum mechanics can lower rates when the the transition state is more highly constrained than the reactant state.

# 9  Acknowledgments

This work was supported in part by the National Science Foundation through grants CHE-9411000 and CHE-9625498. One of us (MAG) also wishes to thank both the National Science Foundation and the Cooperative Research Fellowship Program of AT&T for graduate fellowship support.

This research was sponsored in part by the Phillips Laboratory, Air Force Material Command, USAF, through the use of Maui High Performance Computing Center (MHPCC) under cooperative agreement number F29601-93-2-0001. The views and conclusions contained in this document are those of the authors and should not be interpreted as necessarily representing the official policies or endorsements, either expressed or implied, of Phillips Laboratory or the U.S. Government.

# References

1. P. Fayet, A. Kaldor, and D. M. Cox, J. Chem. Phys. **92**, 254 (1990).
2. J. Völkl and G. Alefeld. *Diffusion in Solids: Recent Developments*, edited by A. S. Nowick and J. J. Burton (Academic Press, New York, 1975), chapter 5.
3. S. W. Rick, D. L. Lynch, and J. D. Doll, J. Chem. Phys. **99**, 8183 (1993).
4. B. Chen, M. A. Gomez, M. Sehl, J. D. Doll, and D. L. Freeman, J. Chem. Phys. **105**, 9686 (1996).
5. M. S. Daw, and M. I. Baskes, Phys. Rev. B. **29**, 6443 (1984).
6. D. L. Lynch, S. W. Rick, M. A. Gomez, B. W. Spath, J. D. Doll, and L. R. Pratt, J. Chem. Phys. **97** 5177 (1992).
7. S. W. Rick, and J. D. Doll, Surface Science Letters. **302**, L305 (1994).
8. B. M. Rice, B. C. Garrett, M. L. Koszykowski, S. M. Foiles, and M. S. Daw, J. Chem. Phys. **92**, 775 (1990).
9. S. E. Wonchoba, and D. G. Truhlar, Phys. Rev. B. **53**, 11222 (1996).
10. A. F. Voter and S. P. Chen, Mater. Res. Soc. Symp. Proc. **82**, 175 (1987)
11. J. Jellinek, in *Metal-Ligand Interactions: Structure and Reactivity*, edited by N. Russo and D. R. Salahub. (Kluwer Academic Publishers, Dordrecht, 1996), p. 325; J. Jellinek and Z. B. Guvenc, in *The Synergy Between Dynamics and Reactivity at Clusters and Surfaces* edited by L. J. Farrugia (Kluwer Academic Publishers, Dordrecht, 1995), p. 217; M. J. Lopez and J. Jellinek, Phys. Rev. A **50**, 1445 (1994); J. Jellinek and Z. B. Guvenc, Z. Phys. D. **26**, 110 (1993).
12. M. S. Stave and A. E. DePristo, J. Chem. Phys **97**, 3386 (1992); R. Fournier, M. S. Stave, and A. E. DePristo, J. Chem. Phys. **96**, 1530 (1992).
13. E. Curotto, A. Matro, D. L. Freeman, and J. D. Doll, J. Chem. Phys. **108**, 729 (1998).
14. H. Gronbeck, D. Tomanek, S. G. Kim, and A. Rosen, Zeitschrift Fur Physik D **40**, 469 (1997)
15. M. W. Lee, R. J. Wolf, and J. R. Ray, J. Alloys Compd. **231**, 343 (1995); R. J. Wolf, M. W. Lee, and J. R. Ray, Phys. Rev. Lett. **73**, 557 (1994); R. J. Wolf, M. W. Lee, and R. C. Davis, Phys. Rev. B. **48**, 12415 (1993); R. J. Wolf, M. W. Lee, and J. R. Ray, Phys. Rev. B. **46**, 8027 (1992).
16. L. Lian, C.-X. Su, and P. B. Armentrout, J. Chem. Phys. **96**, 7542 (1992).
17. D. G. Vlachos, L. D. Schmidt, and R. Aris, J. Chem. Phys. **96**, 6880 (1992).
18. E. K. Parks, L. Zhu, J. Ho, and S. J. Riley, J. Chem. Phys. **100** 7206 (1994).
19. W. H. Press, S. A. Teukolsky, W. T. Vetterling, B. P. Flannery. *Numerical Recipies in FORTRAN. The Art of Scientific Computing. Second Edition* (Cambridge University Press, Cambridge, 1992).
20. A. B. Finnila, M. A. Gomez, C. Sebenik, C. Stenson, and J. D. Doll, Chem. Phys. Lett. **219**, 343 (1994).
21. *Hydrogen in Metals I & II*, edited by G. Alefeld and J. Völkl (Springer-Verlag, Berlin, 1978).
22. K. J. Maynard, A. D. Johnson, S. P. Daley, and S. T. Ceyer, Faraday Discuss. Chem. Soc. **91**, 437 (1991).
23. F. G. Amar and R. S. Berry, J. Chem. Phys. **85**, 5943 (1986).

24. S. W. Rick, D. L. Leitner, J. D. Doll, D. L. Freeman, and D. D. Frantz, J. Chem. Phys. **95**, 6658 (1991).

25. J. D. Doll, D. L. Freeman, and T. L. Beck, Advances in Chemical Physics **78**, 61 (1990).

26. D. D. Frantz, D. L. Freeman, and J. D. Doll, J. Chem. Phys. **93**, 2769 (1990).

27. J. E. Gubernatis, M. Jarrell, R. N. Silver, and D. S. Sivia, Phys. Rev. B. **44**, 6011 (1991).

28. G. A. Voth, D. Chandler, and W. H Miller, J. Chem Phys. **91**, 7749 (1989).

29. A. F. Voter, J. Chem. Phys. **82**, 1890 (1985).

# Metal Cluster – Surface Interaction: Simple Models and *Ab Initio* Calculations

Hannu Häkkinen and Matti Manninen

Department of Physics, University of Jyväskylä,
P.O.Box 35, FIN-40351 Jyväskylä, Finland

**Abstract.** We review recent *ab initio* atomistic calculations on interactions between metal clusters and electronically inert (insulating) substrates. The model system is sodium clusters on the sodium-chloride
(001) surface. This system provides an example of weak cluster-support
interaction (physisorption) which can however be easily modified by
introducing color centers at the surface, resulting in chemisorption of
sodium adatom or cluster. The results obtained from atomistic calculations can be used for constructing simple jellium-type models for the
adsorbed cluster. These models allow for systematic investigations in a
large size-range of clusters on the shell structure, dimensionality, and stability of the cluster as a function of the strength of the cluster-support
interaction.

## 1 Introduction

As has become evident over the past ten years, the physics of small simple metal
clusters is governed by the shell structure of the valence electrons, as shown
by their size-evolutionary properties, including, *e.g.*, the abundance in the mass
spectra, ionization potential, optical absorption spectra, and reactivity. [1, 2]
Deposition of clusters on a supporting surface opens up a way to studies of
a rich variety of phenomena taking place *in* the cluster and *between* the cluster
and the substrate. In these studies the important questions concern the stability,
geometry, and electronic structure of the deposited cluster as a function of the
strength and nature of the cluster-support interaction. [3, 4, 5, 6, 7, 8, 9]

Most metal clusters deposited on a substrate could be expected to be reactive
due to the tendency of their delocalized valence electrons to hybridize with the
available substrate electronic states. This hybridization can be expected to have
a minor role if there exists a large energy gap between the surface and cluster
electronic states, or if the cluster has a closed-shell electronic structure, which in
the case of free clusters is referred to as a "magic" behavior. For larger cluster
sizes, also the closing of a geometric shell in the atomic structure makes the
cluster more stable.[10]

The limit of a very weak cluster-support interaction is an interesting one
because in principle it is possible to create a situation where the clusters' electronic states do not mix with the substrate states. We then have a confinement
of the valence electrons within a spatial region, the dimensionality and size of

which is determined by the arrangement of cluster ions. Particularly, if the ionic arrangement has two-dimensional (2D) nature, shell effects differing from free 3D metal cluster should be detectable. This should lead to analogies to finite, 2D confinements of electrons (or holes) in semiconductors, *i.e.*, quantum dots.[11]

This paper reviews recent *ab initio* studies [12, 13, 14, 15] of small sodium clusters deposited on an alkali-halide NaCl(001) surface. This system provides an example of weak cluster-surface interactions creating the possibility to study shell structure of the cluster both in 3D and 2D, and even dynamical transitions between them. Another interesting aspect with alkali-halide substrates is that the cluster-support interaction can be easily modified by introducing color centers (missing chlorine atoms) which are structural and at the same time electronic defects of the substrate.

We note here that small alkali-halide (AH) clusters (particularly non-stoichiometric clusters $A_m H_n$, $(m \neq n)$) have recently been studied intensively both theoretically and experimentally. [16, 17, 18, 19, 20, 21, 22, 23, 24, 25, 26, 27, 28] An important result from these studies is the surface–segregation of a metallic part in a metal–rich ($n/m < 1$) cluster. The system can then be thought to be composed as $(AH)_n A_{m-n}$, *i.e.*, a metallic "cluster" $A_{m-n}$ attached to an ionic part $(AH)_n$. Recently it was demonstrated [A=Na and H=F, see [20]; A=Na and H=Cl, see [17]] that *e.g.* the $A_{14}H_9$ ($A_{14}H_9^+$) cluster has a surface–segregated component $A_5$ ($A_5^+$) (see Fig. 1) which is thermally considerably stable. Surface–segregation may occur also at surfaces of bulk alkali-halide crystals, and similar aggregation of metal clusters could be expected.

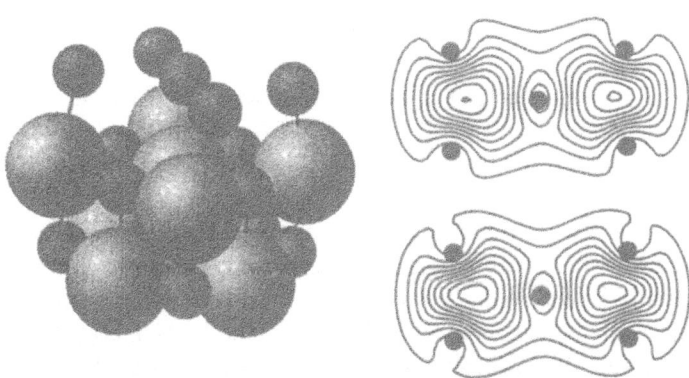

**Fig. 1.** Left: The optimized structure of the neutral $Na_{14}Cl_9$ cluster visualizing how the cluster can be thought to be composed of a $Na_5$ "cluster" attached on stoichiometric $Na_9Cl_9$ part. Right: Excess electron density in the plane of the sodium overlayer on $Na_{14}Cl_9^+$ (top), and electron density of the free $Na_5^+$ cluster (bottom). The spacing between contours is 0.001 a.u. (From [17])

We have studied the geometry, electronic structure and stability of the smallest clusters up to $Na_{12}$ by using a method which incorporates simultaneous solution of the electronic structure and the dynamics of the ions on the Born-Oppenheimer (BO) potential energy surface. Section 2 gives a short account of the method, followed by the discussion of the results in Sect. 3. On the basis of the atomistic calculations we have developed a simplified model (based on the "ultimate" jellium model [29]) for the adsorbed cluster. We discuss the model in Sect. 4. Section 5 gives the results and also discusses some related recent work. [30, 31, 32] Section 6 summarizes the paper.

## 2 BO-LSD-MD Method

The atomistic calculations are done using the BO–LSD–MD (Born-Oppenheimer Local-Spin-Density Molecular Dynamics) method devised by Barnett and Landman, fully documented in [33], to which we refer the reader for details. In the BO–LSD–MD method one solves for the Kohn–Sham (KS) one-electron equations (using a suitable parametrization for the local spin-density approximation to calculate the exchange–correlation part) for the valence electrons of the system corresponding to a given nuclear configuration of the classical ions. From the converged solution the Hellmann-Feynmann forces to ions can be calculated, which together with the classical Coulomb repulsion between the positive ion cores (which are treated as point charges) determine the total forces to ions, according to which one can perform structural optimizations (energy minimizations) or classical molecular dynamics for the ions. The fully converged solution for the electronic structure is obtained for each successive ionic configuration in the course of an optimization or a molecular dynamics run, which ensures that the positions of the ions are advanced in such a manner that the whole system follows dynamics on the Born-Oppenheimer energy surface. The current implementation uses plane waves combined with fast Fourier transform techniques as the basis for the one-electron wave functions and norm–conserving, non–local, separable pseudopotentials by Troullier and Martins [34] to describe the valence electron – ion interaction, and the LSD parametrization by Vosko, Wilks, and Nusair.[35]

The method was initially devised for simulations of finite systems (clusters and molecules) with free boundaries (*i.e.*, without periodic replicas of the ionic system) and has been recently successfully used to study a number of systems, such as pure metal clusters, [36, 37] metal–metal,[38] metal–halogen,[16, 17] and metal–group IV composite clusters,[39] and also computationally demanding molecules, such as water, ammonium and their complexes.[33, 40, 41] The original method has recently been modified [42] for studies of periodic systems like bulk matter or surfaces. The essential modification includes the calculation of the local part of the pseudopotential energy in momentum space by using the structure factor of the ions and the Fourier expansion of the potential and the valence electron density[43] and the use of Ewald summation technique for the ion-ion Coulomb energy. In this work we use the modified version of the method.

In the following we discuss the computational aspects pertinent to our work. The NaCl(001) substrate is constructed parallel to the $xy$-plane of the computational cell, spanned by the elementary directions [100] (x) and [010] (y), with $N_\ell$ layers and $N_{xy}$ sites per layer for ions of alternating charge. Periodic boundary conditions are imposed in each Cartesian direction. We use five different setups (A–E), depending on the size of the adsorbed cluster, and Table 1 summarizes the cell parameters. We have confirmed that the variations in the cell dimensions do not significantly change the calculated binding energies. The pseudopotentials [34] for Cl $3s^2 3p^5$ and Na $3s^1$ valence electrons have been generated and tested in previous studies of alkali-halide clusters,[16, 17] and a typical plane–wave kinetic cutoff energy of $\approx 10$ Ry is found to yield satisfactory accuracy and enable feasible calculations in our studies which span a wide range of systems from about 60 to 300 valence electrons with basis sets of 13000 to 44000 plane waves (Table 1). In calculating the density–related terms in the Kohn–Sham equations we approximate the Brillouin zone sampling by using only the $\Gamma$ point of the supercell.

Table 1. Parameters for the computational cells A–E. $N_\ell$ is the number of substrate layers, $N_{xy}$ is the number of atom positions in a layer, $L_z$ is the cell dimension perpendicular to the (001) surface, and $N^{PW}$ is the size of the plane wave basis set.

| Cell | $N_\ell$ | $N_{xy}$ | $L_z$ (Å) | $N^{PW}$ |
|------|------|------|------|------|
| A | 1 | 16 | 16.1 | 13230 |
| B | 1 | 18 | 13.2 | 25600 |
| C | 2 | 10 | 10.1 | 13230 |
| D | 1 | 32 | 13.4 | 44100 |
| E | 2 | 36 | 16.9 | 32768 |

In the optimization or molecular dynamics runs all the substrate atoms are fixed at their bulk lattice positions corresponding to the lattice parameter of 5.64 Å.[44] The sodium clusters are initially placed within the vacuum region 3.5–4 Å from the substrate, and the total energy of the system is then minimized by allowing the cluster degrees of freedom to relax by a conjugate–gradient method. Energy minimizations are started mainly from the known geometries of the free clusters, but in some cases for $N \geq 5$ variations to the initial geometry are done in repeated minimization procedures. The molecular dynamics runs are done using a fifth–order predictor–corrector method [45] to integrate the equation of motion for cluster ions with a time step of 3–5 fs, which ensures an excellent conservation of the total energy. [17] Equilibration of the system is done when needed at random intervals during the first 0.5–1.0 ps of the run by drawing new velocities of ions from the Maxwell–Boltzmann distribution corresponding to the desired temperature. After the equilibration phase the system is left to evolve at constant total energy.

# 3 BO-LSD-MD Results

## 3.1 Physisorption vs. Chemisorption

As mentioned in the Introduction, the degree of the hybridization, and subsequently the strength of the "bond" created between the adsorbed cluster and the substrate, depends on the relative energy difference of the occupied "active" (valence) electron states in the cluster and in the substrate. For Na clusters deposited on NaCl surfaces the active electron states are those of the $3s$ electron of Na and $3s$, $3p$ electrons of Cl. When the undefected, stoichiometric, NaCl substrate is formed, the valence electrons rearrange themselves to form filled bands of Cl $3s$ and Cl $3p$ type, with a clear band gap to unoccupied levels. The present calculation gives HOMO-LUMO gap of the order of 5 eV, as shown in Fig. 2a.

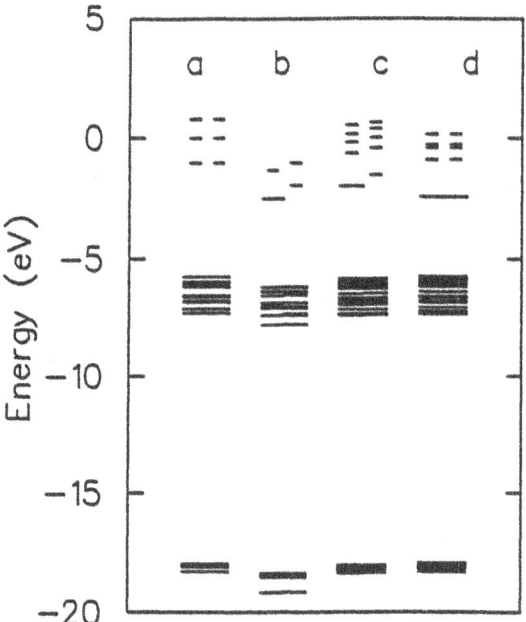

**Fig. 2.** KS one-electron energy levels at the $\Gamma$ point for (a) the pure NaCl substrate, (b) Na/NaCl(001) system, (c) substrate with the F center, and (d) the adatom – F center complex. For each set of levels both spin orientations are shown, the up-spin on the left and the down-spin on the right (when these are degenerate a long horizontal line is drawn). The shortest lines correspond to unoccupied levels and the longer ones to occupied levels.

Bringing a sodium adatom in the vicinity of the surface results in the filling of a state which is created in the band gap of the pure substrate, and is separated from the HOMO of the substrate by 3.6 eV (Fig. 2b). On the basis of the relatively large energy difference between the adatom state and the HOMO of the substrate one could expect a weak adatom-substrate interaction. This turns out to be the case: the calculated adsorption energy (when the adatom is located at the optimal adsorption site atop a substrate Cl ion 2.80 Å above the surface) with LSD approximation is 0.61 eV and no clear chemical bond is formed between the atom and the substrate (Fig. 3a), hence we characterize the process as physisorption.

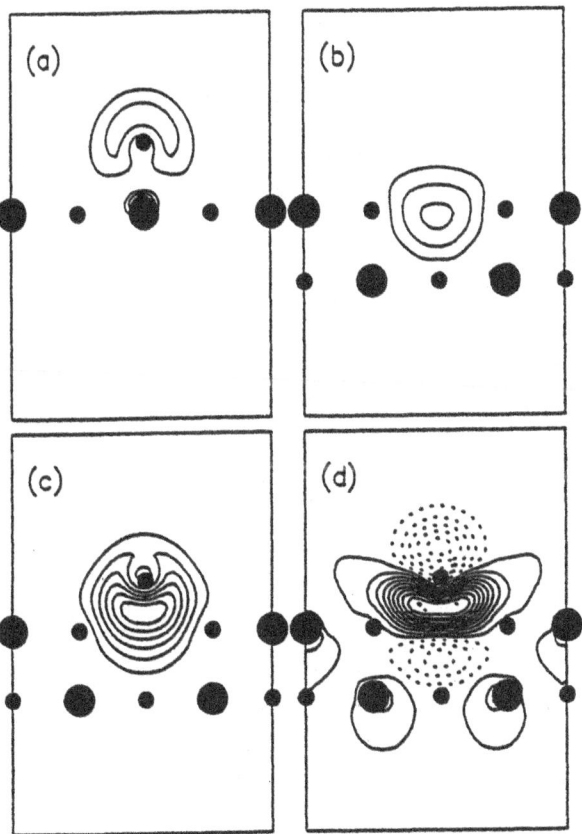

**Fig. 3.** Density contour plot of the KS state of the (a) adatom, (b) F center (FC), and (c) adatom − F center complex. In (a)–(c), contours are drawn evenly with an interval of 0.001 a.u. (d) Difference of total densities $\rho(Na + FC) − (\rho(Na) + \rho(FC))$ showing the bond formation in the region between the adatom and the defect. Accumulation and reduction of the density are denoted by solid and dotted contours, respectively. The contour interval is 0.0002 a.u. in (d). All the contours are calculated on a plane perpendicular to the surface, containing the adatom and the defect. Large and small dots depict chlorine and sodium atoms, respectively.

This picture changes dramatically if we introduce a simple defect on the ideal surface by removing a surface chlorine atom (keeping the system neutral, *i.e.*, with removal of seven valence electrons). This creates a defect which is analogous to F–centers (color centers) in bulk alkali–halides.[46] The "excess electron" (the electron which is not involved in forming the closed shells of the substrate chlorine ions) localizes at the chlorine vacancy, taking as part the role of a negative chlorine ion (Fig. 3b). Similar localization modes of excess electrons are found in halogen–deficient alkali–halide clusters. [16, 17, 18, 19, 20, 21, 22, 23, 24, 25, 26, 27, 28]

The KS eigenvalue associated with the excess electron (Fig. 2c) is located in the gap between the occupied and empty substrate states and happens to be very close to the eigenvalue of the electron of the Na adatom (Fig. 2b). The excess electron forms a spin–pair with the electron of the sodium adatom (Figs. 2d and 3c) which results in a clear chemical bond between the adatom and the F–center (Fig. 3d). As a result, the adatom binding energy increases to 1.58 eV and the optimal adsorption height reduces to 2.17 Å. The change in the nature of bonding is also reflected in the out–of–plane vibrational frequency of the adatom which shifts from 8.8 meV for the physisorbed atom to 6.9 meV for the chemisorbed atom.

When larger clusters are adsorbed on the surface, more states are created in the band gap of the substrate. An example is shown in Fig. 4 for the $Na_8$ cluster. The cluster states in the band gap correspond closely to the states of

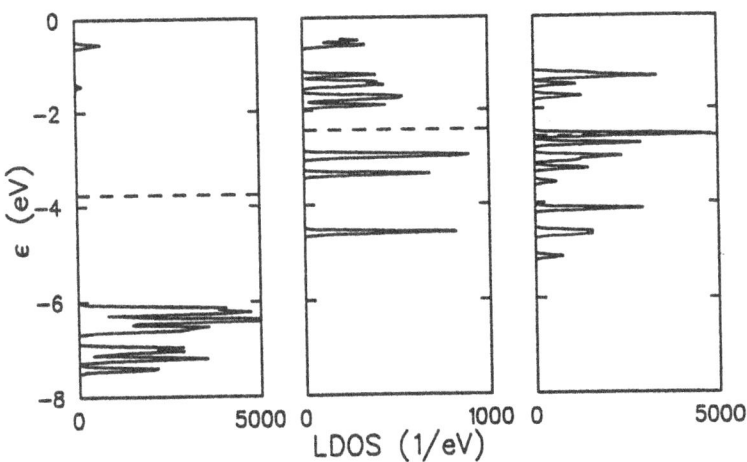

**Fig. 4.** The local density of states (LDOS) at the $\Gamma$ point of the supercell for the pure NaCl substrate (left), the free 3D $Na_8$ cluster (middle), and the Na(110) substrate (right). For NaCl we do not show the Cl $3s$ band which is located at about -18 eV. The discrete KS eigenvalues have been smoothed by 30 meV with Gaussians. The Fermi energy of each system is denoted by the dashed horizontal line.

free clusters, as shown in Sect. 3.2. Na$_8$ happens to be a special case since, having 8 valence electrons, it is "magic" as a free 3D cluster. We show in Sect. 3.3 that a 3D isomer of the adsorbed Na$_8$ behaves even *dynamically* at finite temperatures like a free 3D cluster.

The weak cluster-surface interaction discussed above is made possible by the clear energetic and spatial separation of the substrate and cluster electron states. To stress that this requirement is essential we show in Fig. 4 also the electron structure of a substrate consisting of Na(110) layers. Comparing Figs. 4b and 4c one would expect drastically different behavior if Na$_8$ cluster is deposited on a sodium substrate. In fact, a 3D cluster on top of Na(110) surface is not stable at all but spontaneously collapses on the surface forming an epitaxial adlayer.[12]

We saw that the surface F-center binds strongly the sodium adatom. Similar enhancement of binding of clusters occurs when F-centers are present and is discussed in Sect. 3.4.

## 3.2 Systematics of Geometry and Electronic Structure

The optimized structures for 2D Na$_N$/NaCl(100), $N = 2-8, 12$, are shown in Fig. 5. The mean bond lengths in the cluster and the mean distance of the cluster atoms from the surface plane are given in Table 2. On average, the adsorbed clusters lie 3.0–3.1 Å above the surface plane, which is slightly more than at the adsorption site of a single Na adatom (2.80 Å). It is obvious from Fig. 5 that Na atoms tend to locate themselves on top of substrate Cl$^-$ ions. However, the Cl$^-$–Cl$^-$ distance (3.99 Å) is slightly more than the typical bond length in the free sodium cluster, and the geometry cluster takes on the surface is a result of the competition between the energy cost paid to the corrugations in the substrate potential and the interactions within the cluster. We note that for $N \leq 4$ the symmetry of the surface cluster follows that of the free cluster (optimized with the same pseudopotential): [1] Na$_3$/NaCl(100) is a deformed triangle ($C_{2v}$) with the Na–Na bond length of 3.24 Å (3.15 Å for the free cluster) and the apex angle 95.0° (84.6°), and Na$_4$/NaCl(100) is a rhombus ($D_{2h}$) with the Na–Na bond length of 3.48 Å (3.42 Å) and the sharp angle 57.1° (52.2°). The bond length of the adsorbed dimer is 3.18 Å (2.99 Å). For $N = 5, 6$ the geometry of the adsorbed cluster is still close to that of the free cluster, but the corrugation of the substrate potential starts to play a greater role and the exact symmetries of the free cluster are broken. We note particularly the lowering of the symmetry ($C_{5v} \rightarrow C_{2v}$) for $N = 6$. Beyond $N = 6$, the 2D nature of the adsorbed cluster makes it drastically different from the free cluster.

We compare the electronic structure of the adsorbed clusters to the free ones by showing the Kohn-Sham single-electron levels in Fig. 6. It is seen from Fig. 6 that up to $N = 5$ the KS electronic structure changes very little in the process of adsorption, which correlates with the findings in Fig. 5 for the geometric

---

[1] When discussing the symmetry of the adsorbed 2D clusters we ignore the fact that the substrate induces small distortions in the $z$-direction, i.e., the clusters beyond Na$_3$ are not strictly planar.

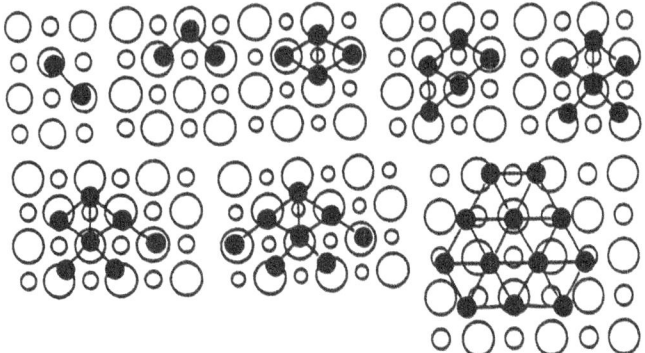

**Fig. 5.** Optimized structures of Na$_2$ (top, left) to Na$_{12}$ (bottom, right) on NaCl(001). Cluster atoms are shaded black, and white large and small circles depict substrate Cl$^-$ and Na$^+$ ions, respectively. In order to aid visualization, "bonds" between cluster atoms are drawn using a cutoff value of 3.7 Å.

**Table 2.** Mean bond length $\bar{d}$ within the adsorbed cluster, mean distance $\bar{z}$ of the cluster atoms from the surface, adsorption energy $E^A$, and binding energy $E^B$ of the cluster to the surface for Na$_N$/NaCl(001), $1 \leq N \leq 12$. The computational cell used for each cluster size is denoted by labels explained in Table 1.

| N | Cell | $\bar{d}$ (Å) | $\bar{z}$ (Å) | $E^A$ (eV) | $E^B$ (eV) |
|---|------|--------|--------|-----------|-----------|
| 1 | A | – | 2.80 | 0.609 | 0.609 |
| 2 | A | 3.18 | 2.82 | 0.574 | 0.983 |
| 3 | B | 3.24 | 3.11 | 0.572 | 0.975 |
| 4 | B | 3.48 | 2.98 | 0.506 | 1.029 |
| 5 | B | 3.49 | 2.99 | 0.512 | 1.082 |
| 6 | B | 3.38 | 2.99 | 0.486 | 1.139 |
| 7 | B | 3.44 | 2.99 | 0.417 | 1.036 |
| 8 | B | 3.46 | 2.97 | 0.414 | 1.051 |
| 12 | D | 3.47 | 3.46 | 0.423 | 1.128 |

structure. For $N = 6$, the lowering of symmetry ($C_{5v} \rightarrow C_{2v}$) splits off the two-fold degenerate HOMO level, although the splitting is hardly visible in Fig. 6. Clear differences become evident for $N = 7, 8$: (i) The free Na$_7$ (pentagonal bipyramid, $D_{5h}$) has degenerate $p$-states in the plane of the pentagon (separately for both spin polarizations), which split off for the 2D structure due to the reduced symmetry. (ii) The HOMO of the free Na$_8$ (dodecahedron, $D_{2d}$) has two-fold degeneracy, which splits off for the 3D adsorbed cluster, correlating with the finding that the initially slightly prolate cluster deforms towards an oblate shape upon adsorption.

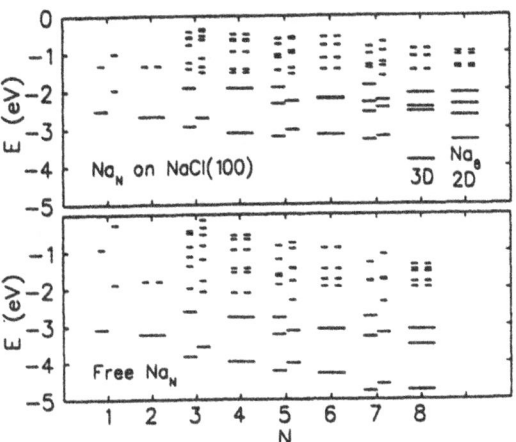

**Fig. 6.** KS one-electron energy levels (top) for $Na_N$/NaCl(001) at the $\Gamma$-point and (bottom) for the free sodium clusters. For $Na_N$/NaCl(001) we do not show the substrate levels, the highest of which are located at about -6 eV.

We show in Table 2 and in Fig. 7 the calculated binding energy $E^B$ of the cluster to the surface (energy per cluster atom to dissociate the adsorbed cluster into noninteracting neutral atoms)

$$E^B(N) = \frac{1}{N}(E^s + NE^a - E^{s+c}),\qquad(1)$$

where $E^s$ is the total energy of the substrate, $E^a$ energy of a free Na atom, and $E^{s+c}$ total energy of the cluster-substrate system. By adding and subtracting the total energy of the cluster, $E^c$, in (1) we rearrange the terms and interpret $E^B$ as the sum of the cluster adsorption energy

$$E^A(N) = \frac{1}{N}(E^s + E^c - E^{s+c})\qquad(2)$$

and the cluster dissociation energy

$$E^D(N) = \frac{1}{N}(NE^a - E^c).\qquad(3)$$

The decomposition of $E^B$ into $E^A$, $E^D$ is also shown in Fig. 7. The most interesting feature in Fig. 7 is the local maximum of $E^B$ at $N = 6$. We interpret this as an electronic shell effect in 2D. Considering the 2D adsorbed $Na_7$ it is clear that the seventh valence electron has to occupy an orbital parallel to the surface having $d$-character, whence in the free 3D $Na_7$ cluster a $p$-type orbital is occupied. Six valence electrons thereby close a $p$-like shell in the 2D shell structure (see Sect. 5.1). We further remark that $Na_8$ which is magic in 3D is now an open-shell cluster and, indeed, is strongly deformed from the circular symmetry, in analogy with open-shell 3D clusters, such as $Na_{14}$, which is far from being spherical.[37] Finally, we remark that while the binding energy per atom has clear odd-even alternations in free clusters,[1, 2] they seem to be absent in $E^B$. The reason for that is seen in Fig. 7: the odd-even staggering in the dissociation energy $E^D$ of the cluster is canceled by the alternations in the cluster adsorption energy $E^A$. The size-dependence of $E^A$ is obviously related to the crystal geometry of the surface and the behavior which almost exactly cancels out the odd-even staggering in $E^D$ is a fortuitous effect of the (001) surface structure.

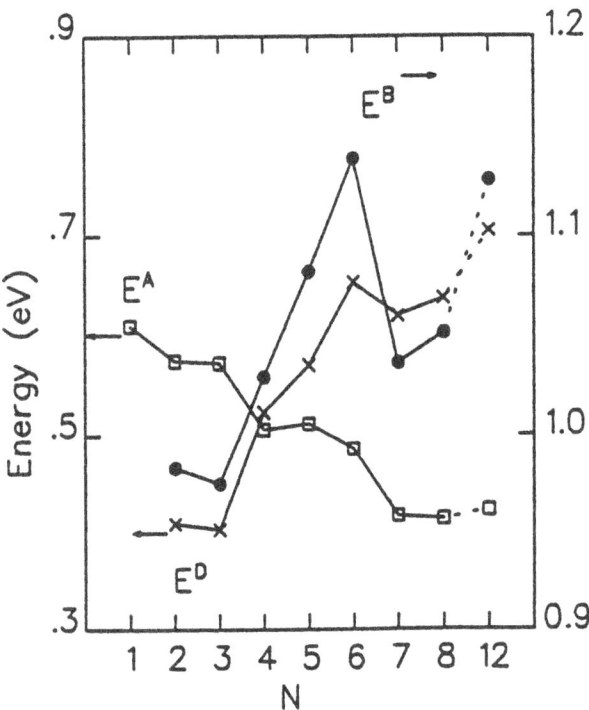

**Fig. 7.** Calculated binding energy $E^B(N)$ (scale on the right) to the surface per cluster atom, and its decomposition into cluster adsorption energy $E^A(N)$ and dissociation energy $E^D(N)$ (scale on the left).

## 3.3   Finite Temperature Effects

Free sodium clusters beyond $Na_6$ are three-dimensional.[1, 2] It is then of interest to study what is the effect of the substrate on the dimensionality and stability of a cluster, deposited initially in a 3D structure at finite temperature.

We have chosen to study Na$_8$ by molecular dynamics at temperatures relevant to experiments.

We select the dodecahedron ($D_{2d}$) as the initial geometry for the cluster,[47] bring the cluster in the vicinity of the surface such that one of its twelve triangular facets is parallel to the surface, and locate the nearest energy minimum of the system by a short quenching molecular dynamics run. The dynamics of the cluster is followed in three microcanonical runs, where the vibrational temperature sets at about 350, 600, and 1100 K, with a duration of 9.5, 9.8, and 6.0 ps, respectively. Figure 8 compares the distribution of Kohn-Sham levels associated to the cluster, time-averaged over the run at 350 K, to those obtained for a free cluster at 550 K. [37] Noting the difference in the cluster temperature, it can be concluded that the thermally averaged structure of the occupied states of the adsorbed cluster is virtually identical to that of the free cluster. A similar agreement of the distributions is seen also in the run at 600 K. Analysis of the dynamical properties of the adsorbed cluster, such as diffusion constant and the fourier spectrum of the velocity autocorrelation function provides further support to the conclusion that the adsorbed 3D Na$_8$ on NaCl(001) behaves at these temperatures very much like a free 3D cluster.

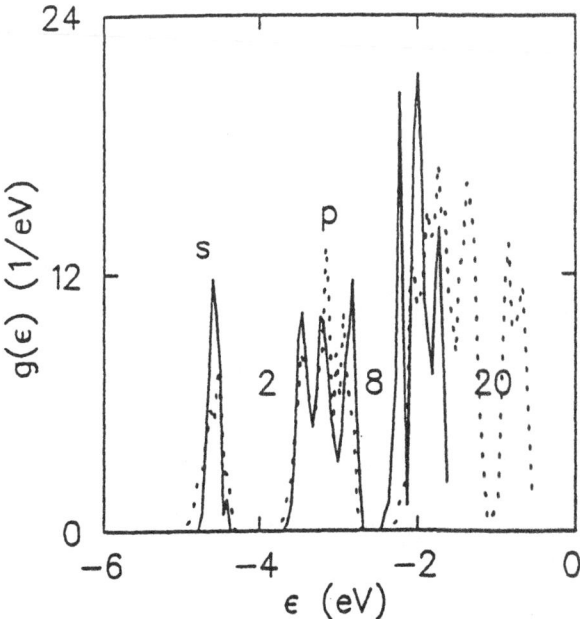

**Fig. 8.** The time-averaged distribution $g(\epsilon)$ of KS states for the adsorbed 3D Na$_8$ cluster at 350 K (solid curve), and the corresponding distribution for a free Na$_8$ at 550 K (dotted curve). $\int g(\epsilon)d\epsilon$ gives the indicated numbers of electrons for the closings of the major shells. The two curves are shifted for comparison such that the $s$-peaks approximately coincide.

338

We wish to remark that we have confirmed the thermal stability of the adsorbed 2D $Na_8$ (shown in Fig. 5) by separate independent molecular dynamics runs. Accordingly, we arrive at a conclusion that a notable energy barrier must exist between the 2D and 3D isomers. This raises an interesting question about the temperature range in which and the mechanism by which a dynamic transition from an initially 3D to a 2D structure would take place. Such a transition is made possible by the fact that a 2D structure is necessarily more open, causing the vibrational spectrum to shift into lower frequencies, which lowers the vibrational contribution of the total free energy. [48] We were able to see the transition in a simulation performed at 1100 K. The transition is characterized by closing of the HOMO–LUMO gap, broadening of the $p$-type states, and decrease of the $s, p$-gap. After the transition the HOMO–LUMO gap opens again which stabilizes the 2D structure. Figure 9 visualizes the process by showing atomic configurations of the cluster before and after the transition.

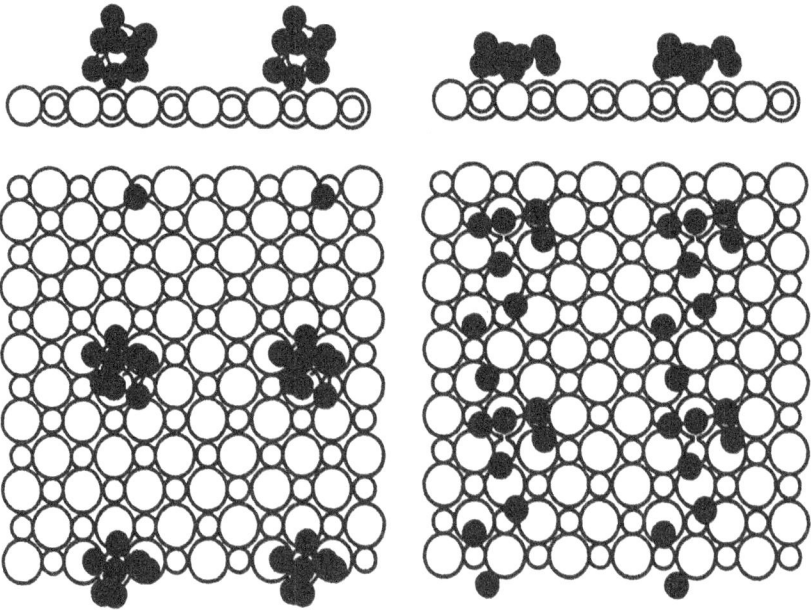

**Fig. 9.** Selected cluster configurations (side and top views) before (left) and after (right) the 3D→2D transition of adsorbed $Na_8$. For better visualization four surface unit cells are shown.

## 3.4   Binding of Metal Clusters to Surface F-centers

As discussed in Sect. 3.1, a surface F center increases significantly (by 1 eV) the binding energy of the sodium adatom to the surface, hence playing an important role in modifying the cluster-surface interaction due to the fact that F centers are one of the most common defect types at real surfaces of alkali-halide crystals.[49] In bulk alkali-halides F centers can form multiple defect complexes, such as M center, R center, and so on.[46] Such defect complexes are seen also in non-stoichiometric alkali-halide clusters.[16, 17] At the surface the excess electrons of an F center complex containing $N$ chlorine vacancies can form spin-pairs with electrons of an $N$-atom sodium cluster, thus enhancing the binding of the cluster to the surface. We have estimated the binding energy of $Na_2$ and $Na_4$ clusters in the presence of multiple F centers by placing the cluster atoms about 2.2 Å above the underlying chlorine vacancies. The resulting binding energies are shown in Table 3. By comparing tables 2 and 3 one can see that F centers enhance the binding by about 0.4 eV/atom. This is in fact a low-limit estimate since we have not allowed relaxations neither within the cluster nor in the surface layer.

**Table 3.** Binding energy $E^B$ of $Na_N$, $N = 1, 2, 4$, to the F center complex having $N$ chlorine vacancies. $4^a$ has a linear chain of F centers in the [11] direction whereas in $4^b$ a square-like arrangement of neighboring F centers is created. Computational cells denoted as in Table 2.

| N | Cell | $E^B$ (eV) |
|---|---|---|
| 1 | C | 1.58 |
| 2 | E | 1.38 |
| $4^a$ | E | 1.40 |
| $4^b$ | E | 1.48 |

# 4   Jellium Model for the Adsorbed Metal Cluster

The BO-LSD-MD studies discussed above, while giving detailed information both about the atomic structure and the whole valence electron structure of the cluster-substrate system, are limited to relatively small systems. For systematic studies of larger size-range of clusters and shell effects in 2D/3D, a simplified approach is necessary. We have chosen to study the electronic states of the cluster by using the jellium model "embedded" in a background potential mimicking the disturbance caused by the surface. The physical reasoning for such an approach is the fact that the nature of interaction between the perfect alkali-halide surface and a metal cluster appears to be weak physisorption.

In the conventional jellium model the positive ions of the metal are replaced with a homogeneous positive background charge[2]. The background density is determined by the bulk density of the metal in question. In the "ultimate" jellium model[29] (UJM) the background density is not fixed but adjusts itself to minimize the total energy. The bulk density of the UJM is close to that of sodium and, in fact, the model gives good results for the deformation parameters and for the binding energy variation as a function of cluster size for sodium clusters[29, 50, 51].

The effect of the surface on the total energy functional of the cluster in the UJM can be included via term [15]

$$E_{SC} = \int d\mathbf{r} n_+(\mathbf{r}) V_S(\mathbf{r}), \qquad (4)$$

where it is assumed that the surface potential interacts with the smooth positive background charge in the cluster. [Note that if the positive background is written as a superposition of delta functions located at the ionic positions $\{\mathbf{r}_i\}$, (4) implies that $E_{SC}$ consists of a sum over pairwise interactions, $E_{SC} = \sum_i V_S(\mathbf{r}_i)$.] Variation of the total energy functional with respect to the background density and the electron density yields the effective potential to electrons

$$V_{\text{eff}} = \frac{\partial E_{xc}[n]}{\partial n(\mathbf{r})} + V_S(\mathbf{r}). \qquad (5)$$

The same effective potential is obtained also in the case that the interaction between the surface and the cluster is assumed to be through the electrons of the cluster. The approach we have chosen does not include any polarization effects. However, it can be argued that a local approximation for the polarization would lead to the same form for the effective potential as given in (5).

The BO-LSD-MD results of $\text{Na}_N/\text{NaCl}(001)$ discussed above provide both a valuable database for constructing the surface potential and a testing ground for UJM results obtained with that potential. We parametrize $V_S$ according to the energy dependence of Na adatom on the position above the NaCl(001) surface. The simplest case is to ignore the spatial corrugation in the plane of the substrate which leads thus to a one-dimensional function $V_S(z)$, $z$ being the distance from the adatom to the surface plane. We select a functional form for $V_S$ as

$$V_S(z) = \min \left\{ \begin{array}{l} A\left[(\frac{\sigma}{z})^6 - (\frac{\sigma}{z})^4\right] \\ V_0 \end{array} \right., \qquad (6)$$

where $A$, $\sigma$ and $V_0$ are parameters deduced from the BO-LSD-MD studies. The exponent 4 in the long-range attractive tail is chosen to model the polarization effect between the atom and the thin substrate slab, [2] and the exponent of the short-range repulsive part is chosen just by practical means to get the best fit to the BO-LSD-MD data.

For the purpose to study the effect of the surface corrugation a sinusoidal term can be added to the attractive part of the potential:

$$V_S(\mathbf{r}) = V_S(z) + [\sin(kx) + \sin(ky)]\frac{C}{z^4}, \qquad (7)$$

---

[2] a semi-infinite substrate would lead to a polarization force $\propto z^{-3}$.

The parameters $k$ and $C$ determine the wave length and amplitude of the corrugation.

After the surface potential $V_S$ is defined the UJM calculations are performed as explained in [29].

# 5 Jellium Results

## 5.1 2D Shapes and Shell Structure of Adsorbed Clusters

The shapes, KS eigenvalues, and total energies have been calculated systematically for the adsorbed jellium clusters of 2–14 electrons in [15]. The main emphasis in [15] was put on studies using the uncorrugated surface potential $V_S(z)$. A recent related work by Kohl and Reinhard [32] studied clusters in a slightly larger size-range and put more emphasis on the spatial corrugation in $V_S$. Here we discuss the results mainly on the basis of [15] but mention also some key ingredients of [32]. Closely related are also recent calculations by Reimann et. al. [31] and Kohl et. al. [30] on *free* jellium clusters, confined in 2D.

Figure 10 shows our calculated shapes for the 2D adsorbed jellium clusters with 5 to 14 electrons, corresponding to the strength of 0.15 eV in the surface potential (i.e., the depth $V_S(z_0)$ of the well in the potential). It should be noted that in most cases for clusters with $N \geq 7$ it was possible to obtain both a 2D and a 3D stable isomer, the energy ordering of which depends on $V_S(z_0)$ (see Sect. 5.2).

The geometries of the smallest clusters can be understood simply by looking at the order how the lowest single particle orbitals would be filled in a 3D spherical jellium model. Up to $N = 6$, 1s and two 1p type orbitals are full, giving rise to a circular shape lying parallel to the surface. The highly symmetric 6-electron cluster is a "magic" one in 2D. The seventh electron starts to fill the next major shell in 2D shell structure, which gets complete in the 12-electron cluster. Figure 11 shows the total energy per electron for the 2D clusters shown in Fig. 10. The dotted line in Fig. 11 connecting the even clusters shows dips at $N = 6, 12$ confirming that these clusters are magic ones. In this context we remark that also the atomistic calculation of $Na_N/NaCl(001)$ gives a local maximum for the binding energy at $N = 6$ (Sect. 3.2). 6 and 12 also come out as magic numbers for free jellium clusters confined in 2D [30, 31].

Figure 11 shows also the odd-even staggering in the total energy, due to the deformation effects. The staggering is similar (albeit slightly weaker) to that of 3D clusters[30, 51, 52].

The effect of the spatial corrugation in the surface potential was studied by setting the amplitude and modulation to correspond to 0.2 eV and the lattice constant, respectively, in (7). For the size-range $N \leq 12$ we did not find any substantial effects in the shape or energy of the clusters due to corrugation. However, in [32] the corrugation was found to affect more the shapes but not energy for larger clusters. This is not conflicting with our findings since a slightly different parametrization for corrugation is used in [32], furthermore, the strongest effects are found for larger clusters than what was considered in [15].

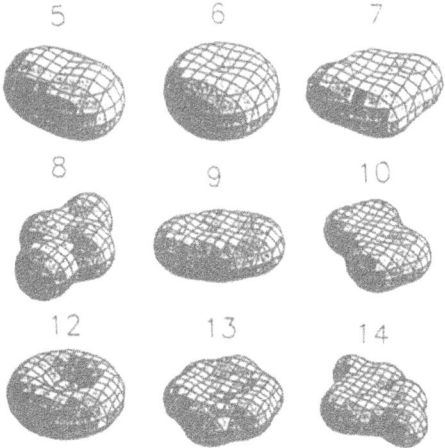

**Fig. 10.** Shapes of adsorbed 2D clusters. The three-dimensional plots are isodensity surfaces with $n = 0.00125a_0^{-3}$ (38 % of the bulk density). The figures do not have an absolute length scale. The 13-electron cluster is a 'saddle-point' geometry. The clusters are viewed from above the surface: The visible flat faces are those on the vacuum side of the cluster. The shapes have been calculated by using only the $z$-dependent part of the surface potential with a strength of 0.15 eV.

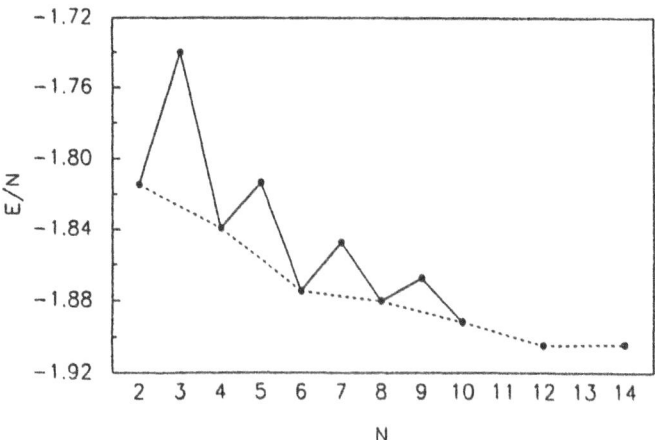

**Fig. 11.** The total energy per electron of adsorbed 2D clusters. Even clusters are connected with a dashed line to clarify the magic numbers 6 and 12

## 5.2 Competition Between 3D and 2D Geometries

Free jellium clusters with seven or more electrons are three-dimensional. Depending on the strength of the surface potential the adsorbed jellium cluster with 7 or more electrons can have either 2D or 3D ground state. In our studies with $V_S(z_0) = 0.15$ eV the only exceptions were 11- and 13-electron clusters. 11-electron cluster did not have at all a 2D isomer but always converged to a 3D geometry, and the 2D geometry of 13-electron cluster (shown in Fig. 10) was in fact a saddle point geometry which converged towards 3D after random perturbations to the effective potential were added.

An interesting case in our size-range is the 8-electron cluster, a well-known magic cluster in 3D. With reasonable variations of $V_S(z_0)$ (from 0.1 eV to 1 eV) this cluster was found to change the ground state from 3D to 2D geometry around a critical strength of about 0.7 eV. In this range of $V_S(z_0)$ the energy difference between the isomers of different dimensionality was of the order of 0.05 eV, which correlates well with the findings from BO-LSD-MD calculations of $Na_8/NaCl(001)$.

Kohl and Reinhard [32] have studied the isomerization in more detail and found that for larger clusters it is even possible to find more than one close-lying 3D isomer. They also developed a simple model for the critical strength of the substrate-cluster interaction. The important factors in the model are the volume and surface energies of the cluster, the interface energy, and the contact area of the cluster-support interface. The model is thus an extension of the conventional Liquid Drop Model (LDM) and can be used in quite general situations to predict the dimensionality of the preferred geometry as a function of the cluster size once the strength of the cluster-substrate interaction is determined.

## 6    Summary and Conclusions

We have reviewed recent studies aimed at contributing to understanding the basic interactions between small metal clusters and surfaces of different electronic properties, particularly insulating alkali-halide surfaces. We have chosen to investigate $Na_N/NaCl(001)$ system using a well-tested *ab initio* total energy method. Previous experimental and theoretical information about interactions between metal clusters and insulating crystal surfaces is scarce. We have studied the adsorption characteristics of a sodium adatom as well as the structure, stability and electronic structure of adsorbed clusters in the size range of $N \leq 12$.

Our results are summarized as follows. The nature of adsorption of a single sodium adatom on the NaCl(001) surface can be characterized as physisorption without any significant spatial overlap between the surface and the adatom one-electron states. The optimal adsorption site is atop a chlorine ion. If the surface contains a chlorine atom vacancy (surface F center) the nature of adsorption changes drastically and can be characterized as chemisorption. Similarly, the binding of clusters to the surface also gets stronger if F centers are present at the surface.

The effect of the undefected surface on the geometry and the electronic structure of the adsorbed clusters is very weak, due to a clear gap between the substrate and the cluster electronic states. The binding energy per cluster atom has a local maximum at $Na_6$ which we interpret to be caused by 2D electronic shell effects. For clusters larger than $Na_7$ both 2D and 3D adsorbed structures seem to be thermally stable. As an example we have studied the dynamics of the 3D $Na_8$ cluster on the surface. At temperatures relevant to experiments the dynamic structural and electronic properties of the cluster are very similar to the free 3D "magic" $Na_8$. A transition to 2D structure can be seen if the simulation is performed at high enough temperature. The structural transition is accompanied by a corresponding change in the electronic shell structure.

We have demonstrated that modern *ab initio* total energy methods can yield valuable information on cluster-surface interactions, which then can be used as a database for simpler models to access phenomena in the larger size-range of clusters. This allows for systematic investigations on the shell structure, stability, and dimensionality as a function of the strength of the cluster-support interaction. We expect that these jellium-based models will be proven useful in future studies of metal clusters on inert supporting substrates.

**Acknowledgement.** During the work reviewed in this paper the authors have benefited from discussions and collaboration with U. Landman and R. N. Barnett. The calculations have been performed using Cray C94 at the Center for Scientific Computing, Espoo, Finland. The collaboration with C. Kohl, S. M. Reimann, M. Koskinen, and J. Kolehmainen in the UJM studies is gratefully acknowledged. This work has been supported by the Academy of Finland and partly (HH) by the Väisälä Foundation.

# References

1. W. A. de Heer, Rev. Mod. Phys. **65**, 611 (1993).
2. M. Brack, Rev. Mod. Phys. **65**, 677 (1993).
3. *Physics and Chemistry of Finite Systems: ¿From Clusters to Crystals*, Vols. I-II, ed. by P. Jena, S. N. Khanna, and B. K. Rao (Kluwer, 1992).
4. H. Haberland, Z. Insepov, and M. Moseler, Z. Phys. D **26**, 229 (1993).
5. H.-P. Cheng and U. Landman, Science **260**, 1304 (1993).
6. P. M. St. John, R. D. Beck, and R. L. Whetten, Z. Phys. D **26**, 226 (1993).
7. A. N. Patil, D. Y. Paithankar, N. Otsuka, and R. P. Andres, Z. Phys. D **26**, 135 (1993).
8. T. Moriwaki, H. Shiromaru, and Y. Achiba, Z. Phys. D **26**, S320 (1993).
9. I. Moullet, Surf. Sci. **331-333**, 697 (1995).
10. T. P. Martin, Phys. Rep. **273**, 199 1996.
11. For review, see T. P. Smith III, Surf. Sci. **229**, 232 (1990).
12. H. Häkkinen and M. Manninen, Phys. Rev. Lett. **76**, 1599 (1996).
13. H. Häkkinen and M. Manninen, Europhys. Lett. **34**, 177 (1996).
14. H. Häkkinen and M. Manninen, J. Chem. Phys. **105**, 10565 (1996).
15. J. Kolehmainen, H. Häkkinen and M. Manninen, Z. Phys. D **40**, 306 (1997).

16. R. N. Barnett, H.-P. Cheng, H. Häkkinen, and U. Landman, J. Phys. Chem. **99**, 7731 (1995); H. Häkkinen, R. N. Barnett, and U. Landman, Chem. Phys. Lett. **232**, 79 (1995).

17. H. Häkkinen, R. N. Barnett, and U. Landman, Europhys. Lett. **28**, 263 (1994).

18. V. Bonačič-Koutecký, C. Fuchs, J. Gaus, J. Pittner, and J. Koutecký, Z. Phys. D **26**, 192 (1993).

19. U. Landman, D. Scharf, and J. Jortner, Phys. Rev. Lett. **54**, 1860 (1985); D. Scharf, U. Landman, and J. Jortner, J. Chem. Phys. **87**, 2716 (1987).

20. G. Rajagopal, R. N. Barnett, and U. Landman, Phys. Rev. Lett. **67**, 727 (1991); U. Landman, R. N. Barnett, C. L. Cleveland, and G. Rajagopal, in Ref. [3].

21. G. Galli, W. Andreoni, and M. P. Tosi, Phys. Rev. A **34**, 3580 (1986).

22. G. Rajagopal, R. N. Barnett, A. Nitzan, U. Landman, E. Honea, P. Labastie, M. L. Homer, and R. L. Whetten, Phys. Rev. Lett. **64**, 2933 (1990).

23. P. W. Weiss, C. Ochsenfeld, R. Ahlrichs, and M. M. Kappes, J. Chem. Phys. **97**, 2553 (1992).

24. T. Bergmann, H. Limberger, and T. P. Martin, Phys. Rev. Lett. **60**, 1767 (1988).

25. E. C. Honea, M. L. Homer, P. Labastie, and R. L. Whetten, Phys. Rev. Lett. **63**, 394 (1989); E. C. Honea, M. L. Homer, and R. L. Whetten, Phys. Rev. B **47**, 7480 (1993).

26. S. Pollack, C. R. C. Wang, and M. M. Kappes, Chem. Phys. Lett. **175**, 209 (1990); Z. Phys. D **12**, 241 (1989).

27. Y. A. Yang, C. W. Conover, and L. A. Bloomfield, Chem. Phys. Lett. **158**, 279 (1989).

28. P. Xia and L. A. Bloomfield, Phys. Rev. Lett. **70**, 1779 (1993); P. Xia, A. J. Cox, and L. A. Bloomfield, Z. Phys. D **26**, 1841 (1993).

29. M. Koskinen, P. O. Lipas, and M. Manninen, Z. Phys. D **35**, 285 (1995).

30. C. Kohl, B. Montag, and P.-G. Reinhard, Z. Phys. D **38**, 81 (1996).

31. S. M. Reimann, M. Koskinen, P. E. Lindelof, and M. Manninen, Z. Phys. D **40**, 310 (1977).

32. C. Kohl, Z. Phys. D **39**, 225 (1997).

33. R. N. Barnett and U. Landman, Phys. Rev. B **48**, 2081 (1993).

34. N. Troullier and J. L. Martins, Phys. Rev. B **43**, 1993 (1991).

35. S. H. Vosko, L. Wilks, and M. Nusair, Can. J. Phys. **58**, 1200 (1980); S. Vosko and L. Wilks, J. Phys. C **15**, 2139 (1982).

36. C. Brechignac, P. Cahuzac, F. Carlier, M. de Frutos, R. N. Barnett, and U. Landman, Phys. Rev. Lett. **72**, 1636 (1994); R. N. Barnett, U. Landman, and G. Rajagopal, Phys. Rev. Lett. **67**, 3058 (1991).

37. H. Häkkinen and M. Manninen, Phys. Rev. B **52**, 1540 (1995).

38. H.-P. Cheng, R. N. Barnett, and U. Landman, Phys. Rev. B **48**, 1820 (1993).

39. S. Wei, R. N. Barnett, and U. Landman, Bull. Am. Phys. Soc. **41**, 10 (1996).

40. H.-P. Cheng, R. N. Barnett, and U. Landman, Int. J of Quantum Chem. **29**, 615 (1995); Chem. Phys. Lett. **237**, 161 (1995); R. N. Barnett and U. Landman, Phys. Rev. Lett. **70**, 1775 (1993).

41. R. N. Barnett and U. Landman, J. Phys. Chem. **100**, 13950 (1996).

42. R. N. Barnett, H. Häkkinen, and U. Landman, unpublished.

43. J. Ihm, A. Zunger, and M. L. Cohen, J. Phys. C **12**, 4409 (1979).

44. N. W. Ashcroft and N. D. Mermin, *Solid State Physics* (Holt, Rinehart and Winston 1976).

45. C. W. Gear, *Numerical initial value problems in ordinary differential equations* (Prentice-Hall, NJ 1971).

46. W. B. Fowler (ed.), *Physics of color centers* (Academic Press, New York 1968).

47. V. Bonačič-Koutecký, P. Fantucci, and J. Koutecký, Chem. Rev. **91**, 1035 (1991); U. Röthlisberger and W. Andreoni, J. Chem. Phys. **94**, 8129 (1991).

48. T. P. Martin, Phys. Rep. **95**, 167 (1983).

49. U. Malaske, H.Pfnür, M. Bässler, M. Weiss, and E. Umbach, Phys. Rev. B **53**, 13115 (1996); S. Fölsch and M. Henzler, Surf. Sci. **247**, 269 (1991).

50. S. Bjørnholm, J. Borggreen, H. Busch, and F. Chandezon, in *Large Clusters of Atoms and Molecules*, ed. by T. P. Martin (Kluwer 1996).

51. C. Yannouleas and U. Landman, Phys. Rev. B **51**, 1902 (1995).

52. M. Manninen, J. Mansikka-aho, H. Nishioka, Y. Takahashi, Z. Phys. D **31**, 259 (1994).

# Experimental Studies of the Structures and Isomerization of Atomic Clusters

Ph. Dugourd,[1] R. R. Hudgins, A. A. Shvartsburg, and M. F. Jarrold

Department of Chemistry, Northwestern University
2145 Sheridan Road, Evanston, IL 60208, USA

**Abstract** Ion mobility measurements can be used to separate structural isomers of atomic clusters and to provide information about their geometries and isomerization processes. The principles behind ion mobility measurements and the methods used to calculate mobilities for comparison with the experimental data are briefly reviewed. With the development of high resolution ion mobility measurements, it is now possible to separate many more structural isomers than could be resolved using conventional techniques. Some recent results for carbon and silicon clusters are described. For sodium chloride nanocrystals several families of structural isomers have been resolved and the results show that dramatic shape transformations can occur at room temperature for these species.

## 1 Introduction

Since all the physical and chemical properties of an atomic cluster ultimately depend on its geometry, information about the geometric structure of atomic clusters is of critical importance. However, this information has been excruciatingly difficult to obtain for clusters in the critical 4-100 atom size regime. The presence of isomers makes the problem of determining structural information even more problematic. Spectroscopic information is difficult to interpret if there is a distribution of isomers present. And since naked atomic clusters have coordinatively unsaturated surfaces, it is quite common to find several different geometric structures with quite similar energies for clusters with more than a few atoms. The issue of isomerization and the activation energies associated with interconversion between the different geometries then becomes an important issue. In the last few years, ion mobility measurements have emerged as a technique that can provide information about the geometries and the isomerization processes of medium-sized atomic clusters. In this article we describe some of the recent developments in this area, including advances made in the methods used to calculate mobilities for comparison with the experimental data, and experimental advances that now make it possible to resolve many more structural isomers.

---

[1] Present and permanent address: Laboratoire de Spectrometrie Ionique et Moleculaire (UMR n°5579), CNRS et Universite Lyon I, 43 bd du 11 novembre 1918, 69622 Villeurbanne Cedex, France

The mobility of a gas phase ion is a measure of how rapidly it moves through a buffer gas under the influence of an electric field. Mobility measurements are performed in a drift tube, which contains the buffer gas and has a series of electrodes to provide a uniform electric field. The electric field accelerates the ions, while collisions with the buffer gas decelerate them, leading to a constant drift velocity, $v_D$. The mobility, $K$, is the ratio of the drift velocity to the electric field, $K = v_D/E$. For atomic ions, the mobility depends on the electronic state [1]. For a polyatomic ion, the mobility depends on the average collision integral or collision cross section. An ion with a large average collision cross section has more collisions with the buffer gas and travels more slowly than an ion with a small cross section. Thus mobility measurements can be used to separate cluster ions with different geometries [2]. Information about the resolved geometries can be obtained by comparing the measured mobilities to mobilities calculated for trial geometries. The structural information obtained from these studies can be ambiguous because more than one geometry can have the same mobility, and it is not as detailed as that available from high-resolution spectroscopic studies. But in many cases the types of spectroscopic studies needed to provide structural information cannot be applied to gas phase clusters with more than a few atoms.

In the last few years ion mobility measurements have been used to examine carbon clusters [3,4,5,6], silicon clusters [7], germanium clusters [8], aluminum clusters [9], and a variety of metal-containing carbon clusters [10]. The results for carbon clusters, first studied by Bowers and collaborators, are particularly noteworthy because so many different structural isomers have been resolved. These include linear chains, monocyclic rings, bicyclic and polycyclic rings, roughly planar graphite sheets, and fullerenes.

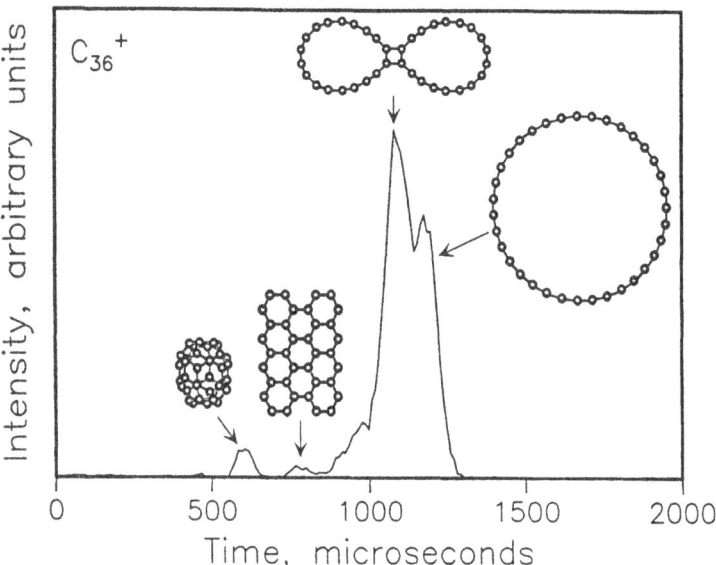

**Fig. 1.** Drift time distribution for $C_{36}^+$

Figure 1 shows a drift time distribution measured for $C_{36}^+$ clusters generated by laser vaporization of graphite. The drift time distribution shows the amount of time it takes for the cluster ions to travel across the drift tube. Slow moving ions with larger collision cross sections have longer drift times. For $C_{36}^+$ at least four isomers are resolved. With increasing drift time, the resolved isomers are a fullerene at around 600 μs, a roughly planar graphite sheet which is a minor isomer at around 800 μs, a bicyclic ring at around 1100 μs, and a monocyclic ring at around 1200 μs.

Information about the structure of the cluster ions generated by the source is only part of the story because the clusters may not be in their lowest energy geometry. This is particularly true for a laser vaporization source where the clusters are generated under kinetic rather than thermodynamic control. A simple annealing technique makes it possible to use ion mobility measurements to examine the isomerization processes of a cluster ion [11]. If the ions are injected into the drift tube at elevated kinetic energies, collisions with the buffer gas lead to a transient heating cycle. While hot, the ions can isomerize and at high injection energies they may fragment. Since the transient heating cycle occurs close to the entrance of the drift tube, the rest of the drift tube can be used to probe the geometries of the annealed parent ion or the fragments. This approach has been used to provide important information about isomerization processes in carbon clusters and the mechanism of fullerene formation [12,13]. For $C_{36}^+$ shown in Figure 1, the bicyclic rings convert into monocyclic rings as the injection energy is raised. This process can be understood in terms of strain relief: the monocyclic ring is less strained than the bicyclic ring. For slightly larger carbon clusters, some of the polycyclic ring isomers convert into fullerenes, and the efficiency of fullerene formation increases with increasing cluster size.

## 2 Mobility Calculations

Structural information is obtained from ion mobility measurements by calculating the mobilities for trial geometries and comparing them to the measured mobilities. In the low field limit, where the mobility is independent of the drift field, the mobility is given by [14,15]

$$K = \frac{(18\pi)^{\frac{1}{2}}}{16} \left[ \frac{1}{m} + \frac{1}{m_B} \right]^{\frac{1}{2}} \frac{ze}{(k_B T)^{\frac{1}{2}} \Omega_{avg}^{(1,1)}} \frac{1}{N} . \tag{1}$$

In this expression, m is the mass of the ion, $m_B$ is the mass of a buffer gas atom, N is the buffer gas number density, ze is the ion's charge, and $\Omega^{(1,1)}_{avg}$ is the average collision integral or collision cross section. Assuming that there is no alignment in the drift tube, which is a reasonable assumption for mobilities determined in the low field limit where the drift velocity is small compared to thermal velocities, the average collision cross section can simply be obtained by averaging over all possible collision geometries. Treating the polyatomic ion as a collection of hard spheres, one for each atom, and assuming hard sphere interactions between the ion and buffer gas atom, the average cross section is obtained by averaging the geometric cross section over all possible orientations in space. We refer to this as the hard sphere projection approximation because the geometric cross section is simply the area of the shadow cast by the trial geometry in collisions with the buffer gas [16,17]. This type of model has been widely used in the last few years [4,7,9,18,19,20].

While it is obvious that the hard sphere projection approximation ignores the long-range interactions between the ion and buffer gas, this approach also ignores all the details of the scattering process between the polyatomic ion and buffer gas atom. $\Omega^{(1,1)}_{avg}$ in Equation 1 is really a collision integral that should be calculated by averaging the momentum transfer cross section over relative velocity and collision geometry [14,15]. The momentum transfer cross section depends on the scattering angle, the angle between the incoming and outgoing trajectory in a collision between the polyatomic ion and a buffer gas atom. The projection approximation ignores all these details. We have recently developed an exact hard spheres scattering model [21] and have shown that the cross section obtained using the hard sphere projection approximation is equal to the collision integral only for a body with no concave surfaces. With concave surfaces, multiple scattering occurs and the projection approximation deviates from the true collision integral. Large deviations occur for some geometries, particularly those with grossly concave surfaces such as cups, and geometries with large-scale surface roughness.

Studies of mobilities as a function of temperature show that the long-range interactions between the buffer gas and the ion should also not be ignored [22,23]. Accounting for the long-range interactions correctly is not trivial. First an effective potential, consisting

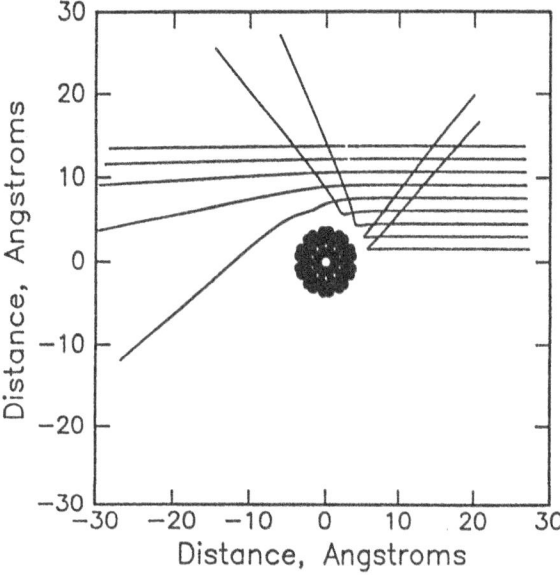

**Fig. 2.** Trajectories calculated for $C_{60}^{+}$/He collisions

of a sum of interactions between the buffer gas atom and all the atoms in the ion, must be defined and then trajectories are run in this potential to determine the scattering angle. Many trajectories must be run to average over the impact parameter, the relative velocity, and the collision geometry. Calculations along these lines have recently been reported by Mesleh et al. [23] employing a potential consisting of a sum of two body Lennard-Jones interactions and ion induced-dipole interactions. Figure 2 shows examples of trajectories calculated for collisions between He and $C_{60}^+$ using this potential. The trajectories were calculated with a collision energy of $k_BT$ with a temperature of 298 K. They clearly show the effects of the long range interactions between the helium buffer gas atom and the fullerene.

To illustrate the magnitude of the deviations between the values obtained using the different models described above, Table 1 shows inverse mobilities (the inverse mobility is proportional to the collision integral) calculated for $(C_{60})_2^{+/-}$ and icosahedral $(C_{60})_{13}^{+/-}$ [24] using the projection approximation, the exact hard spheres scattering model, and trajectory calculations with parameters determined from fitting the measured mobility of $C_{60}^{+/-}$. The projection approximation underestimates the inverse mobilities because it ignores the details of the scattering process. The difference between the values determined from the exact hard spheres scattering model and from trajectory calculations are smaller and result primarily from the difference in the effective potential between the buffer gas atom and the clusters. The effective potential is assumed to be given by a sum of two-body C-He interactions, and the effective potential experienced by the helium atom is different for $He/C_{60}^{+/-}$, $He/(C_{60})_2^{+/-}$, and $He/(C_{60})_{13}^{+/-}$. The effective potential depends on both the cluster size and geometry even if the same parameters are used for all C-He interactions. The differences between the predictions of the different models

**Table 1** A comparison of the inverse mobilities calculated for $(C_{60})_2^{+/-}$ and $(C_{60})_{13}^{+/-}$ using trajectory calculations, the exact hard spheres scattering model, and the hard sphere projection approximation. The calculations were performed with a temperature of 298 K. The parameters for all three models were obtained by fitting the measured mobility for $C_{60}^+$. The numbers in parenthesis show the ratio of the calculated inverse mobility to that determined by trajectory calculations.

| Fullerene Cluster | (Mobility)$^{-1}$ $m^{-2}Vs$ | | |
|---|---|---|---|
| | Trajectory Calculations | Exact Hard Spheres Scattering | Projection Approximation |
| $(C_{60})_2^{+/-}$ | 4222 | 4204 (1.00) | 4032 (0.95) |
| $(C_{60})_{13}^{+/-}$ | 15190 | 14690 (0.97) | 12750 (0.84) |

shown in Table 1 are large enough to make the wrong structural assignments, even for the dimer [25]. The measured inverse mobility of the $(C_{60})_2{}^{+/-}$ dimer generated by laser desorption from fullerene films is 4070 Vsm$^{-2}$, which is in reasonable agreement with the inverse mobility in Table 1 obtained using the projection approximation. However, the geometry used to calculate the inverse mobilities in Table 1 was for an isomer known as a 5-6 stick [26]. Numerous theoretical studies indicate that the lowest energy geometry for a chemically bound $C_{60}$ dimer is a [2+2] adduct [26,27]. And the inverse mobility from trajectory calculations for this geometry is 4047 Vsm$^{-2}$, in close agreement with the measured quantity, while the projection approximation yields a value of 3883 Vsm$^{-2}$. This is consistent with the expectation that the dimer is a [2+2] adduct rather than a 5-6 stick.

It should be clear from the preceeding examples that the evaluation of mobilities for comparison with experimental data is still not a completely solved problem. With the recent improvements in resolution, more isomers are being resolved with more subtle structural differences, placing even more stringent demands on the theoretical methods. All the methods described above assume a rigid geometry and elastic collisions. Lin et al. [17] have considered the effects of rotation of the polyatomic ion during a collision with a buffer gas atom. For a helium buffer gas the effect is small [28]. Book et al. [20] have reported a study of the effects of vibrational motion on mobilities calculated using the projection approximation. This was done by averaging over a Boltzmann ensemble of geometries generated by molecular dynamics simulations. There is no reason why this approach cannot be employed with the more sophisticated methods for calculating mobilities, including trajectory calculations. There have not yet been any studies of the effects of inelastic collisions on the calculated mobility of a polyatomic ion. While on average the collisions are elastic, because the ion is in thermal equilibrium with the buffer gas, individual collisions can be inelastic. A full molecular dynamics treatment is required to examine whether inelastic collisions significantly affect the mobilities.

## 3 Experimental Considerations

The minimum requirements for an ion mobility measurement are a source of ions, a drift tube containing a buffer gas that the ions travel across under the influence of a uniform electric field, and an ion detector. The mobility is measured by injecting a short packet of ions, and determining the amount of time it takes for them to travel across the drift tube. Since the mobility depends on the buffer gas number density, reduced mobilities, scaled to the number density at STP, are usually reported. The reduced mobility is given by:

$$K_o = \frac{L^2}{t_D V} \frac{273.2}{T} \frac{p}{760} \tag{2}$$

where V is the voltage drop across the drift tube, L is the length of the drift tube, $t_D$ is the drift time, p is the buffer gas pressure in torr, and T is the temperature of the buffer gas.

Since all the parameters in Equation 1 can be determined with an accuracy of better than 1%, it is not difficult to obtain an accuracy of a few percent in ion mobility measurements, and a reproducibility of better than 1%. The parameter that determines an ion's energy in a drift tube is E/N, where E is the electric field and N is the buffer gas number density. At low E/N, where the drift velocity is small compared to thermal velocities, the mobility is independent of the field. This is called the low field limit. In the high field limit, where the drift velocity is much larger than thermal velocities, the mobility is no longer constant and depends on E/N. Ion mobility measurements performed to obtain structural information should be done in the low field regime. It is easier to calculate the mobility in the low field limit, and in the intermediate and high field regimes, the ions may align to some extent in the drift field.

Diffusion limits the resolution in ion mobility experiments. If a short packet of ions is injected into a drift tube, the packet expands by diffusion, as it travels through the drift tube. If two isomers have mobilities that differ by less than the ion packet expands, they will not be resolved. As shown by Revercomb and Mason [29] the resolving power is given by:

$$\frac{t_D}{\Delta t} = \left[ \frac{L\, E\, z\, e}{16\, k_B T \ln 2} \right]^{1/2} \tag{3}$$

where ze is the charge and $\Delta t$ is the width of the peak at half height. It is apparent from this expression that in order to increase the resolving power it is necessary to lower the temperature, increase the length of the drift tube, or increase the drift field. Lowering the temperature to 77 K increases the resolving power by a factor of two. The length is limited by expansion of the ion packet by diffusion as it travels through the drift tube. If the drift tube is too long the ion packet expands and hits the walls and very few ions exit the drift tube. If the drift field is increased there must be a corresponding increase in the buffer gas pressure in order to keep the mobilities in the low field limit. However, for pressures above around 10 torr it becomes increasingly difficult to inject intact polyatomic ions into the drift tube from an external source, since they must overcome the buffer gas flowing out of the drift tube.

Previous ion mobility measurements for atomic clusters were performed using an injected ion drift tube configuration, that was originally developed by Hasted and collaborators [30]. With this configuration mass-selected ions are injected into the drift tube from an external source. With the buffer gas pressure limited to less than 10 torr, the drift field is limited to around 10 V/cm and the length, which is limited by diffusional loss of the ions, is limited to tens of centimeters. So the resolving power for this configuration is around 10-20. The results shown in Figure 1 were recorded with an injected ion drift tube apparatus, and the resolution is not sufficient to fully separate the isomers for $C_{36}{}^+$. In particular, the monocyclic ring isomers at around 1100 μs are not completely resolved from the bicyclic ring isomers at around 1200 μs. While carbon clusters have isomers with very different shapes, and hence very different mobilities, metal clusters and ionic clusters generally have isomers with similar shapes. The application of ion mobility measurements to these species is hindered by the low resolution available in injected ion drift tube experiments.

In order to significantly increase the resolution it is necessary to increase the electric field in the drift tube and increase the length of the drift tube. With the higher electric field it is necessary to increase the buffer gas pressure to keep the drifting ions in the low field regime, and with higher buffer gas pressures it is no longer possible to inject intact cluster ions into the drift tube. Thus the source must be attached directly to the drift tube. A buffer gas pressure of over an atmosphere can then be employed. With this buffer gas pressure, the drift field can be increased to hundreds of volts per centimeter, and the drift tube can be made around a meter long. The resolving power for this configuration is around 200-400, which is over an order of magnitude better than with the conventional injected ion drift tube configuration.

## 4 High Resolution Ion Mobility Measurements

Figure 3 shows a schematic diagram of the high resolution ion mobility apparatus recently constructed at Northwestern University [31]. The apparatus consists of four

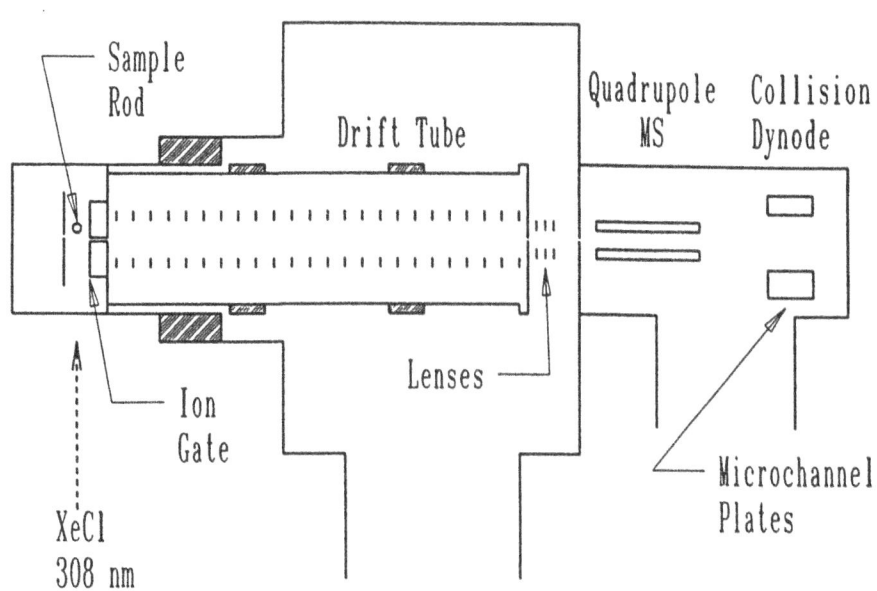

**Fig. 3.** Schematic diagram of high resolution ion mobility apparatus

main regions: 1) the source, where the clusters are produced; 2) the ion gate, which connects the source to the drift tube; 3) the drift tube; and 4) the mass spectrometer and ion detector. The drift tube and source region contain helium buffer gas at a pressure of around 500 torr. The source is a laser vaporization/desorption source with a near static buffer gas. The cluster ions are produced near the target rod and then guided to the entrance of the ion gate by a shaped electric field. The function of the ion gate is to allow cluster ions to pass from the source region to the drift tube while preventing neutral clusters from entering the drift tube. It is important that neutral clusters do not enter the drift tube, because if a cluster ion charge transfers in the drift tube to a neutral cluster with a different number of atoms, it will spend part of its time travelling through the drift tube as one cluster and the rest as another cluster, and so the drift time will not be characteristic of the cluster ion that is detected. The ion gate consists of a 2.5 cm long by 0.5 cm diameter cylindrical channel with a uniform electric field along its length to carry ions through. A counter-flow of buffer gas, from the drift tube to the source region, prevents neutral species from entering the drift tube from the source. The source region is pumped by a small mechanical pump to generate the buffer gas flow. The drift tube is 63 cm long and has 43 drift guard rings separated by ceramic spacers to generate a uniform electric field along its length. Drift voltages of up to 14 kV can be employed with a buffer gas pressure of 500 torr. At the end of the drift tube some of the cluster ions exit through a 0.13 mm diameter aperture into the vacuum chamber. The ions are then accelerated and focused by a set of electrostatic lenses into a differentially pumped chamber that houses a quadrupole mass spectrometer and ion detector.

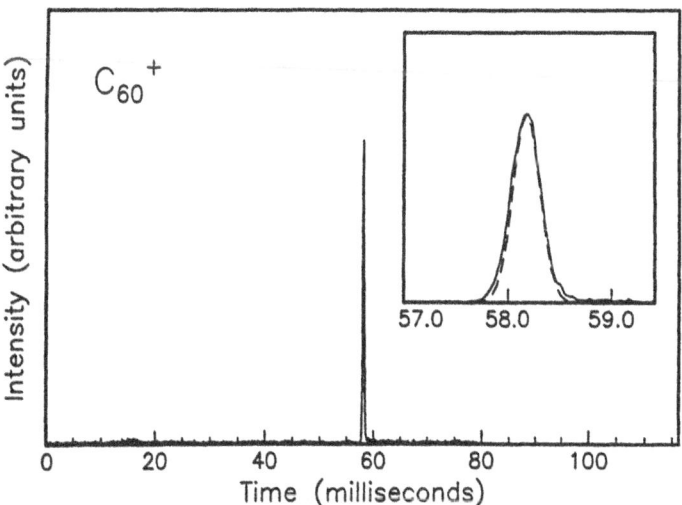

**Fig. 4.** Drift time distribution recorded for $C_{60}^+$ fullerene

Figure 4 shows the drift time distribution recorded for $C_{60}^+$ fullerene. The $C_{60}^+$ was generated by laser desorption from a fullerene film deposited on a copper rod. Assuming that all the ions are produced in the source at the same time, the shape of the drift time distribution can be calculated from [14]:

$$\Phi(t) = \frac{\Phi_o}{(4\pi Dt)^{1/2}} \left(1 - \exp(-r_o^2/4Dt)\right) \exp(-(L-v_Dt)^2/4Dt) \tag{4}$$

where $\Phi(t)$ is the flux of ions leaving the drift tube as a function of time, $D$ is the diffusion constant which under low field conditions is given by $eK/k_BT$, and $r_o$ is the radius of the source of ions. The calculated peak is shown by the dashed line in the insert in Figure 4. The measured peak is slightly broader than the calculated one. This is probably due to the initial spatial distribution of the ions. While the ions are generated by the laser pulse at essentially the same time, they are not produced at exactly the same position. Ions generated at slightly different positions are subjected to slightly different electric fields in the source, and so they enter the drift tube at slightly different times. The width at half height of the peak shown in Figure 4 is $\Delta t = 0.34$ ms, and the resolving power, $t_D/\Delta t = 172$. Isomers with mobilities that differ by more than 0.5% are easily separated in this apparatus.

## 5. Carbon Clusters

Figure 5 shows a high resolution drift time distribution recorded for $C_{24}^-$ clusters that were generated by laser vaporization of a graphite target. The most abundant isomer,

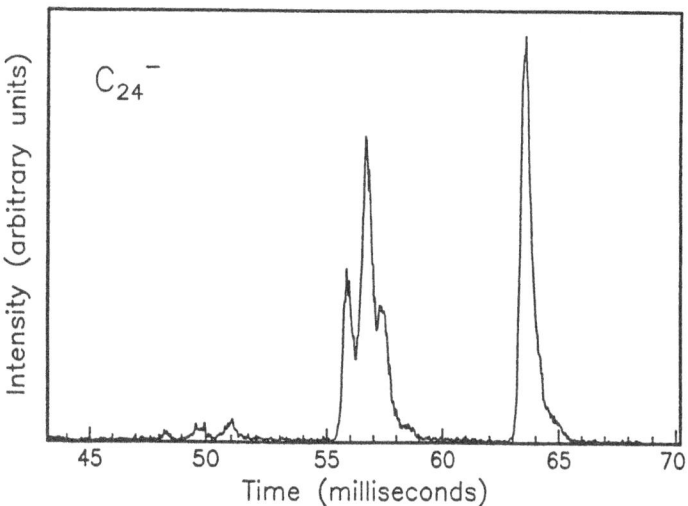

**Fig. 5.** High resolution drift time distribution measured for $C_{24}^-$.

which occurs at a drift time of around 64 ms has a mobility close to that expected for a linear chain. This isomer is not present in the drift time distribution for $C_{36}^+$ shown in Figure 1. The linear chains persist to larger cluster sizes for anions than for cations [32]. A group of three large peaks at around 57 ms in the drift time distribution for $C_{24}^-$ have mobilities close to that expected for a monocyclic ring. In the low resolution drift time distribution for $C_{36}^+$ shown in Figure 1, the peak at around 1200 µs has been attributed to a monocyclic ring. This peak is not completely resolved from the peak at 1100 µs which is attributed to a bicyclic ring. In the high resolution distribution shown in Figure 5 the bicyclic ring isomers are expected to occur at around 50 ms. There are three small peaks in this region of the drift time distribution which are presumably different types of bicyclic ring isomers. The presence of several bicyclic ring isomers is not unexpected because it is possible to generate a number of different bicyclic rings. The bicyclic rings are believed to result from the coalescence of two smaller monocyclic rings and there are a number of ways that this can occur. On the other hand, the resolution of several monocyclic ring isomers was not expected. There should be only one monocyclic ring isomer resolved in these studies, and the other isomers must be due to different geometries. A good candidate geometry for the other isomers is a ring with a short chain attached [33]. Peaks would then be expected in the drift time distribution for different combinations for an n-atom cluster with (n-m) atoms in the ring and m atoms in the chain. It is easy to see how these geometries could be produced in a laser vaporization plasma. Chains are the preferred conformation for small carbon clusters while rings become important for the larger clusters. For the cations the linear chains disappear around $C_{10}$ and then the rings dominate [4]. As noted above the linear chains persist to larger cluster sizes for the anions. The relative stability of the chain and ring isomers is determined by a balance between the strain energy for bending the chain into a ring, and the energy associated with making a new bond to form the ring. In the anions the extra electrons effectively tie-up one of the dangling bonds at the ends of the chain and the energy associated with making the bond to form a ring is less than for a cation [32]. So the chain persists to larger sizes. The ring+chain geometry is also expected to be more stable for anions than for cations because the extra electrons in the anion can tie up the dangling bonds at the end of the chain. Experiments have not yet been performed to examine this issue. Calculations indicate that the ring+chain geometry is relatively low in energy [33]. The only problem with assigning the additional "monocyclic ring" isomers to a ring+chain geometry is that mobilities calculated for these geometries are not in particularly good agreement with the measured mobilities. On the other hand, it is difficult to imagine another family of geometries that could account for the observed isomers, and so this may indicate that there is a problem with the calculated mobilities for these structures.

## 6 Silicon Clusters

Previous ion mobility measurements for silicon cluster cations resolved two main families of isomers [7]. For clusters with 10-30 atoms the relative mobilities systematically decreased, and the results were interpreted as indicating that silicon clusters in this size regime followed a prolate growth sequence to give elongated geometries. Starting at around $Si_{24}$ another family of isomers emerged with more spherical geometries. Combined ion mobility and chemical reactivity studies have shown that more than two isomers were present for many of the clusters [34]. At least some of these isomers have now been resolved. Figure 6 shows a high resolution drift time distribution measured for $Si_{30}^-$. At least five isomers are clearly resolved in the insert in

358

the Figure. In fact more than five isomers are present for this cluster because many of the peaks in the drift time distribution are broader than expected for a single isomer. Similar results were obtained for other cluster sizes. The large number of isomers resolved here may explain why optical absorption spectra and photoelectron spectra [35,36] recorded for silicon clusters in this size regime are broad and do not show many sharp features. We have attempted to perform laser annealing studies of the silicon clusters in the drift tube. While irradiation of the clusters with a pulsed laser apparently affected the relative abundances of some of the isomers, it was not possible to reduce the number of isomers present. Some of the effects of laser irradiation of these ions could be due to photodetachment. It is clearly desirable to develop an annealing method that will convert all of the isomers into the lowest energy geometry. This is not easy for silicon, because the directional covalent bonding leads to substantial activation barriers for isomerization. It is also possible that at the high temperatures necessary to cause isomerization on a reasonable timescale, the lowest energy geometry may not dominate because a higher energy geometry has a larger entropy. Thus it may not be physically possible to generate the lowest energy structure. So the best strategy to examine the

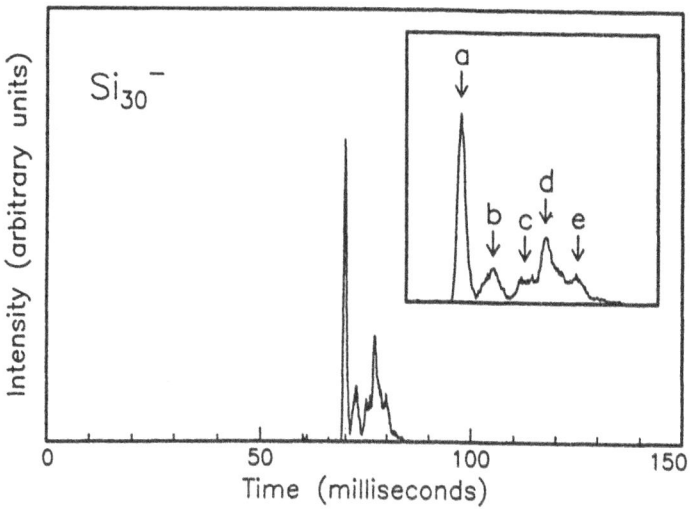

**Fig. 6.** Drift time distribution measured for $Si_{30}$

properties of these clusters is probably to separate out the different structural isomers and examine their properties individually. These types of experiments can be performed by coupling ion mobility measurements to other experimental techniques such as photoelectron spectroscopy and chemical reactivity studies. Using this approach, information can be obtained for not only the most stable geometry, but for other geometries as well.

## 7. Alkali Halide Nanocrystals

Alkali halide clusters are interesting model systems because the bulk fcc structure emerges at very small cluster sizes [37,38,39,40,41,42,43]. Mass spectrometry studies and ionization energy measurements indicate that alkali halide cluster ions, $(MX)_n X^-$ or $(MX)_n M^+$, with complete fcc nanocrystal structures, such as a 3x3x3 cube and the 5x3x3 and 5x5x3 cuboids, are particularly stable [37,38,39,40]. Cuboids with one vacancy also appear to be stable structures and it has been suggested that geometries with complete terraces are stable as well [40]. However, except for the "magic" number cuboid clusters, the determination of structural information from ionization energy, absorption cross section, or mass spectrometry measurements is not straightforward, particularly if isomers are present [41].

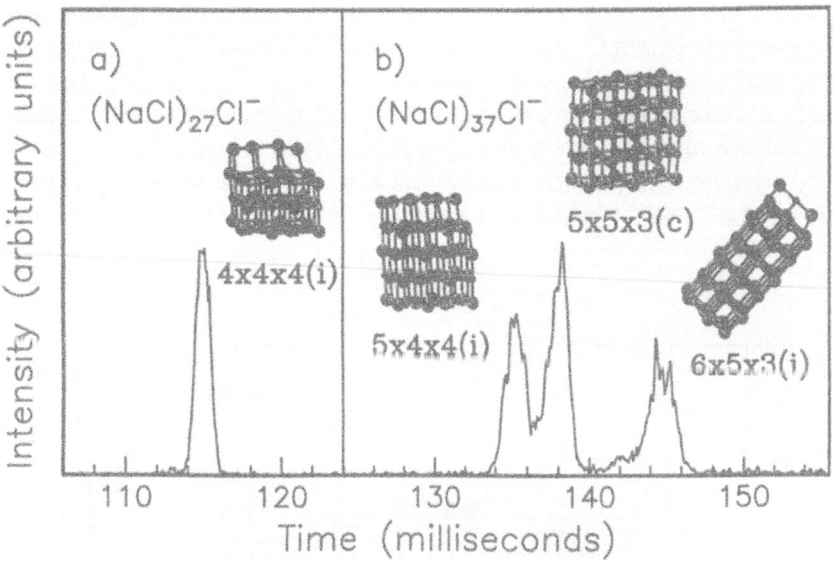

**Fig. 7.** Drift time distributions measured for $(NaCl)_{27}Cl^-$ and $(NaCl)_{37}Cl^-$

Figure 7 shows portions of the drift time distributions measured for $(NaCl)_{27}Cl^-$ and $(NaCl)_{37}Cl^-$ cluster ions. For $(NaCl)_{27}Cl^-$ only a single peak is observed indicating that all clusters have the same mobility. For $(NaCl)_{37}Cl^-$, three peaks are present in the drift time distribution. As will be described below, the observed peaks are assigned to specific structures by comparing measured mobilities to those calculated for geometries derived using an ionic potential. For $(NaCl)_{27}Cl^-$ the observed peak is assigned to an incomplete 4x4x4 geometry. For $(NaCl)_{37}Cl^-$ the three peaks are assigned to an incomplete 5x4x4 structure, a perfect 5x5x3 geometry, and an incomplete 6x5x3 structure. These different geometries are shown in the figure. Multiple geometries are present for almost all clusters with more than 30 NaCl units. The relative stabilities of these isomers were

360

investigated by recording drift time distributions as a function of the buffer gas temperature, from -7°C to +77°C. The number of peaks present, and their relative intensities depend on the temperature. At low temperatures, multiple isomers are observed for some of the small clusters, n<30, where only a single isomer is observed at room temperature, and even more isomers, up to four, are observed for the larger clusters, n>30, where multiple isomers were observed at room temperature. At high temperatures only a single isomer is present or dominant in the drift time distributions for most of the clusters. The buffer gas pressure in the drift tube in these experiments is around 500 torr, and the ions spend around 100 ms in the drift tube where they experience around $10^9$ collisions with the buffer gas. *So there is absolutely no question that the cluster ions are brought into thermal equilibrium with the buffer gas, and that the changes that occur as the drift tube temperature is raised result from the isomerization of higher energy isomers into lower free energy geometries.* The changes in the drift time distributions cannot be due to dissociation or electron detachment because the activation energies for these processes are too large for them to occur with the temperatures employed [39,44]. Figure 8 shows a plot of the relative inverse mobilities of the $(NaCl)_nCl^-$ clusters with up to 49 NaCl units. The inverse mobilities systematically increase with increasing cluster size because of the change in the physical size of the cluster. In order to remove this systematic change, and emphasize the changes in the mobilities that are related to structural changes, relative inverse mobilities [45] are plotted in Figure 8. The isomers that dominate at high temperatures are represented by the filled circles, while additional isomers that are observed at lower temperatures are

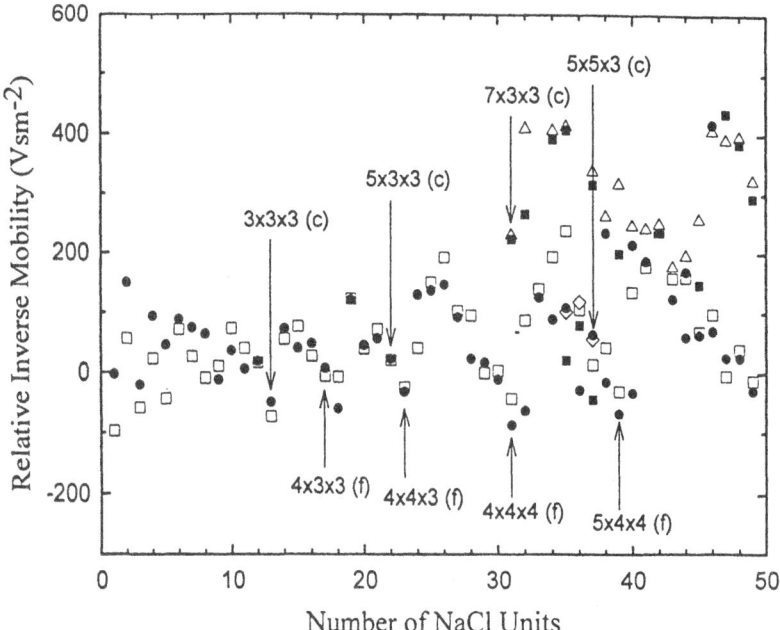

**Fig. 8.** Relative inverse mobilities for $(NaCl)_nCl^-$ clusters

represented by filled squares. Except for two clusters, $(NaCl)_{35}Cl^-$ and $(NaCl)_{37}Cl^-$, the most stable isomer is the most compact one, with the smallest collision cross-section, and smallest inverse mobility. The relative inverse mobilities show a series of steps where the inverse mobility increases sharply on going from one cluster size to the next. After each step the relative inverse mobilities gradually decrease.

The open points in Figure 8 show the mobilities calculated using the exact hard spheres scattering model for geometries obtained using an ionic potential that includes polarization effects [46,47]. The parameters used in the potential were taken from Welch et al. [48]. For clusters with less than 28 NaCl units, the most stable geometries were determined using a metropolis algorithm and then optimized using a conjugate gradient method [47]. For clusters with more than 28 NaCl units, a global optimization was not performed because of the enormous number of possible geometries. Geometries for the larger clusters were obtained by building a structure on an NaCl lattice and then optimizing it using the ionic potential. Calculations were performed with several different cuboid dimensions for each cluster size. Overall, there is good agreement between the measured mobilities and those calculated for the geometries determined from the ionic potential. For clusters with more than five NaCl units the root mean square deviation is 1.3%.

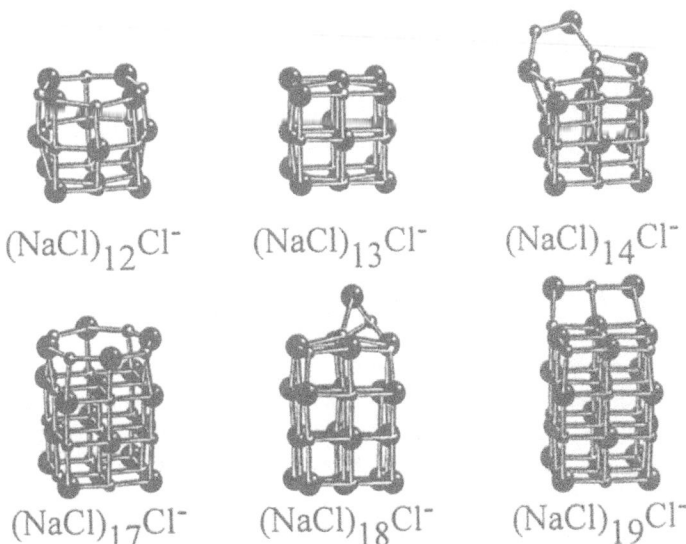

$(NaCl)_{12}Cl^-$   $(NaCl)_{13}Cl^-$   $(NaCl)_{14}Cl^-$

$(NaCl)_{17}Cl^-$   $(NaCl)_{18}Cl^-$   $(NaCl)_{19}Cl^-$

**Fig. 9.** Optimized geometries for some $(NaCl)_nCl^-$ clusters

As can be seen from Figure 8, the relative inverse mobilities for the larger clusters show a number of steps. The first step occurs between $(NaCl)_{13}Cl^-$ and $(NaCl)_{14}Cl^-$. Geometries optimized with the ionic potential for $(NaCl)_{12}Cl^-$, $(NaCl)_{13}Cl^-$, and $(NaCl)_{14}Cl^-$ are shown in Figure 9. $(NaCl)_{13}Cl^-$ is a 3x3x3 cuboid, and the step is due to the completion of this cuboid. While $(NaCl)_{12}Cl^-$ and $(NaCl)_{13}Cl^-$ have approximately the same collision cross sections, $(NaCl)_{14}Cl^-$ has two atoms on the surface of the cuboid, which causes a significant increase in its collision cross section relative to the $(NaCl)_{13}Cl^-$ cuboid. For clusters larger than n = 14 the relative inverse mobilities decrease as the cluster grows from a 3x3x3 geometry to a 4x3x3 and the next step occurs at n = 18. Optimized geometries are shown in Figure 9 for $(NaCl)_{17}Cl^-$, $(NaCl)_{18}Cl^-$, and $(NaCl)_{19}Cl^-$. For $(NaCl)_{18}Cl^-$, which has one atom in excess of a 4x3x3 cuboid, the additional atom is close to one face of the cuboid and does not cause a significant increase in the cross section. A substantial increase in the cross-section is then observed for $(NaCl)_{19}Cl^-$, which has three atoms in excess of the 4x3x3 cuboid. The next step occurs at n = 23 and corresponds to the completion of the 4x4x3 cuboid. However, going from the 4x3x3 cuboid to the 4x4x3 cuboid is not a simple progression because according to the calculations with the ionic potential, the 5x3x3 geometry is more stable than an incomplete 4x4x3 geometry for n = 21 and 22. For n = 22 the 5x3x3 geometry is a perfect cuboid, and like n = 13 this is a "magic" number in mass spectra of these clusters. Between n = 24 and n = 32 the relative inverse mobilities decrease as the clusters progress from a 4x4x3 cuboid to a 4x4x4 cuboid. A 4x4x4 cuboid with a single vacancy occurs at n = 31. $(NaCl)_{31}Cl^-$ is the first cluster for which more than one isomer is observed at room temperature. The isomer with the smaller inverse mobility is probably a 4x4x4 cuboid with a single vacancy. The inverse mobility of the second isomer corresponds almost exactly to the inverse mobility estimated for an elongated 7x3x3 cuboid. For n = 31 the 7x3x3 geometry is a perfect cuboid, like the 3x3x3 for n = 13 and the 5x3x3 for n = 22. However, the complete 7x3x3 cuboid is not the most stable geometry for $(NaCl)_{31}Cl^-$ because at higher drift tube temperatures the other isomer, which is probably a 4x4x4 cuboid with a single defect, dominates. The compact 4x4x4 structure is more stable than the elongated 7x3x3 cuboid because of the finite range of the ionic potential. The next step at n = 40 appears to be associated with completion of the 5x4x4 cuboid. Multiple isomers are observed for most of the clusters with more than 30 NaCl units, and three other geometries appear to be important for clusters with 32-40 NaCl units. Elongated isomers that appear to be 8x3x3 geometries are observed for n = 32, 34, and 35. The other two geometries that are important in this size regime are the 5x5x3 and the 6x5x3. The perfect 5x5x3 cuboid is completed at n = 37, and this is a "magic" number cluster like n = 13 and 22. The drift time distribution measured for n = 37 at room temperature (see Figure 7) has three peaks which are assigned to an incomplete 5x4x4, a perfect 5x5x3, and an incomplete 6x5x3. At higher drift tube temperatures the relative abundance of the 5x5x3 cuboid increases, suggesting that this is the most stable geometry. Although, a significant abundance of the other geometries remains at the high drift tube temperatures. For clusters with 35 and 36 NaCl units we observe geometries that appear to be incomplete 5x5x3 structures, but these are not the most stable geometries for these clusters. For $(NaCl)_{35}Cl^-$ there are three isomers which we have assigned to an incomplete 5x5x3, an incomplete 5x4x4, and an 8x3x3 with a single defect. At high drift tube temperatures only the incomplete 5x4x4 geometry remains, indicating that this is the most stable isomer. For clusters with more than 40 NaCl units 5x5x4 and 6x5x3 geometries appear to be important. The 6x5x3 cuboid is completed at n = 44 and at this point the 7x5x3 geometry emerges. The 5x5x4 cuboid is completed at n = 49.

Previous studies have emphasized the special stability of the "magic" number cuboid clusters. The first perfect cuboid is 3x3x3 and there is little doubt that this is the structure of the $(NaCl)_{13}Cl^-$ cluster. Calculated mobilities for the $(NaCl)_{22}Cl^-$ (5x3x3), $(NaCl)_{31}Cl^-$ (7x3x3), and $(NaCl)_{37}Cl^-$ (5x5x3) perfect cuboids are also in excellent agreement with the measured mobilities. However, the 5x3x3, 7x3x3, and 5x5x3 geometries only emerge at or near the point where there are the right number of atoms to generate the perfect cuboids. The steps in the observed mobilities correspond to completion of 4x3x3, 4x4x3, 4x4x4, and 5x4x4 cuboids, and most of the clusters have these geometries. These more compact fcc geometries are more stable than incomplete "magic" number cuboids because of the finite range of the ionic potential.

## 9 Structural Transformations of Alkali Halide Nanocrystals

As described above, some of the $(NaCl)_nCl^-$ geometries isomerize as the drift tube temperature is raised. By measuring drift time distributions as a function of the drift voltage it is possible to determine rate constants for the isomerization processes. Since the buffer gas pressure in these experiments is close to 500 torr, and the isomerization

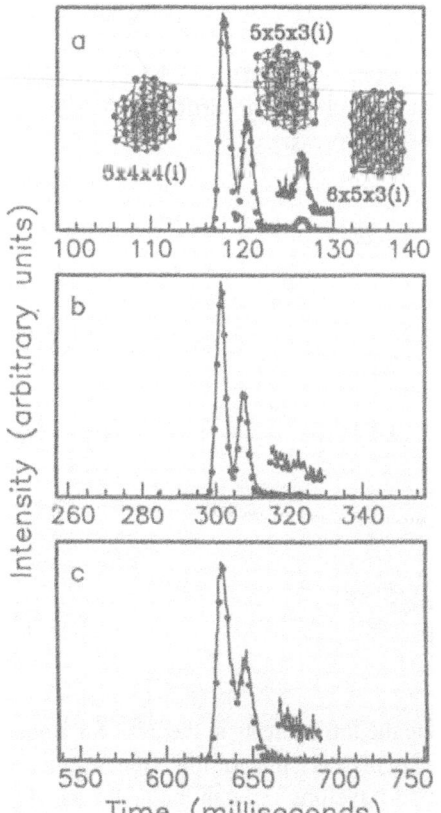

**Fig. 10.** Drift time distributions measured for $(NaCl)_{36}Cl^-$ with drift voltages af a) 10.5 kV, b) 4.0 kV, and c) 1.95 kV.

processes occur on a millisecond timescale there is no question that these are thermally-activated isomerization processes. Thus accurate activation energies can be determined from the rate constants measured as a function of temperature. This is the most direct and most accurate way to determine the activation energies for these isomerization processes. Injected ion drift tube techniques have been used to examine the isomerization processes of silicon [11,34,49] and carbon clusters [13,50]. However, these studies were performed by collisional annealing, where the injection energy is increased to collisionally heat the clusters as they enter the drift tube.

Figure 10 shows drift time distributions measured for $(NaCl)_{36}Cl^-$ at room temperature with drift voltages of 10.5 kV, 4kV, and 1.95 kV. There are three peaks in the drift time distribution measured with a drift voltage of 10.5 kV. These have been assigned to an incomplete 5x4x4 fcc geometry, an incomplete 5x5x3 geometry, and an incomplete 6x5x3 geometry. As the drift voltage is lowered the peak at long drift times, which is assigned to an incomplete 6x5x3 geometry, disappears. The ions spend longer travelling through the drift tube as the drift voltage is lowered. The average drift times are around 120 ms, 300 ms, and 650 ms at drift voltages of 10.5 kV, 4.0 kV, and 1.95 kV, respectively. These results indicate that the incomplete 6x5x3 geometry is isomerizing into one or both of the other two geometries as it travels through the drift tube. This isomerization process occurs at room temperature. Experiments performed at higher drift tube temperatures suggest that the 6x5x3 geometry isomerizes into the 5x4x4 geometry.

Rate constants for the isomerization of the 6x5x3 geometry can be obtained by simulation of the drift time distributions shown in Figure 10. The drift time distribution for two isomers, A and B, can be described by the expression:

$$\Phi(t) = \int \rho(t) \, g(t,t') \, dt' \qquad (5)$$

with

$$\rho(t) = A_o \delta(t - L/v_A) + B_o \delta(t - L/v_B) \, e^{-kt} + \int_0^t B_o \delta\left( t - \tau - \frac{L - v_B \tau}{v_A} \right) e^{-k\tau} \, k \, d\tau \qquad (6)$$

and

$$g(t,t') = \frac{v_A}{(Dt)^{1/2}} \, e^{-\frac{v_A^2 (t - t')^2}{4Dt}} . \qquad (7)$$

The function $\rho(t)$ has three terms, that describe the ion intensity at the peak for isomer A and isomer B, and the area between the two peaks that results from isomer B converting into isomer A during the time spent travelling through the drift tube. The function $g(t,t')$ accounts for the diffusional broadening of the ion packet as it travels through the drift tube. The broadening due to the initial spatial distribution of the ion packet is also accounted for, though this is not shown in the equations. The distribution is thus described in terms of the drift tube length L, the drift velocities $v_A$ and $v_B$ of isomers A and B, the initial ion intensities $A_o$ and $B_o$, the diffusion constant for the ions D, and the isomerization rate constant k. D, $v_A$, and $v_B$ are extracted from the drift time distributions directly. Thus $A_o$, $B_o$, and k are the only variables. For situations where there is more than one interconverting isomer, additional terms, analogous to the last two terms in Equation 6, can be added to $\rho(t)$ to account for the additional isomers.

The fit to the data for $(NaCl)_{36}Cl^-$ shown in Figure 10 is represented by the dotted lines. The rate constant employed was $k = 7$ $s^{-1}$ and this value successfully accounts for the relative intensity of the peak assigned to the incomplete 6x5x3 geometry at all drift voltages. Similar simulations have been performed for $(NaCl)_{30}Cl^-$, $(NaCl)_{35}Cl^-$, $(NaCl)_{36}Cl^-$, and $(NaCl)_{37}Cl^-$ at various temperatures between -7°C and 77°C. Figure 11 shows drift time distributions for $(NaCl)_{35}Cl^-$ at three temperatures at a fixed drift voltage of 7.0 kV. The three peaks in the drift time distribution at low temperature have been assigned to an incomplete 5x5x3 fcc geometry, an incomplete 5x4x4 geometry, and an 8x3x3 geometry with a single defect. As the temperature is raised the peaks due to the 5x5x3 and 8x3x3 geometries disappear leaving only the 5x4x4 geometry. The intensity between the middle and rightmost peaks increases and develops a steeper slope as the

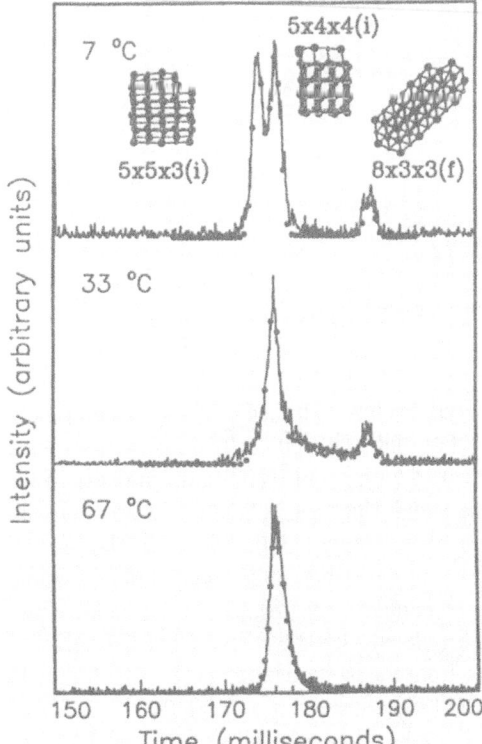

**Fig 11.** Drift time distributions for $(NaCl)_{35}Cl^-$ as a function of temperature.

temperature increases. This reflects an increase in the isomerization rate constant at the higher temperature. The dotted lines show the fits to the experimental data used to derive isomerization rate constants. For the 8x3x3 to5x4x4 transformation the following rate constants were obtained: $0.9\pm0.3$ s$^{-1}$ at 7°C, $2.7\pm0.5$ s$^{-1}$ at 33°C, and $60\pm3$ s$^{-1}$ at 67°C. Since the isomerization processes occur under thermally activated conditions, the rate constants should follow the Arrhenius expression, $k = A \exp(-E_A/k_BT)$ where $E_A$ is the Arrhenius activation energy. Figure 12 shows an Arrhenius plot, ln k plotted against $1/k_BT$, for the 8x3x3 to 5x4x4 transformation for $(NaCl)_{35}Cl^-$. The slope of this plot yields an activation energy of $0.57\pm0.05$ eV. Activation energies for some other isomerization processes for the $(NaCl)_nCl^-$ clusters are given in Table 2.

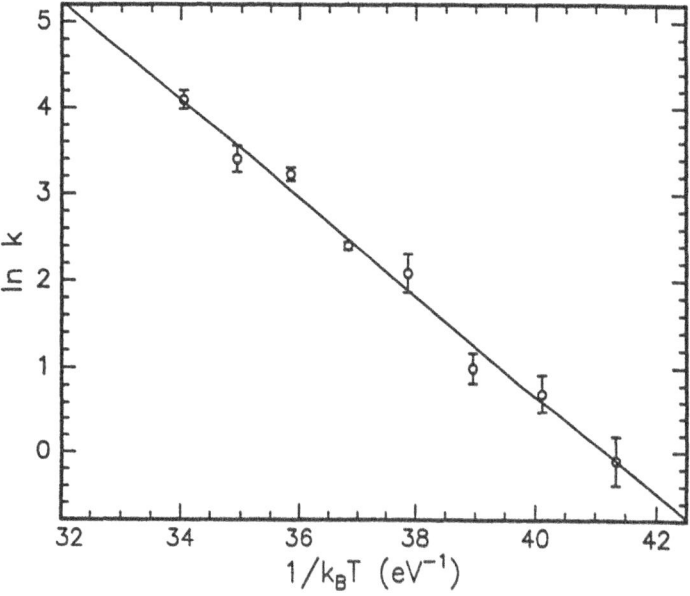

**Fig. 12.** Arrhenius plot for 8x3x3 to 5x4x4 transformation of $(NaCl)_{35}Cl^-$

The isomerization of NaCl and other alkali halide clusters has been studied theoretically by several groups [43,51,52,53,54,55,56]. For NaCl clusters, activation energies have only been reported for clusters with less than 10 atoms. For $(NaCl)_4$, the upper limit for the activation energy for isomerization of a planar ring to a 2x2x2 cuboid was found to be 0.56 eV by Martin [51]. A value of 0.33 eV has been reported by Heidenreich et al. [52]. The magnitude of these values is similar to the activation energies shown in Table 2. However, it may not be appropriate to compare the results for the nanocrystal geometries shown in Table 2 to those for the molecule-sized systems. What distinguishes the structural transformations studied here from the isomerization processes studied in smaller clusters is the scale of the structural transformation. For the 8x3x3 to 5x4x4 transformation of $(NaCl)_{35}Cl^-$ close to half the atoms in the nanocrystal must be

**Table 2**     Activation energies of isomerization of some $(NaCl)_nCl^-$ clusters. The letters following the *jxkxl* descriptions indicate the following: c corresponds to a complete cuboid, f indicates a cuboid with one vacancy, and i corresponds to a cuboid with an incomplete face.

| Cluster | Converting Cluster Geometry | Product Isomer Geometry | Activation Energy, eV |
|---------|------------------------------|--------------------------|------------------------|
| $(NaCl)_{30}Cl^-$ | 7x3x3 i | 4x4x4 i | $0.57 \pm 0.05$ |
| $(NaCl)_{35}Cl^-$ | 8x3x3 f | 5x4x4 i | $0.57 \pm 0.05$ |
| $(NaCl)_{35}Cl^-$ | 5x5x3 i | 5x4x4 i | $0.53 \pm 0.05$ |
| $(NaCl)_{36}Cl^-$ | 6x5x3 i | 5x4x4 i | $0.44 \pm 0.05$ |
| $(NaCl)_{37}Cl^-$ | 6x5x3 i | 5x5x3 c | $0.36 \pm 0.05$ |

rearranged to give a nanocrystal with a dramatically different overall shape, but the same fcc packing motif. It seems unlikely that this would occur in a single concerted step, and so it probably involves a closely choreographed sequence of transformations, each involving the movement of one or two atoms. For bulk NaCl, the activation energies for the migration of an ion vacancy through the solid is 0.85 eV [57]. Note that all the activation energies in Table 1 are all less than 0.6 eV. Thus the isomerization mechanism for NaCl clusters must be less energetically hindered than the full dislocation of atoms in the bulk. This suggests a mechanism which involves a restructuring of the cluster through a series of small-scale atomic rearrangements at the surface of the cluster. Theoretical studies predict activation energies for lateral and diagonal diffusion on KCl(100) of 0.52 eV and 0.26 eV, respectively [58]. Analogous values for NaCl are expected to be similar. Thus it appears that surface diffusion is the most plausible mechanism for the structural transformations observed for the $(NaCl)_nCl^-$ clusters.

## 10. Concluding Remarks

Ion mobility studies have revealed the presence of structural isomers for many different types of atomic clusters. It some cases it has been possible to examine the structural transitions between the different isomers and even obtain activation energies for these transformations. The presence of multiple isomers for medium sized clusters hinders efforts to examine their properties. It seems likely that in the future experiments must be performed on shape selected clusters by combining ion mobility measurements with other experimental techniques. For example the combination of ion mobility measurements with photoelectron spectroscopy is feasible and would provide access to the photoelectron spectra of specific isomers. Spectroscopic studies would also aid in assigning geometries to the features resolved in the ion mobility measurements, because structural assignments based on just the mobility can be ambiguous.

## Acknowledgements

We gratefully acknowledge the support of the National Science Foundation (CHE-9306900) and the Petroleum Research Fund (administered by the American Chemical Society). Ph. Dugourd also acknowledges financial support from CNRS and NATO.

## References

1.  B. R. Rowe, D. W. Fahey, F. C. Fehsenfeld, and D. L. Albritton, J. Chem. Phys. 73, 194 (1980).
2.  D. F. Hagen, Analyt. Chem. 51, 870 (1979).
3.  G. von Helden, M.-T. Hsu, P. R. Kemper, and M. T. Bowers, J. Chem. Phys. 93, 3835 (1991).
4.  G. von Helden, M.-T. Hsu, N. Gotts, and M. T. Bowers, J. Phys. Chem. 97, 8182 (1993).
5.  K. B. Shelimov, J. M. Hunter, and M. F. Jarrold, Int. J. Mass Spectrom. Ion Proc. 138, 17 (1994).
6.  J. M. Hunter and M. F. Jarrold, J. Am. Chem. Soc. 117, 103 (1995).
7.  M. F. Jarrold and V. A. Constant, Phys. Rev. Lett. 67, 2994 (1992).
8.  J. M. Hunter, J. L. Fye, M. F. Jarrold, and J. E. Bower, Phys. Rev. Lett. 73, 2063 (1994).
9.  M. F. Jarrold and J. E. Bower, J. Chem. Phys. 98, 2399 (1993).
10. D. E. Clemmer, J. M. Hunter, K. B. Shelimov, and M. F. Jarrold, Nature 372, 248 (1994).
11. M. F. Jarrold and E. C. Honea, J. Am. Chem. Soc. 114, 459 (1992).
12. J. M. Hunter, J. L. Fye, and M. F. Jarrold, Science 260, 784 (1993).
13. G. von Helden, N. G. Gotts, and M. T. Bowers, Nature 363, 60 (1993).
14. E. A. Mason and E. W. McDaniel, *Transport Properties of Ions in Gases* (Wiley; New York, 1988).
15. J. O. Hirschfelder, C. F. Curtiss, and R. B. Bird, *Molecular Theory of Gases and Liquids* (Wiley; New York, 1954).
16. E. Mack, J. Amer. Chem. Soc. 47, 2468 (1925).
17. S. N. Lin, G. W. Griffin, E. C. Horning, and W. E. Wentworth, J. Chem. Phys. 60, 4994 (1974).
18. G. von Helden, N. G. Gotts, P. Maitre, and M. T. Bowers, Chem. Phys. Lett. 227, 601 (1994).
19. S. Lee, T. Wyttenbach, and M. T. Bowers, J. Am. Chem. Soc. 117, 10159 (1995).
20. L. D. Book, C. Xu, and G. E. Scuseria, Chem. Phys. Lett. 222, 281 (1994).
21. A. A. Shvartsburg and M. F. Jarrold, Chem. Phys. Lett. 261, 86 (1996).
22. G. von Helden, T. Wyttenbach, and M. T. Bowers, Int. J. Mass Spectrom. Ion Proc. 146/147, 349 (1995).

23. M. F. Mesleh, J. M. Hunter, A. A. Shvartsburg, G. C. Schatz, and M. F. Jarrold, J. Phys. Chem. 100, 16082 (1996).

24. J. P. Doye and D. J. Wales, Chem. Phys. Lett. 262, 167 (1996).

25. A. A. Shvartsburg, R. R. Hudgins, Ph. Dugourd, and M. F. Jarrold, J. Phys. Chem. A 101, 1684 (1997).

26. D. L. Strout, R. L. Murry, C. Xu, W. C. Eckhoff, G. K. Odom, and G. E. Scuseria, Chem. Phys. Lett. 214, 576 (1993).

27. G. E. Scuseria, Chem. Phys. Lett. 257, 583 (1996); and references therein.

28. M. F. Mesleh, G. C. Schatz, and M. F. Jarrold, (unpublished).

29. H. E. Revercomb and E. A. Mason, Analyt. Chem., 47, 970 (1975).

30. Y. Kaneko, M. R. Megill, and J. B. Hasted, J. Chem. Phys., 45, 3741 (1966).

31. Ph. Dugourd, R. R. Hudgins, D. E. Clemmer, and M. F. Jarrold, Rev. Sci. Instrum. 68, 1122 (1997).

32. G. von Helden, P. R. Kemper, N. Gotts, and M. T. Bowers, Science, 259, 1300 (1993); N. G. Gotts, G. von Helden, and M. T. Bowers, Int. J. Mass Spectrom. Ion Proc. 149/150, 217 (1995).

33. D. L. Strout, L. D. Book, J. M. Millam, C. Xu, and G. E. Scuseria, J. Chem. Phys. 98, 8622 (1994).

34. M. F. Jarrold and J. E. Bower, J. Chem. Phys. 96, 9180 (1992).

35. K. D. Rinnen and M. L. Mandich, Phys. Rev. Lett. 69, 1823 (1992).

36. G. Gantefor, (private communication).

37. J. E. Campana, T. M. Barlak, R. J. Colton, J. J. Decorpo, J. R. Wyatt, and B. I. Dunlap, Phys. Rev. Lett. 47, 1046 (1981).

38. R. Pflaum, P. Pfau, K.. Sattler, and E. Recknagel, Surf. Sci. 156, 165 (1985); R. Pflaum, K. Sattler, and E. Recknagel, Chem. Phys. Lett. 138, 8 (1987).

39. E.C. Honea, M.L. Homer, P. Labastie, and R.L. Whetten, Phys. Rev. Lett. 63, 394 (1989).

40. Y. J. Twu, C. W. S. Conover, Y. A. Yang and L. A. Bloomfield, Phys. Rev. B 42, 5306 (1990).

41. P. Labastie, J.M. L'Hermite, Ph. Poncharal, and M. Sence, J. Chem. Phys. 103, 6362 (1995).

42. R.L. Whetten, Acc. Chem. Res. 26, 49 (1993).

43. J. Luo, U. Landman, and J. Jortner, in Physics and Chemistry of Small Clusters, P. Jena, B. K. Rao, and S. N. Kanna, Eds. (Plenum, New York, 1987) p 201.

44. D. O. Welch, O. W. Lazereth, G. J. Dienes, and R. D. Hatcher, J. Chem. Phys. 68, 2159 (1978).

45. Ph. Dugourd, R. R. Hudgins, and M. F. Jarrold, Chem. Phys. Lett. 267, 186 (1997).

46. J. Diefenbach and T. P. Martin, J. Chem. Phys. 83, 4585 (1985).

47. N. G. Phillips, C. W. S. Conover, and L. A. Bloomfield, J. Chem. Phys. 94, 4980 (1991).

48. D. O. Welch, O. W. Lazareth, G. J. Dienes, and R. D. Hatcher, J. Chem. Phys. 64, 835 (1975).

49. M. F. Jarrold, J. Phys. Chem. 99, 11 (1995).

50. J.M. Hunter, J.L. Fye, E.J. Roskamp, and M.F. Jarrold, J. Phys. Chem. 98, 1810 (1994).

51. T.P. Martin, J. Chem. Phys. 72, 3506 (1980).

52. A. Heidenreich, J. Jortner, and I. Oref, J. Chem. Phys. 97, 197 (1992).

53. U. Landman, D. Scharf, and J. Jortner, Phys. Rev. Lett. 54, 1860 (1985).

54. D. Scharf, J. Jortner, and U. Landman, J. Chem. Phys. 87, 2716 (1987).

55. J.P. Rose and R.S. Berry, J. Chem. Phys. 96, 517 (1992).

56. V.K.W. Cheng, J.P. Rose, and R.S. Berry, Z. Phys. D 26, 195 (1993).

57. H.W. Etzel and R.J. Maurer, J. Chem. Phys. 18, 1003 (1950).

58. V. K. W. Cheng, B. A. W. Coller, and E. R. Smith, J. Chem. Soc. Faraday Trans. I 84, 899 (1988).

# Silver Clusters
# and Silver Cluster/Ammonia Complexes

David M. Rayner, Kalliopi Athanassenas, Bruce A. Collings,
Steven A. Mitchell and Peter A. Hackett

Steacie Institute for Molecular Sciences, National Research Council of Canada,
100 Sussex Drive, Ottawa, Ontario K1A 0R6, Canada

**Abstract.** The chemistry and physics of small silver clusters are reviewed. Particular attention is directed towards establishing the electronic and geometric structure of silver clusters and linking these with their chemical properties in order to understand structure/reactivity correlations at a molecular level. In this regard we focus on silver cluster/ammonia complexes as a prototype for cluster association complexes. We include new data on the optical absorption spectra of silver clusters and their cations, $Ag_n$ and $Ag_n^+$ ($n = 4, 7, 9, 10$ and $12$) and on the infrared absorption spectra of silver cluster/ammonia complexes, $Ag_n(NH_3)_m$ ($n = 4$ - $18$, $m = 1, 2$).

## 1 Introduction

While experimentalists were learning to prepare and study discrete metal clusters, theorists were already exploiting cluster models as an approach to understanding surface chemistry. As methods for producing and investigating isolated metal clusters have been refined, we have had the opportunity to bring theory and experiment much closer together in this field. The study of reactions on small metal clusters clearly has the potential to reveal intimate details about the binding site specific, *short range* interactions which lead to the catalytic activity of certain metals. This was recognized early on, following the discovery of marked cluster size dependencies in transition metal cluster reactivities [1]. Progress since then has been slower than one would have hoped - for instance there is still active discussion of how the addition of a single Nb atom to $Nb_{13}$ results in the three orders of magnitude change observed in reactivity with $H_2$ [2]. However, understanding and exploiting cluster promoted chemistry is a goal which can only be reached through a thorough understanding of more basic aspects of cluster science. Here the central topics remain electronic and geometric structure and characterizing the interaction of metal clusters with small ligands.

Rather than attempt to review the complete field of cluster reactivity, this report concentrates one of the most comprehensively studied metal systems, silver clusters, $Ag_n$, and their interaction with a single ligand, ammonia. In addition to reviewing published work we include previously unpublished results on the spectroscopy of $Ag_n$ clusters and $Ag_n NH_3$ complexes. In many ways this system shows how far we have come and how far we have to go in the use of clusters to understand surface reactivity. The synergistic roles theory and experiment have played and will continue to play are evident.

# 2 Silver Cluster Structure

The attention silver clusters have received is partly due to their importance in photography and partly because, in electronic complexity, the coinage metals fall between the alkali metals and transition metals with their partially filled $d$-shells. To the theoreticians they represent a stepping stone to more complex systems. In terms of structure they have provided a forum where the validity of the concepts of molecular structure (as understood by the chemist in terms of a defined geometric structure, associated with an electronic structure which can be rationalized, at the simplest level, in terms of molecular orbitals) can be contrasted with those of the essentially structureless shell models which arise from the jellium description.

## 2.1 Ionization Potentials

The ionization potentials of $Ag_n$ have been measured by electron impact (EI) ionization [3] for $n \leq 36$ and by single photon photoionization (PI) threshold measurements [4] for $n \leq 100$. As the EI results represent an upper limit and as the PI threshold measurements are consistently lower than the EI measurements (by 0.5 eV to 1 ev in the range $n = 3 - 20$) the PI measurements are considered the most reliable. Both show the same qualitative size dependence evident in Figure 1 where the results derived from PI measurements are reproduced. A marked even-odd variation, with even clusters having the higher IPs, and discontinuities consistent in first order with shell-closings at $n = 2$, 8 and 20 in the spherical jellium model are apparent.

## 2.2 Photoelectron Spectroscopy and Electron Affinities

The photoelectron spectroscopy of silver cluster anions, $Ag_n^-$, has been studied by several groups. Electron affinities for neutral $Ag_n$ obtained by this approach are shown in Figure 1. There is good agreement between groups [5 - 7]. As in the IPs there is a strong even-odd variation and evidence for shell-closings at $n = 8$, 20, 34 and 58. The spheroidal jellium model predicts the features in the size dependence associated with shell and sub-shell closings [7] but does not account fully for the even-odd oscillations. It has recently been shown that for sodium, potassium and copper clusters the even-odd oscillations in both EAs and IPs can be accounted for almost quantitatively by an ellipsoidal jellium model in which triaxial deformations are allowed [8]. Calculations have not been reported on $Ag_n$ but the indications are that, in this case too, the frontier orbital electronic properties of Ag clusters are closely linked to their shapes.

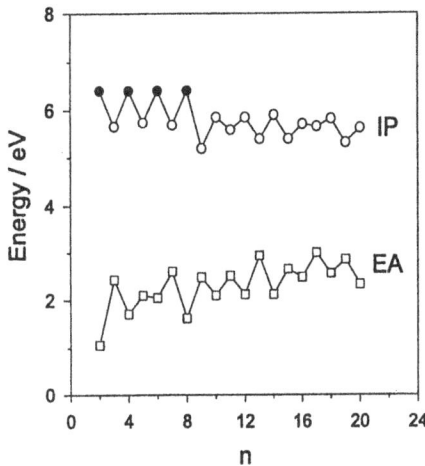

**Fig. 1.** Ionization potentials(o, after Ref. 4) and electron affinities (□, after Ref. 5 and Ref. 6). Values which are only lower limits are indicated by filled symbols.

Quantum chemical calculations offer the chance to detail the role of geometric structure in more detail. For $Ag_n$ clusters this has been achieved for $n$ up to 9 for the neutral, anionic and cationic forms using an effective core potential-configuration interaction Hartree-Fock approach [9, 10]. The experimental trends in both IP and EA are quantitatively reproduced. In addition the calculations account for much of the observed $Ag_n^-$ photodetachment spectra and lead to assignments for electronic excited states in neutral $Ag_n$. Recently, more detailed photoelectron spectra of $Ag_n^-$ clusters have been obtained [11] and have been used to contrast the quantum mechanical approach with the electronic shell model. The quantum chemical approach provides a quantitative description and allows most of the features of the observed spectra to be assigned to electronic states of the neutral clusters in the geometry of the anion. The agreement between experiment and theory establishes the ground state geometries of small $Ag_n$ cluster anions unambiguously for $n \leq 9$. The exception is $Ag_7$ where the calculations predict two low lying candidates for the ground state geometry. A $D_{3h}$ form is found only 0.05 eV above the $C_{3v}$ lowest energy structure. The $Ag_7$ photoelectron spectra fit neither of these forms, possibly due to their presence as a mixture or because the calculation does not find the global minimum in this case. The results are contrasted to the spheroidal shell model by assigning single particle orbitals to electronic subshells. By adjusting the shape of the cluster it is possible to account qualitatively for the features of the photoelectron spectra beyond the range of $n$ covered by the quantum chemical calculations. For $n \leq 9$ there is a correspondence between the symmetries of the exact single particle orbitals and those of the subshells.

## 2.3 Optical Spectroscopy

Optical absorption spectra of small $Ag_n$ clusters were first reported for $Ag_n^+$ (n=9, 11, 15, 21) cations studied by photofragmentation of mass selected $Ag_n^+$ cluster ion beams [12,13] and for neutral $Ag_n$ (n = 5, 7 – 9, 11, 13, 15 – 21) by conventional absorption spectroscopy performed on mass selected clusters embedded in rare-gas matrices [14, 15]. The relatively broad spectra observed in both these studies were interpreted in terms of collective dipole resonances. Splittings were observed which were in general accord with the shape predictions of the jellium model. Shifts relative to the Mie plasma frequency predicted by considering only the $s$-electrons have been attributed to the screening influence of the $d$-electrons [13, 15, 16]. Absorption cross sections reported for both neutral and cation species were also consistent with the plasma resonance description, in that a significant portion of the total oscillator strength appeared to be exhausted in the observed transitions.

Recently we have applied a photodepletion technique to $Ag_n$ clusters which has the advantage that it gives *absolute* and *size-specific* cross sections without the need to mass select the cluster beam. The technique, which was developed concurrently by Knickelbein and Menezes [17, 18] and us [19] involves photodepleting van der Waals complexes formed between the clusters and rare gas atoms. The photodepletion event, loss of the rare gas, can be followed quantitatively by laser ionization time-of-flight mass spectrometry and has the significant advantage over direct cluster photodissociation spectroscopy of being free of interferences due to fragments from dissociation of larger complexes. The spectroscopy of *both* neutral and cationic species is accessible in this way, simply by adjusting the relative timing of the photodepletion and photoionization lasers [19, 20].

Using this approach we have measured the absorption spectra of the $Ag_n$ clusters and their cations for n = 4, 7, 9, 10 and 12. Results for the neutral clusters are shown in Figure 2 and for the cations in Figure 3. The spectra for $Ag_7$, $Ag_9^+$ and $Ag_9$ have been communicated previously [21]. These are the clusters for which we could make direct comparison with the selected cluster ion photodissociation and matrix adsorption studies. The similar bandshapes and positions found validate the photodepletion technique. Our observed spectra, taken in the gas phase at a temperature of 100K, are sharper than the previous ones. At this resolution, complex structure is apparent. In addition we find significantly lower oscillator strengths compared to the previous studies, even when the acknowledged large error limits of the three techniques are taken into account. A recent spectrum of $Ag_9^+$, obtained by mass selected photofragmentation in a Penning trap, is in good agreement with our photodepletion spectrum of $Ag_9Kr^+$, both in terms of band structure and oscillator strength [22]. Our oscillator strengths are more in line with distinct single particle excitations rather than a giant plasmon resonance. We

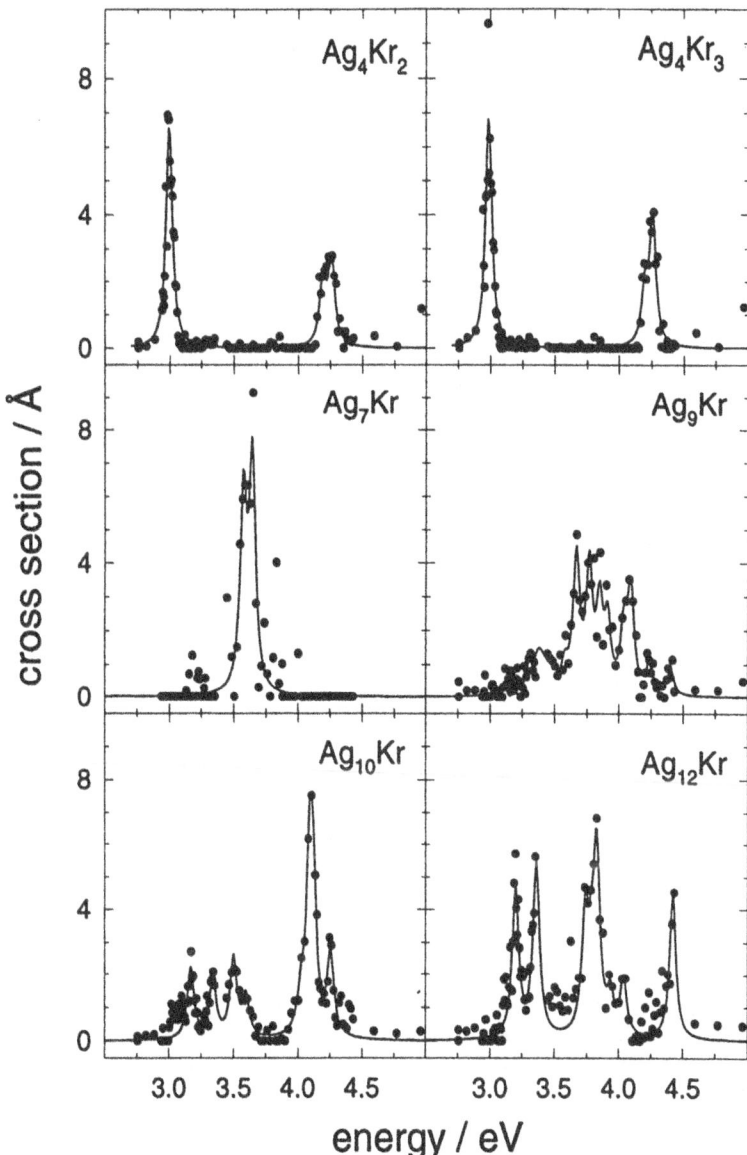

**Fig. 2.** Photodepletion spectra of $Ag_nKr_m$. The solid lines are to guide the eye. They are result of empirical fits using a minimum number of Lorentzians, each of 0.08 eV FWHM (the width of the $Ag_4Kr_2$ transition at 3.0 eV, taken to correspond to a single particle transition).

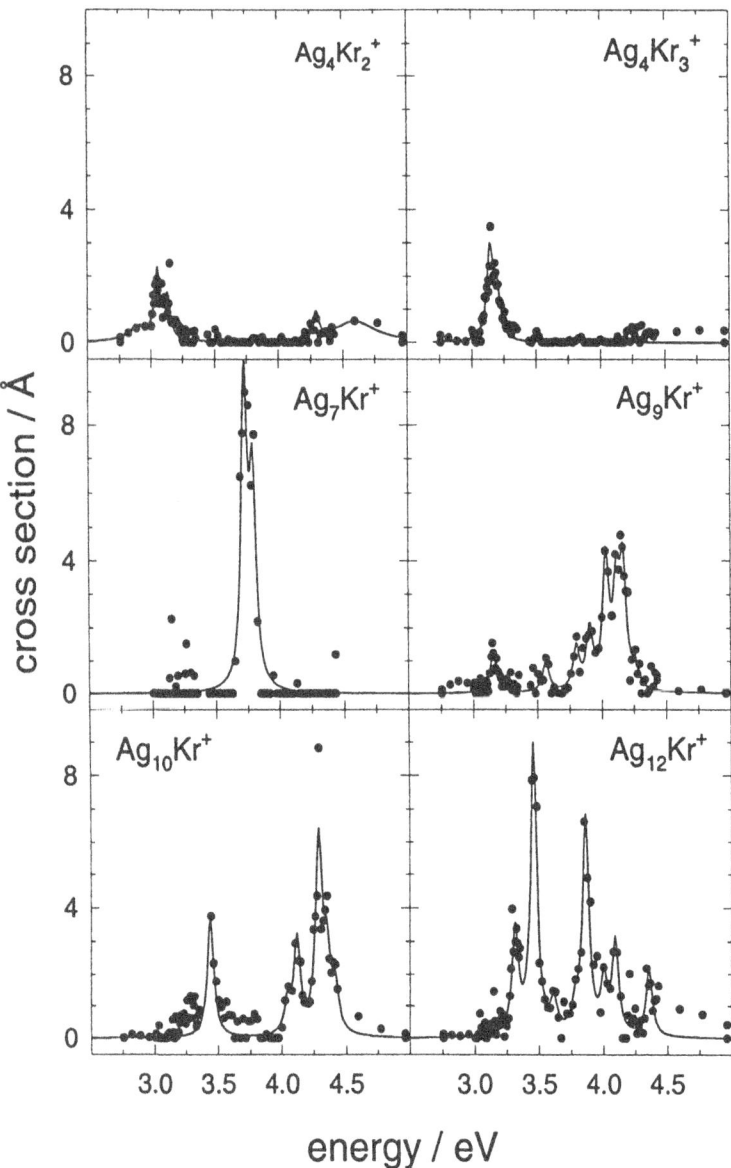

**Fig. 3.** Photodepletion spectra of $Ag_n Kr_m^+$. The solid lines are to guide the eye (see caption for Figure 2.

take our results to demonstrate the influence of geometric structure on the electronic properties of $Ag_n$ clusters and to imply that clusters as large as $Ag_{12}$ will be well described in molecular terms.

Similar implications have recently been reported for small sodium clusters cations, $Na_n^+$, cooled to 35 K [23]. At higher temperatures $Na_n^+$ and $Na_n$ clusters exhibit spectra which are in accord with jellium models, presumably because they are melted under these conditions. We can conclude for small $Ag_n$ clusters that they are solid and retain a definitive molecular structure at least to 100 K. The parallel with $Na_n$ clusters is strong. Our differences with the results of cluster ion beam studies may well lie in the temperature of the clusters in that case which was estimated at 2000 to 3000 K [12]. At this temperature it is sure that $Ag_n$ clusters are liquid, ideally suiting them for description under the jellium model.

Although the spectra of matrix deposited clusters are obtained at 10 K they are still broad compared to our 100 K spectra [15]. This is due to the importance of site-specific matrix effects. Our results for one, two or three Kr atom van der Waals complexes with $Ag_n$ may be regarded as the simplest models for such effects. The effect attaching a rare gas atom has on the electronic structure of a cluster is apparent in a lowering in the IP of 0.1 eV per rare gas atom [24]. This lowering can be understood in terms of a simple electrostatic model in which an additional charge induced dipole interaction stabilizes the ionic complex. For $Ag_4$ it explains why, ionizing at 193 nm (6.42 eV), we observe $Ag_4Kr_2$ and $Ag_4Kr_3$ but not $Ag_4$ and $Ag_4Kr$. Comparison of the absorption spectra of $Ag_4Kr_2^+$ and $Ag_4Kr_3^+$ shows the optical transition at 3.1 eV to shift 0.1 eV to the blue on adding a third Kr atom (Figure 3). This shows that the spectral shift is also largely attributable to the destabilization of the LUMO in the ion. Consistent with this interpretation the equivalent neutral complexes do not exhibit an equivalent shift in their spectra.

The resolved electronic spectra reported for $Ag_n$ clusters in Figures 2 and 3 offer the hope that structures can be established by comparison with high level theory. Unfortunately, except for some information on $Ag_4$, the effective core potential-CI calculations for $Ag_n$ and $Ag_n^+$ of Bonačić-Koutecký [9, 10] do not yet extend to excited states with allowed transitions in the visible/uv. Work in this area, involving a new 11-electron relativistic core potential for Ag, is presently under way [25]. In the case of $Ag_4$, calculated excited state levels are available to high enough energy but only from a calculation restricted to the rhombic geometry of the $Ag_4^-$ anion. These calculations are consistent, as far as they go, with our observed spectra and would lead to the assignment of the 3.0 eV transition in $Ag_4$ to excitation from the $^1A_g$ ground state to the $1^1B_{2u}$ state. However the next calculated allowed transition is at only 0.3 eV higher whereas in the depletion spectra it is 1.2 eV to the blue higher. Better calculations with relaxed geometries are clearly necessary to match the absorption spectra from $Ag_4$ up. Until then our knowledge of $Ag_n$ structures must rely on the calculations. Confidence in the one-valence-electron CI predictions is raised by the overall agreement for the anions with PES results. In addition, a recently obtained Raman spectrum of matrix isolated $Ag_5$ confirms the planar trapezoidal structure predicted by these calculations [26]. It must be noted, however, all-electron density functional calculations on $Ag_n$ for $n = 1 - 6$ report a difference in the ground state geometry for the hexamer and in the energies of other, low-lying isomers for $n = 4 - 6$ [27]. It is hoped that our spectra will eventually be used to resolve this conflict and establish the structures of larger clusters.

# 3 Formation of Silver Cluster $NH_3$ Complexes

Notwithstanding the above, we know more about the structure of small silver clusters than many other systems and they form one of the best systems to develop our ideas of molecular surface science. For the same reasons that they are attractive to theorists, in being intermediate between alkali metal clusters and open $d$-shell metal clusters, they are attractive to experimentalists concerned with reactivity. Although an eventual goal of molecular surface science is to understand chemistry taking place on a cluster surface we need to start with systems which focus on part of the problem. $Ag_n$ clusters are relatively benign and much of their chemistry is dominated by formation of what, in molecular terms, would be called an association complex. Here we have the opportunity to study mechanisms of association, surface transport and dissociation from the cluster without the complication of ligand dissociation. These are key processes which precede and follow any chemical transformation on a surface, be it cluster or extended. In this review we focus on these processes through the ammonia ligand.

We have used a fast flow cluster reactor [28 – 31] to study the reactivity of silver clusters in the gas phase under well thermalized conditions at pressures in the range 0.4 to 5 Torr and temperatures in the range 270 - 370 K. The dimer, $Ag_2$, has been monitored by laser induced fluorescence and higher clusters by photoionization TOFMS. In the case of the dimer these flow reactor studies have shown that $Ag_2$ is significantly less reactive than its coinage metal counterparts, $Cu_2$ and $Au_2$, especially towards ligands which require backdonation to form stable complexes (e.g. CO, $C_2H_4$) [32]. This is due to the relative stability of the Ag $d$-orbitals which makes $d$-electrons unavailable for backdonation. $Ag_2$ does react with ammonia and is shown to come to equilibrium with the complex $Ag_2NH_3$ in the gas phase at room temperature. The $Ag_2NH_3$ complex is unreactive towards the addition of a further ligand.

In the case of larger clusters we are unable to monitor $Ag_3$, $Ag_4$ and $Ag_6$ by photoionization TOFMS due to their high IPs. Otherwise we find all clusters with $n > 3$ form complexes with $NH_3$. Using TOFMS we are able to monitor the appearance of $Ag_n(NH_3)_m$ in addition to the disappearance of $Ag_n$. Although $Ag_4$ is not detectable, the addition of one $NH_3$ ligand lowers the IP sufficiently for the reaction to be followed through product appearance. Unlike $Ag_2$, the larger clusters can add more than one $NH_3$ ligand in a sequential manner. From the dependence of the ratio $Ag_n(NH_3)_m/Ag_n(NH_3)_{m-1}$ on $p(NH_3)$ it is shown that the $NH_3$ complexes are relatively weakly bound and are at equilibrium in the gas phase under our conditions [33].

# 4  Bonding in Silver Cluster $NH_3$ Complexes

## 4.1  Thermodynamics of $NH_3$ Complex Formation

Under conditions where the equilibrium

$$Ag_n(NH_3)_m + NH_3 \rightleftharpoons Ag_n(NH_3)_{m-1} + NH_3 \qquad (1)$$

can be established, one can obtain the associated equilibrium constant, $K_m$, from the dependence of the relative concentrations of $Ag_n(NH_3)_{m-1}$ and/or $Ag_n(NH_3)_m$ on $p(NH_3)$. Measurement of $K_m$ as a function of temperature allows the van't Hoff relationship to be applied to obtain binding enthalpies, $\Delta H_T^\circ$, and entropies, $\Delta S_T^\circ$, as a function of $n$ and $m$. Values of $\Delta H_T^\circ$ and $\Delta S_T^\circ$ obtained in this way have been published for $Ag_2NH_3$ [34], $Ag_{10}NH_3$ and $Ag_{16}NH_3$ [33]. Figure 4 shows $\Delta H_T^\circ$ and $\Delta S_T^\circ$ for the addition of a single $NH_3$ ligand as a function of $n$, including preliminary data for the other accessible clusters in the size range up to $n = 18$. The results are tabulated in Table 1. As expected from our ability to bring the reaction to equilibrium in the gas phase, $Ag_n$ clusters are weakly bound, with $\Delta H_T^\circ$ in the range $-7$ to $-17$ kcal mol$^{-1}$. The complexes divide into two categories. Clusters with $n = 2, 4, 16$ and $17$ have relatively high binding enthalpies ($\Delta H_T^\circ < -14$ kcal mol$^{-1}$) and a relatively large loss of entropy on binding ($\Delta S_T^\circ < -23$ cal mol$^{-1}$). The rest have relatively low binding enthalpies ($\Delta H_T^\circ > -10$ kcal mol$^{-1}$) and small entropy losses ($\Delta S_T^\circ > -14$ cal mol$^{-1}$), with the exception of $n = 18$ which is an intermediate case in both senses. The distinct correlation between the dependence of binding enthalpy and of entropy on $n$ indicates that two, size-dependent types of binding operate in $Ag_nNH_3$ complexes.

This difference has been examined in detail for the representative complexes $Ag_{10}NH_3$ and $Ag_{16}NH_3$ [33]. The binding entropy for $Ag_{16}NH_3$ of $-23.5$ cal mol$^{-1}$ K$^{-1}$ is in the range expected for a molecular association complex where the ligand forms a localized dative bond with the cluster. A prototype for this type of interaction is the dimer complex, $Ag_2NH_3$, which has been studied in some detail [32, 34, 35]. It is known from theoretical studies and from photodissociation experiments (see below) that $NH_3$ is bound end-on in this complex. In this configuration there is a strong mixing interaction between the empty $\sigma^*$-orbital on $Ag_2$ and the $n$-orbital on $NH_3$. Net dative bonding is obtained as the stabilization of the $n(NH_3)$-orbital is significantly less than the destabilization of the $\sigma(Ag_2)$-orbital. Mixing of the $\sigma$ and

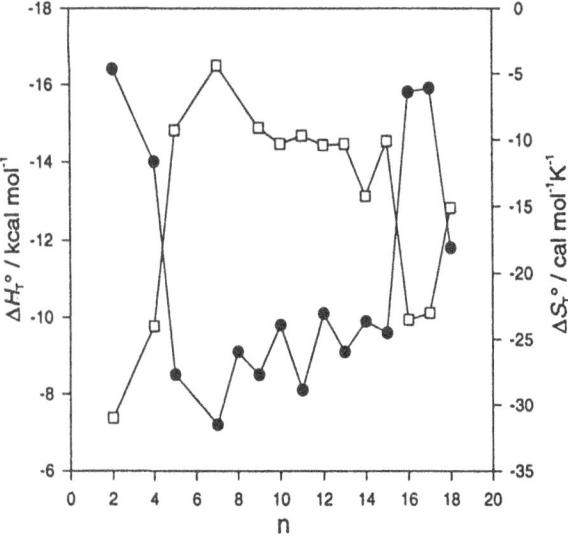

**Fig. 4.** Binding enthalpies, $\Delta H_T^\circ$ (•), and binding entropies, $\Delta S_T^\circ$ (□),for $Ag_n NH_3$ complexes.

$\sigma^*$-orbitals allows polarization to alleviate repulsion between the $\sigma$-electrons and the lone pair. This ability to polarize is critical to bond formation and results in an electrostatic contribution to the bonding. The measured entropy change in forming $Ag_2NH_3$ is $-31$ cal mol$^{-1}$ K$^{-1}$ which can be compared to a value of $-27$ cal mol$^{-1}$ K$^{-1}$ derived from statistical mechanics using molecular constants provided by density functional theory (DFT) calculations [35]. The large change in entropy finds its roots in the loss of three translational degrees of freedom in forming the complex.

It is clear that $Ag_{10}NH_3$, with $\Delta S_T^\circ = -9.8$ cal mol$^{-1}$ K$^{-1}$, must be much more fluctional than $Ag_2NH_3$ and $Ag_{16}NH_3$. A simple statistical mechanics model which accounts for this difference remarkably well, in that it finds $\Delta S_T^\circ = -10.2$ cal mol$^{-1}$ K$^{-1}$, is one which treats the ligand as a 2-dimensional gas, i.e. $NH_3$ is free to move on the surface of the cluster. The model has a parallel in surface science in the lattice gas model for surface adsorption in the $kT \gg V_0$ limit. The low binding enthalpies in $Ag_{10}NH_3$ and other complexes in the range $n = 5$ to $15$ are consistent with this interpretation. On these clusters there is apparently a non-site specific binding mechanism operating. A candidate for this mechanism is the dipole-induced dipole interaction which helps stabilize the $Ag_2NH_3$ complex [34, 35]. If the ability to polarize $Ag_n$ clusters becomes largely insensitive to direction once the cluster has reached a certain size, then a purely electrostatic interaction

Table 1. Binding enthalpies (kcal mol$^{-1}$) and binding entropies (cal mol$^{-1}$ K$^{-1}$) for Ag$_n$NH$_3$ complexes.

| $n$ | Ag$_n$NH$_3$ | |
| --- | --- | --- |
| | $\Delta H^{\circ}_{300}$ | $\Delta S^{\circ}_{300}$ |
| 2 | -16.4±3.0[a] | -31±6[a] |
| 4 | -14 | -24 |
| 5 | -8.5 | -9.3 |
| 7 | -7.2 | -4.4 |
| 8 | -9.1 | |
| 9 | -8.5 | -9.1 |
| 10 | -9.8±1.0[b] | -10.3±3.0[b] |
| 11 | -8.1 | -9.7 |
| 12 | -10.1 | -10.4 |
| 13 | -9.1 | -10.3 |
| 14 | -9.9 | -14.2 |
| 15 | -9.6 | -10.1 |
| 16 | -15.8±1.0[b] | -23.5±3.0[b] |
| 17 | -15.9 | -23 |
| 18 | -11.8 | -15.1 |

[a]From Ref. 35
[b]From Ref. 33

will be insensitive to the location of the ligand on the surface. This will allow free motion in two degrees of freedom. The development of this ability is seen in DFT calculations on very small coinage metal cluster NH$_3$ complexes. In the dimer complex the two-fold bridge binding site is non-bonding but in the trimer complex already shows net bonding as the cluster develops the ability to polarize charge away from the ligand in this configuration [36].

In contrast, the entropy changes on the formation of Ag$_{16}$NH$_3$ and Ag$_{17}$NH$_3$ are as expected for a complex in which the ammonia is locally bound. An idea of the depth of the local well which would be required to constrain motion of NH$_3$ can be estimated from the difference in $\Delta S$ between the two types of complex. At 300 K to have 90% of the ligands locally bound would require a local well of the order of 4 kcal mol$^{-1}$. The total binding enthalpies for the strongly bound clusters are larger than those for the weakly bound clusters by this amount or more. Ag$_{16}$ and Ag$_{17}$ are thought to present favourable sites for dative bond formation involving some degree of orbital overlap, a mechanism similar to that operating in Ag$_2$NH$_3$. Stabilization at these sites is pictured as operating in addition to the electrostatic mechanism which dominates binding for the range $n = 5$ to 15. In these systems local binding can be the result of quite subtle local variations in the potential energy surface.

It has been proposed, in reference to $Cu_n^+CO$ cluster complexes, that electronic shell closings can account for the extra stability of certain complexes if the electrons donated by the ligand are included in the count, so that $Cu_7^+CO$ and $Cu_{17}^+CO$, with electron counts of 8 and 18 were found to be exceptionally stable [37]. Although $Ag_{16}NH_3$ has a combined count of 18, $Ag_{17}NH_3$, which has similar thermodynamics, does not so our results are not consistent with this mechanism for stabilization. Our entropy results show that the $NH_3$ ligands are locally bound in these complexes requiring that geometries and site-specific interactions be considered in discussing details of the bonding. It is hoped that extension of calculations on $Ag_n$ and $Ag_nNH_3$ to higher $n$ will develop this understanding in the near future.

## 4.2   $Ag_2NH_3$ Photodissociation

Many of the ideas presented above concerning the interaction of $Ag_n$ with $NH_3$ have been developed from an in depth understanding of the simplest member of the series, the $Ag_2NH_3$ complex. The thermodynamics of its formation and theoretical approaches using both simple molecular orbital ideas and high level DFT theory have been discussed above. A third, revealing, source of information on $Ag_2NH_3$ has been its dissociation dynamics.

Laser UV photodissociation of $Ag_2NH_3$ in the region of 310 nm results in emission of light identified as the $A \rightarrow X$ transition of the silver dimer [34, 35]. Modeling the emission spectrum indicates that the dissociation produces $Ag_2$ $(A^1\Sigma_u)$ with a vibrational temperature of $600 \pm 100$K. The action spectrum for this photodissociation is shifted to the blue in comparison to the $A \leftarrow X$ transition in $Ag_2$. In the dimer this transition is associated with promotion from the $s\sigma$ to $s\sigma^*$-orbital. In the complex, $Ag_2NH_3$, the molecular orbital picture has the $s\sigma^*$-orbital substantially destabilized compared to the $s\sigma$-orbital when $NH_3$ approaches end-on, predicting the observed blue shift. The $s\sigma*$ state is repulsive in the $Ag - N$ co-ordinate, and correlates to the electronic excited $A^1\Sigma_u$ state of $Ag_2$, hence the observed $Ag_2$ emission. Energy disposal in the photodissociation is also consistent with an end-on structure and a dissociative excited state. Following excitation at 308 nm there is an excess of 13 kcal mol$^{-1}$ to produce $Ag_2$ $(A^1\Sigma_u)$ from ground state $Ag_2NH_3$. Conservation of energy and momentum in a limiting impulsive model applied to an end-on geometry predicts a vibrational temperature of 710 K in $Ag_2$ $(A^1\Sigma_u)$. The difference between this value and the measured value, if it is significant, suggests the involvement of bending modes of the complex. Even then the energy disposal is largely accounted for by the simple model.

We have found similar results in the case of photodissociation of the related $Cu_2NH_3$ complex, although this case proves to be complicated by the additional involvement of excited states of the complex which correlate to the $B$ and $C$ states of $Cu_2$ as well as the $A$. The observed emission is from the thermally equilibrated $B$ and $C$ states, populated, depending on the excitation wavelength, either by curve crossing from a state of $Cu_2NH_3$ which correlates to $Cu_2$ $(A)$ or directly to the state which correlates to $Cu_2$ $(B)$ [34]

## 4.3   Infrared Photodepletion Spectroscopy of $Ag_n(NH_3)_m$

Vibrational spectroscopy is one of the most successful approaches to probing the adsorption of molecules to surfaces. Unfortunately one cannot apply the direct techniques of transmission and reflectance Fourier transform infrared and electron energy loss spectroscopy to cluster complexes. The requirements for mass selection have led us to apply infrared laser photodepletion techniques to measure the infrared spectra of $Ag_n(NH_3)_m$. Infrared multiphoton decomposition (IRMPD) photodepletion spectroscopy was first applied to cluster complexes by Cox and co-workers [38] where they concluded that methanol undergoes dissociative chemisorption on small iron clusters. This system has recently been revisited by Knickelbein, who was able to carry out deuterium substituted experiments and concluded that methanol adsorbs non-dissociatively, at least at 77 K [39].

We have used the approach to measure the infrared spectrum of $Ag_2NH_3$ [35] in the region of the $NH_3$ umbrella mode frequency which is accessible to the wavelengths of the TEA $CO_2$ laser. We have recently published preliminary spectra for $Ag_nNH_3$, n = 8, 11, 14 [40].

Molecular IRMPD has received significant attention in the past. In molecules with sufficient density of states the process is understood in terms of a kinetic model where absorption of the first few photons is the rate determining step [41]. Consequently the action spectrum for IRMPD is expected to be centred close to the peak of the small-signal IR absorption spectrum even in systems where tens of photons have to be absorbed to cause dissociation. We, and others, have used this assumption to assign IRMPD ligand dissociation spectra to fundamental modes of cluster ligand complexes. Closer consideration of the mechanism operating in IR ligand photodepletion indicates that the correspondence with the small-signal spectrum may be even closer than the "traditional" IRMPD picture allows in certain circumstances. These circumstances prevail when weakly bound species are prepared under thermal equilibrium conditions prior to expansion into a molecular beam. This means that a significant population of molecules already have internal energy greater than the binding energy as they expand from the the reactor, where they are part of a canonical ensemble, onto the molecular beam where they form a collection of isolated microcanonical ensembles. Some, highly excited complexes will loose their ligand spontaneously so that as the beam reaches the detection region the internal energy distribution is non-Boltzmann and peaked around the dissociation energy. Simulations based on unimolecular reaction theory show that a significant number of complexes can dissociate on the absorption of a single IR photon, if binding energies are in the range of those found for $Ag_nNH_3$ complexes. Detailed discussion of this IR photodepletion mechanism will appear elsewhere [42]. Here we want to point out that, in this case, the correspondence of the depletion action spectrum with the IR absorption spectrum is likely even better than can be assumed under the traditional IRMPD mechanism.

Infrared photodepletion for $Ag_nNH_3$ and $Ag_n(NH_3)_2$ cluster complexes with $n$ in the range 4 to 18 are shown in Figure 5. Again they have been obtained using a TEA $CO_2$ laser to drive the photodissociation and an ArF excimer laser photoionization TOFMS to monitor the depletion. Details of the experiment are essentially the same as described in our preliminary study [40]. The points are the slopes of plots of $\ln(I/I_0)$, the $NH_3$ photodepletion rate, against IR fluence obtained at individual $CO_2$ laser frequencies. Here $I$ is the depleted complex MS peak height and $I_0$ is the non-depleted height. When studying mono-ligated complexes, reactor conditions were adjusted so that the related di-ligated complex was not present to distort the measurement by photolysing to produce the mono-ligated complex. Absorption was found in the region of the IR laser's 9 $\mu m$ P and R branches. The discontinuous nature of the spectra in Figure 5 is due to the gap between these branches where no laser lines are available. No absorption was observed in the 10 $\mu m$ laser bands which cover the 923−956 and 965−988 $cm^{-1}$ regions of the spectrum.

The absorption features observed for $Ag_n(NH_3)_m$ complexes in Figure 5 are all attributable to the $NH_3$ umbrella mode. In $Ag_2NH_3$ $\nu$-$NH_3$(umbrella) is found to be 1056 $cm^{-1}$ by depletion spectroscopy compared to 1102 $cm^{1-}$ from DFT theory [35]. Binding to $Ag_2$ produces a ~100 $cm^{-1}$ shift to the blue compared to the gas phase, similar to that found on binding to the Ag(110) surface [43]. Figure 6 and Table 2 summarize the peak absorption positions found in the $Ag_n(NH_3)_m$) spectra. A shift of the same order is observed for one and two $NH_3$ ligands binding to larger clusters. Observed peaks are in the range 1060 - 1090 $cm^{-1}$ with the proviso that certain clusters ($n = 4$, 9 and 11, $m =1$) have appreciable absorption at the high frequency limit of our experiment and may have maxima outside the accessible range. There are no other normal mode candidates for absorption at this relatively high frequency. $Ag_n$ frame modes are expected to be significantly lower. The Ag − N stretch and $NH_3$ rocking mode are calculated to be 305 and 513 $cm^{-1}$ respectively in the dimer complex. These modes will be even softer in the less tightly bound complexes formed by larger $Ag_n$ clusters. Retaining the umbrella mode on binding is further confirmation that $NH_3$ binds with the N atom directed towards the cluster, as presumed from the dimer studies and from electrostatic considerations.

While it is clear that $\nu$-$NH_3$(umbrella) is size sensitive there is no striking correlation with known properties of the bare clusters. Comparison of Figure 6 with the IPs and EAs given in Figure 1 shows that the odd-even behavior linked with electronic properties does not show up in $\nu$-$NH_3$(umbrella).

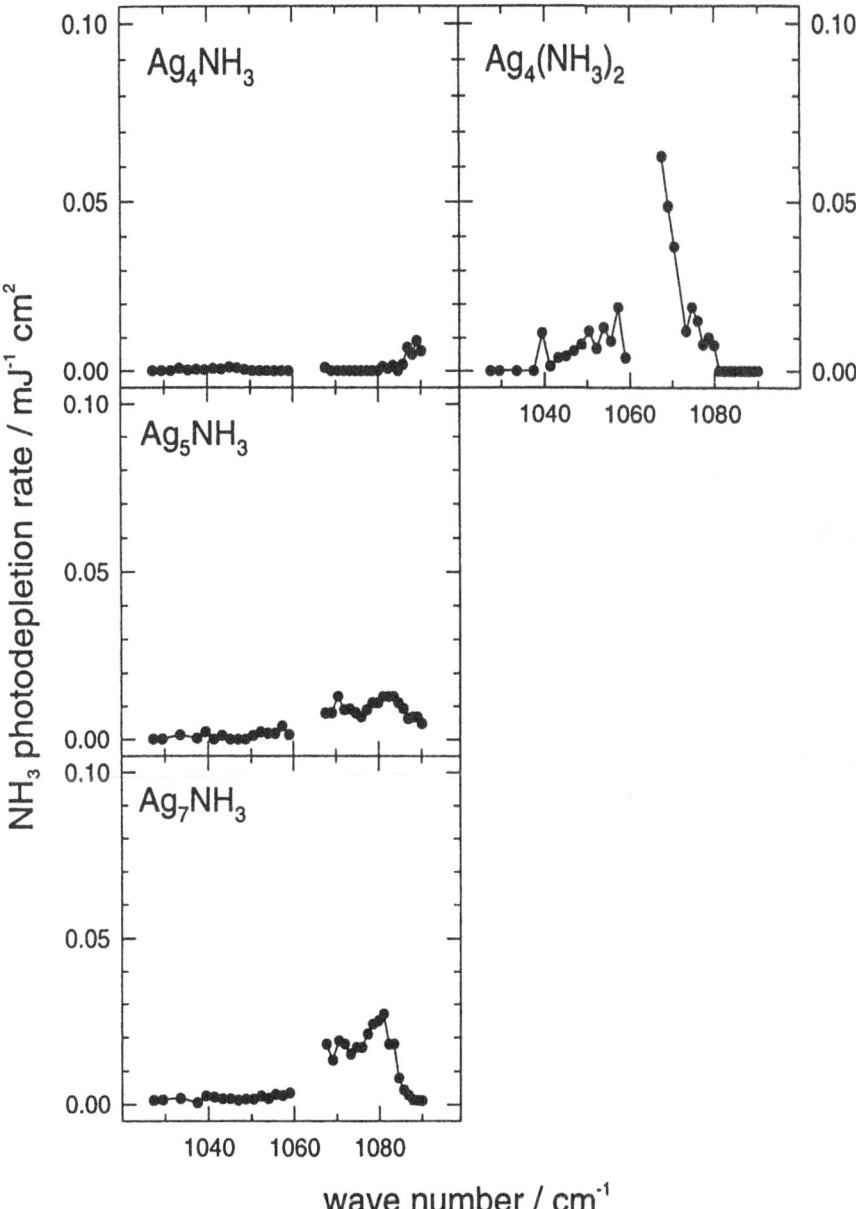

**Fig. 5.** Infrared photodepletion spectra of $Ag_nNH_3$ and $Ag_n(NH_3)_2$ cluster complexes in the frequency region of the umbrella vibrational mode of bound $NH_3$. The cross section for photodepletion implied by an $NH_3$ photodepletion rate of 0.1 $mJ^{-1}$ $cm^2$ is $2 \times 10^{-18}$ $cm^2$. (*Continues overleaf*)

386

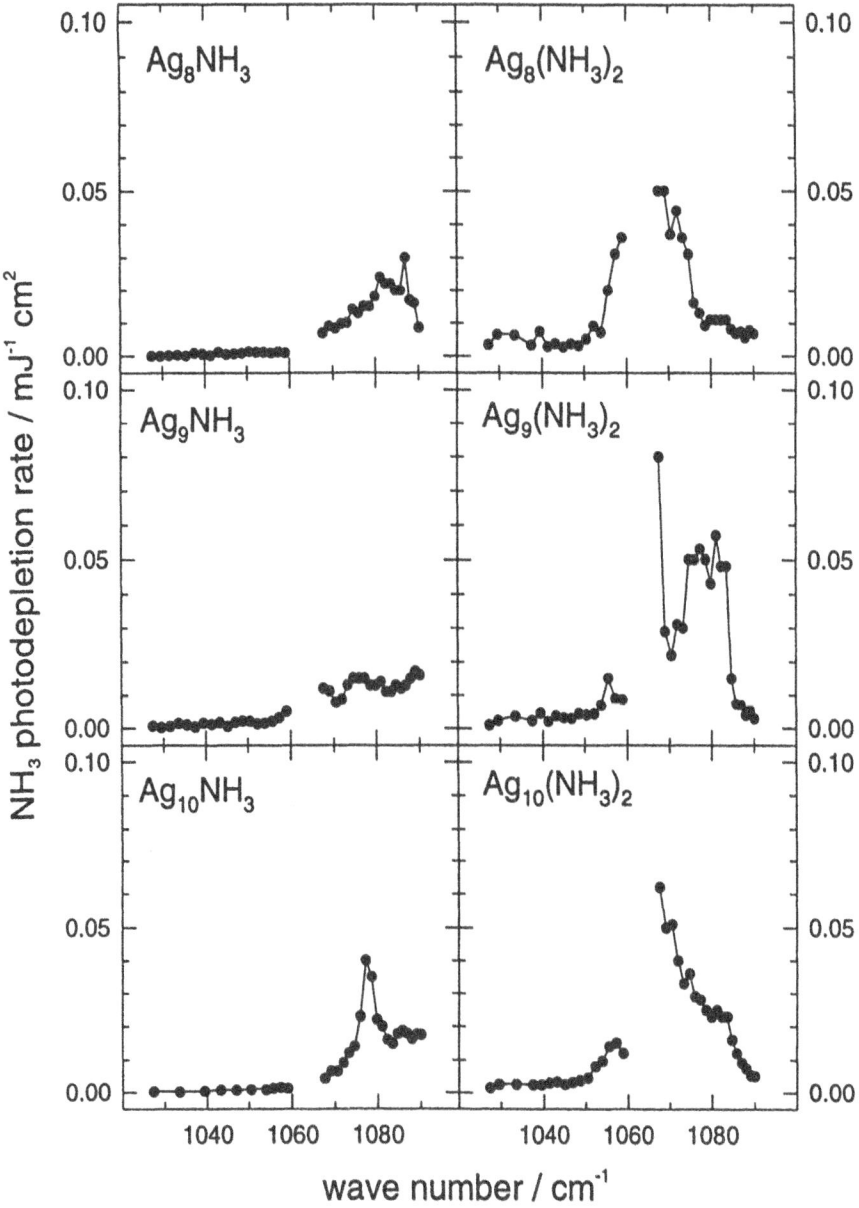

**Fig. 5.** (*Continued*) Infrared photodepletion spectra of $Ag_n NH_3$ and $Ag_n(NH_3)_2$ cluster complexes in the frequency region of the umbrella vibrational mode of bound $NH_3$. (*Continues overleaf*)

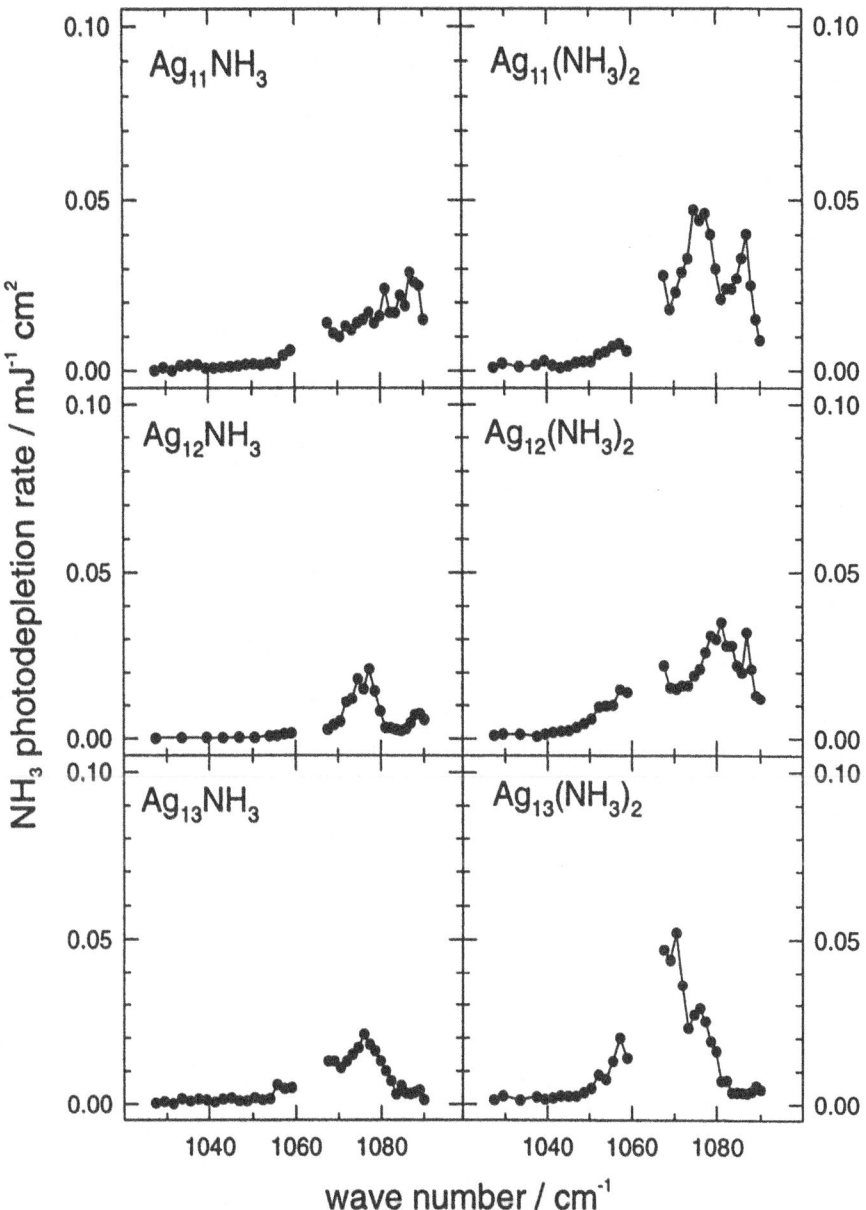

**Fig. 5.** (*Continued*) Infrared photodepletion spectra of $Ag_nNH_3$ and $Ag_n(NH_3)_2$ cluster complexes in the frequency region of the umbrella vibrational mode of bound $NH_3$. (*Continues overleaf*)

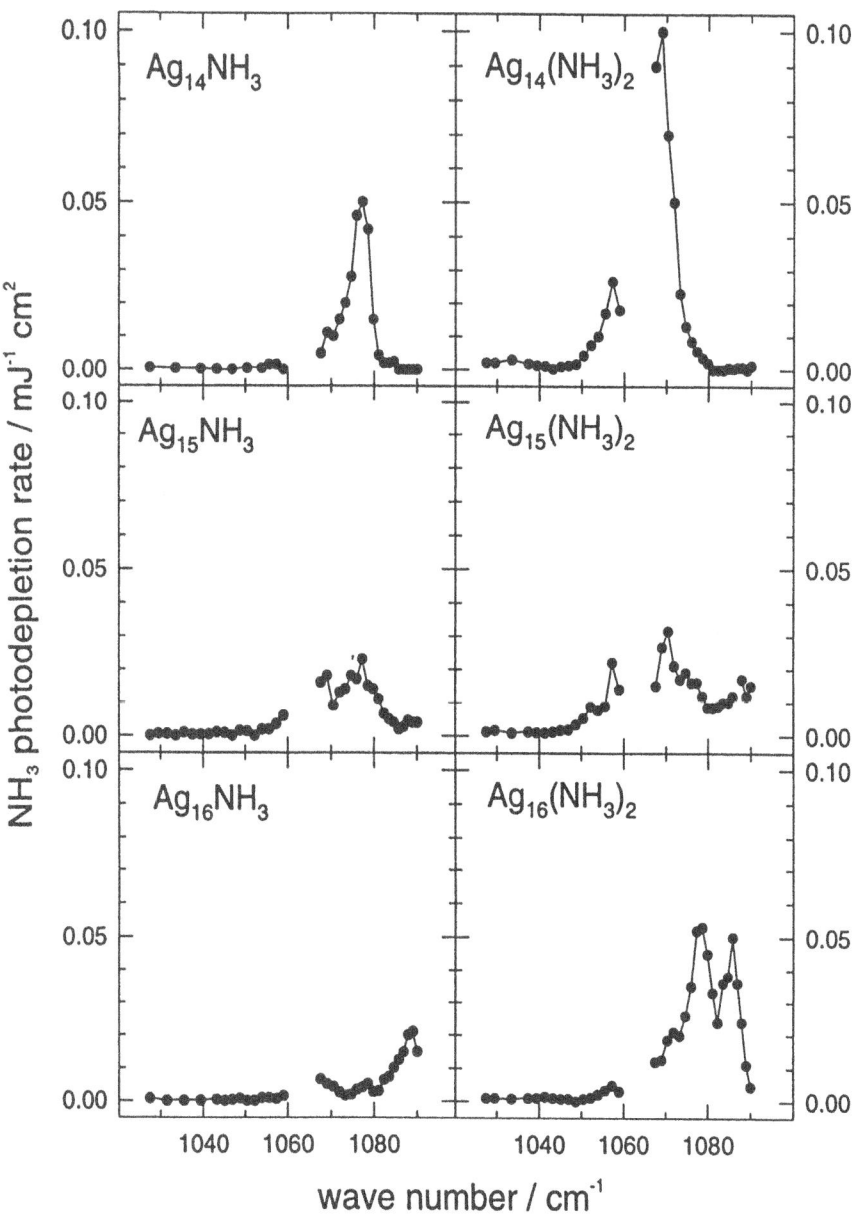

Fig. 5. (*Continued*) Infrared photodepletion spectra of $Ag_n NH_3$ and $Ag_n(NH_3)_2$ cluster complexes in the frequency region of the umbrella vibrational mode of bound $NH_3$. (*Continues overleaf*)

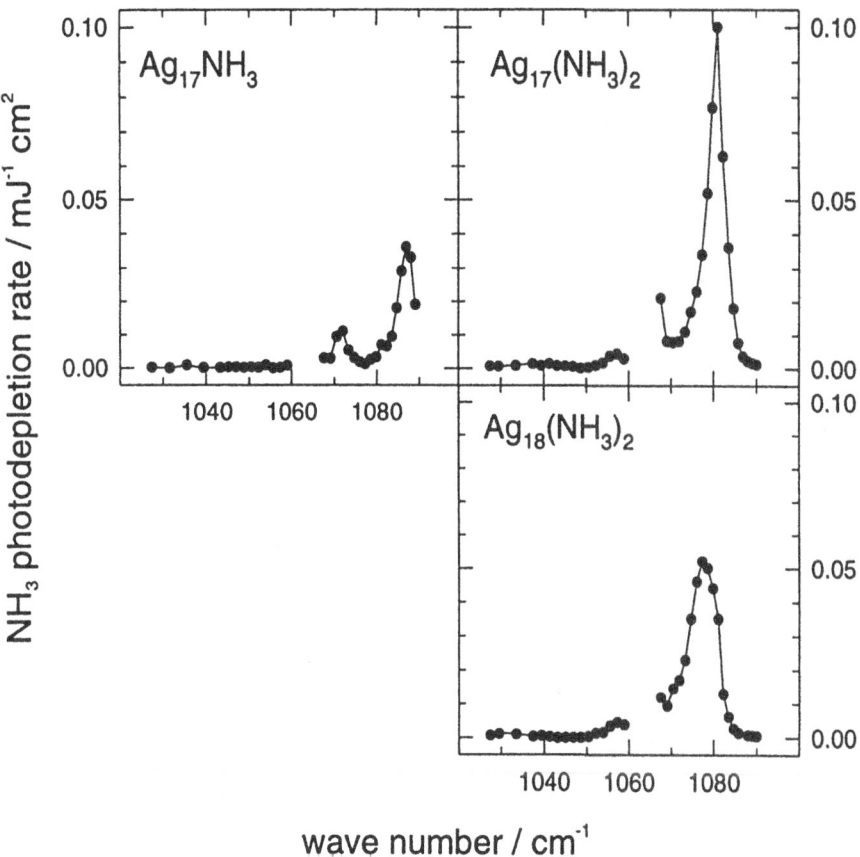

Fig. 5. (*Continued*) Infrared photodepletion spectra of $Ag_n NH_3$ and $Ag_n(NH_3)_2$ cluster complexes in the frequency region of the umbrella vibrational mode of bound $NH_3$.

Nor is there any apparent correlation with the thermodynamic properties of the cluster shown in Figure 4. This is somewhat disappointing as DFT theory carried out on $NH_3$ complexes of coinage metal dimers and trimers found $\nu$-$NH_3$(umbrella) to correlate to the $NH_3$ binding energy [36]. Apparently these findings do not extrapolate to larger clusters.

Addition of a second $NH_3$ ligand to $Ag_n NH_3$ complexes increases the strength of the infrared absorption and, generally, results in a red shift of the spectrum. Characteristics of the second addition vary with cluster size. Evidence for co-operative binding is seen, most strongly when $n = 8$ and 14. In these cases addition of a second $NH_3$ results in a shift of the original

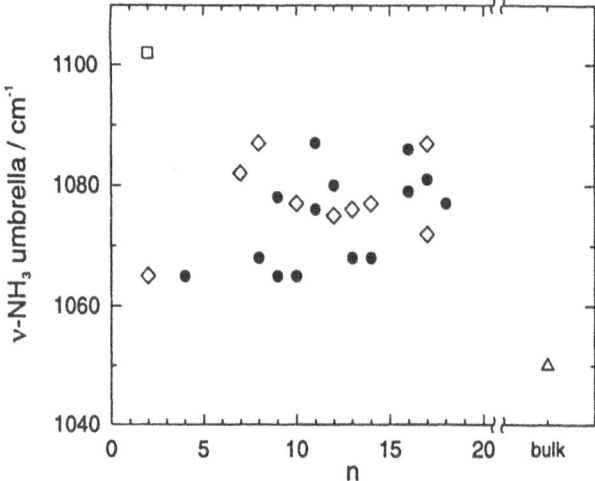

**Fig. 6.** Positions of the $\nu$-NH$_3$ umbrella frequency as a function of Ag$n$ cluster size in Ag$_n$NH$_3$ ($\diamond$) and Ag$_n$(NH$_3$)$_2$ ($\bullet$) cluster complexes. The positions for Ag$_2$NH$_3$ calculated from DFT theory ($\square$) (Ref. 35) and for NH$_3$ on an Ag(110) surface ($\triangle$) (Ref. 43) are also shown.

Ag$_n$NH$_3$ $\nu$-NH$_3$(umbrella) resonance to the red. It is not clear at this time what mechanism is responsible for this co-operativity. Whether it is due to direct interaction between two ligands or a through the cluster effect is not known. Of the clusters which are thought to be locally bound, Ag$_{16}$(NH$_3$)$_2$ exhibits two distinct peaks of similar intensity. This is either due to the two ligands occupying different sites and being largely uncoupled or to two strongly coupled ligands which may be in similar sites but exhibit two peaks associated with their in-phase and out-of-phase umbrella motion. In contrast the spectrum of Ag$_{17}$(NH$_3$)$_2$ is dominated by one strong peak, indicating sites that are at least equivalent in their effect on $\nu$-NH$_3$(umbrella) and that are relatively weakly coupled.

Several complexes, most notably Ag$_{10}$NH$_3$, Ag$_{14}$NH$_3$ and Ag$_{17}$(NH$_3$)$_{1,2}$, exhibit relatively narrow peaks with FWHM $\sim$7 cm$^{-1}$. Such single absorption peaks might be expected when a locally bound complex exists in a single configuration. Ag$_{17}$NH$_3$, shown in Section 4.1 to be locally bound, would fit this case. Single peaks are also expected in the extreme of non-local binding where the ligand cluster complex interaction is insensitive to the position on

**Table 2.** $NH_3$ umbrella mode frequencies $(cm^{-1})$ of $Ag_nNH_3$ and $Ag_n(NH_3)_2$ complexes.[a,b]

| $n$ | $Ag_nNH_3$ | $Ag_n(NH_3)_2$ |
|---|---|---|
| 2 | $1065^c$, *$1102^c$* | |
| 4 | >1090 | 1065 |
| 5 | 1082 | |
| 7 | 1082 | |
| 8 | 1087 | 1068 |
| 9 | >1090 | 1065, 1078 |
| 10 | 1077 | 1065 |
| 11 | >1090 | 1076, (1087) |
| 12 | 1075 | 1080 |
| 13 | 1076 | 1068 |
| 14 | 1077 | 1068 |
| 15 | 1077 | 1070 |
| 16 | ≥1089 | 1079, 1086 |
| 17 | 1089, (1072) | 1081, (1065) |
| 18 | | 1077 |
| Ag(110) | $1050^d$ | |

[a]Values in *italics* are from DFT theory.
[b]Values in brackets indicate minor peaks.
[c]From Ref. 35
[d]From Ref. 43

the surface. $Ag_{10}NH_3$ and $Ag_{14}NH_3$ appear to fall in this category and on these clusters either all positions are equivalent or exchange is so rapid that a single average peak is detected. From uncertainty principle considerations, free translation on the surface at room temperature would be sufficient to collapse bands separated by 30 $cm^{-1}$ or less. Other complexes which are also thought to be non-locally bound, most notably those with $n = 5$, 8 9 and 11, exhibit relatively broad absorbances. One explanation is that some degree of residual localization still holds in these complexes, slowing exchange down sufficiently to broaden the absorption. From the thermodynamic point of view, some degree of localization is possible within the bounds of the experimental uncertainty in the binding entropies reported in Section 4.1. An alternative explanation of the broad peaks, where they occur, is that they are due to the presence of isomers of the $Ag_n$ clusters themselves. Laser vapourization sources are known to produce mixtures of metal cluster isomers. Perhaps the best documented case is that of $Nb_n$ clusters [44, 45]. Isomers are almost certainly ubiquitous in metal clusters. They are most apparent in the size range where a marked difference in metal atom coordination between two low lying isomers results in distinctive physical or chemical properties, i.e. close to where the transitions from planar to 3-dimensional geometries and from single shell geometries to geometries having a central atom occur [46]. This is the range covered by these experiments on $Ag_n$ clusters. In the case of $Ag_{17}NH_3$ a subsidiary peak at 1072 cm$^{-1}$ is resolved from the main peak at 1087 cm$^{-1}$. Both are narrow and in the isomer picture would be assigned to two forms of $Ag_{17}NH_3$ differing in the underlying structure in the metal cluster itself. However, as $NH_3$ is locally bound on this cluster (see Section 4.1), it is also possible that two different local sites are available on this cluster.

Infrared desorption of $NH_3$ from $Ag_nNH_3$ complexes has parallels in surface science. Laser induced thermal desorption from surfaces has been well studied and resonant desorption is the subject of experimental and theoretical work [47]. As in molecular IRMPD the process is statistical with the sequential absorption of IR photons accompanied by rapid transfer of energy from the pumped mode [48]. The time-scale and dynamics of transfer of vibrational energy from the adsorbate to the substrate is an important issue which may be addressed by extension of work on the IR photodepletion of $Ag_nNH_3$ complexes to measure desorption rates and intensity effects. Clusters are different from surfaces in that they are finite heat baths. Their heat capacities tune with size which could be invaluable in testing statistical models.

## 5  Other Silver Cluster Complexes

Recently the study of silver cluster complexes has been widened to include other adsorbates. Infrared photodepletion spectroscopy has been applied to complexes of small silver clusters with benzene [49] and with ethylene and ethylene oxide [50]. This work was similar in concept to that described above for $Ag_nNH_3$ but used a CW $CO_2$ probe laser. Vibrational bands observed in the region of 10 $\mu$m are consistent with the formation of molecular association complexes in all cases. Relaxation of infrared selection rules is associated with reduced symmetry on adsorption and gives an indication of the adsorbate geometry in relation to the gas phase and on single crystal surfaces.

The dynamics of cluster mediated chemistry has been approached through the study of the photochemistry of silver cluster complexes. Near UV photodepletion experiments carried out on small silver cluster/carbonyl sulfide complexes, $Ag_nOCS$, show a remarkable odd-even dependence on the number of Ag atoms [51]. OCS photodesorbs non-dissociatively from even numbered clusters but, on odd silver clusters a dissociative channel leading to $Ag_nS$ predominates. Comparison of the photoaction spectra with the photodepletion spectra of $Ag_n Kr$ complexes (see above) demonstrates that electronic excitation of the metal cluster initiates the chemistry in either case, analogous to substrate mediated photochemistry and photodesorption on extended surfaces. A charge-transfer mechanism is thought to be involved in the OCS dissociation channel, with the accessibility of the intermediate ion-pair state of $Ag_n OCS$ governing the size-dependence.

# 6  Conclusions

Significant progress has been made in understanding silver cluster/ammonia complexes and the underlying properties of silver clusters. Experimental information on the structure and reactivity other silver cluster complexes is also emerging. However, there is still a long way to go to a full description of these systems and their application to understanding more complex chemical systems. Theory has made a large contribution in establishing the structures of small silver clusters, although discrepancies between CI and DFT ground state structures still remain to be settled. Theory has been particularly powerful when combined with PES experiments in establishing the structure of anionic clusters up to $n = 9$. It is challenged to do the same for neutral clusters by predicting the optical absorption spectra presented here, a harder problem because calculations must be taken to higher excited states.

Silver cluster/ammonia complexes have intriguing experimental size-dependent properties. These include binding enthalpies and entropies which indicate local bonding on some particular clusters compared to non-local bonding on others and infrared spectra which show the frequency of the $NH_3$ umbrella mode to depend on cluster size and on $NH_3$ coverage. If we could understand these size-dependencies in molecular terms we would have a greater understanding of the factors at play in the formation of association complexes at metal clusters and at metal surfaces. It is important to understand such complexes because of the role they play as precursors to dissociative chemistry in more reactive systems. We look to theoreticians to make the next steps forward in understanding this important prototype system for metal cluster association complexes. At present we are not aware of any theoretical studies of $Ag_n NH_3$ complexes for $n > 2$ or for the related coinage metal complexes, $Cu_n NH_3$, for $n > 3$. Links between cluster reactivity and cluster structure are surely there. In the near future we can expect that they will be revealed and understood at a fundamental level.

# References

1 Morse M.D., Geusic M.E., Heath J.R. and Smalley R.E., J. Chem. Phys. **83**, 2293 (1883)
2 Bérces A., Hackett P.A., Lian .L., Mitchell S.A. and Rayner D.M., J. Chem. Phys. **108**, 5476 (1998)

394

3 Jackschath C., Rabin I. and Schulze W., Z. Phys. D Atoms, Molecules and Clusters **22**, 517 (1992)

4 Alameddin G., Hunter J., Cameron D. and Kappes M.M., Chem. Phys. Lett. **192**, 122 (1992)

5 Ho J., Ervin K.M. and Lineberger W.C., J. Chem. Phys. **93**,6987 (1990)

6 Ganteför G., Gausa M., Meiwes-Broer K.H. and Lutz H.O., J. Chem. Soc. Faraday Trans. **86**, 2483 (1990)

7 Taylor K.L., Pettiette-Hall C.L., Cheshnovosky O. and Smalley R.E., J. Chem. Phys. **96**, 3319 (1992)

8 Yannouleas C. and Landman U., Phys. Rev. B **51**, 1902 (1995)

9 Bonačić-Koutecký V., Češpiva L., Fantucci P. and Koutecký J., J. Chem. Phys. **98**, 7981 (1993)

10 Bonačić-Koutecký V., Češpiva L., Fantucci P., Pittner J. and Koutecký J., J. Chem. Phys. **100**, 490 (1994)

11 Handschuh H., Cha C.-Y., Bechthold P.S., Ganteför G. and Eberhardt W., J. Chem. Phys. **102**, 6406 (1995)

12 Tiggesbäumker J., Köller L., Lutz H.O. and Meiwes-Broer K.H., Chem. Phys. Lett. **190**, 42 (1992)

13 Tiggesbäumker J., Köller L., Meiwes-Broer K.H. and Liebsch A., Phys. Rev. A **48**, R1748 (1993)

14 Harbich W., Fedrigo S. and Buttet J., Chem. Phys. Lett. bf 195, 613 (1992)

15 Fedrigo S., Harbich W. and Buttet J., Phys. Rev. B **47**, 10706 (1993)

16 Kresin V.V., Phys. Rev. B **51**, 1844 (1995)

17 Knickelbein M.B. and Menezes W.J.C., Phys. Rev. Lett. **69**, 1046 (1993)

18 Menezes W.C.J. and Knickelbein M.B., J. Chem. Phys. **98**, 1867 (1993)

19 Collings B.A., Athanassenas K., Rayner D.M. and Hackett P.A., Z. Phys. D **26**, 36 (1993)

20 Collings B.A., Athanassenas K., Lacombe D.M., Rayner D.M. and Hackett P.A., J. Chem. Phys. **101**, 3506 (1994)

21 Collings B.A., Athanassenas K., Rayner D.M. and Hackett P.A., Chem. Phys. Lett. **227**, 490 (1994)

22 Lindinger M., Dasgupta K., Dietrich G.,Krückeberg S., Kuznetsov S., Lützenkirchen K., Schweikhard L., Walther C. and Ziegler J., Z. Phys. D: Atoms Mol. Clusters **40**, 347 (1997)

23 Ellert C., Schmidt M., Schmitt C., Reiners T. and Haberland H. Phys. Rev. Lett. **75**, 1731 (1995)

24 Knickelbein M.B. and Menezes W.J.C., Chem. Phys. Lett. **184**, 436 (1991)

25 Bonačić-Koutecký V., private communication

26 Haslett T.L., Bosnick K.A. and Moskovits M., J. Chem. Phys. **108**, 3453 (1998)

27 Santamaria R., Kaplan I.P. and Novaro O., Chem. Phys. Lett. **218**, 395 (1994)

28 Lian L., Akhtar F., Hackett P.A. and Rayner D.M., Chem. Phys. Lett. **205**, 487 (1993)

29 Lian L., Akhtar F., Parsons J.M., Hackett P.A. and Rayner D.M., Z. Phys. D: Atoms Mol. Clusters **26S**, S168 (1993)

30 Lian L.,Mitchell S.A. and Rayner D.M., J. Phys. Chem. **98**, 11637 (1994)

31 Mitchell S.A., Lian L., Rayner D.M. and Hackett P.A., J. Chem. Phys. **103**, 5539 (1995)

32 Lian L., Hackett P.A. and Rayner D.M., J. Chem. Phys. **99**, 2583 (1993)

33 Lian L., Michell S.A., Hackett P.A. and Rayner D.M., J. Chem. Phys. **104**, 5338 (1996)

34 Mitchell S.A., Lian L., Rayner D.M. and Hackett P.A., J. Phys. Chem.**100**, 15708 (1996)

35 Rayner D.M., Lian L., Fournier R., Mitchell S.A. and Hackett P.A., Phys. Rev. Lett **74**, 2070 (1995)

36 Fournier R., J. Chem. Phys. **102**, 5396 (1995)

37 Nygren M.A., Siegbahn P.E.M., Jin C., Guo T. and Smalley R.E., J. Chem. Phys. **95**, 6181 (1991)

38 Zakin M.R., Brickman D.M., Cox D.M., Reichmann K.C., Trevor D.J. and Kaldor A., J. Chem. Phys. **85**, 1198 (1986)

39 Knickelbein M.B., Chem. Phys. Lett. **239**, 11, (1995)

40 Rayner D.M., Lian L., Athanassenas K., Collings B.A., Fournier R., Mitchell S.A. and Hackett P.A., Surface Rev. Lett. **3**, 649 (1996)

41 Black J.G., Yablonovitch E., Bloembergen N. and Mukamel S., Phys. Rev. Lett. **38**, 1131 (1977)

42 Athanassenas K., Collings B.A., Hackett P.A. and Rayner,D.M., to be published.

43 Thornburg D.M. and Madix R.J., Surf. Sci. **220**, 268 (1989)

44 Hamrick Y.M. and Morse M.D., J. Phys. Chem. **93**, 6494 (1989)

45 Knickelbein M.B. and Yang S., J. Chem. Phys. **93**, 1476 (1990);**93**, 5760 (1990)

46 Athanassenas K., Kreisle D., Collings B.A., Rayner D.M. and Hackett P.A., Chem. Phys. Lett. **213**, 105 (1993)

47 Ertl G. and Neumann M., Z. Naturforsch. **27A**, 1607 (1972)

48 Hussla I., Seki H., Chuang T.J., Gortel H.J., Kreuzer H.J. and Piercy P., Phys. Rev. B **32**, 3489 (1985)

49 Koretsky G.M. and Knickelbein M.B., Chem. Phys. Lett. **267**, 485 (1997)

50 Koretsky G.M. and Knickelbein M.B., J. Chem. Phys. **107**, 10555 (1997)

51 Brown L.A. and Rayner D.M., J. Chem. Phys. **109**, 2474 (1998)

# Laser-Femtochemistry of Small Clusters

Elmar Schreiber*

Max-Born-Institut, Rudower Chaussee 6, D-12489 Berlin, Germany;
email: eschreib@mbi-berlin.de

**Abstract.** An overview on the opportunities of 'Laser-Femtochemistry' applied to small alkali and silver clusters is presented. The experimental technique of femtosecond real-time spectroscopy is briefly described. By means of several model systems phenomena like the observation of vibrational wave packets and their revivals on perturbed potential energy surfaces, the control of molecular dynamics, selective state preparation, ultrafast internal vibrational redistribution, photodissociation and femtosecond structural relaxation are exemplarily introduced.

## 1 Introduction

The study of molecular dynamics in the femtosecond time domain (1 fs= $10^{-15}$ s) can be regarded as the ultimate achievement in half a century of the development of techniques for exploration of the most elementary motion of atoms bound by chemical forces [1]. The outstanding books of A.H. Zewail 'Femtochemistry Vol.I&II' [2] and J. Manz & L. Wöste 'Femtosecond Chemistry Vol.I&II' [3] as well as the presented contributions [4, 5] at the two 'Femtochemistry' conferences in Berlin (1993) and Lausanne (1995) organized by J. Manz and M. Chergui, respectively, document how femtosecond probing of molecular dynamics allows the viewing of new phenomena and the reaching of new frontiers. Applying real-time femtosecond spectroscopy to molecules or clusters enables us to make a 'movie' of their molecular dynamics revealing such phenomena as wave packet propagation, coherent control, internal vibrational redistribution, and ultrafast photodissociation [6].

The central idea of this ultrafast spectroscopy of molecules and clusters is the preparation of molecular wave packets followed by the detection of their motion in real-time (see Fig. 1a). Vibrational periods of small molecules and clusters are of the order of $10^{-14}...10^{-12}s$ [7]. With the availability of lasers generating pulses in the sub-100 fs [8–11] regime, the real-time spectroscopy of these systems became feasible. By his advent of femtosecond pump&probe techniques [12–16] A.H. Zewail then established the direct observation of ultrafast molecular dynamics, named 'Laser-Femtochemistry' [17].

Since then the ultrafast dynamics of a few molecular, especially dimer systems, have been studied. Zewail observed the vibration and rotational revival

---

* The experiments presented here have been carried out at the Institut für Experimentalphysik, Freie Universität Berlin, in the laboratories of L. Wöste during the stay of the author in his group.

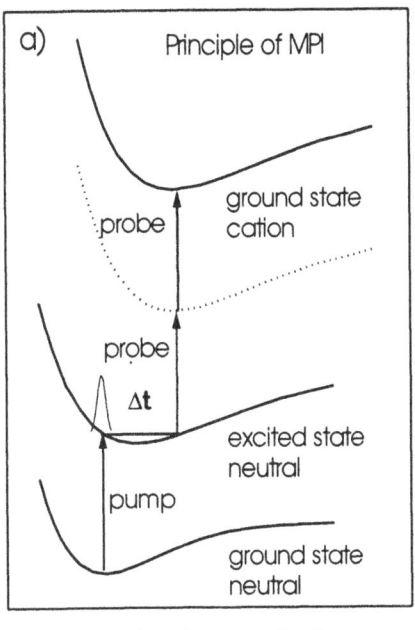
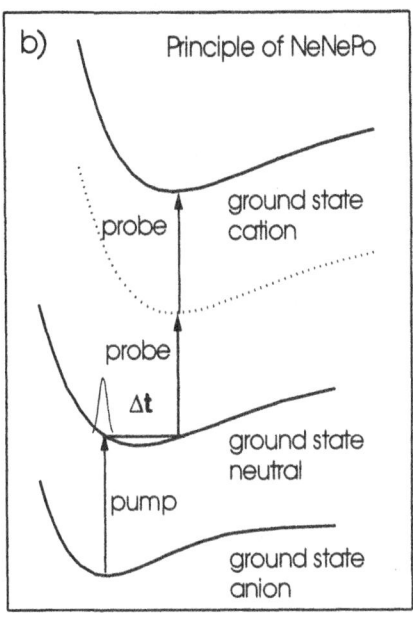

**Fig. 1.** Principle scheme of the transient MPI (a) and NeNePo (b) spectroscopic technique. a) Principle of time-resolved MPI spectroscopy: A wave packet is prepared in an excited state of the neutral system by a pump pulse. Since in general the transition probability to the ion state is a function of the wave packets' location on the potential energy surface, the evolution of the wave packet can be probed by a second time-delayed pulse. b) Principle of the NeNePo process: Starting in the anion's potential energy surface an ultrashort pump pulse detaches an electron and prepares a wave packet in the neutral. After a certain delay time $\Delta t$ a probe pulse photoionizes the neutral. The time-dependent signal of the cation's intensity is detected. For convenience, this method is named NeNePo, **Ne**gative-to-**Ne**utral-to-**Po**sitive

of excited $I_2$ [18–20]. A. Stolow studied the same system applying femtosecond pump&probe zero-kinetic-energy (ZEKE) photoelectron spectroscopy [21, 22]. G. Gerber observed fascinating features in the ultrafast dynamics of the sodium dimer's multiphoton ionization (MPI) [23–27] being well described by theoretical simulation [28–31]. Recently, B. Girard presented a theory and an experiment of one-color coherent control for the closely related $Cs_2$ system [32]. In $Li_2$ S.R. Leone observed vibrational and rotational recurrences with single rovibronic control of an intermediate state [33].

In Sec. 3.1 for similar systems, two isotopes of $K_2$ excited to its $A\,^1\Sigma_u^+$ state, the wave packet propagation is investigated by real-time three-photon ionization (3PI) spectroscopy. The high resolution in time reveals pronounced differences of the isotopes' dynamics (see also [34–42]). The coherence of the femtosecond pulse transferred to the dimer systems enables the recovery of nearly all of the spectral information. Strongly different energetic shifts due to different spin orbit coupling of the A state with a crossing 'dark' b $^3\Pi_u$ state for the isotopes show up. The achieved accuracy in time is close to 1 fs, the spectral resolution is less than $0.1\,\text{cm}^{-1}$. Similar results are obtained for $Na_2$ [42–44].

The general complementarity of sensitivities in continuous wave and femtosecond spectroscopy has been anticipated by A.H. Zewail [45] and it is verified for the $Na_3$ molecule excited to its electronic B state (see Sec. 3.2). The effect of selective state preparation and internal vibrational redistribution (IVR) can be achieved by variation of the exciting pump pulse duration [46–50]. The possibility of preparing specific vibrational modes by ultrafast pulses provides a detailed understanding on how the induced molecular oscillation is built up. Starting in the dominantly prepared symmetric stretch mode the sodium trimer finds its well-known pseudorotational rhythm after a few picoseconds due to an ultrafast IVR process [46].

Concerning the observation of wave packet propagation phenomena the $K_3$ molecule is another promising candidate. Theoretical calculations [51] predicted an electronic state comparable to the $Na_3$ B state at about 800 nm. Different, highly sensitive methods, such as stationary MPI and depletion spectroscopy were applied but failed [52]. As presented in Sec. 3.3 with femtosecond real-time spectroscopy, however, it is possible to observe both the vibrational and dissociation dynamics of this system [40, 53, 54].

The $K_3$ system represents a limiting case, where both, wave packet propagation phenomena and dissociation dynamics of the molecule can be analyzed. This leads to a new chapter (Sec. 4) of the presented investigations.

The photodissociation, especially of small molecules and clusters, can be regarded as the motor for many important chain reactions. To determine the characteristics of this "motor" the exploration of the real-time dissociation of small elemental clusters induced by ultrashort light pulses is of great help. It will give a fundamental insight into the stability and chemical forces of these species. Due to the inherent instability of clusters, general information about their fundamental properties have become available only recently using new experimental techniques. Here, catalysis [55–57], solvation [58], reactivity [12, 14, 15, 59], energy transfer [60], and fragmentation [61–65] processes are some of the main areas of interest. One essential lack of empirical knowledge is the stability of these clusters itself. To investigate their stability, real-time studies of the fragmentation probability are required. The lifetime of excited clusters is expected in the range of $10^{-10} \ldots 10^{-14}\,\text{s}$ [7, 66, 67].

To observe ultrafast fragmentation of excited alkali clusters the appropriate tool is real-time MPI spectroscopy. This technique allows the mass-selected detection of the ultrafast photodissociation with high sensitivity. In 1992 G. Gerber and coworkers presented the first femtosecond time-resolved experiments in cluster physics [28, 68, 69] showing differences in the fragmentation behavior of $Na_{n\leq21}$ clusters dependent on the excitation at different wavelength. In Sec. 4 the real-time photodissociation dynamics of small sodium ($Na_{n=3\ldots10}$) is studied as a function of cluster size as well as excitation wavelength [70–76].

The ultrafast dissociation, however, often prevents the obtaining of information concerning the vibrational dynamics of the studied system. Here a new approach – called NeNePo – opens a window to obtain a deeper insight into the dynamics of a cluster (see Fig. 1b). This technique (mind the analogy to the time-resolved MPI spectroscopy) enables the direct preparation of a wave packet in the ground state of a cluster. An ultrashort pump pulse interacts with an anion to produce by photodetachment a neutral cluster excited to several vibrational levels of its ground state. Hence, the neutral's pure ground state dynamics can be probed by a subsequent laser pulse. First results on $Ag_3$ are presented in Sec. 5 revealing information about ultrafast structural relaxation times of the prepared molecule or cluster [77–79]. The results are in excellent agreement with an elegant theoretical description given by K.H. Bennemann and coworkers [80–82]. They combine an electronic theory with molecular dynamics calculations to study the ultrafast structural response of optically excited small clusters.

With these introducing aspects I like to enter the fascinating world of 'Laser-Femtochemistry' of small clusters by giving first a little information on the experimental set-ups used for the presented investigations.

# 2 Experimental Set-ups

Here, the principles of two set-ups for the spectroscopy of ultrafast processes in clusters are briefly described. In Sec. 2.1 the set-up for the presented real-time MPI experiments is sketched. With this arrangement the investigations discussed in Sec. 3 and Sec. 4 are carried out. The set-up for the NeNePo experiment (see Fig. 1 and Sec. 5) is presented in Sec. 2.2. Further details about the experimental set-ups are given in [6].

## 2.1 Set-up for Real-Time Multiphoton Ionization

In this paper, different alkali molecules and clusters are investigated with real-time MPI spectroscopy. The time evolution of the MPI signals are obtained by means of femtosecond pump&probe technique followed by mass-selective detection of the alkali ions. Since several modifications of the employed ultrafast laser system are used to perform the real-time investigations on the different clusters, here the configuration for the $K_2$ studies is briefly sketched (see Fig. 2). It neatly introduces all principles of the utilized technique. Special details on the other set-ups are given in [34] for $K_2$, in [83] for $Na_3$, in [84] for $K_3$, in [73] for $Na_n$.

Femtosecond laser pulses of 90 fs duration (FHWM, assuming $sech^2$ pulse shape) were generated in an argon-ion laser-pumped modelocked titanium sapphire laser (Spectra Physics model 2080, all lines visible, 8 W, and model 3960). The spectral width of the femtosecond pulse spectrum at the wavelength 833.7 nm was measured, covering $190\,cm^{-1}$ (FWHM), so the pulses reached 1.6 times the Fourier limit. The pulse repetition rate was $\sim 80\,MHz$. A Michelson-like arrangement was used to split the laser beam and to realign it collinearly with identical polarization, using single stack dielectric beam splitters. The length of one of the

Michelson branches was controlled by a computer-driven DC-motor translation stage. Its position defines the delay time $\Delta t$ between pump and probe pulse and was read out by an optical encoder. The resolution of the delay time $\Delta t$ was about 0.1 fs and the minimal step width amounted 0.3 fs. Each of the laser pulse trains had an average power of about 200 mW.

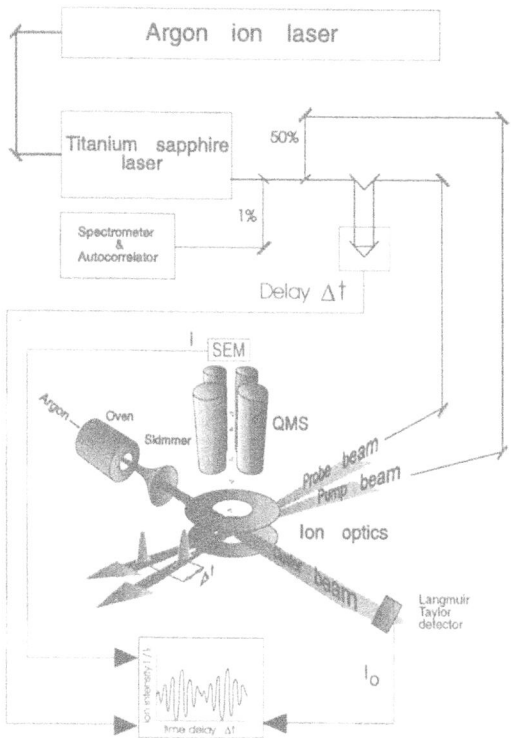

**Fig. 2.** An argon ion laser pumps a modelocked titanium:sapphire laser. 1% of the laser output is used to obtain the laser parameters by means of a spectrometer and an autocorrelator. The laser beam is splitted into two identical parts, one passing through the delay unit. Both laser beams are recombined in the interaction region of the cluster chamber. Alkali metal vapor is produced in the oven and coexpanded with argon. The cluster beam is collimated using a skimmer. Cluster ions produced by interaction of the laser beams with the cluster beam are focused by the ion optics into the quadrupole mass filter (QMS) followed by a secondary electron multiplier (SEM). A Langmuir-Taylor detector (LTD) measures the cluster beam intensity $I_0$. The relative intensity $I/I_0$ of ions is recorded as a function of the delay time $\Delta t$

In order to produce a molecular beam of high stability, pure potassium was evaporated at a temperature of 850 K in a TZM (Titanium-Zirconium-Molybde-num) oven. The alkali metal vapor was coexpanded with an inert carrier gas (3...5 atm) through a 70 μm nozzle. In this continuous supersonic beam source the rotational and vibrational temperatures amounted to about 10 K and 50 K, respectively. Therefore, the $K_2$ molecules, arriving from the oven, were in the ground state $X^1\Sigma_g^+$. Their vibrational quantum number was $v = 0$ and their rotational quantum number is $J \approx 20$.

The laser beams were focused on the molecular beam by means of a 400 mm lens. Each pulse reached a peak power of about 0.5 GW. Photoionized potassium dimers were mass-selectively detected by means of a quadrupole mass spectrometer with a resolution of $\frac{m}{\Delta m} > 240$, being sufficient to distinguish between $^{39,39}K_2$ and the heavier $^{39,41}K_2$ isotope. The ion intensity $I$ was continuously recorded as a function of the delay time $\Delta t$ between pump and probe pulse. The result is the so-called real-time spectra $I(\Delta t)$ of the MPI process.

We chose a typical time step for the pump&probe experiments for $K_2$ of $\Delta t_{step} = 50$ fs. For this step width the Nyquist critical frequency $\omega_c = \frac{1}{2\Delta t_{step}c} = 333$ cm$^{-1}$ is even larger than four times the expected frequency $\omega_0 \approx 65$ cm$^{-1}$. Therefore, an aliasing of realistic frequency components larger than $\omega_c$ is not expected in a Fourier analysis. The time to record a transient spectrum up to 200 ps was about two hours, a time where the molecular beam is not a priori stable. While in our experiments the intensity of the laser was stable within 3% the Langmuir-Taylor detector controlled intensity of the molecular beam varied over a range of about ±20%. We normalized the transient data to obtain data points oscillating around the zero-line to perform a correct real Fourier analysis.

## 2.2 The Set-up for the Real-Time NeNePo Experiments

As described in detail in Sect. 5 a real-time NeNePo (see also Fig. 1b)) experiment starts with a photodetachment process followed by a photoionisation step, each induced by an ultrashort laser pulse. Hence, besides a source of femtosecond pulses for NeNePo investigations negatively charged molecules and clusters (here silver) are necessary. They are generated in a sputtering ion source by bombarding targets of elemental silver with fast Xe$^+$ ions. Vaporizing cesium metal onto the targets lowers the work function of the silver surface and enables a drastic increase of the anions' yield.

An ion lens collects the generated anions and leads them into a first quadrupole ion guide of large cross-section (see Fig .3). Here, the cluster ions are cooled and moderated by collisions with a background gas of He ($p_{He} \approx 10^{-2}$ mbar) [85]. The cations then have about room temperature. Next the anions pass a quadrupole mass filter, where the species of interest can be selected. A beam of well thermalized mass selected metal cluster ions is created by this method.

To increase the density of the anions, they are stored in a linear quadrupole ion guide, which is filled with gas and operates as an ion trap [86]. The entrance lens of the ion guide is kept at a potential being slightly below the kinetic energy of the ions, while the exit lens is on a higher potential. Hence, the anions can enter the ion guide, but are reflected back at the exit lens. Traveling through the gas cell, the ions lose kinetic energy by collisions and are no longer able to escape via the entrance lens. The number of stored ions can be monitored by pulsing the exit lens open and recording the magnitude of the ion-current pulse. Without the detachment laser beam, more than $10^8$ ions can be stored in the trap.

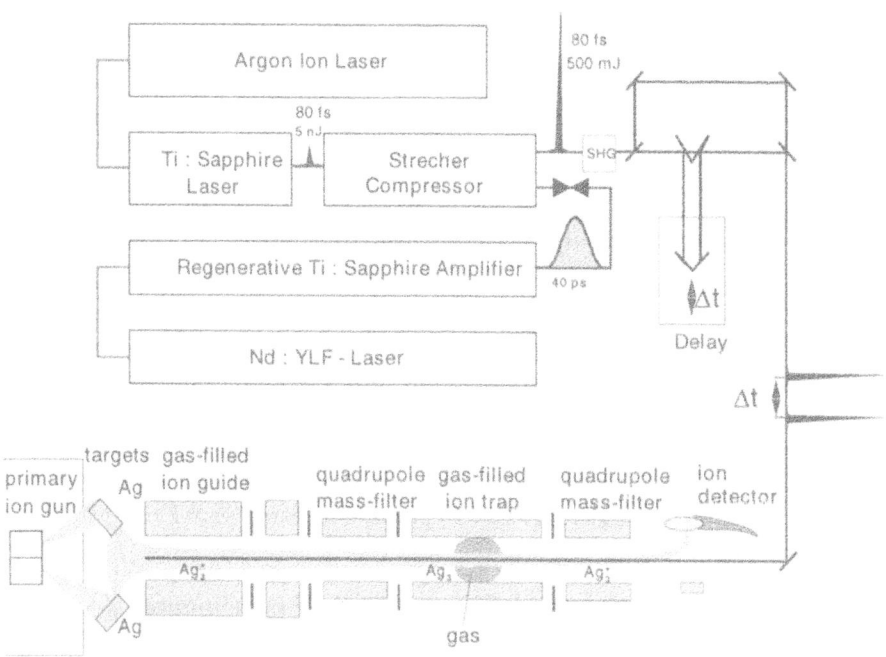

**Fig. 3.** Experimental set-up for the NeNePo investigations. Schematic of the ion trap used for the real-time NeNePo experiment. The first quadrupole prepares a beam of mass-selected negative ions (here $Ag_3^-$). They are stored in the central quadrupole ion trap. Here, detachment (pump) and ionization (probe) laser pulses interact with the stored clusters. The positively charged ions subsequently are accelerated to a third quadrupole, where again mass filtering can be carried out

With the detachment laser, an equilibrium between the continuous filling and the depletion of the trap by the detachment process is reached. In this case, an ensemble of about $10^6$ mass selected anions is stored and can interact with the subsequent detachment and ionization pulses. The cations created by the laser pulses are extracted by the exit lens. Having passed a second quadrupole mass filter the cations are detected by means of a secondary electron multiplier detector. This allows the detection of possible cluster fragments during or after the interaction with the laser pulses.

To analyze the time evolution of the neutral trimer's configuration a laser system consisting of a titanium sapphire oscillator (Spectra Physics Tsunami) which is pumped by a 12 W argon ion laser and of a Nd:YLF-pumped regenerative amplifier [87] (Quantronix 4800) is used. This configuration enables us to produce ultrashort ($t < 100fs$, 500 mJ/pulse) laser pulses at a repetition rate of 1 kHz. The pulses are efficiently ($\approx 40\%$) frequency doubled in a BBO crystal. The second harmonic is split into pump and probe pulses, with the probe pulse delayed with respect to the pump pulse by a computer controlled translation stage. Pump and probe laser beams are imaged into the trap collinear with the ion trajectories and overlap throughout the whole length of the trap (see Fig.3). The electrons of the stored cluster anions are detached by the pump pulse and, subsequently, after a certain delay time $\Delta t$, the just created neutrals are ionized by the probe pulse. Now the trap no more acts as a trap, but preferably accelerates the cations to the ion detector. To obtain the real-time spectra the cation's intensity now is detected as a function of the delay time $\Delta t$. In Sec. 5 first fascinating results measured with this technique are presented.

# 3  Wave Packet Propagation Phenomena

A central idea to investigate the molecular dynamics in real-time is to prepare a wave packet on a potential energy surface (PES), i.e. exciting coherently several energy eigenstates of the molecule. Since the spectral width of ultrashort laser pulses is broad compared to the level spacing of the molecular eigenenergies, this coherent superposition of eigenstates can be well managed with ultrafast laser systems. Applying a fs pump pulse, a wave packet will be prepared in the excited state of the cluster. This wave packet will propagate on the PES. Since in general the transition probability to the ion state is a function of the wave packet's location on the PES the evolution of the wave packet propagation can be probed by a second time delayed pulse. High transition probability will result in a maximum in the detected ion signal. As a consequence, a more or less pronounced oscillation will appear. Three fascinating examples of model systems demonstrating the expressiveness of wave packet propagation studies are given in this section. Further examples and details are presented in [6].

## 3.1  K$_2$: Isotopic Effects and Revival Structure

From the viewpoint of classical spectroscopic experimental techniques the electronic $A^1\Sigma_u^+$ state of $^{39,39}$K$_2$ has been studied e.g. by laser-induced fluorescence, optical-optical double resonance and Fourier transform spectroscopy [88–93]. Rovibrational levels could be numbered and the spectroscopic constants are calculated by a Dunham fit. An RKR (Rydberg-Klein-Rees) analysis was used to deduce the potential energy curves. A strong spin-orbit-coupling between the $A^1\Sigma_u^+$ state and the $b^3\Pi_u$ state was observed around $v = 8$ and $v = 12$, respectively. In $^{39,41}$K$_2$ a couple of fluorescence lines could be identified [92]. The influence of inter system crossing (ISC) processes on the molecular dynamics has been pointed out theoretically for some examples [94–96]. Special interest is focused on the intersection of the intersystem crossing of the K$_2$ $A^1\Sigma_u^+$ state with the $b^3\Pi_u$ state. Applying one-color pump&probe spectroscopy with a wavelength of 833.7 nm we studied the wave packet dynamics directly around $v = 12$

404

in $^{39,39}K_2$. The isotope-selective detection allows the comparison with the isotope $^{39,41}K_2$.

The real-time spectra for $^{39,39}K_2$ and $^{39,41}K_2$ were recorded for delay times between $-5$ ps and more than 180 ps (see also [34–38, 97, 98]). The temporal evolution of the ion signal's intensity for both isotopes are shown in Fig. 4 for delay times between 0 and 200 ps. In both pump&probe spectra a fine oscillatory structure with an oscillation period $T_A \approx 500$ fs — being the full $2\pi$ oscillation time of the wave packet in the A state — is present over the whole range. The first maximum of the oscillation appears at a delay time of 250 fs being half ($\pi$) of the A state period. This fine oscillation is superimposed on a long-time evolution which reveals totally different features for the two isotopes: in the transient of the $^{39,39}K_2$ (Fig. 4a) a beat structure with a period $T_{BS} \approx 10$ ps dominates in the measured temporal region. A double structure of the beat oscillation maxima appears at $T_{BD,1} \approx 10$ ps and at $T_{BD,2} \approx 60$ ps. For the $^{39,41}K_2$ (Fig. 4b)) the long-time structure includes different features: A regular dephasing and some fractional revivals [99] are observed. The main revivals appear at 38 ps, 60 ps and 82 ps.

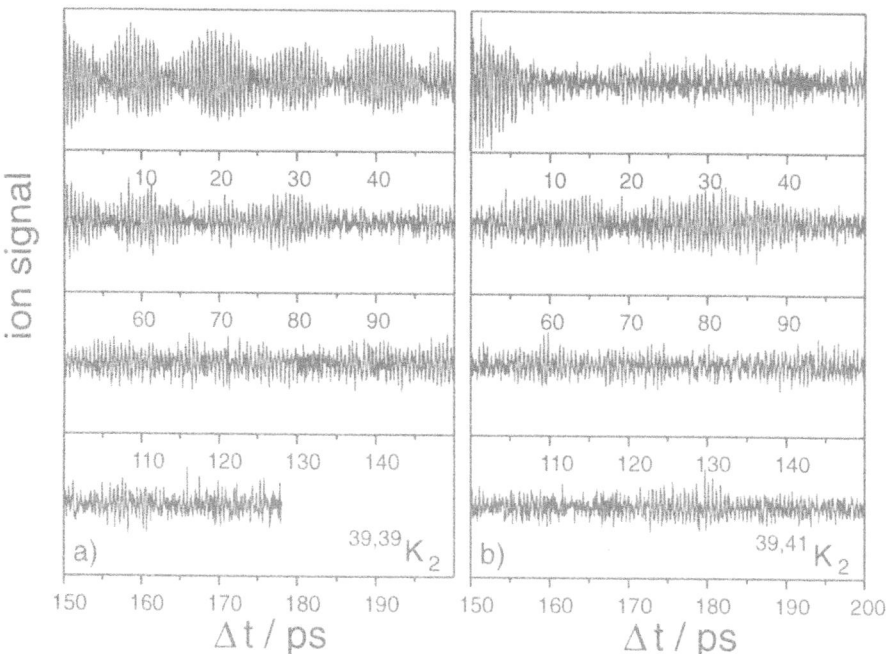

**Fig. 4.** Temporal evolution of the 3PI signal for the two $K_2$ isotopes (a) $^{39,39}K_2$ and (b) $^{39,41}K_2$

Quantum dynamical calculations of the real-time spectra for the two isotopes were performed for delay times up to 40 ps. In agreement with the experimental data the short time dynamics of the theoretical signal show the 500 fs oscillation period of the wave packet prepared in the $A^1\Sigma_u^+$ state (centered around $v = 11$) and the long time dynamics reflects the totally different beat structures of the two isotopes. However, the (sequence) oscillation period of the pronounced and regular beat structure of $^{39,39}K_2$ as well as the faded and irregular beat structure of $^{39,41}K_2$ are somewhat shorter for the theoretical signal.

The frequency components involved in the transient spectra are obtained by Fourier analysis of the normalized real-time data (see Fig. 5). Both Fourier spectra are dominated by a group of frequencies around $w_0^{(1)} \approx 65\,\mathrm{cm}^{-1}$. Two additional frequency groups with lower amplitudes appear at $w_0^{(2)} \approx 130\,\mathrm{cm}^{-1}$ and at $w_0^{(3)} \approx 195\,\mathrm{cm}^{-1}$. An additional peak is observed at $w_x \approx 90\,\mathrm{cm}^{-1}$, where the relative intensity is slightly larger in case of the lighter isotope (Fig. 5a).

The frequency group around $w_0^{(1)}$ is illustrated in the insets of Fig. 5 a,b for the case of the studied isotopes of $K_2$. The inset of Fig. 5 a shows the Fourier components in the real-time spectrum of $^{39,39}K_2$. Two main frequencies at $63.8\,\mathrm{cm}^{-1}$ and $67.2\,\mathrm{cm}^{-1}$ dominate this spectrum. The component at $67.2\,\mathrm{cm}^{-1}$ seems to be broadened by a component at $66.8\,\mathrm{cm}^{-1}$. Further frequency components can be observed at $64.7\,\mathrm{cm}^{-1}$, $65.2\,\mathrm{cm}^{-1}$, $65.6\,\mathrm{cm}^{-1}$ and $67.9\,\mathrm{cm}^{-1}$. Hence, the wave packet consists of a non-monotonic frequency distribution of the contributing vibrational levels, which manifests as a spectral hole. Instead of the frequency values corresponding vibrational level pairs are given. These pairs of vibrational levels are found by introducing an energetic shift to the RKR levels of Ref. [93] and by comparing the resulting energy spacings between neighboring levels with the Fourier analysis data (see Table 1). Introducing a shift for $v = 12$ and $v = 13$ of $1.2\,\mathrm{cm}^{-1}$ and $2.1\,\mathrm{cm}^{-1}$, respectively, results in a good agreement. The two dominant frequencies can clearly be proved to be responsible for the beat oscillation period of $T_{BS} \approx 10\,\mathrm{ps}$ being observed in the pump&probe spectrum: overlaying the frequencies $w_{13,14} = 63.8\,\mathrm{cm}^{-1}$ and $w_{12,13} = 67.2\,\mathrm{cm}^{-1}$ leads to a frequency $w_{BS} = w_{12,13} - w_{13,14} = 3.4\mathrm{cm}^{-1}$, which gives a corresponding period of $T_{BS} = 9.8\,\mathrm{ps}$. The spectrum around $65\,\mathrm{cm}^{-1}$ for the isotope $^{39,41}K_2$ is presented in the inset of Fig. 5 b). There, we observe five distinct components in the frequency group at $64.3\,\mathrm{cm}^{-1}$, $64.7\,\mathrm{cm}^{-1}$, $65.5\,\mathrm{cm}^{-1}$, $65.9\,\mathrm{cm}^{-1}$ and $66.4\,\mathrm{cm}^{-1}$. The numbering here is not included due to missing spectroscopic data for this isotope.

The Fourier spectra were calculated as well for the theoretical data. Since the simulations were performed up to 40 ps, only, two main frequencies $w_{calc\,1,2}$ are solely resolved (see Table 1). Experimental and theoretical Fourier spectra reveals the effect of the spin-orbit coupling to the $b^3\Pi_u$ state. In the region of the pump laser pulse the perturbation is most effective for $^{39,39}K_2$. Two vibrational levels of the $b^3\Pi_u$ state ($v = 23, 24$) are close [92], nearly energetically degenerated, to the two vibrational levels of the $A^1\Sigma_u^+$ state ($v = 12, 13$). Thus the perturbation is very pronounced in these two vibrational levels and they are shifted (as seen in the inset of Fig. 5 a apart. The resulting frequencies $w_{calc\,1} = 66.8\,\mathrm{cm}^{-1}$ and $w_{calc\,2} = 62.5\,\mathrm{cm}^{-1}$ lead to the period $T_{BS} = 7.8\,\mathrm{ps}$. For the $^{39,41}K_2$ the situation is different. The perturbing levels in the excitation range of the pump pulse are

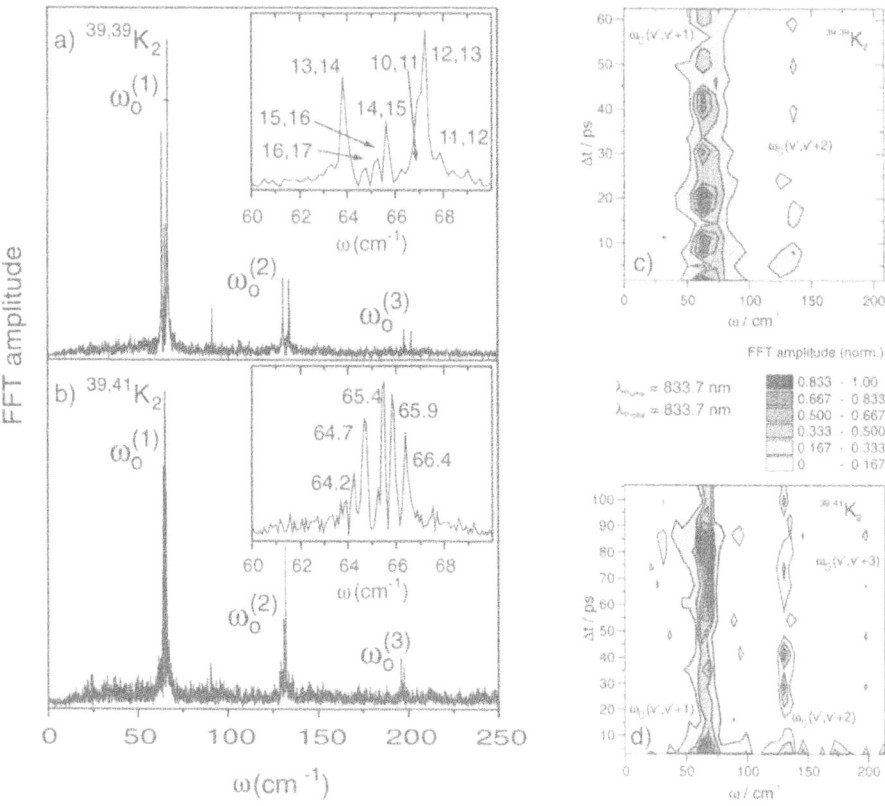

**Fig. 5.** Fourier analysis calculated of the normalized real-time data, (a) for $^{39,39}K_2$ and (b) for $^{39,41}K_2$. The insets describe details of the Fourier spectra between 60 and $70\,\mathrm{cm}^{-1}$. The inset of part (a) contains the vibrational level pairs belonging to the frequency components, while in inset (b) the frequency values are given. The corresponding spectrograms $I(t,\omega)$ of both isotopes are presented in panels c) and d) using contours plots

not energetically close and all the vibrational levels contributing in the wave packet are perturbed, but only to a small amount (see Fig. 5 b). The regular pattern of an unperturbed spectrum (see e.g. [19]) is conserved.

To get even deeper insight into the induced wave packet dynamics, we use the visualization by spectrograms $I(\Delta t, \omega)$ [100, 101]. This technique nicely enables the direct observation of the time dependence ($\Delta t$) of the different frequency ($\omega$) components originating from the propagating wave packet, including its relative contribution $I$(see Fig. 5 c,d). Especially, the interplay of the involved frequency groups can be seen at first glance. Revivals, total and fractional [99], are emphasized in the spectrograms. For further details see Refs. [6, 97, 102].

**Table 1.** Vibrational RKR level pairs reached by the excitation laser and respective energy spacings without and with energy level shift for $v = 12, 13$ of $1.2\,\mathrm{cm}^{-1}$ and $2.1\,\mathrm{cm}^{-1}$, respectively, in comparison with experimental and theoretical found data for $^{39,39}K_2$

| $v, v+1$ | $w_{RKR}/\mathrm{cm}^{-1}$ [a] | $w_{shift}/\mathrm{cm}^{-1}$ [b] | $w_{FFT}/\mathrm{cm}^{-1}$ [c] | $w_{calc}/\mathrm{cm}^{-1}$ [d] |
|---|---|---|---|---|
| 9,10 | 67.35 | e | f | f |
| 10,11 | 67.03 | e | 66.8 | f |
| 11,12 | 66.71 | 67.91 | 67.9 | f |
| 12,13 | 66.40 | 67.30 | 67.2 | 66.8 |
| 13,14 | 66.08 | 63.98 | 63.8 | 62.5 |
| 14,15 | 65.76 | e | 65.6 | f |
| 15,16 | 65.44 | e | 65.2 | f |
| 16,17 | 65.11 | e | 64.7 | f |
| 17,18 | 64.47 | e | f | f |

[a] Vibrational level spacings of Ref. [93].
[b] Level spacings with introduced shifts for $v = 12, 13$.
[c] Fourier components of pump&probe data.
[d] Main Fourier components of theoretical simulation.
[e] Not changed.
[f] Not observed.

## 3.2 Controlled Molecular Dynamics of Na$_3$: Mode Selective Excitation

As exemplarily demonstrated in Sect. 3.1 the ultrafast pump&probe technique has been established as a powerful instrument to analyze the real-time wave packet dynamics in molecules. The goal of this technique is, however, not only to investigate the induced dynamics of a system, but also to control the molecular dynamics. Starting from the theoretically developed pump and control concept by Tannor, Rice and Kosloff [103, 104], first examples now exist where those ideas have been realized experimentally [26, 45, 98]. The control parameters used in these examples are the delay time between pump- and probe-pulse [26, 45], the intensity of the laser pulse [98, 105] and the pulse duration [106], respectively. Most of these pump&control experiments were carried out on diatomic molecules, because the dynamics can be controlled relatively easy in such simple systems with only one vibrational degree of freedom. In larger molecular systems with three or more vibrational degrees of freedom, however, the situation becomes much more complicated and it is an interesting question, whether the concept of "controlled molecular dynamics" can still be realized or whether the signal will lose its characteristics.

The sodium trimer is one of the best known metal clusters and can be considered as a model system for these investigations. In particular, its excited electronic B state presents a characteristic pseudorotation between obtuse and acute triangular geometry inside the rather weakly localized state. Several research groups have investigted the B state both experimentally [46, 50, 69, 107– 112]

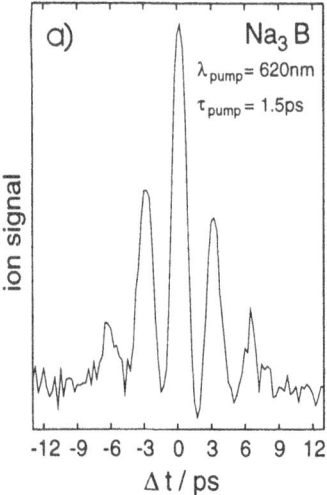
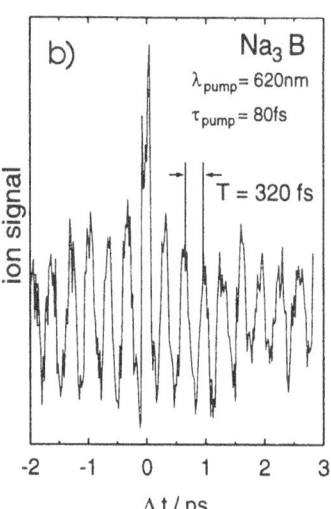

**Fig. 6.** Temporal evolution of the TPI signal of $Na_3$ excited to the electronic B state with **a)** a 1.5 ps and **b)** an 80 fs laser pulse

and theoretically [46, 49, 83, 113–120]. Next to the traditional cw or nanosecond (ns) absorption spectroscopy [49, 107, 115], in the last few years the ultrafast pump&probe technique was mainly applied to investigate this system [49, 50, 69, 83, 107–109, 113, 117– 120]. These investigations can be divided into two groups: experiments that have been carried out with a ps laser pulse [50, 108] and others in which a fs laser pulse was used [69, 83, 109, 113]. Different pulse intensities were used in the fs pump&probe experiments. Both lead to comparable results concerning the dominant oscillatory structure in the pump-probe signal. The high intensities used in the experiment of the Gerber group [69, 109] mean, that stimulated Raman processes are as well involved. These additional processes complicate a full analysis of the experiment. For this reason, here ps-[50, 108] and fs-pump&probe [83, 113] experiments at low or moderate laser intensities are presented. As a consequence impulsive stimulated Raman processes [105] do not interfere with the wave packet propagation of the excited electronic state.

The pseudorotation can be visualized in a first approximation as vibronic coupling between the bending $Q_x$ and asymmetric stretch $Q_y$ mode. It astonished, however, that the experimentally obtained spectra did not present any hint on the symmetric stretch mode $Q_s$. Recently performed theoretical calculations [83, 113] demonstrated that the use of ultrafast spectroscopy might open a temporal window to even observe this mode. Hence, preparing a wave packet in the B state's PES and probing the propagating wave packet with an ultrashort laser pulse by ionizing the trimer should enable to make a real-time movie of the described complex molecular dynamics.

First, we measured the temporal evolution of the B state's two photon ionization (TPI) employing tunable dye laser (620nm, rhodamine 6G) synchronously pumped by a modelocked $Ar^+$ laser. The dye laser provides pulses with a duration of 1.25 ps and a repetition rate of about 82 MHz. The temporal evolution of the TPI signal is shown in Fig. 6 a. A clear beat structure, symmetrical to the zero-of-time point is observed. The period of these oscillations amounts $\approx$ 3 ps and can be assigned to the pseudorotation. The oscillation is damped with a time constant of about 3.5 ps. A constant offset is due to an efficient TPI process caused by the pump pulse itself. The second experiment was performed with 80 fs laser pulses at 620 nm produced by a frequency doubled optical parametric oscillator synchronously (Spectra Physics OPAL) pumped by a modelocked titanium:sapphire laser. In this case a fast oscillation of the ion signal with a period of $\approx$ 310 fs is clearly observable for the first 4 − 5 ps (see Fig. 6 b).

Theoretical simulations on the basis of the first ever done 3d (3 normal modes) quantum *ab initio* calculations [46, 48, 83, 113] are in excellent agreement with these results. Within the pulsewidth of the 1.5 ps pulse a wave packet is prepared at one of the local minima of the B state PES. Here the wave packet represents the obtuse geometry of the trimer. After 1.5 ps the wave packet has moved to the other side of the PES trough which represents the acute geometry now. From this region a transition to the ion state cannot take place due to an extremely low Frank-Condon factor. This is nicely reflected by the minimum in the experimental result (see Fig. 6). The wave packet then turns back and reaches again the local minimum after a pump-probe delay of 3 ps. Here a maximum of ionization probability is given. Hence, the second maximum of the experimental results represents the high ionization probability in the case of obtuse shape of the trimer. The observed temporal evolution of the ion signal therefore mirrors the oscillatory changes between obtuse and acute triangular shape i.e. the pseudorotation of the trimer.

In contrast, the specific 120 fs pump pulse excites due to significantly different equilibrium distances in the $Q_s$-coordinate nearly selectively the symmetric stretch (Fig. 6b) between the X- and the B-state [83]. The 120 fs excitation drives the excited wave packet along $Q_s$. Due to specific Franck-Condon windows for the selected pump and probe pulses the B ⟵ X and $X^+$ ⟵ B transitions can be monitored at specific values of $Q_s$ for obtuse angle geometries of $Na_3$. For longer times an IVR-process to the pseudorotations as well takes place so that after 5 ps these pseudorotations dominate again. This behavior correlates with the time evolution of $\theta_B(t)$ as calculated by [113] . It explains that only the fs pump-probe technique is able to detect the $Q_s$-mode with such a large selectivity.

This nicely demonstrates that cw/ps and fs spectroscopy have different sensitivities for excitations of different vibrational modes, thus confirming the original conjecture of A. H. Zewail [45]. The results demonstrate how sensitively the reaction of different vibrational modes depends upon the applied pulse length and as well shows that it is possible to excite different vibrational modes selectively during an electronic excitation with ultrashort laser pulses. For this reason, it should in future be possible to control subsequent reactions.

410

## 3.3  K₃: Ultrafast Dynamics on a Repulsive Potential Energy Surface

Although theoretical calculations [51] predicted an excited electronic state for the potassium trimer in the spectral region of 790 to 830 nm many trials failed to observe this electronic state by multi photon ionization (TPI, 3PI) spectroscopy [52]. The reason for this was assumed to be ultrafast photodissociation which depopulates the excited electronic state, before a second photon can ionize the excited trimer. Depletion spectroscopy [121] which avoids this problem similary did not show any reasonable absorption. Hence, the experimental proof of this electronic state was in doubt.

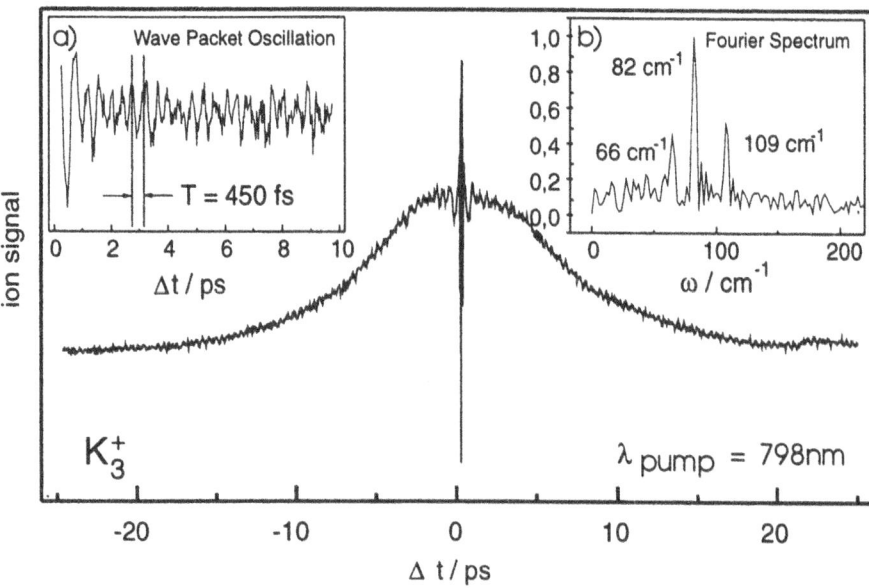

**Fig. 7.** Temporal evolution of the 3PI signal for K₃ excited at $\lambda = 798$ nm. The symmetric shape is due to the applied pump and probe pulse being of similar wavelength. Therefore, at the zero-of-time both pulses interchange their role. Besides the ultrafast decay a superimposed oscillation with $T = 450$ fs is visible. Around the zero-of-time the interferometric autocorrelation is seen. Inset: a) The first 10 ps with a clear oscillation of ∼ 450 fs is visible. The original data are deconvoluted to overcome the decay. b) Fourier spectra of the 3PI signal of K₃. Three dominant lines are visible. They are assigned to three normal modes of the trimer

For several excitation wavelengths the femtosecond real-time dynamics of the trimer is observed. In Fig. 7, a representative time evolution of the 3PI signal for excitation of K₃ with 70 fs laser pulses at a central wavelength of 798 nm is shown. In the center, which marks the zero-of-time, the interferometric autocorrelation peak is clearly observable. The transient ion signal is symmetric with respect to this peak as it is expected for a one-color 3PI experiment. A fast decay is clearly seen and is superimposed by a 450 fs oscillation. Two processes are involved, wave packet propagation on the PES as well as ultrafast dissociation.

To describe the fast decay of the ion signal, a simple energy level model is applied similar to that described in Sect. 4. Since radiative decay of the trimer is in the time domain of nanoseconds, the observed change in population can be established either by photodissociation or by intersystem crossing to an electronic state, from which, under these experimental conditions, no ionization can take place. However, in the second case this behavior should be as well directly visible in drastic modulation of the observed wave packet propagation [34, 122, 123]. But this is not found. Apart from this, first theoretical calculations of the PES of $K_3$ gave no evidence to the existence of any "dark state" in this energy regime. Hence, we believe ultrafast photodissociation causes the fast decay of the ion signal. We believe that the fast photodissociation process coupled with a rather low oscillator strength is the reason why cw spectroscopy was unable to detect this electronic state. Femtosecond spectroscopy with high peak power and broad spectral width of the exciting laser pulses combined with the probing of the excited electronic state within a few picoseconds, however, opens a time window to efficiently detect this state and its dynamics. Further details on the photodissociation process are given in [53].

Superimposed on the just described decay a pronounced wave packet dynamics is clearly visble (see Fig. 7). From the time evolution of the ion signal it can be seen that the wave packet loses intensity with each oscillation period. This can be explained by the dissociative character of this excited state, which is visible in the overall decay of the ion signal. With higher resolution, the inset a) of Fig. 7 presents this oscillation with its dominant period of $T \sim 450$ fs. Since it is not at all a pure 450 fs vibration, a mixture of several modes of the trimer might cause the observed oscillation pattern. For a detailed analysis it is, therefore, necessary to perform a Fourier analysis of the transient spectra. We normalized these data to obtain data points oscillating around a zero-line. In Fig. 7 b the corresponding Fourier spectrum is depicted. It reveals three frequencies with wavenumbers $\bar{\nu}_1 = 66\,\mathrm{cm}^{-1}$, $\bar{\nu}_2 = 82\,\mathrm{cm}^{-1}$ and $\bar{\nu}_3 = 109\,\mathrm{cm}^{-1}$. In Table 2 the corresponding vibrational periods and intensities are listed.

**Table 2.** Wavenumber $\bar{\nu}$, vibrational period T, and relative intensity I (compared to $Q_s^*$) of the symmetric stretch $Q_s^*$ of the excited as well as the asymmetric $Q_y^X$, and symmetric stretch $Q_s^X$ modes of $K_3$ ground state

| $K_3$ modes | $\bar{\nu}/\mathrm{cm}^{-1}$ | T/fs | I/a.u. |
|---|---|---|---|
| $Q_y^X$ | 66 | 505 | 0.45 |
| $Q_s^*$ | 82 | 406 | 1 |
| $Q_s^X$ | 109 | 306 | 0.52 |

By comparison with data for ground (X) [124] and excited (*) electronic B state [26, 109, 114] of $Na_3$, we assign these values to normal modes of the trimer's ground and excited state. Since, as a first approximation, $\bar{\nu} \propto \sqrt{1/m}$ with m the mass of a single atom of the trimer a ratio $\kappa := \bar{\nu}_{K_3}/\bar{\nu}_{Na_3}$ can be introduced to compare the vibrational modes of the alkali trimers. With $m_K = 39$ amu and $m_{Na} = 22.9$ amu one obtains: $\kappa = 0.78$.

The dominant line at $82\,\mathrm{cm}^{-1}$ can be assigned to the symmetric stretch mode $Q_s^*$ of the excited system. Comparing this frequency with the data in [26, 109, 114], where $\bar{\nu}(Q_s^{Na_3B}) = 105\,\mathrm{cm}^{-1}$ is determined, the ratio is $\kappa = 0.78$. This value is in excellent agreement with our estimate, demonstrating its expressiveness. However, both the weaker lines at $66\,\mathrm{cm}^{-1}$ and $109\,\mathrm{cm}^{-1}$ have no correspondings to known vibrational frequencies of $Na_3$ excited to its bound B state. Since the peak intensities of the applied 70 fs pulses with $\geq 1\,\mathrm{GWcm}^{-2}$ are rather high we believe that both lines document the ground state dynamics of the trimer. Similar to the results of Gerber [26, 109] for $Na_3$ and de Vivie-Riedle et al. [98] for $K_2$ via impulsive stimulated Raman scattering (see [98] and refs. therein) a wave packet is generated in the ground state. This dynamics as well is reflected in the real-time one color 3PI signal. By comparison with the vibrational frequencies of the $Na_3$ X system [124], we believe that the line at $66\,\mathrm{cm}^{-1}$ can be assigned to the asymmetric bending mode $Q_y^X$ of the ground state. With $\bar{\nu}(Q_y^{Na_3X}) = 87\,\mathrm{cm}^{-1}$ from [124] the value of $\kappa$ amounts 0.76 which again is in reasonable agreement with our approximated $\kappa = 0.78$. The line at $109\,\mathrm{cm}^{-1}$ belongs similarly to the symmetric stretch mode $Q_s^X$ of $K_3$. Compared with $\bar{\nu}(Q_s^{Na_3X}) = 140\,\mathrm{cm}^{-1}$, taken from [124], it is $\kappa = 0.78$.

For a more detailed picture of the dynamics, CI ab initio calculations of the PES combined with time dependent quantum dynamical simulations, as performed for the sodium trimer, are essential.

# 4   Ultrafast Photodissociation

The ultrafast photodissociation dynamics of excited electronic states of $Na_3$ up to $Na_{10}$ has been studied by means of femtosecond real-time pump&probe spectroscopy. Two-color-TPI technique with laser energies of 1.48 eV and 2.96 eV (second harmonic generation of the output pulses of the used titanium:sapphire laser) is employed. In Fig. 8 a for different cluster sizes the real-time evolution of the ion signal is presented. The intensity of the ion signal at $E_e = 2.96$ eV (positive delay times) is for all sizes of the $Na_n$ clusters higher than at $E_e = 1.48$ eV (negative delay times). This corresponds to the higher density of electronic states in the region around 3 eV. The decay of the ion yield in each curve might have two reasons: first the ultrafast fragmentation of the excited cluster and second a reduction due to a rapid internal vibrational redistribution (IVR) of the excited system. However, since no wave packet propagation indicated by oscillations superimposed on the decay curves (as is seen e.g. for the dissociative $K_3$ molecule excited to its B state discussed in this paper in Sec. 3.3 [40, 53]) are observed, we believe that fragmentation is the only reason for the ultrafast decay.

To describe this ultrafast photodissociation process, we like to use a simple energy level model which we successfully applied to first investigations on $Na_3$ [75], $Na_n$ [73], and $K_n$ [72]. The mechanism of pump photon ($E_e = h\nu_e$) ab-

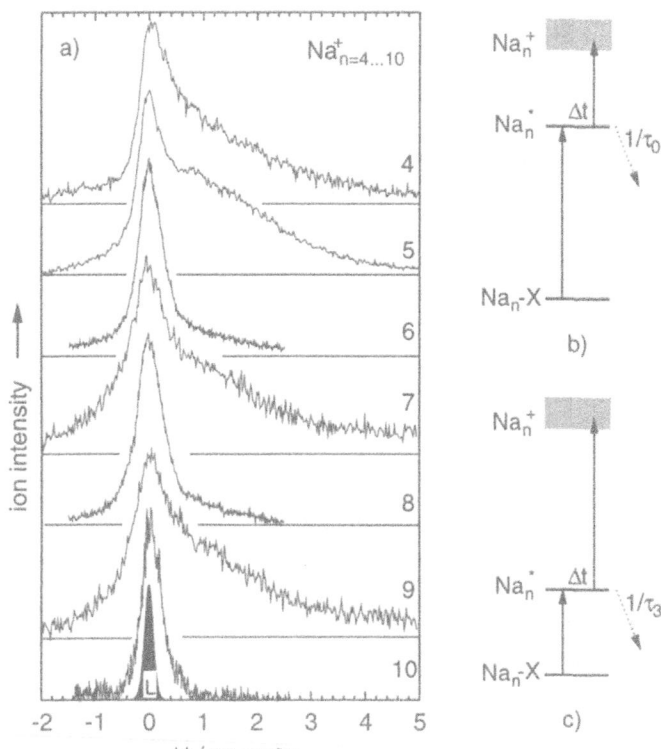

**Fig. 8.** Real-time evolution of the ion signals for $Na_{n=4...10}$ (a) and energy level schemes to describe the temporal evolution of the $Na_n^+$-signal (X = ground state) for $\Delta t > 0$ $E_e = 2.96\,\text{eV}$ and $E_i = 1.48\,\text{eV}$ (b) and for $\Delta t < 0$ $E_e = 1.48\,\text{eV}$ and $E_e = 2.96\,\text{eV}$ (c). $\frac{1}{\tau_0}$ and $\frac{1}{\tau_3}$ indicate the fragmentation probabilities of an excited state $Na_n^*$ for different excitation $(E_e)$ and ionization energies $(E_i)$

sorption of an observed $Na_n$ cluster simply is given by eqn.(1). Then, either the excited state (∗) can dissociate and two fragment products are found (2) or by absorbing a probe photon $(E_i = h\nu_i)$, being more or less delayed in time, the cluster can be ionized (3).

$$Na_n + h\nu_e \longrightarrow Na_n^* \tag{1}$$
$$Na_n^* \longrightarrow Na_{n-m}^* + Na_m \tag{2}$$
$$Na_n^* + h\nu_i \longrightarrow Na_n^+ + e^-. \tag{3}$$

For photoexcited $Na_n$ $(n \leq 10)$ it is supposed [73], that under these excitation conditions clusters with even number of sodium atoms split off a monomer $(m = 1)$ whereas odd-numbered clusters separate a dimer $(m = 2)$.

For the interpretation of the recorded transients from $Na_3$ to $Na_{10}$ one might expect – with respect to the model presented in Fig. 8 b,c – for positive as well as negative delay times $\Delta t$ a single exponential decay convoluted with the overall system response $L(t)$ to the laser pulses:

$$I(t) \propto l(t) * e^{-\frac{t}{\tau}}. \tag{4}$$

Here the time constant $\tau$ is given by the inverse of the fragmentation probability of the excited state. The ion signal is assumed to be a direct measure of the population density of excited clusters. But, as being obviously seen in Fig. 8 and as an example in Fig. 9 (topmost curve: Na$_3$ excited to its D-state), the shapes of the transient signals can even at a first glance not all be described by this simple fragmentation model.

As has been shown elsewhere [73, 75], this unexpected feature, however, can easily be explained: The cluster beam is composed of an ensemble of cluster sizes. Hence, the detected ion signal contains besides the ion signal of the excited cluster Na$_n^*$ $(:= Type I)$ contributions caused by ionization of larger clusters' fragments $(:= Type II)$ being of the same mass (Na$_n$). Therefore, the observed ion signal is the sum of ionized Na$_n$ clusters of different origin, namely $Type$ $I$ and $II$. In [73] it was supposed that excited sodium clusters break into a monomer or dimer and the respective daughter clusters. Splitting off a monomer or a dimer, an energy of $0.6\,$eV or $0.9\,$eV, respectively, is necessary [125]. Hence, taking into account the energy balance of the $Type\ II$ processes the ionization of the fragments needs – in nearly all of the examined cases – two photons of the probe pulse. Starting with the excited cluster the detailed $Type\ II$ processes studied here are listed below (5-11). The processes calculated for $E_e = 2.96\,$eV base on the data given in [125, 126]. The results modify (1-3).

$$Na_3^* \longrightarrow Na_2^* + Na \xrightarrow{2h\nu} Na_2^+ \tag{5}$$

$$Na_4^* \longrightarrow Na_3^* + Na \xrightarrow{h\nu} Na_3^+ \tag{6}$$

$$Na_5^* \longrightarrow Na_4^* + Na \xrightarrow{2h\nu} Na_4^{+*} \longrightarrow Na_3^+ + Na \tag{7}$$

$$Na_6^* \longrightarrow Na_4^* + Na_2 \xrightarrow{2h\nu} Na_4^{+*} \longrightarrow Na_3^+ + Na \tag{8}$$

$$Na_7^* \longrightarrow Na_6^* + Na \xrightarrow{2h\nu} Na_6^{+*} \longrightarrow Na_5^+ + Na \tag{9}$$

$$Na_8^* \longrightarrow Na_6^* + Na_2 \xrightarrow{2h\nu} Na_6^{+*} \longrightarrow Na_5^+ + Na \tag{10}$$

$$Na_9^* \longrightarrow Na_8^* + Na \xrightarrow{2h\nu} Na_8^{+*} \longrightarrow Na_7^+ + Na \tag{11}$$

Equations 5-11 nicely demonstrate that $Type\ II$ processes will be detected in the detection channel of Na$_3^+$, Na$_5^+$, and Na$_7^+$. Therefore, the ion signal off the studied odd numbered clusters should deviate more or less from a single expo-nential decay, whereas for even numbered clusters a single exponential decay is expected. To describe this more complicate behavior, we use the fragmen-tation model given in [75]. It takes into account the population density and fragmentation characteristics of $Type\ I$ and $Type\ II$ clusters. The model contains four different fragmentation processes with four time constants $\tau_0$, $\tau_1$, $\tau_2$ and $\tau_3$. Here, $\tau_0$ characterizes the fragmentation behavior of the relevant $Type\ I$ clusters, whereas $\tau_1$ and $\tau_2$ are due to *Type II* clusters. $\tau_3$ describes the fragmentation while pump and probe pulses are exchanged ($\Delta t < 0$ in Fig. 9).

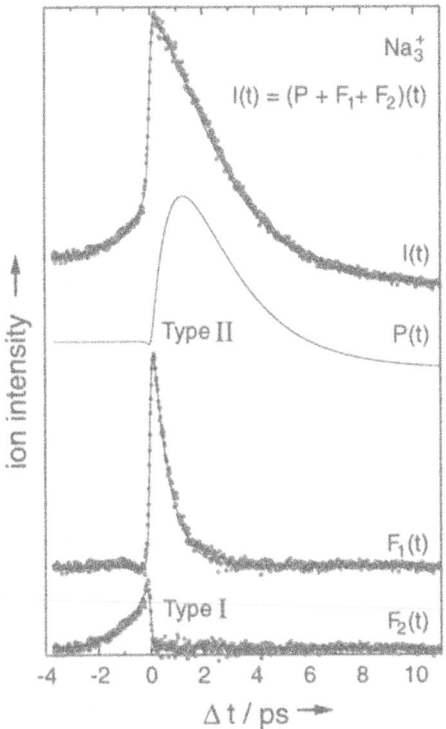

**Fig. 9.** Recorded TPI-signal $I(t)$ for $Na_3$ excited with $E_e = 2.96$ eV ($\Delta t \geq 0$) and $E_e = 1.48$ eV ($\Delta t < 0$). Due to the different excitation, ionization, and fragmentation processes of the cluster ensemble in the beam, $I(t)$ is the sum of three processes: $I(t) = P(t) + F_1(t) + F_2(t)$. $F_1(t)$ and $F_2(t)$ describe the photofragmentation of the *Type I* clusters whereas $P(t)$ represents *Type II* fragmentation. Dots: experimental data; lines: fitted function due to the fragmentation model explained in the text ($\Delta t$: delay time)

Due to this model the transient intensity of the detected ions is proportional to the convolution of the population density $n(t)$ of excited clusters (*Type I* and *II*) with the overall system response $l(t)$ to the laser pulses

$$I(t) \propto n(t) * l(t) = (n_+(t) + n_-(t)) * l(t) \tag{12}$$

For $t \geq 0$ ($E_e = 2.96$ eV, $E_i = 1.48$ eV) the temporal dependence of the excited population density is given by $n_+(t)$ described by the sum of three exponential functions [75]:

$$n_+(t) = \overbrace{N_0 e^{-\frac{t}{\tau_0}}}^{Type I} + \overbrace{M_0(e^{-\frac{t}{\tau_1}} - e^{-\frac{t}{\tau_2}})}^{Type II} = n_1(t) + n_2(t) \tag{13}$$

with weighting constants $N_0$ and $M_0$. The first term of this sum ($n_1(t)$) represents the temporal evolution of the relevant population density of *Type I* clusters, whereas the second term ($n_2(t)$) is ascribed to the *Type II* ions.

For $t < 0$ ($E_e = 1.48$ eV, $E_i = 2.96$ eV) no *Type II* fragmentation could be observed. Hence the transient population density for this two photon process is given by a single exponential decay, namely

$$n_-(t) = N_3 e^{-\frac{t}{\tau_3}}. \tag{14}$$

$N_3$ is a weighting factor, reflecting the initial population density of excited states for $E_e = 1.48$ eV.

To determine the fragmentation probability of *Type I* clusters the ion signal of *Type II* ions has to be separated and subtracted from the observed signal. Therefore, one has first of all to deconvolute the transient ion signal $I(t)$. This is done by convoluting the overall system response with four exponential functions using a least square fit algorithm. Then the contribution of the *Type II* ions has to be subtracted. The last step is to separately obtain the transient population densities $n_1(t)$ and $n_-(t)$ of the *Type I* clusters for $t \geq 0$ and $t < 0$ by subtracting the other contribution, respectively. The result of this procedure are two curves each decreasing with a single exponential decay. The decay constants are $\tau_0$ and $\tau_3$.

Fig. 9 presents as an overview the separated contributions of $I(t)$ obtained by this algorithm:
For $t \geq 0$ (*Type II*)

$$P(t) := l(t) * n_2(t) \tag{15}$$

For $t \geq 0$ (*Type I*)

$$F_1(t) := l(t) * n_1(t) \propto l(t) * e^{-\frac{t}{\tau_0}} \tag{16}$$

and for $t < 0$ (*Type I*)

$$F_2(t) := l(t) * n_-(t) \propto l(t) * e^{-\frac{t}{\tau_3}}. \tag{17}$$

So, $I(t)$ is given by

$$I(t) = P(t) + F_1(t) + F_2(t). \tag{18}$$

The comparison of fit-functions and experimental data (Fig. 10) presents excellent agreement. We like to point out that although the fit routine works with four exponential functions, for negative delay times only one of these is essential. The remaining three describe the behavior for positive delay times, only. Here, again only one exponential function is necessary in the case of clusters with even number of atoms. For clusters with odd number of atoms we find a more or less strong amount of *Type II* clusters. The results are in excellent agreement with those predicted by (5-11).

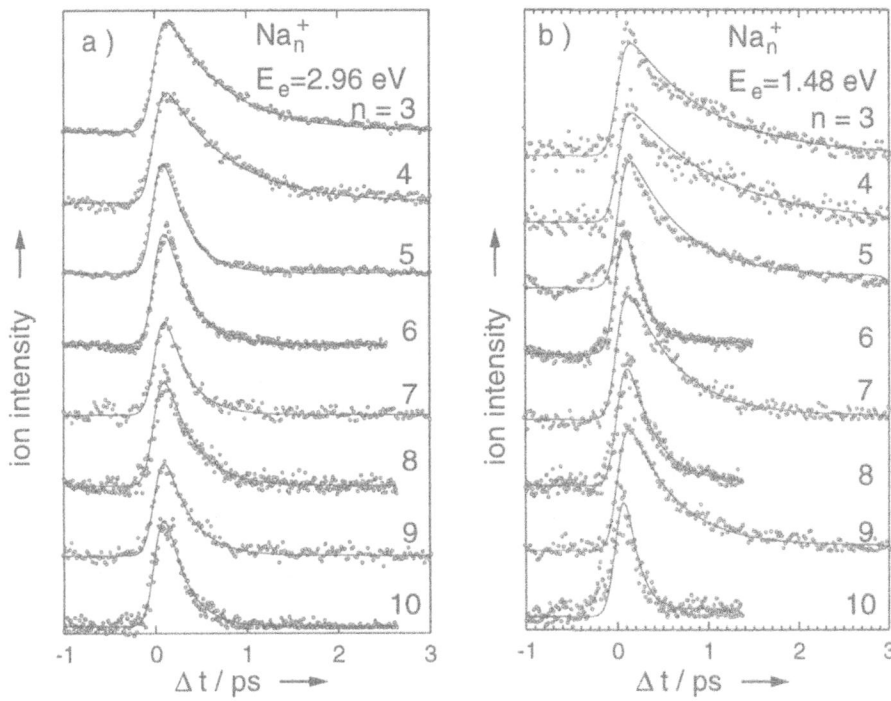

**Fig. 10.** Temporal evolution of the photofragmentation a)$F_1(t)$ for $Na^+_{n=3...10}$ excited at $E_e = 2.96$ eV and b) $F_2(t)$ for $Na^+_{n=3...10}$ excited at $E_e = 1.48$ eV. Dots: experimental data; lines: fitted single exponential function

The transients $F_1(t)$ and $F_2(t)$ directly present the temporal evolution of the (*Type I*) cluster photodissociation after excitation with photon energies of $E_e = 2.96$ eV and $E_e = 1.48$ eV, respectively. It has to be emphasized, that in case of $Na_3$ –depicted as an example in Fig. 9– the amount of $P(t)$ is rather large compared to that of $F_1(t)$. This ratio changes drastically for larger cluster sizes and is discussed elsewhere [73].

**Table 3.** Lifetimes $\tau_0$ and $\tau_3$ of $Na_n$ clusters after photoexcitation with energy $E_e = 2.96$ eV and $E_e = 1.48$ eV, respectively. The error on determining the lifetimes is $\Delta\tau = \pm 10$ fs

| n | 3 | 4 | 5 | 6 | 7 | 8 | 9 | 10 |
|---|---|---|---|---|---|---|---|----|
| $\tau_0$/fs | 600 | 690 | 260 | 270 | 240 | 310 | 280 | 220 |
| $\tau_3$/fs | 970 | 990 | 580 | 170 | 550 | 270 | 510 | 120 |

By this mathematical procedure we obtained for the different cluster sizes the relevant fragmentation time constants $\tau_0$ and $\tau_3$ listed in Tab.3. These values mirror the lifetimes of the photoexcited clusters for two different excitation

418

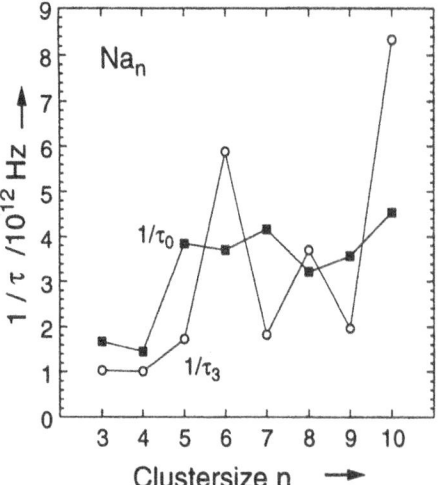

**Fig. 11.** Fragmentation probabilities $\frac{1}{\tau_0}$ and $\frac{1}{\tau_3}$ for different cluster sizes $n$

energies $E_e$. Fig. 11 summarizes these results by presenting the fragmentation probabilities $\frac{1}{\tau_0}$ and $\frac{1}{\tau_3}$ with respect to the number $n$ of atoms of the observed cluster. For $E_e = 1.48$eV an even-odd alternation (except for $n = 4$) of the photodissociation probability is clearly visible. Even numbered clusters tend to dissociate faster, whereas the odd numbered clusters are the more stable candidates with respect to this excitation energy.

Besides the distinct even-odd alternation for both excitation energies an obvious change with $n = 5$ appears. For $E_e = 1.48$ eV the even-odd alternation is drastically enhanced. Here, compared to $Na_3$ the fragmentation probability of $Na_{10}$ is about 8 times bigger. For $E_e = 2.96$ eV the fragmentation probability reveals at $n = 5$ a sudden increase to about 2 times the value of $n = 3$. Hence, for this excitation energy the stability of clusters larger than $Na_4$ is drastically smaller than that of the smaller ones.

It has to be stated that this simple fragmentation model can not explain any energy dependence of the recorded data. The model as well does not take into account any propagation of wave packets being prepared in the excited states of the clusters as are observed e.g. in the B-state of $Na_3$ [49]. On the other hand, investigations of ultrafast fragmentation processes of potassium clusters $K_{n=3...10}$ seem to confirm our model to be the right ansatz [72]. For other excitation conditions, e.g. as Gerber and coworkers used in their experiments on $Na_n$, the model might lose its validity. Gerber [68, 69] excited e.g. $Na_8$ close to four surface plasmon resonances at 2.39 eV (518 nm). Therefore, several ultrashort decay processes are involved simmultaneously, instead of one in our case. For further details see [6].

# 5 Ultrafast Structural Relaxation

In this section a new approach, used to investigate small silver molecules and clusters as $Ag_n$, $(n = 3, 5, 7$ and $9)$, is described. The method starts with a beam of mass-filtered, negatively-charged clusters (for the used set-up see Sect. 2.2). These anions are subjected to photodetachment. Subsequently, after a variable but selected temporal delay they are photoionized. Then the produced cations are mass-analyzed and collected. Their intensity as a function of the delay interval between the pump and probe is a measure of the Franck-Condon factor for photoionization of a neutral prepared by a vertical detachment process from a low-lying vibrational state of the anion. Hence, the vibrational motion of the neutral is probed. As an abbreviation for this process NeNePo is used, Negative-to-Neutral-to-Positive (see also [77–79]). Fig.1 displays the principle of the NeNePo process. Here, the results for $Ag_3$ will be presented, only. Further results are found in [6].

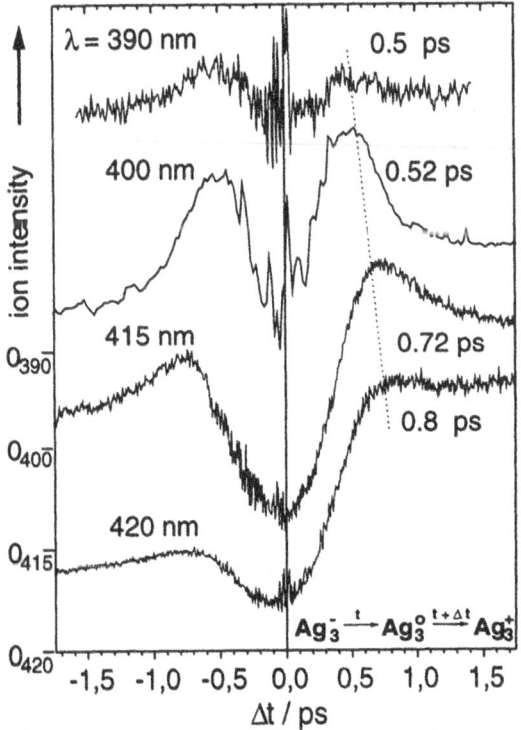

**Fig. 12.** NeNePo spectra of the silver trimer taken with wavelength of $\lambda = 390$ nm, 400 nm, 415 nm, and 420 nm. Note that each curve has its own axis of zero signal, and that the time-independent background increases steadily with decreasing wavelength. The fine structure around $\Delta t = 0$ is due to interference of pump and probe pulses. 0 indicates the zero line of the ion intensities for the different excitation wavelengths

The detected ion signal gives the yield of mass selected cations produced from mass selected anions as a function of the delay time between pump and probe pulse at fixed wavelength and fluence of these pulses. In first experiments the silver trimer is studied. Mass selected $Ag_3^-$ ions were produced with an intensity of about 2 nA, and stored in the ion trap (see Fig. 3). For the detachment process laser pulses with a central wavelength of 420 nm, 415 nm, 400 nm, and 390 nm were applied. This allowed one-photon detachment of the anions. The ionization was performed nonresonantly using two photons of the same wavelength. Since the energy of two photons of 420 nm is just slightly above the ionization potential of the silver trimer a very soft ionization is realized. Employing a wavelength of 415 nm, positive ions are mainly detected when there is a nonzero time delay between pump and probe laser pulses. This confirms that sequential processes of detachment and ionization are involved in the creation of the cations.

In Fig. 12 the yield of $Ag_3^+$ is displayed as a function of the delay time $\Delta t$ for various wavelengths of the detachment and ionization laser. At $\Delta t = 0$, pump and probe laser pulses exchange their role. However, the traces are not symmetrical since the respective fluences of pump and probe pulse are slightly different. For the longer wavelengths, the ion yield rises from almost zero to a maximum around $\Delta t = 750 fs$, and then decays to a constant value at longer time delays. There it stays constant for more than 100 ps which is the longest time delay used in the experiment. If light of shorter wavelength is used this phenomenon is progressively washed out. Applying pulses at a central wavelength of 390 nm the ionization efficiency is almost independent of the delay time $\Delta t$. The time required to reach the maximum grows with increasing wavelength from 500 fs to about 800 fs. This is consistent with the notion that the extra energy goes at least in part into the bending vibration.

The dependence of the cation yield on the power of the pump and probe laser pulses shows that the detachment process depends linearly, but the ionization process quadratically on the respective light intensity. This is in good agreement with the creation process of the cations as discussed above. As theoretical predictions [127, 128] indicate the most stable geometry of the trimer is linear for the anion, obtuse isosceles for the neutral and equilateral for the cation. The neutral trimer is presumably generated in a linear and therefore highly vibrationally excited configuration, at a saddle point, from which it bends slowly at first and then faster, comes through the geometry of the obtuse isosceles minimum and then decelerates until it approaches equilateral geometry, where its overlap with the positive ion is greatest.

In fact the results obtained with the silver trimers do not show the multiply-periodic behavior of a simple vibrational spectrum measured in the time domain. Rather, some of the results reveal less of the vibrational spectra but more of the dynamics of the internal rearrangements of these species, as shown by the time-dependent currents of positive ions in Fig. 12. Hence, the following hypothesis is assumed to interpret the behavior of the trimer signal: initially, the neutral is produced in a linear configuration by the vertical Franck-Condon detachment process. The Franck-Condon overlap factor of the linear neutral with the equilateral positive ion is so low that virtually no positive ions are generated. However, the neutral starts to bend and after passing through the obtuse equilibrium geometry of the neutral it approaches the equilateral equilibrium geometry of the cation after a certain delay time $\Delta t$. During this time the cation signal grows continuously reaching a maximum when the system is close to the classical turning point near the equilateral triangular geometry. Then the system rebounds and the signal decreases. However, the vibrational excitation is high enough that the modes mix, and after the rebound, the still-unionized neutral trimers are left with enough energy to pseudorotate through their three equivalent obtuse-triangular equilibrium structures, going around the trough of their "Mexican hat" potential energy surface [129]. In so doing, they remain at a roughly constant distance from the equilateral geometry, so that the Franck-Condon factor also remains nearly constant, and therefore so does the positive ion signal. The signal at short times is more pronounced at relatively long wavelength of the ionization laser, as then the ionization probability is strongly dependent on the vertical ionization potential in the momentary configuration of the neutral.

A first promising theoretical approach was done by the group of K.H. Bennemann [80–82]. By performing molecular dynamics simulations based on a microscopic electronic theory they studied the ultrafast dynamics of the neutral silver trimer immediately after the photodetachment of the anion ($Ag_3^-$). They could nicely determine the time scale for the change from the linear to the triangular structure of the ground state of $Ag_3$ being in excellent agreement with the presented experimental results.

To sum up, the NeNePo technique can be regarded as a quite general scheme to investigate the time evolution of a coherent nonequilibrium state in neutral molecules. Further investigations will be carried out using different molecular systems (e.g. isomerization of $Ag_5$) and different ionization pathways. NeNePo might also be used to study larger molecules and reactive compounds (i. e. ligand systems) where a chemical reaction or molecular rearrangement starts after the neutralization. The associated dynamics could eventually deduced by detecting and energy analyzing the photo electrons of the probe process as a function of $\Delta t$.

# 6   Conclusion and Forward View

In this work an overview has been given of the amazing opportunities of femtosecond real-time spectroscopy applied to small molecules and clusters. Fascinating phenomena like wave packet propagation, control of molecular dynamics, selective state preparation, ultrafast IVR, femtosecond photodissociation, and ultrafast structural relaxation with unexpected and sometimes exceptional features have been introduced.

Wave packet phenomena are examined for several model systems ($K_2$, $Na_3$ and $K_3$). The applied pump&probe technique enables an analysis of the cluster dynamics with high temporal ($\approx 1\,\mathrm{fs}$) and via Fourier analysis as well with high spectral resolution ($\approx 0.05\,\mathrm{cm}^{-1}$).

For two isotopes of the potassium dimer the strong influence of perturbation of electronic states due to spin-orbit coupling with a crossing triplet state on the wave packet propagation is presented. The applied spectrogram technique opens a direct view on the changing dynamics of the photo-excited systems.

In certain cases it even is possible to come closer to the control of molecular dynamics by means of tailored ultrashort laser pulses. Selectively two vibrational modes, the symmetric stretch and the pseudorotation mode of the sodium trimer could be excited by the right choice of the laser pulse duration.

The complemantarity of cw and fs spectroscopy is nicely demonstrated in the case of potassium trimer. While stationary spectroscopic techniques due to ultrafast photodissociation up to now failed to observe a theoretically predicted electronic state fs spectroscopy opened a temporal window to study the dynamics of this repulsive state in detail. Superimposed on an ultrafast decay – caused by photodissociation – the laser-induced wave packet dynamics is clearly observable. By Fourier analysis for the first time three normal modes of the trimer ground and excited state could be estimated experimentally.

For larger size sodium aggregates the characteristics of the real-time MPI spectra drastically change. Here ultrafast dissociation dominates the femtosecond cluster dynamics. Lifetimes in the region of 200 fs up to 1 ps show a clear even-odd alternation with cluster size. Detailed theoretical calculations of the real-time photodissociation based e.g. on the potential energy surfaces of the examined excited states are necessary but have not been performed up to now. With femtosecond laser sources today sufficient temporal resolution allows the measurement of extremely fast processes induced by the interaction of light with clusters. This can be regarded as a great challenge to the theoretical physicists and chemists working e.g. in the field of femtochemistry.

A new approach (NeNePo) to investigate the molecular dynamics of the ground state is successfully applied to the silver trimer and allows the direct observation of the ultrafast change of the molecular geometry during the NeNePo process. The technique gives direct access to structural relaxation times of clusters and complexes. A prototype for this might be the $Ag_5$ cluster with its 2- and 3-dimensional isomers. In future chemical reaction as well as solvation real-time dynamics might be successfully studied with this new promising experimental method.

**Acknowledgment.** The author gratefully acknowledges continuous financial support by the Deutsche Forschungsgemeinschaft. He is indebted to Ludger Wöste. All experiments were performed at his laboratory. His great enthusiasm were an important basis of the performed experiments. Holger Kühling carried out the measurements on the $Na_n$ clusters and Harald Ruppe those on $K_3$. The experiment on the silver trimer is due to a fruitful cooperation with R. Stephen Berry (University of Chicago), Thomas Leisner and Sebastian Wolf. The author especially likes to thank Soeren Rutz, who was involved in nearly every of the presented experiments.

# References

1. *Femtosecond Chemistry*, eds.: J. Manz and L. Wöste (VCH Verlagsgesellschaft, Weinheim, 1995), Vol. 1, Chap. 1, pp. 3.

2. *Femtochemistry: Ultrafast Dynamics of the Chemical Bond Vol. 1 & 2, World Scientific Series in 20th Century Chemistry*, eds.: A.H. Zewail (World Scientific, Singapore, 1994).

3. *Femtosecond Chemistry*, eds.: J. Manz and L. Wöste (VCH Verlagsgesellschaft, Weinheim, 1995), Vol. 1 & 2.

4. *Femtosecond Chemistry*, Vol. 97 *Special Issue of the Journal of Physical Chemistry*, eds.: J. Manz and Jr. A.W. Castleman (The American Chemical Society, USA, 1993), pp. 12423.

5. *Femtochemistry: Ultrafast Chemical and Physical Processes in Molecular Systems*, eds.: M. Chergui (World Scientific, Singapore, 1996).

6. E. Schreiber, *Femtosecond Real-Time Spectroscopy of Small Molecules and Clusters, Springer Tracts in Modern Physics, Vol.143* (Springer, Berlin, Heidelberg, 1998).

7. V. Brückner, K.-H. Feller and U.-W. Grummt, *Applications of Time-Resolved Optical Spectroscopy*, Vol. 66 *Studies in Physical and Theoretical Chemistry* (Elsevier, Amsterdam, 1990).

8. W.H. Knox, R.L. Fork, M.C. Downer, R.H. Stolen, C.V. Shank and J.A. Valdmanis, Optical pulse compression to 8 fs at a 5-kH repetition rate, Appl. Phys. Lett. **46**, 1120 (1985).

9. R.L. Fork, C.H. Brito Cruz, P.C. Becker and C.V. Shank, Compression of optical pulses to six femtoseconds by using cubic phase compensation, Opt. Lett. **12**, 483 (1987).

10. *Ultrashort Laser Pulses and Applications*, Vol. 60 *Topics in Applied Physics*, eds.: W. Kaiser (Springer Verlag, Berlin, 1988)

11. G.H.C. New, Femtofascination, Physics World **7**, 33 (1990).

12. M.J. Rosker, M. Dantus and A.H. Zewail, Femtosecond real-time probing of reactions. I. The technique, J. Chem. Phys. **89**, 6113 (1988).

13. M. Dantus, M.J. Rosker and A.H. Zewail, Femtosecond real-time probing of reactions. II. The dissociation reaction of ICN, J. Chem. Phys. **89**, 6128 (1988).

14. L.R. Khundkar and A.H. Zewail, Ultrafast molecular reaction dynamics in real-time: Progress over a decade, Ann. Rev. Phys. Chem. **41**, 15 (1990).

15. M. Dantus, M.H.M. Janssen and A.H. Zewail, Femtosecond probing of molecular dynamics by mass-spectrometry in a molecular beam, Chem. Phys. Lett. **181**, 281 (1991).

16. A.H. Zewail, Femtochemistry, J. Phys. Chem. **97**, 12427 (1993).

17. A.H. Zewail, Laser femtochemistry, Science **242**, 1645 (1988).

18. M. Dantus, R.M. Bowman and A.H. Zewail, Femtosecond laser observations of molecular vibration and rotation, Nature **343**, 737 (1990).

19. M. Gruebele, G. Roberts, M. Dantus, R.M. Bowman and A.H. Zewail, Femtosecond temporal spectroscopy and direct inversion to the potential: Application to iodine, Chem. Phys. Lett. **166**, 459 (1990).

20. M. Gruebele and A.H. Zewail, Femtosecond wavepacket spectroscopy: Coherences, the potential, and structural determination, J. Chem. Phys. **98**, 883 (1993).

21. I. Fischer, D.M. Villeneuve, M.J.J. Vrakking and A. Stolow, Femtosecond wavepacket dynamics studied by time-resolved zero-kinetic energy photoelectron spectroscopy, J. Chem. Phys. **102**, 5566 (1995).

424

22. M.J.J. Vrakking, I. Fischer, D.M. Villeneuve and A. Stolow, Collisional enhancement of rydberg lifetimes observed in vibrational wave packet experiments, J. Chem. Phys. **103**, 4538 (1995).

23. T. Baumert, B. Bühler, R. Thalweiser and G. Gerber, Femtosecond spectroscopy of molecular autoionization and fragmentation, Phys. Rev. Lett. **64**, 733 (1990).

24. T. Baumert, B. Bühler, M. Grosser, R. Thalweiser, V. Weiss, E. Wiedenmann and G. Gerber, Femtosecond time-resolved wave packet motion in molecular multiphoton ionization and fragmentation, J. Phys. Chem. **95**, 8103 (1991).

25. T. Baumert, M. Grosser, R. Thalweiser and G. Gerber, Femtosecond time-resolved molecular multiphoton ionization: The $Na_2$ system, Phys. Rev. Lett. **67**, 3753 (1991).

26. T. Baumert and G. Gerber, Fundamental interactions of molecules ($Na_2$, $Na_3$) with intense femtosecond laser pulses, Isr. J. Chem. **34**, 103 (1994).

27. A. Assion, T. Baumert, V. Seyfried, V. Weiss, E. Wiedemann and G. Gerber, Femtosecond spectroscopy of the (2) $^1\Sigma_u^+$ double minimum state of $Na_2$: time domain and frequency spectroscopy, Z. Phys. D. **36**, 265 (1996).

28. T. Baumert, V. Engel, C. Röttgermann, W.T. Strunz and G. Gerber, Femtosecond pump – probe study of the spreading and recurrence of a vibrational wave packet in $Na_2$, Chem. Phys. Lett. **191**, 639 (1992).

29. T. Baumert, V. Engel, C. Meyer and G. Gerber, High laser field effects in multiphoton ionization of $Na_2$. experiment and quantum calculations, Chem. Phys. Lett. **200**, 488 (1992).

30. C. Meier and V. Engel, Electron kinetic energy distributions from multiphotonionization of $Na_2$ with femtosecond laser pulses, Chem. Phys. Lett. **212**, 691 (1993).

31. Ch. Meier and V. Engel, in *Femtosecond Chemistry*, eds.: J. Manz and L. Wöste (VCH Verlagsgesellschafft, Weinheim, 1995), Vol. 1, Chap. 11, pp. 369.

32. V. Blanchet, M.A. Bouchene, O. Cabrol and B. Girard, One-color coherent control in $Cs_2$. observation of 2.7 fs beats in the ionization signal, Chem. Phys. Lett. **233**, 491 (1995).

33. J.M. Papanikolas, R.M. Williams, P. Kleiber, J.L. Hart, C. Brink, S.D. Price and S.R. Leone, Wave-packet dynamics in the $Li_2(^1\Sigma_g^+)$ shelf state: Simultaneous observation of vibrational and rotational recurrences with single rovibronic control of an intermediate state, J. Chem. Phys. **103**, 7269 (1995).

34. S. Rutz, R. de Vivie-Riedle and E. Schreiber, Femtosecond wave packet propagation in spin-orbit coupled electronic states of $^{39,39}K_2$ and $^{39,41}K_2$, Phys. Rev. A **54**, 306 (1996).

35. R. de Vivie-Riedle, B. Reischl, S. Rutz and E. Schreiber, Femtosecond study of multiphoton ionization processes in $K_2$ at moderate laser intensities, J. Phys. Chem. **99**, 16829 (1995).

36. S. Rutz, E. Schreiber and L. Wöste, in *Fast Elementary Processes in Chemical and Biological Systems*, Vol. 364 *AIP Conference Proceedings*, eds.: A. Tramer (AIP Press, Woodbury, New York, 1996), pp. 652.

37. E. Schreiber, S. Rutz and R. de Vivie-Riedle, in *Laser Laser in Forschung und Technik, Laser in Research and Engineering*, eds.: W. Waidelich, H. Hügel, H. Opower, H. Tiziani, R. Wallenstein and W. Zinth (Springer Verlag, Berlin, 1996), pp. 203–212.

38. E. Schreiber, S. Rutz and L. Wöste, in *Fast Elementary Processes in Chemical and Biological Systems*, Vol. 364 *AIP Conference Proceedings*, eds.: A. Tramer (AIP Press, Woodbury, New York, 1996), pp. 645.

39. S. Rutz, E. Schreiber and L. Wöste, in *Ultrafast Processes in Spectroscopy*, eds.: O. Svelto, D. De Silvestri and G. Denardo (Plenum Publ., New York, 1996), pp. 127–131.

40. S. Rutz, E. Schreiber and L. Wöste, Femtosecond vibrational dynamics of the potassium dimer, Surf. Rev. and Lett. **3**, 475 (1996).

41. S. Rutz and E. Schreiber, in *Ultrafast Phenomena IX*, Vol. 60 *Springer Series in Chemical Physics*, eds.: P.F. Barbara, W.H. Knox, G.A. Mourou and A.H. Zewail (Springer Verlag, Berlin, 1994), pp. 312.

42. E. Schreiber, in *Proceedings of the International Conference on LASERS '95*, eds.: V.J. Corcoran and T. Goldman (Society for Optical and Quantum Electronics, McLean, 1996), pp. 53.

43. E. Schreiber and S. Rutz, in *Femtochemistry: Ultrafast Chemical and Physical Processes in Molecular systems*, eds.: M. Chergui (World Scientific, Singapore, 1996), pp. 217.

44. S. Rutz, S. Greschik, E. Schreiber and L. Wöste, Femtosecond wave packet propagation in spin-orbit coupled electronic states of the $Na_2$ molecule, Chem. Phys. Lett. **257**, 365 (1996).

45. A.H. Zewail, in *Femtosecond Chemistry*, eds.: J. Manz and L. Wöste (VCH Verlagsgesellschaft, Weinheim, 1995), Vol. 1, Chap. 2, pp. 15.

46. B. Reischl, R. de Vivie-Riedle, S. Rutz and E. Schreiber, Ultrafast molecular dynamics controlled by pulse duration: The $Na_3$ molecule, J. Chem. Phys. **104**, 8857 (1996).

47. S. Rutz, H. Ruppe, E. Schreiber and L. Wöste, Femtosecond wave packet dynamics in alkali trimers, Z. Phys. D **40**, 25 (1997).

48. R. de Vivie-Riedle, J. Gaus, V. Bonačić-Koutecký, J. Manz, B. Reischl, S. Rutz, E. Schreiber and L. Wöste, in *Femtochemistry: Ultrafast Chemical and Physical Processes in Molecular Systems*, eds.: M. Chergui (World Scientific, Singapore, 1996), pp. 319.

49. J. Gaus, K. Kobe, V. Bonačić-Koutecký, H. Kühling, J. Manz, B. Reischl, S. Rutz, E. Schreiber and L. Wöste, Experimental and theoretical approach to the pseudorotating $Na_3$ (B), J. Phys. Chem. **97**, 12509 (1993).

50. K. Kobe, H. Kühling, S. Rutz, E. Schreiber, J.P. Wolf, L. Wöste, M. Broyer and Ph. Dugourd, Time-resolved observation of molecular pseudorotation in $Na_3$, Chem. Phys. Lett. **213**, 554 (1993).

51. V. Bonačić-Koutecký and J. Gaus, private communication.

52. L. Wöste, private communications.

53. H. Ruppe, S. Rutz, E. Schreiber and L. Wöste, Femtosecond wave packet propagation dynamics in the dissociative $K_3$ molecule, Chem. Phys. Lett. **257**, 356 (1996).

54. H. Ruppe, Y.U. Rutz, S. Rutz and E. Schreiber, in *Ultrafast Phenomena X*, Vol. 62 *Springer Series in Chemical Physics*, eds.: P.F. Barbara, J. Fujimoto, W.H. Knox and W. Zinth (Springer Verlag, Berlin, 1996), pp. 192.

55. M.R. Zakin, R.O. Brickman, D.M. Cox and A. Kaldor, Dependence of metal cluster reaction kinetics on charge state. II. chemisorption of hydrogen by neutral and positively charged iron clusters, J. Chem. Phys. **88**, 6605 (1985).

56. P. Fayet, F. Granzer, G. Hegenbart, E. Moisar, B. Pischel and L. Wöste, Latent-image generation by deposition of monodisperse silver clusters, Phys. Rev. Lett. **55**, 3002 (1985).

426

57. G. Delacrétaz, P. Fayet, J.P. Wolf and L. Wöste, in *Proc. of the Intern. School of Physics "Enrico Fermi", Course CVII*, eds.: G. Scoles (North Holland, Amsterdam, 1990), pp. 359–396.

58. N. Lee, R.Ĝ. Keessee and Jr. A.W. Castleman, On the correlation of total and partial enthalpies of ion solvation and the relationship to the energy barrier to nucleation, J. Colloid Interface Sci. **75**, 555 (1980).

59. N.F. Scherer, L.R. Khundkar, R.B. Bernstein and A.H. Zewail, Real-time picosecond clocking of the collision complex in a bimolecular reaction: The birth of OH from $H+CO_2$, J. Chem. Phys. **87**, 1451 (1987).

60. A. Amirav, U. Even and J. Jortner, Electronic-vibrational excitations of aromatic molecules in large argon clusters, J. Phys. Chem. **86**, 3345 (1982).

61. L. Bewig, U. Buck, C. Mehlmann and M. Winter, Ionization induced fragmentation of size selected neutral sodium clusters, J. Chem. Phys. **100**, 2765 (1994).

62. J. Tiggesbäumker, L. Köller, H.O. Lutz and K.H. Meiwes-Broer, Giant resonances in silver-cluster photofragmentation, Chem. Phys. Lett. **190**, 42 (1992).

63. C. Bréchignac, Ph. Cahusac, J. Leygnier and J. Weiner, Dynamics of unimolecular dissociation of sodium cluster ions, J. Chem. Phys. **90**, 1492 (1989).

64. C. Bréchignac, Ph. Cahusac, R. Pflaum and J.-Ph. Roux, Adiabatic unimolecular dissociation of heterogeneous alkali clusters, J. Chem. Phys. **88**, 3732 (1988).

65. C. Bréchignac, Ph. Cahusac, J.-Ph. Roux, D. Pavolini and F. Spiegelmann, Adiabatic decomposition of mass-selected alkali clusters, J. Chem. Phys. **87**, 5694 (1987).

66. R. Schinke, *Photodissociation Dynamics, Cambridge Monographs on Atomic, Molecular and Chemical Physics* (University Press, Cambridge, 1993).

67. D.M. Willberg, M. Gutmann, J.J. Breen and A.H. Zewail, Real-time dynamics of clusters. I. $I_2X_n$ ($n = 1$), J. Chem. Phys. **96**, 198 (1992).

68. T. Baumert, R. Thalweiser, V. Weiß and G. Gerber, in *Ultrafast Phenomena VIII*, Vol. 55 *Springer Series in Chemical Physics*, eds.: J.-L. Martin, E.P. Ippen, G.A. Mourou and A.H. Zewail (Springer Verlag, Berlin, 1993), pp. 83.

69. T. Baumert, R. Thalweiser, V. Weiß and G. Gerber, Time-resolved studies of neutral and ionized $Na_n$ clusters with femtosecond light pulses, Z. Phys. D **26**, 131 (1993).

70. E. Schreiber, K. Kobe, A. Ruff, S. Rutz, G. Sommerer and L. Wöste, Ultrafast fragmentation probability of the $Na_3$ C-state, Chem. Phys. Lett. **242**, 106 (1995).

71. E. Schreiber, S. Rutz, S. Wolf, T. Leisner and L. Wöste, in *Structures and Dynamics of Clusters*, Vol. 16 *Frontiers Science Series*, eds.: T. Kondow, K. Kaya and A. Terasaki (Universal Academy Press, Inc., Tokyo, 1996), pp. 199, in: Proceedings of the YAMADA Conference, Tokyo 1995.

72. A. Ruff, S. Rutz, E. Schreiber and L. Wöste, Ultrafast photodissociation of $K_{n=3...9}$ clusters, Z. Phys. D **37**, 175 (1996).

73. H. Kühling, S. Rutz, K. Kobe, E. Schreiber and L. Wöste, Odd-even alternation of femtosecond fragmentation processes of excited $Na_{3-10}$ clusters, J. Phys. Chem. **98**, 6697 (1994).

74. H. Kühling, K. Kobe, S. Rutz, E. Schreiber and L. Wöste, Time-resolved spectroscopy of $Na_n$-cluster fragmentation, Z. Phys. D **26**, 33 (1993).

75. H. Kühling, S. Rutz, K. Kobe, E. Schreiber and L. Wöste, Femtosecond fragmentation of the $Na_3$ D-state, J. Phys. Chem. **97**, 12500 (1993).

76. S. Rutz, K. Kobe, H. Kühling, E. Schreiber and L. Wöste, Time-resolved tpi spectroscopy of $Na_3$-clusters, Z. Phys. D **26**, 276 (1993).

77. S. Wolf, G. Sommerer, S. Rutz, E. Schreiber, T. Leisner and L. Wöste, Spectroscopy of size-selected neutral clusters: Femtosecond evolution of neutral silver trimers, Phys. Rev. Lett. **74**, 4177 (1995).

78. S. Wolf, G. Sommerer, E. Schreiber, S. Rutz, T. Leisner, L. Wöste and R.S. Berry, in *Femtochemistry: Ultrafast Chemical and Physical Processes in Molecular Systems*, eds.: M. Chergui (World Scientific, Singapore, 1996), pp. 225.

79. E. Schreiber, R.S. Berry, T. Leisner, S. Rutz, S. Wolf and L. Wöste, in *Ultrafast Processes in Spectroscopy*, eds.: O. Svelto, D. De Silvestri and G. Denardo (Plenum Publ., New York, 1996), pp. 133–137.

80. H.O. Jaeschke, M.E. Garcia and K.H. Bennemann, Analysis of the ultrafast dynamics of the silver trimer upon photodetachment, J. Phys. B: At. Mol. Opt. Phys. **29**, L545 (1996).

81. H.O. Jeschke, M.E. Garcia and K.H. Bennemann, Theory for the ultrafast structural response of optically excited small clusters: Time dependence of the ionization potential, Phys. Rev. A **54**, R4601 (1996).

82. M. Garcia and K.H. Bennemann, this volume.

83. B. Reischl, Qantum dynamical three-dymensional *ab-initio* approach to a femtosecond pump-probe ionization spectrum of Na$_3$ (B) at low laser field intensities, Chem. Phys. Lett. **239**, 173 (1995).

84. E. Schreiber, in *Proceedings of the International Conference on LASERS '94*, eds.: V.J. Corcoran and T. Goldman (Society for Optical and Quantum Electronics, McLean, 1995), pp. 490.

85. L. Hanley, S.A. Ruatta and S.L. Anderson, Collision-induced dissociation of aluminum cluster ions: Fragmentation patterns, bond energies, and structures for $Al_2^+$-$Al_7^+$, J. Chem. Phys. **87**, 260 (1987).

86. G.G. Dolnikowski, M.J. Kristo, C.G. Enke and J.T. Watson, Ion-trapping technique for ion/molecule reaction studies in the center quadrupole of a triple quadrupole mass spectrometer, Int. J. Mass Spectr. Ion Proc. **82**, 1 (1988).

87. F. Salin, J. Squier, G. Mourou and G. Vaillancourt, Multikilohertz Ti:Al$_2$O$_3$ amplifier for high-power femtosecond pulses, Opt. Lett. **16**, 1964 (1991).

88. P. Kusch and M.M. Hessel, An analysis of the $B^1\Pi_u$ - $X^1\Sigma_g^+$ band system of Na$_2$, J. Chem. Phys. **68**, 2591 (1978).

89. J.B. Atkinson, J. Becker and W. Demtröder, Experimental observation of the $A^3\Pi_u$ state of Na$_2$, Chem. Phys. Lett. **87**, 92 (1982).

90. G. Gerber and R. Möller, Optical-optical double resonance spectroscopy of high vibrational levels of the Na$_2$ A $^1\Sigma_u^+$ state in a molecular beam, Chem. Phys. Lett. **113**, 546 (1984).

91. C. Effantin, O. Babaky, K. Hussein, J. d'Incan and R.F. Barrow, Interactions between the $A^1\Sigma_u^+$ and $b^3\Pi_u$ states of Na$_2$, J. Phys. B **18**, 4077 (1985).

92. A.J. Ross, P. Crozet, C. Effantin, J. d'Incan and R.F. Barrow, Interactions between the a(1) $^1\Sigma_u^+$ and b(1) $^3\Pi_u$ states of K$_2$, J. Phys. B **20**, 6225 (1987).

93. A.M. Lyyra, W.T. Luh, L. Li, H. Wang and W.C. Stwalley, The $A^1\Sigma_u^+$ state of the potassium dimer, J. Chem. Phys. **92**, 43 (1990).

94. G. Stock and W. Domcke, Femtosecond spectroscopy of ultrafast nonadiabatic excited-state dynamics on the basis of *ab-initio* potential-energy surfaces: The S$_2$ state of pyrazine, J. Phys. Chem. **97**, 12466 (1993).

95. L. Seidner and W. Domcke, Microscopic modelling of photoisomerization and internal-conversion dynamics, Chem. Phys. **186**, 27 (1994).

428

96. C. Daniel, M.-C. Heitz, J. Manz and C. Ribbing, Spin-orbit induced radiationless transitions in organometallics: Quantum simulation of the $^1E \rightarrow ^3A_1$ intersystem crossing process in $HCO(CO)_4$, J. Chem. Phys. **102**, 905 (1995).

97. S. Rutz and E. Schreiber, Fractional revivals of wave packets in the $A^1\sigma_u^+$ state of $K_2$. a comparison of two different pump&probe cycles by spectrograms, Chem. Phys. Lett. **269**, 9 (1997).

98. R. de Vivie-Riedle, K. Kobe, J. Manz, W. Meyer, B. Reischl, S. Rutz, E. Schreiber and L. Wöste, Femtosecond study of multiphoton ionization processes in $K_2$: from pump-probe to control, J. Phys. Chem. **100**, 7789 (1996).

99. I.Sh. Averbukh and N.F. Perelmann, Fractional revivals: Universality in the long-term evolution of quantum wavepackets beyond the correspondence principle dynamics, Phys. Lett. A **139**, 449 (1989).

100. M.J.J. Vrakking, D.M. Villeneuve and A. Stolow, Observation of fractional revivals of a molecular wave packet, Phys. Rev. A **54**, R37 (1996).

101. I. Fischer, M.J.J. Vrakking, D.M. Villeneuve and A. Stolow, Femtosecond time-resolved zero kinetic energy photoelectron and photoionization spectroscopy studies of $I_2$ wavepacket dynamics, Chem. Phys. **207**, 331 (1996).

102. J. Heufelder, H. Ruppe, S. Rutz, E. Schreiber and L. Wöste, Fractional revivals of vibrational wave packets in the NaK $A^1\sigma^+$ state, Chem. Phys. Lett. **269**, 1 (1997).

103. D.J. Tannor and S.A. Rice, Control of selectivity of chemical reactions via wave packet evolution, J. Chem. Phys. **83**, 5013 (1985).

104. D.J. Tannor, R. Kosloff and S.A. Rice, Coherent pulse sequence induced control of selectivity of reactions: Exact quantum mechanical calculations, J. Chem. Phys. **85**, 5805 (1986).

105. U. Banin, A. Bartana, S. Ruhman and R. Kosloff, Impulsive excitation of coherent vibrational motion ground surface dynamics induced by intense short pulses, J. Chem. Phys. **101**, 8461 (1994).

106. S.A. Rice, in *Mode Selective Chemistry*, eds.: J. Jortner, R.D. Levine and B. Pullman (Kluwer Academic Publishers, Dordrecht, 1991), pp. 485.

107. G. Delacrétaz, E.R. Grant, R.L. Whetten, L. Wöste and J. Zwanziger, Fractional quantization of molecular pseudorotation in $Na_3$, Phys. Rev. Lett. **56**, 2598 (1986).

108. E. Schreiber, H. Kühling, K. Kobe, S. Rutz and L. Wöste, Time-resolved tpi-spectroscopy of the B- and D-state of $Na_3$-clusters, Ber. Bunsenges. Phys. Chem. **96**, 1301 (1992).

109. T. Baumert, R. Thalweiser and G. Gerber, Femtosecond two-photon ionization spectroscopy of the B-state of $Na_3$-clusters, Chem. Phys. Lett. **209**, 29 (1993).

110. W.E. Ernst and S. Rakowsky, Integer quantization of the pseudorotational motion in $Na_3$ B, Phys. Rev. Lett. **74**, 58 (1995).

111. S. Rakowsky, R.F.W. Herrmann and W.E. Ernst, High resolution laser spectroscopy of the $Na_3$ B – X system, Z. Phys. D **26**, 273 (1993).

112. W.E. Ernst and S. Rakowsky, Is the B state of $Na_3$ a case of Berry's phase?, Z. Phys. D **26**, 270 (1993).

113. B. Reischl, Ph.D. thesis, Freie Universität Berlin, Berlin-Dahlem, 1995.

114. F. Cocchini, T.H. Upton and W. Andreoni, Excited states and Jahn-Teller interactions in the sodium trimer, J. Chem. Phys. **88**, 6068 (1988).

115. R. Meiswinkel and H. Köppel, A pseudo-Jahn-Teller treatment of the pseudorotational spectrum of $Na_3$, Chem. Phys. **144**, 117 (1990).

116. R. Meiswinkel and H. Köppel, A pseudo-Jahn-Teller treatment of the B system of Na$_3$, Z. Phys. D **19**, 63 (1991).

117. J. Schön and H. Köppel, Femtosecond time-resolved ionization spectroscopy of Na$_3$ (B) and the question of the geometric phase, Chem. Phys. Lett. **231**, 55 (1994).

118. J. Schön and H. Köppel, Geometric phase effects and wave packet dynamics on intersecting potential energy surfaces, J. Chem. Phys. **103**, 9292 (1995).

119. A.J. Dobbyn and J.M. Hutson, Wavepacket calculations of femtosecond pump-probe experiments on the sodium trimer, J. Phys. Chem. **98**, 11428 (1994).

120. A.J. Dobbyn and J.M. Hutson, The influence of the ionisation potential on the simulated ion signal from femtosecond pump–probe spectroscopy, Chem. Phys. Lett. **236**, 547 (1995).

121. M. Broyer, G. Delacrétaz, P. Labastie, J.P. Wolf and Wöste, Size-selective depletion spectroscopy of predissociated states of Na$_3$, Phys. Rev. Lett. **57**, 1851 (1986).

122. T.S. Rose, M. Rosker and A.H. Zewail, Femtosecond real-time observation of wave packet oscillations (resonances) in dissociation reactions, J. Chem. Phys. **88**, 6672 (1988).

123. M. Rosker, T.S. Rose and A.H. Zewail, Femtosecond real-time dynamics of photofragment-trapping resonances on dissocative potential energy surfaces, Chem. Phys. Lett. **146**, 175 (1988).

124. M. Broyer, G. Delacrétaz, P. Labastie, J.P. Wolf and Wöste, Vibronic structure of the Na$_3$ ground state by stimulated emission spectroscopy, Phys. Rev. Lett. **62**, 2100 (1989).

125. C. Bréchignac, Ph. Cahuzac, F. Carlier, M. de Frutos and J. Leygnier, Simple metal clusters, Z. Phys. D **19**, 1 (1991).

126. M.M. Kappes, M. Schär, U. Röthlisberger, C. Yeretzian and E. Schumacher, Sodium cluster ionization potentials revisited: higher resolution measurements for Na$_x$ ($x < 23$) and their relation to bonding models, Chem. Phsy. Lett. **143**, 251 (1988).

127. V. Bonačić-Koutecký, L. Češpiva, P. Fantucci and J. Koutecký, Effective core potential-configuration interaction study of electronic structure and geometry of small neutral and cationic Ag$_n$ clusters: predictions and interpretation of measured properties, J. Chem. Phys. **98**, 7981 (1993).

128. V. Bonačić-Koutecký, L. Češpiva, P. Fantucci, J. Pittner and J. Koutecký, Effective core potential-configuration interaction study of electronic structure and geometry of small anionic Ag$_n$ clusters: predictions and interpretation of photodetachment spectra, J. Chem. Phys. **100**, 490 (1994).

129. M. Broyer, G. Delacrétaz, P. Labastie, J.P. Wolf and L. Wöste, Spectroscopy of vibrational ground-state levels of Na$_3$, J. Phys. Chem. **91**, 2626 (1987).

# Springer
# and the
# environment

At Springer we firmly believe that an
international science publisher has a
special obligation to the environment,
and our corporate policies consistently
reflect this conviction.
We also expect our business partners –
paper mills, printers, packaging
manufacturers, etc. – to commit
themselves to using materials and
production processes that do not harm
the environment. The paper in this
book is made from low- or no-chlorine
pulp and is acid free, in conformance
with international standards for paper
permanency.

The manufacturer's authorised representative in the EU is Springer
Nature Customer Service Centre GmbH, Europaplatz 3, 69115 Heidelberg,
Germany. If you have any concerns regarding our products, please
contact ProductSafety@springernature.com

Printed and bound by CPI Group (UK) Ltd, Croydon, CR0 4YY

28/04/2026

02098503-0002